Max-Planck-Institut für ausländisches
öffentliches Recht und Völkerrecht

Beiträge zum ausländischen öffentlichen Recht und Völkerrecht

Begründet von Viktor Bruns

Herausgegeben von
Armin von Bogdandy · Anne Peters

Band 247

Jürgen Friedrich

International Environmental "soft law"

The Functions and Limits of Nonbinding Instruments in International Environmental Governance and Law

Springer

ISSN 0172-4770 ISSN 2197-7135 (electronic)
ISBN 978-3-662-52525-8 ISBN 978-3-642-44946-8 (eBook)
DOI 10.1007/978-3-642-44946-8
Springer Heidelberg New York Dordrecht London

Die Deutsche Nationalbibliothek verzeichnet diese Publikation in der Deutschen Nationalbibliografie; detaillierte bibliografische Daten sind im Internet über http://dnb.d-nb.de abrufbar.

Gedruckt auf säurefreiem und chlorfrei gebleichtem Papier

Springer ist Teil der Fachverlagsgruppe Springer Science+Business Media (www.springer.com)

Für meine Eltern, Heidi und Lutz Friedrich

Foreword

This book reflects my research and thinking on this topic from my time as a research fellow at the Max Planck Institute for Comparative Public Law and International Law in Heidelberg and as a visiting scholar to Columbia University Law School from 2004-2008. It also includes some insights gained during my work as a legal officer for international environmental law for the Federal Ministry for Environment, Nature Conservation and Nuclear Safety of Germany in 2009-2010.

I foremost thank my parents and my sister for their encouragement and support over all these years and my friend Isabel Feichtner for inspiring discussions and her insightful comments on the manuscript. I am grateful to Rüdiger Wolfrum for supervising this work and for giving me both the opportunity as well as the freedom to work independently in the intellectually stimulating environment of the Max Planck Institute. I would also like to thank Stefan Oeter for acting as the second reviewer, José Alvarez for engaging with my thoughts during my time at Columbia University and Armin von Bogdandy for thoughtful discussions and feedback during our workshops on the exercise of public authority through international institutions. Last but not least, the book could not have been written nor finished without the dedicated support of the staff of the Max Planck Institute, in particular Sara von Skerst and the library team as well as Verena Schaller-Soltau for finalizing the manuscript.

Nairobi, July 2013

Jürgen Friedrich

Outline

Table of Contents

Abbreviations

ABS	Access and Benefit Sharing
BIAC	Business and Industry Advisory Committee
CCRF	Code of Conduct for Responsible Fisheries
CIME	Committee on International Investment and Multinational Enterprises
CITES	Convention on International Trade in Endangered Species of Wild Fauna and Flora
COFI	Committee for Fisheries
COP	Conference of the Parties
CSD	Commission on Sustainable Development
DNA	Designated National Authorities
DRC	Democratic Republic of Congo
ECJ	European Court of Justice
EPA	Economic Partnership Agreement
EU	European Union
FAO	Food and Agriculture Organization
GATS	General Agreement on Trade in Services
GATT	General Agreement on Tariffs and Trade
GEF	Global Environment Facility
GIFAP	National Associations of Manufacturers of Agrochemical Products
IAEA	International Atomic Energy Agency
IC	Investment Committee
ICAO	International Civil Aviation Organization
ICCAT	International Commission for the Conservation of Atlantic Tunas
ICJ	International Court of Justice
ILA	International Law Association
ILC	International Law Commission
ILO	International Labour Organization
IMDG Code	International Maritime Dangerous Goods Code

IMF	International Monetary Fund
IMO	International Maritime Organization
IPOA	International Plan of Action
ISM Code	International Management Code for the Safe Operation of Ships and for Pollution Prevention
ISO	International Organization for Standardization
ITLOS	International Tribunal for the Law of the Sea
IUU	Illegal, unreported and unregulated
KPCS	Kimberley Process Certification Scheme for Rough Diamonds
MEA	Multilateral Environmental Agreement
MNC	Multinational Corporation
MOP	Meeting of the Parties
MoU	Memorandum of Understanding
NAFO	Northwest Atlantic Fisheries Organization
NCP	National Contact Point
NGO	Non-Governmental Organisation
NPOA	National Plan of Action
OECD	Organisation for Economic Co-operation and Development
OSPAR	Convention for the Protection of the Marine Environment of the North-East Atlantic
PAN	Pesticide Action Network
PIC	Prior Informed Consent
SAICM	Strategic Approach to International Chemicals Management
SARPs	Standards and Recommended Practices
SOLAS	International Convention for the Safety of Life at Sea
SPS	Sanitary and Phytosanitary Measures
TBT	Technical Barriers to Trade
TUAC	Trade Union Advisory Committee
UNCLOS	United Nations Convention on the Law of the Sea
UNCTAD	U.N. Conference on Trade and Development
UNEP	United Nations Environment Programme

UNFCCC	United Nations Framework Convention on Climate Change
UNGA	United Nations General Assembly
WHO	World Health Organization
WTO	World Trade Organization

Introduction

1. The increasing use of nonbinding instruments

The development of international relations and international law over the last few decades has been defined by the ongoing move from a law of coexistence defining interstate relations and rules of abstention to a law of positive cooperation.[1] With states becoming more dependent on each other in a world of globalising ecological and social problems as well as economic processes, an increasing number of issue areas require international regulatory activity and cooperation. The field of international environmental law is one of the key examples of this development. The transboundary and global nature of environmental issues, the linkage of these issues with questions of global justice and trade as well as the increasing ability of globalising business actors to eschew tight domestic regulation make international cooperation necessary and inevitable.

This need for international cooperation and order has resulted in another major structural development: the rise of permanent international institutions. International organisations, treaty regimes and informal institutions today engage in the regulation and administration of an increasing number of activities formerly within the domestic realm of states.[2] International institutions often do not only act as facilitators of intergovernmental negotiations and treaty-making. The traditional image of international institutions acting simply as agents of states must often give way to one where international institutions act more independently.[3]

These two trends coincide with a third one that is the main focus of this study: the increasing use of legally nonbinding instruments, often re-

---

1 W. Friedmann, The Changing Structure of International Law, 1964, 60 et seq. and 152 et seq.

2 G. Dahm/J. Delbrück/R. Wolfrum, Völkerrecht, Band I/1, Die Grundlagen. Die Völkerrechtssubjekte, 1989, 13 et seq.

3 J.E. Alvarez, "International Organizations: Then and Now", American Journal of International Law 100 (2006), 324-347; J. Klabbers, "The changing image of international organizations" in: J.-M. Coicaud/V. Heiskanen (eds.), The Legitimacy of International Organizations, 2001, 221-255.

ferred to as "soft law".[4] Such nonbinding instruments – which lack the specific form of the known and recognised sources of international law but nevertheless have legal and behavioural effects[5] – have become ubiquitous in international relations and law. International institutions and states cooperating internationally increasingly resort to various forms of nonbinding instruments instead of or in addition to binding ones to pursue their objectives.[6] This is particularly true for environmental matters. In fact, there is hardly any forum which does not resort to nonbinding instruments and hardly any issue where states do not at one point use nonbinding instruments as the basis for their common efforts.

Not only the number, but also the role and functions of nonbinding instruments are ever expanding. Nonbinding instruments are not any longer simply a precursor to subsequent treaty making or customary law. Nor do they only define general objectives for inter-governmental cooperation as in traditional declarations. Resolutions, declarations, codes of conduct, guidelines and action plans are today used to define concrete measures and best practices to be taken by states as well as private actors. And nonbinding instruments increasingly establish international procedures with direct implications for state administrators and private actors. Furthermore, nonbinding instruments are supported by international institutions, sometimes created only for that purpose irrespective of the nonbinding character of the instrument. And established

[4] The term "soft law" is attributed to Lord McNair for a transitional stage in the development of norms. One of the first to use the term in the context of nonbinding resolutions of international organizations of a programmatory character was R.J. Dupuy, "Declaratory Law and Programmatory Law: From Revolutionary Custom to 'Soft Law'" in: R.J. Akkerman/P.J. v. Krieken/C.O. Pannenborg (eds.), Declarations on principles: a quest for universal peace. Liber Röling, 1977, 247-257 (252). The term "soft law" is however not used in this study for the reasons outlined further below in this introduction at 4.

[5] Similarly e.g. G.F. Handl, "A Hard Look at Soft Law", American Society of International Law Proceedings 82 (1988), 371-393 (371). Some authors sceptical of the category of "soft law" argue that the term should be reserved for treaty provisions that are not clear and specific but of more general and vague content and otherwise avoided, see P. Weil, "Towards Relative Normativity in International Law", American Journal of International Law 77 (1983), 413-442 (414); J. Klabbers, "The Redundancy of Soft Law", Nordic Journal of International Law 65 (1996), 167-182 (168).

[6] E. Brown Weiss (ed.), International Compliance with Nonbinding Accords, 1997, 4.

international institutions use their resources and standing directly to promote and enhance the implementation of these instruments. Thus, they establish mechanisms designed to enhance compliance with non-binding instruments such as reporting mechanisms, capacity building and other forms of subtle pressure and persuasion.

In light of these developments, it is the purpose of this study to have a closer look at the specific role of these instruments in practice, and at their limitations. Many questions remain: Given the lack of enforcement in international law and in particular in international environmental law, one wonders if it matters at all whether an environmental norm is legally binding or not. What difference does it make if such instruments, as is increasingly the case, are supported through institutional underpinnings and follow-up mechanisms? And could these instruments indeed provide a useful alternative to treaty making, or how else do they contribute exactly? These and other questions will be central themes in this study.

The use of nonbinding instruments or "soft law" in international law has for a long time been the subject of legal research[7] and has recently received renewed interest,[8] in particular in the context of legal research

7 The literature is broad, see e.g. M. Bothe, "Legal and Non-Legal Norms – A Meaningful Distinction in International Relations?", Netherlands yearbook of international law 11 (1980), 65-95; A.E. Boyle, "Some Reflections on the Relationship of Treaties and Soft Law", International and Comparative Law Quarterly 48 (1999), 901-913; C. Chinkin, "The Challenge of Soft Law: Development and Change in International Law", International and Comparative Law Quarterly 38 (1989), 850-866; R.J. Dupuy, "Declaratory Law and Programmatory Law: From Revolutionary Custom to 'Soft Law'" in: R.J. Akkerman/P.J. v. Krieken/C.O. Pannenborg (eds.), Declarations on principles: a quest for universal peace. Liber Röling, 1977, 247-257; W. Heusel, "Weiches" Völkerrecht: Eine vergleichende Untersuchung typischer Erscheinungsformen, 1991; J.J. Kirton/M.J. Trebilcock (eds.), Hard Choices, Soft Law: Voluntary Standards in Global Trade, Environment and Social Governance, 2004; J. Klabbers, "The Undesirability of Soft Law", Nordic Journal of International Law 67 (1998), 381-391; L. Senden, Soft law in European Community law, 2004; K. Raustiala, "Form and Substance in International Agreements", American Journal of International Law 99 (2005), 581-614. W. Lang, "Die Verrechtlichung des internationalen Umweltschutzes: Vom 'soft law' zum 'hard law'", Archiv des Völkerrechts 22 (1984), 283-305; D. Shelton (ed.), Commitment and Compliance: The Role of Non-Binding Norms in the International Legal System, 2000.

8 See e.g. M. Knauff, Der Regelungsverbund: Recht und Soft Law im Mehrebenensystem, 2010 and the recent debate in the European Journal of International Law: J. D'Aspremont, "Softness in International Law: A Self-

on the changing role of international institutions[9] and so-called global administrative law.[10] One recurrent question in these debates is whether one should acknowledge the existence of a source called "soft law". With the rising significance of these instruments in practice, there is considerable scholarly support for reconsidering the sources doctrine of international law in that respect.[11] The heart of this debate is whether legal doctrine should reflect a continuum of normativity or "relative normativity" that possibly exists in a sociological reality,[12] or if one

---

Serving Quest for New Legal Materials", European Journal of International Law 19 (2008), 1075-1093; A.A. D'Amato, "Softness in International Law: A Self-Serving Quest for New Legal Materials: A Reply to Jean d'Aspremont", European Journal of International Law 20 (2009), 897-910; J. D'Aspremont, "Softness in International Law: A Self-Serving Quest for New Legal Materials; a rejoinder to Tony D'Amato", European Journal of International Law 20 (2009), 911-917.

[9] J.E. Alvarez, International Organizations as Law-makers, 2005, 217 et seq.; A. von Bogdandy/R. Wolfrum/J. von Bernstorff/P. Dann/M. Goldmann (eds.), The Exercise of Public Authority by International Institutions: Advancing International Institutional Law, 2010.

[10] B. Kingsbury/N. Krisch/R. Stewart, "The Emergence of Global Administrative Law", Law & Contemporary Problems (2004-2005), 15-62.

[11] G.F. Handl, "A Hard Look at Soft Law", American Society of International Law Proceedings 82 (1988), 371-393 (373); Alvarez suggests to form a subsidiary source of international obligation, see J.E. Alvarez, "International Organizations: Then and Now", American Journal of International Law 100 (2006), 324-347 (328-329, 333); Fastenrath calls for the incorporation of soft law into the existing legal methodology and doctrine in order for international legal doctrine to reflect the changes within international relations, see U. Fastenrath, "Relative Normativity in International Law", European Journal of International Law 4 (1993), 305-340 (340); see also E. Riedel, "Standards and Sources. Farewell to the Exclusivity of the Sources Triad in International Law?", European Journal of International Law 2 (1991), 58-84.

[12] K.W. Abbott & D. Snidal, "Hard and Soft Law in International Governance", 54 International Organization (2000), 421 et seq. (424); Guzman also considers that "some obligations are more binding than others", compare A.T. Guzman, "A Compliance-Based Theory Of International Law", California Law Review 90 (2002), 1823-1887 (1882 and 1887); according to Fastenrath, law necessarily depends on soft law to solve its problems of divergence and varying modes of interpretation, since it provides common understandings and agreement on (linguistic) conventions regarding the existence, formulation and interpretation of law, compare U. Fastenrath, "Relative Normativity in International Law", European Journal of International Law 4 (1993), 305-340 (323-324, 315);

should maintain a binary perspective of binding law and non-binding non-law. Defenders of the traditional view fear that the doctrinal acceptance of a scale of normativity contributes to a pathological "relative normativity" where it is not clear "where the legal norm begins and where it ends".[13] Without a clear conception of what is law and what is not, one automatically challenges the ability of the legal order to fulfil its functions and purpose.[14] Giving up the binary conception is feared to undermine legal certainty about what is law and what not, and thus to risk the proper functioning of law itself.[15] A clear understanding of what is law is needed in order to distinguish between the legal and political spheres.[16]

This study does not aim to renew this debate, nor does it seek to develop a new doctrine. By looking at the role and functions of nonbinding instruments and their practical application, it adopts a more pragmatic perspective. The study however takes the above-mentioned arguments of the "soft law" debate into account. It does acknowledge the existence and important role of nonbinding instruments in both political and legal spheres.

---

for a detailed critique of this view see J. Klabbers, "The Redundancy of Soft Law", Nordic Journal of International Law 65 (1996), 167-182 (178-179).

13 P. Weil, "Towards Relative Normativity in International Law", American Journal of International Law 77 (1983), 413-442 (417-418).

14 P. Weil, "Towards Relative Normativity in International Law", American Journal of International Law 77 (1983), 413-442 (418-419); G.M. Danilenko, "Sources of International Law in a Changing International Community: Theory and Practice" in: W. E. Butler (ed.), Perestroika and International Law, 1990, 61-80 (61).

15 P. Weil, "Towards Relative Normativity in International Law", American Journal of International Law 77 (1983), 413-442 (417).

16 P. Weil, "Towards Relative Normativity in International Law", American Journal of International Law 77 (1983), 413-442 (441); A. Fischer-Lescano/P. Liste, "Völkerrechtspolitik. Zu Trennung und Verknüpfung von Politik und Recht der Weltgesellschaft", Zeitschrift für internationale Beziehungen 12 (2005), 7-48 (20); P. Malanczuk, Akehurst's modern introduction to international law, 1997, 54-55.

16 P. Malanczuk, Akehurst's modern introduction to international law, 1997, 54-55.

The study starts from the assumption that the formal distinction between binding and nonbinding should not be watered down,[17] but must to the contrary be upheld as a distinguishing device between two broad types of distinguishable instruments that both fulfil important functions in international relations. The binary distinction of legally binding and legally nonbinding is not theoretical but to a significant extent reflects state practice and the realities of international negotiations. States clearly distinguish between binding and nonbinding norms both in the negotiation of instruments as well as in their implementation.[18] They carefully choose between non-binding and binding instruments, and often fiercely negotiate over that particular issue.[19] Just to name a few examples, the question of binding or nonbinding status was an issue in the negotiations leading up to the adoption of the Agreement for the Implementation of the Provisions of the United Nations Convention on the Law of the Sea of 10 December 1982 Relating to the Conservation and Management of Straddling Fish Stocks and Highly Migratory Fish Stocks of 1995 (hereinafter Fish Stocks Agreement)[20];[21] it was one of the most hotly debated issues in the negotiations on an instrument defining the conditions of access and benefit sharing under the Biodiversity Convention;[22] and has played a prominent role in the decade-long ne-

---

17 Similarly R. Wolfrum, "Introduction" in: R. Wolfrum/V. Röben (eds.), Developments of International Law in Treaty Making, 2005, 1-13; K. Raustiala, "Form and Substance in International Agreements", American Journal of International Law 99 (2005), 581-614 (586); H. Hillgenberg, "A Fresh Look at Soft Law", European Journal of International Law 10 (1999), 499-515 (508); W. Heusel, "Weiches" Völkerrecht: Eine vergleichende Untersuchung typischer Erscheinungsformen, 1991.

18 P. Weil, "Towards Relative Normativity in International Law", American Journal of International Law 77 (1983), 413-442 (417).

19 K. Raustiala, "Form and Substance in International Agreements", American Journal of International Law 99 (2005), 581-614 (587).

20 Agreement for the Implementation of the Provisions of the United Nations Convention on the Law of the Sea of 10 December 1982 Relating to the Conservation and Management of Straddling Fish Stocks and Highly Migratory Fish Stocks (4 August 1995), 2167 UNTS 88, 34 ILM 1542.

21 See D.A. Freestone/Z. Makuch, "The new international environmental law of fisheries: the 1995 United Nations Straddling Stocks Agreement", Yearbook of international environmental law 7 (1996), 3-51.

22 The nonbinding Bonn guidelines on access to genetic resources and the fair and equitable sharing of the benefits arising from their utilization were adopted in 2002, 10 years after the Convention, but were only of a provisional

gotiations of an instrument on the protection of forests. Remarkably for the present context, the negotiations on forests have recently produced a *Non-legally Binding* Instrument on Sustainable Forest Management of all Types of Forests.[23]

Neglecting the difference between binding and nonbinding as a distinguishing feature of various instruments as sometimes suggested[24] would not only remove legal doctrine from the actual realities of international affairs. It would also be detrimental to the status and the functioning of international law. The concept of "bindingness" refers to the presumption that the norm in question is and will be handled in accordance with a number of well-established legal procedures and rules such as consent requirements, ratification procedures, the law of treaties et cetera.[25] Although it is necessary to advance our understanding and doctrine to capture nonbinding phenomena, one should not in doing so too easily give up the achievements of the past which lies exactly in a common

---

nature. Only in 2009 did the Working Group on ABS of the Convention decide on a draft Protocol text on Access and Benefit Sharing which was adopted as the Nagoya Protocol on Access to Genetic Resources and the Fair and Equitable Sharing of Benefits Arising from their Utilization to the Convention on Biological Diversity COP 10 in 2010; for the debate over the issue of "bindingness", where proposals ranged from "legally binding" to a "mix of legally binding and legally nonbinding instruments" to a "non-binding instrument", compare e.g. Conference of the Parties to the Convention on Biological Diversity, Decision IX/12, Annex.

23 Non-legally Binding Instrument on Sustainable Forest Management of all Types of Forests, U.N. Doc. A/Res/62/98 of 17 December 2007. Originally aimed at the development of a binding Forest Convention, negotiations failed in 2005, and states could only agree to finalise a non-legally binding instrument in short term and to develop a forest convention by 2015. On these developments, compare K. Kunzmann, "The Non-legally Binding Instrument on Sustainable Management of All Types of Forests – Towards a Legal Regime for Sustainable Forest Management?", German Law Journal 9 (2008), 981-1005.

24 M. Goldmann, "Inside Relative Normativity: From Sources to Standard Instruments for the Exercise of International Public Authority" in: A. von Bogdandy/R. Wolfrum/J. von Bernstorff/P. Dann/M. Goldmann (eds.), The Exercise of Public Authority by International Institutions: Advancing International Institutional Law, 2010, 661-711 (709).

25 For a similar critique in which Jan Klabbers upholds the distinction of binding and nonbinding see J. Klabbers, "Goldmann Variations" in: A. von Bogdandy/R. Wolfrum/J. von Bernstorff/P. Dann/M. Goldmann (eds.), The Exercise of Public Authority by International Institutions, 2010, 713-725 (722-724).

understanding of what comes with a legally binding obligation. Moreover, without a clear distinction one would also risk depriving nonbinding instruments of their potential and role as a more flexible and more informal form of cooperation, and possibly drive states into even less formal ways.

Rather than adding another layer to the theoretical "hard versus soft" and "binding versus nonbinding" debate, the study thus attempts to show how binding and nonbinding instruments, both in distinct ways, occupy an important place in the ordering of international relations today. As the experience from domestic law and administration teaches us, social processes have always been a mixture of formal and informal elements that operate side by side.[26] In focussing on the differences between binding and nonbinding instruments but also on their interrelation and respective potentials, including their limits, the study aims to contribute to a clearer distinction between both forms of instruments and to a better understanding of their respective utility in practice. In the long run, I believe that maintaining a clear distinction will help preserve the usefulness and functionality of both binding and nonbinding instruments.

2. Purpose and focus of the study

On a more concrete level, the study pursues a dual objective. On the one hand it aims to analyse the role and function of nonbinding instruments in terms of their utility for effective cooperation and regulation, and on the other hand analyses the legitimacy of their use.

First, regarding utility, this study aims to identify the specific role and functions of nonbinding instruments, including the comparative advantages of nonbinding instruments as compared to treaty law and the comparative advantages of treaty law as compared to nonbinding instruments.[27] The analysis of the specific weaknesses and limits of non-

[26] E. Schmidt-Aßmann, "Die Herausforderung der Verwaltungsrechtswissenschaft durch die Internationalisierung der Verwaltungsbeziehungen", Der Staat 3 (2006), 315-338 (325); S.J. Toope, "Formality and Informality" in: D. Bodansky/J. Brunnée/E. Hey (eds.), The Oxford Handbook of International Environmental Law, 2007, 107-124 (124).

[27] The need for research on the comparative advantages of binding and nonbinding instruments has been stressed as one of the first scholars by K. Raustiala, "Compliance & effectiveness in international regulatory coopera-

binding instruments is significant, because these often make them attractive for states.[28] Overall, this analysis may help decision makers in taking optimal decisions with respect to regulatory choice. The study thereby hopes to contribute to what has always been the task of legal scholarship; namely to develop suitable legal frameworks, procedures and mechanisms which provide practitioners with effective instruments.[29]

Second, nonbinding instruments, if and to the extent that they are used to exercise authority, must be sufficiently legitimated for the sake of long-term effectiveness and accountability. Legitimacy questions are often discussed in general terms. This study attempts to explore the specific challenges which accompany the move to nonbinding instruments.

Pursuing an inductive approach, the study takes the area of environmental protection and sustainable development law as reference. This issue area is one where nonbinding instruments are especially flourishing. This may have to do with some of the particular conditions of international environmental problems and law, such as uncertainty about the regulatory problem; a particular need for the protection of common goods that require the integration of social, economic and environmental issues; the need for flexibility due to fast technological and scientific advancement; and the relative absence of reciprocity as a basis for enforcement.[30]

Nonbinding instruments are not only often used, but also highly diverse. In order to come to meaningful results, this study therefore mainly focuses only on particular nonbinding instruments. It thus does not attempt to analyse all types of nonbinding instruments from the field of environmental protection and sustainable development. In par-

---

tion", Case Western reserve journal of international law 32 (2000), 387-440 (427).

28 Similarly A. von Bogdandy, "Lawmaking by International Organisations: Some Thoughts on Non-Binding Instruments and Democratic Legitimacy" in: R. Wolfrum/V. Röben (eds.), Developments of International Law in Treaty Making, 2005, 171-182 (172).

29 U. Beyerlin, Umweltvölkerrecht, 2000, 627; similarly for administrative law scholarship E. Schmidt-Aßmann, "Die Herausforderung der Verwaltungsrechtswissenschaft durch die Internationalisierung der Verwaltungsbeziehungen", Der Staat 3 (2006), 315-338 (324).

30 Similar reasons are given by P.-M. Dupuy, "Soft law and the international law of the environment", Michigan Journal of International Law 12 (1990–1991), 420-435 (420-422).

ticular it is not another study on general resolutions and declarations and their role in the development of international law, even though this issue is also treated.[31] Instead its main attention is on instruments containing *norms.* As a *norm*, I understand the prescription of certain appropriate behaviour; in contrast to mere descriptions of facts or behavioural regularities.[32] Therefore the study is concerned with instruments which contain commands, requests or recommendations through which certain behaviour is expected to be achieved, and which could very well also be contained in a legally binding document.[33] Furthermore, I am concerned here only with those instruments which are supported by *institutionalised follow-up mechanisms*, because these instruments reflect a more direct attempt by international institutions to shape the behaviour of states and private actors.[34] Moreover, by concentrating on *public international* institutions, i.e. multilateral institutions where states are members, the study is not concerned with the increased standard-setting activities by private actors such as environmental NGOs, business associations or so-called multi-stakeholder initiatives.[35]

3. Way of proceeding

This study proceeds in three steps. As a first step, Part 1 introduces, systematises and analyses nonbinding instruments in the field of envi-

---

31 See on this issue instead of many already J.A. Frowein, "Der Beitrag der internationalen Organisationen zur Entwicklung des Völkerrechts", Zeitschrift für ausländisches öffentliches Recht und Völkerrecht 36 (1976), 147-167.

32 D. Bodansky, The art and craft of international environmental law, 2010, 87-88.

33 Norms thus are here understood to be those of a deontic nature.

34 The study is thus concerned with external and not internal instruments. Internal are those instruments which regulate the functioning of the organization, e.g. by establishing rules for its bodies or staff. Where internal instruments have external effects, for example in constellations in which they become standards for external practice as illustrated by the World Bank Operational Standards, they are considered to be external and are included in this study.

35 See on this issue G. Teubner, "'Global Bukowina': Legal Pluralism in the World Society" in: G. Teubner (ed.), Global Law without a state, 1996, 3-28; J. Morrison/N. Roht-Arriaza, "Private and Quasi-Private Standard Setting" in: D. Bodansky/J. Brunnée/E. Hey (eds.), Oxford Handbook of International Environmental Law, 2007, 498-527.

ronmental protection and sustainable development. In this part, I will first give an overview of the variety of nonbinding instruments used by international organisations and in treaty regimes, and provide a first systematisation of such instruments. Three case studies of particularly interesting and diverse instruments follow. These case studies are intended to give a more profound background to the regulatory challenges, the instruments and the institutions, but also to identify cross-cutting characteristics and differences of nonbinding instruments. A number of parameters are then introduced which reflect these cross-cutting characteristics. These parameters can serve to assess the effectiveness and legitimacy of any nonbinding instrument.

Part 2, by drawing upon the material and analysis of Part 1, analyses the utility and limits of nonbinding instruments in the area of sustainable development. As nonbinding instruments have an impact on and develop their influence through cooperative processes at different interrelated levels of governance and law, the study adopts a multilevel approach. The first section of Part 2 looks at the impact of nonbinding instruments on the international level, notably with respect to international law on the one hand and international institutions on the other hand. It stresses the interrelation of binding and nonbinding instruments, as well as inter-institutional cooperation on the basis of nonbinding instruments. In the past, only limited attention has been given to the interplay of binding and nonbinding instruments.[36] By focussing on inter-institutional interplay on the basis of nonbinding instruments, this study aims to contribute to scholarship that considers institutional approaches as one possible way to mitigate sectoral fragmentation.[37] The study then turns to the impact of nonbinding instruments at state level. For this purpose, the study first analyses the potential of the various characteristics, institutional mechanisms and limitations of nonbinding instruments in inducing state compliance. This is followed by an analysis of the various possible modes and limits of implementation

---

[36] One exception in which the lack of systematic attention to the interaction of soft and hard law is deplored is the contribution by J.B. Skjaerseth/O.S. Stokke/J. Wettestad, "Soft Law, Hard Law, and Effective Implementation of International Environmental Norms", Global Environmental Politics 6 (2006), 104-120.

[37] On institutional coordination as a means to address conflicts see N. Matz, Wege zur Koordinierung völkerrechtlicher Verträge: völkervertragsrechtliche und institutionelle Ansätze, 2005, 340 et seq.; M. Schroeder, Die Koordinierung der internationalen Bemühungen zum Schutz der Umwelt, 2005.

in national legal systems. The third section in Part 2 considers the rather novel attempts of international organisations to address private actors directly, and looks at mechanisms that enhance the direct impact of such norms. Again, particular attention will be devoted to the limitations of these attempts.

Part 3 addresses the issue of legitimacy. It first questions whether and to which extent a legitimacy issue may arise at all, given the fact that the norms are of voluntary nature. Secondly, some specific challenges for the legitimacy of nonbinding instruments and the institutions who adopt them are identified, again with a focus on the particularities of these instruments as compared to treaty law. Part 3 concludes with some proposals for responding to the identified challenges. This responds to the need for legal research to develop adequate strategies which can ensure the long-term legitimacy of norm development and norm implementation processes.

Finally, this study will end with a concluding summary in Part 4.

4. On terminology

A final word is needed on the terminology used throughout the study. The study will use the term *instrument* for the document in which norms are contained.[38] Further, the term *nonbinding instruments* refers to instruments that do not give rise to legal obligations irrespective whether they are politically binding. The choice of the term *nonbinding instrument* deliberately avoids the commonly used term *soft law*. The term *soft law* appears inadequate in two respects. First of all, and most importantly, using this term for nonbinding instruments implies that these instruments constitute or give rise to law. Secondly, the distinction between "hard" and "soft" carries the connotation that "soft" is less strong. As this study starts from the assumption that both have important but different roles to play, the prioritisation implicit in "soft" versus "hard" is avoided. The distinction between legally nonbinding and binding is a merely formal one and does not imply any normative statement.

---

[38] Similar D. Shelton, "Law, Non-Law and the Problem of 'Soft Law'" in: D. Shelton (ed.), Commitment and Compliance: The Role of Non-Binding Norms in the International Legal System, 2000, 1-18 (5).

While making the distinction between binding and nonbinding, the study will employ the term *norm* for provisions which prescribe certain appropriate behaviour, i.e. provisions of a deontic nature that contain a request or a recommendation. The core of the concept of a *norm* considered to be the desideratum to behave in a certain way. It does not matter whether a provision is binding or nonbinding for it to prescribe certain behaviour. Regardless of whether we are concerned with binding or nonbinding norms, both types of prescriptions may express a preference for one particular conduct of the addressee.

Further, I will use the term *compliance* not only with respect to binding, but also with respect to nonbinding norms. Although this term has been traditionally used for behaviour conforming to binding law, it is also useful for the assessment of whether an actor follows a nonbinding norm. This approach is supported by recent academic works on nonbinding instruments.[39] The difference between nonbinding and binding is sufficiently expressed in the inability of the former to be the basis of an assessment of illegality.

The notion of *compliance* is different from mere *implementation* which is also used with respect to nonbinding norms.[40] *Implementation* is used to describe those measures that are taken to adapt domestic legal systems through legislation, judicial decision, executive decree or administrative processes to international norms.[41] In contrast, *compliance* is wider and refers to the matching of the international norm and actual behaviour of actors. *Compliance* includes *implementation*, but also depends on whether a government ultimately follows through on its im-

---

[39] Compare e.g. by E. Brown Weiss (ed.), International Compliance with Nonbinding Accords, 1997; D. Shelton (ed.), Commitment and Compliance: The Role of Non-Binding Norms in the International Legal System, 2000; J.J. Kirton/M.J. Trebilcock (eds.), Hard Choices, Soft Law: Voluntary Standards in Global Trade, Environment and Social Governance, 2004.

[40] D. Shelton, "Law, Non-Law and the Problem of 'Soft Law'" in: D. Shelton (ed.), Commitment and Compliance: The Role of Non-Binding Norms in the International Legal System, 2000, 1-18 (5).

[41] H.K. Jacobson/E. Brown Weiss, "A Framework for Analysis" in: H.K. Jacobson/E. Brown Weiss (eds.), Engaging Countries: Strengthening Compliance with International Environmental Accords, 1998, 1-18; D. Shelton, "Law, Non-Law and the Problem of 'Soft Law'" in: D. Shelton (ed.), Commitment and Compliance: The Role of Non-Binding Norms in the International Legal System, 2000, 1-18 (5).

plementing acts.[42] Finally, *compliance* is not identical with *effectiveness* of the instrument. Although *compliance* with an instrument may be perfect, the international instrument may not be *effective* in resolving or at least addressing the problems it was meant to address because of the inadequateness of its approach or its prescriptions.[43]

---

[42] H.K. Jacobson/E. Brown Weiss, "A Framework for Analysis" in: H.K. Jacobson/E. Brown Weiss (eds.), Engaging Countries: Strengthening Compliance with International Environmental Accords, 1998, 1-18 (4); D. Shelton, "Law, Non-Law and the Problem of 'Soft Law'" in: D. Shelton (ed.), Commitment and Compliance: The Role of Non-Binding Norms in the International Legal System, 2000, 1-18 (5); R. Wolfrum, "Means of Ensuring Compliance with and Enforcement of International Environmental Law", Recueil des cours de l'Académie de Droit International de La Haye 272 (1998), 13-154 (29-30).

[43] H.K. Jacobson/E. Brown Weiss, "A Framework for Analysis" in: H.K. Jacobson/E. Brown Weiss (eds.), Engaging Countries: Strengthening Compliance with International Environmental Accords, 1998, 1-18 (5); D.G. Victor/K. Raustiala/E.B. Skolnikoff, "Introduction and Overview" in: D. G. Victor/K. Raustiala/E.B. Skolnikoff (eds.), The Implementation and Effectiveness of International Environmental Commitments: Theory and Practice, 1998, 1-30 (7).

Part 1

Nonbinding instruments in international practice

A. The variety of nonbinding instruments: an overview

Nonbinding instruments are as ubiquitous in international relations as they are diverse. The following overview attempts to present some of the most important instruments prescribing norms of behaviour in the field of sustainable development.[1] As will be seen, this overview goes hand in hand with a preliminary categorization of these instruments.

The categorization is based on differences in types of norms contained in the instruments, the actors that develop them and the addressees of the instrument. The resulting categories can often overlap and only represent ideal types. For instance, just like treaty law, nonbinding instruments may contain various types of norms.[2] And the distinction between various types of norms may sometimes be blurred, for instance as between technical standards and policy-oriented norms.

Moreover, the categorization does not relate simply to the title of instruments. The name of an instrument only serves as a weak indicator of its actual content and function. The same type of nonbinding instrument may be called "code of conduct" by one organisation and "guidelines" by another. Instruments that are called "codes of conduct" often contain guiding principles just as guidelines or declarations do,

---

[1] For other overviews of nonbinding instruments and categorizations see e.g. M. Knauff, Der Regelungsverbund: Recht und Soft Law im Mehrebenensystem, 2010, 257–295; J.E. Alvarez, International Organizations as Lawmakers, 2005, 217 et seq.; for the specific context of environmental law and sustainable development short overviews are given in U. Beyerlin/T. Marauhn, International Environmental Law, 2011, 289-297; S. Fritz, Integrierter Umweltschutz im Völkerrecht, 2009, 116-138; A.C. Kiss/D. Shelton, International environmental law, 2004, 89 et seq.

[2] Often an instrument contains various types of norms, as exemplified by the general principles and prescription of measures to be undertaken by states and private actors in the FAO Code of Conduct for Responsible Fisheries.

and "guidelines" may contain specific standards and best practices or remain relatively vague and lay down broad principles.

I. Memoranda of Understanding

Memoranda of Understanding are nonbinding documents which are most often negotiated and adopted bilaterally by two actors to clarify issues of common political interest and to set out general agreement on cooperation.[3] However, as illustrated by the Memorandum of Understanding on the Conservation of Migratory Sharks adopted in 2010 by Parties of the Convention on the Conservation of Migratory Species of Wild Animals,[4] sometimes states cooperating multilaterally also adopt so-called Memoranda of Understanding in a multilateral setting.

One can distinguish between Memoranda of Understanding between institutions and between states. With respect to the first group, Memoranda of Understanding are commonly used by international institutions to formalise and enhance their cooperation. Memoranda of Understanding are in this manner commonly employed between treaty bodies,[5] between organs of international organisations[6] and between in-

---

[3] For a similar definition compare R. Wolfrum/N. Matz, Conflicts in international environmental law, 2003, fn. 354.

[4] The Memorandum of Understanding on the Conservation of Migratory Sharks was adopted on 12 February 2010 at the 8th Conference of the Parties to the Convention on the Conservation of Migratory Species of Wild Animals (23 June 1979), 1651 UNTS 333, 19 ILM 15 (1980). The Memorandum is available at http://www.cms.int/species/sharks/MoU/Migratory_Shark_MoU_Eng.pdf.

[5] The numbers of Memoranda of Understanding between various treaty bodies and treaty secretariats is overwhelming, instead of many see e.g. from the area of biodiversity protection, the Memorandum of Understanding between the secretariat of the Convention on Biological Diversity and the Bureau of the Convention on Wetlands of International Importance, especially as Waterfowl Habitat, UN Doc.UNEP/CBD/COP/3/Inf.38; the Memorandum of Understanding between the secretariat of the Convention on Biological Diversity of the Convention on Biological Diversity and the secretariat of the Convention on International Trade in Endangered Species, UNEP/CBD/COP/3/Inf.39; the Memorandum of Understanding between the secretariat of the Convention on Biological Diversity and the secretariat of the Convention on the Conservation of Migratory Species of Wild Animals, UN Doc. UNEP/CBD/COP/3/Inf.40; all available at http://www.cbd.int.

ternational institutions and treaty bodies.[7] They are important instruments for addressing conflicts of norms and avoiding overlap of activities through horizontal coordination and cooperation.[8] Memoranda of Understanding are also commonly used as a basis for environmental cooperation between states[9] or between states and organisations.[10] As this study is concerned with multilateral instruments established by international institutions which are addressed to states or private actors, Memoranda of Understanding in their bilateral form will not be the main focus of this study.

---

[6] E.g. the Memorandum of Understanding concerning establishment of the Inter-Organization Programme for the Sound Management of Chemicals between the United Nations Environment Programme, the International Labour Organization, the Food and Agriculture Organization of the United Nations, the World Health Organization, the United Nations Industrial Development Organization and the Organisation for Economic Co-operation and Development, available at http://www.who.int/iomc/participants/iomc-mou.pdf.

[7] E.g. the Memorandum of Understanding between the Council of the Global Environment Facility and the Conference of the Parties of the Convention on Biological Diversity, UN Doc. UNEP/CBD/COP/2/11; Memorandum of Understanding between the Council of the Global Environment Facility and the Conference of the Parties with the Framework Convention on Climate Change, UN Doc. FCCC/CP/1996/9; Memorandum of Understanding between the secretariat of the Convention on International Trade in Endangered Species and the Food and Agriculture Organisation, COFI:FT/X/2006/3.

[8] R. Wolfrum/N. Matz, Conflicts in international environmental law, 2003, 173-174.

[9] Germany for example has signed a great number of Memoranda of Understanding on environmental cooperation with other states on matters that are not addressed in treaties, e.g. in January 2009 the Memorandum of Understanding on Cooperation in Combating Climate Change between the Government of the Federal Republic of Germany and the Government of the People's Republic of China; the Memorandum of Understanding on cooperation regarding Clean Development Mechanism project activities and climate policy issues between Germany and Peru signed in 2008; at www.bmu.de.

[10] E.g. the Memorandum of Understanding signed between the World Bank and Poland at Poznan in 2008 to enable a Green Investment Scheme transaction of emission credits tradable under Article 17 of the Kyoto Protocol, at www.worldbank.org.

II. International programmes

International programmes, often also called action plans, are here referred to as those documents that outline future activities of international institutions and states. Oftentimes these instruments additionally contain a number of recommendations, as described in the next section. These programmes often shape institutional policies but also general international environmental policy and legal development.

One example is the Programme for the Development and Periodic Review of Environmental Law (Montevideo Programme) adopted by the United Nations Environment Programme (UNEP).[11] The Montevideo Programme mainly delineates and guides the activities of UNEP in the field of environmental law and governance.

The Action Plan for the Human Environment, adopted by the 1972 Stockholm Conference on the Human Environment alongside the Stockholm Declaration not only lays down recommendations for concrete actions to be taken by states, but also outlines a programme for international measures considered to be necessary for the implementation of the principles of the Stockholm declaration. The Action Plan has provided the foundation for a number of international measures, including a global environment assessment plan (Earthwatch), a number of global conventions and UNEP's regional seas programme.

International Programmes, often called action plans, are also frequently employed by the Conferences of the Parties (COPs) of Multilateral Environmental Agreements (MEAs). The Bali Action Plan[12] adopted in 2007 by the 13th COP of the Framework Convention on Climate Change[13] served as a guiding framework for negotiations for a successor agreement to the Kyoto Protocol.

---

[11] The first Programme for the Development and Periodic Review of Environmental Law (Montevideo I) was adopted in 1982 by UNEP GC 10/5/Add.2, Annex, Ch. 11 (1981); successor programmes were adopted in 1993 [Montevideo Programme II, UNEP/GC.17/5 (1993)], in 2001 [Montevideo Programme III, UNEP GC 21/23 (2001)]. A Montevideo Programme IV is currently being developed, compare for the respective report to the 25th UNEP Governing Council UNEP/GC.25/11 (2009).

[12] Bali Action Plan, Decision 1/CP.13 of the Report of the Conference of the Parties on its thirteenth session, held in Bali from 3 to 15 December 2007, UN Doc. FCCC/CP/2007/6/Add.1 of 14 March 2008.

[13] United Nations Framework Convention on Climate Change, U.N. Doc. A/AC.237/18 (Part II)/Add.1, 1771 UNTS 107, 31 ILM 849 (1992).

III. Declarations of principles and action plans adopted at international conferences

Nonbinding declarations refer to documents containing norms in the form of principles and more concrete rules that are adopted at international conferences. In contrast to international recommendations discussed in the next section, these instruments are not adopted by the body or organ of an international institution but are in principle one-time decisions.

At least in United Nations practice, declarations of principles are adopted on special occasions to formulate the fundamental values and principles which are considered of great and lasting importance by all convened states.[14] Even though not formally binding, these declarations may nevertheless be influential on the development of subsequent legal rules and principles, on the mandates, work and acts of international organisations, on international jurisprudence, or even directly on the practice of states.[15]

Some of the most influential declarations of principles have been adopted at intergovernmental conferences.[16] The 1972 Declaration of Principles[17] and the 1992 Rio Declaration on Environment and Development[18] constitute hallmarks in the development of international environmental principles, law and institutions.[19] In particular the Rio Decla-

---

[14] R. Higgins, "The Role of Resolutions of International Organizations in the Process of Creating Norms in the International System" in: W.E. Butler (ed.), International Law and the International System, 1987, 21-30 (26).

[15] A.C. Kiss/D. Shelton, International environmental law, 2004, 93.

[16] Principles and concepts of environmental governance and law are almost always included in any relevant nonbinding instruments adopted by international institutions.

[17] Declaration of the United Nations Conference on the Human Environment (16 June 1972), U.N. Doc. A/CONF.48/14/Rev.1 (1973), 11 ILM. 1416 (1972); generally on the conference A. Kiss/J.-D. Sicault, "La Conférence des Nations Unies sur l'Environnement", Annuaire Français du Droit International 18 (1972), 603-628.

[18] United Nations Declaration on Environment and Development (13 June 1992), UN Doc. A/CONF.151/26 (vol. I); 31 ILM 874 (1992); generally on the conference and its results A. Kiss/S. Doumbé-Billé, "La Conférence des Nations-Unies sur l'Environnement et le Développement", Annuaire Français du Droit International 38 (1992), 823-843.

[19] P. Sands, Principles of International Environmental Law, 2003, 40 et seq.

ration, as an expression of a delicately balanced consensus of developed and developing states, has greatly contributed to the codification and development of principles of international environmental law and the concept of sustainable development.[20] The influence of the concepts and principles of the above mentioned documents on international jurisprudence is illustrated in, for example, the judgment of the International Court of Justice (ICJ) in *Gabčikovo-Nagymaros Project,* where the Court held that the need to reconcile economic development with environmental protection as prescribed by the concept of sustainable development is one factor that requires the Parties to the dispute to enter into fresh negotiations.[21] Another early example is the Declaration of Principles Governing the Sea-Bed and the Ocean Floor, and the Subsoil Thereof, beyond the Limits of National Jurisdiction of the United Nations General Assembly.[22] It first stipulated the principle of the common heritage of mankind in relation to the law of the sea, and formed the basis of the rules of the United Nations Convention on the Law of the Sea on the resources of the deep seabed beyond national jurisdiction.[23]

A more recent example of an international declaration of principles is the Dubai Declaration on International Chemicals Management adopted by the International Conference on Chemicals Management in 2006. Together with the so-called Overarching Policy Strategy and the more concrete Global Plan of Action, it forms the basis of the completely nonbinding initative of the Strategic Approach to International Chemicals Management (SAICM). As with other nonbinding instru-

---

[20] On the declarations and the subsequent development of sustainable development, including its normative status M. Fitzmaurice, Contemporary issues in international environmental law, 2009, 61-109; P. Birnie/A. Boyle, International law and the environment, 2002, 82 et seq.; P. Sands, Principles of International Environmental Law, 2003, 143.

[21] International Court of Justice, Gabčikovo-Nagymaros Project (Hungary/Slovakia), Judgment, ICJ Reports 1997, p. 7, para. 140; generally on the application of the concept of sustainable development by international courts P. Sands, "International courts and the application of the concept of 'sustainable development'", Max Planck Yearbook of United Nations Law 3 (1999), 389-405.

[22] UNGA Res. 2749 (XXV) of 17 December 1970.

[23] Compare UNCLOS, Articles 136 and 137; on common heritage and the deep seabed regime, see R. Wolfrum, Die Internationalisierung staatsfreier Räume: die Entwicklung einer internationalen Verwaltung für Antarktis, Weltraum, Hohe See und Meeresboden, 1984, 389-395.

ments, one can in this case observe the development of institutions to underpin the nonbinding form, including in this case an Executive Board, a trust fund and a secretariat.[24]

International conferences do not only adopt declarations of principles, but often supplement these with more concrete action plans. These prescribe measures to be taken at international, regional and national levels for reaching a particular goal, such as sustainable development as in the case of the Agenda 21 or better chemical management as in the case of the Global Plan of Action adopted in 2006 as part of the overall nonbinding SAICM.[25] Sometimes, action plans are programmatic in nature as discussed in the previous section, but oftentimes action plans also contain norms directed at states, organisations and private actors. They are in this sense very similar to international recommendations as discussed in the next section, with the exception that they are often not adopted by established institutions. However, in many cases they are supplemented with some institutional underpinnings, so that the differences are further diminished.

Perhaps the best known and most influential of such action plans in the area of sustainable development are the Agenda 21[26] and the 2002 Johannesburg Plan of Implementation.[27] Adopted by the 1992 Rio Conference, the Agenda 21 in particular has played a significant role by setting out a policy programme for implementing the concept of sustainable development at the national level.[28] This process was enhanced by the creation of the Commission on Sustainable Development (CSD). Created by the United Nations General Assembly in 1992 through a nonbinding General Assembly Resolution, its mandate is, *inter alia*, to examine the progress of the implementation of Agenda 21 at the national, regional and international levels.

Two aspects of this development are remarkable. First, and similarly to the Strategic Approach to Chemicals Management mentioned earlier, it

---

[24] For details, see www.saicm.org.

[25] All of the main documents of the Strategic Approach to International Chemicals Management are available at http://www.saicm.org/documents/saicm%20texts/SAICM_publication_ENG.pdf.

[26] Agenda 21: Programme of Action for Sustainable Development, UN Doc A/Conf.151/26 (1992).

[27] Plan of Implementation of the World Summit on Sustainable Development, UN Doc. A/CONF.199/20 (2002), Resolution 2, Annex.

[28] A.C. Kiss/D. Shelton, International environmental law, 2004, 96.

is remarkable that an institution is created to promote the implementation of a *nonbinding* instrument. And second, the CSD is also authorized to monitor state compliance with the "commitments" and "agreed objectives" of Agenda 21 through a periodic reporting procedure even though the underlying instruments are of nonbinding nature. As will be seen throughout this study, such institutionalised follow up mechanisms are increasingly established in international environmental affairs irrespective of the nonbinding status of the instrument.

These two aspects, institutional underpinnings and follow-up procedures, are distinct elements of support for nonbinding instruments that are increasingly used today. Both are central and recurring themes in this study.

IV. International recommendations adopted by international institutions

International recommendations are nonbinding instruments adopted in decision making bodies of international institutions which prescribe norms for state and/or private actors. The terms used as titles of these instruments vary, but most often one will them being labelled as resolutions, guidelines, action plans and codes of conduct. The term 'code of conduct' is most often used for instruments that contain norms addressed to non-state actors, either exclusively or in addition to norms directed at states.[29]

International recommendations constitute one of the principal means of expression for most international institutions. One of the reasons is that the constitutions or founding treaties only rarely authorise their organs and treaty bodies to adopt instruments that are directly binding for states. In particular in the field of environmental protection, secondary law-making competencies of international organisations remain excep-

[29] J. Friedrich, "Codes of Conduct", in: R. Wolfrum (ed.), Max Planck Encyclopedia of Public International Law, 2010, online at: www.mpepil.com; for a similar approach of identifying codes of conduct as instruments characterised by a particular set of features and thus distinguishable from other instruments see already R.E. Lutz/G.D. Aron, "Codes of Conduct and Other International Instruments" in: G. Handl/R.E. Lutz (eds.), Transferring Hazardous Technologies and Substances: The International Legal Challenge, 1989, 129-151 (153 et seq.).

tional.[30] Therefore, many of the increasingly significant contributions of international institutions in the field of environmental protection and sustainable development are achieved through these nonbinding instruments. The following section will provide an overview of such instruments of the most important international institutions in the field of environmental protection. The overview will include both international organisations as well as, in an exemplary fashion, Multilateral Environmental Agreements (MEAs.).

1. International Organisations

a) United Nations General Assembly

The United Nations General Assembly has the broad power to address any issue within the scope of the UN Charter.[31] Although the environment is not specifically mentioned in the Charter, the broad competences included in Articles 1 and 55 of the Charter, in particular the references to improvement of the standard of living in Article 55 (a) and to the "solution of economic, social health and related problems" in Article 55 (b), can be interpreted to give the UN competence in environmental matters.[32] On those issues within its competence, the General Assembly may make recommendations to Member states and promote progressive development of international law and its codification.[33] While lacking the competence to adopt binding instruments,[34] the reso-

---

[30] The WHO, ICAO and WMO are among the few international organisations which have the power to adopt binding norms, albeit with an opt-out possibility. Generally on secondary law by international organizations J.D. Aston, Sekundärgesetzgebung internationaler Organisationen zwischen mitgliedstaatlicher Souveränität und Gemeinschaftsdisziplin, 2005; M. Benzing, "International Organizations or Institutions, Secondary Law", in: R. Wolfrum (ed.), Max Planck Encyclopedia of Public International Law 2010, online at: www.mpepil.com

[31] UN Charter, Article 10.

[32] Similarly, mentioning in addition the implied powers doctrine, P. Birnie/A. Boyle, International law and the environment, 2002, 48; on Article 55 (a) and (b) of the UN Charter R. Wolfrum, "Article 55 (a) and (b)" in: B. Simma (ed.), The Charter of the United Nations: A Commentary, 2002, 897-917.

[33] Charter of the United Nations, Articles 10 and 13 (1).

[34] J. Brunnée, "International Legislation", in: R. Wolfrum (ed.), Max Planck Encyclopedia of Public International Law, 2010, para. 40, online at: www.mpepil.com; P. Sands, Principles of International Environmental Law, 2003, 81.

lutions adopted by the General Assembly under this competence have often proven to be highly influential, especially when adopted by a broad majority of states or by consensus.[35]

In the environmental field, the General Assembly has directly or indirectly contributed significantly to the changes in the environmental policies of States and of other organisations. Important resolutions on environmental principles included the one on the historical responsibility of states for the preservation of nature of 1979,[36] the resolution by which the General Assembly adopted the World Charter for Nature[37] or the one by which it endorsed the Brundtland Report.[38] The Brundtland Report has significantly shaped the programmatic orientation of the Rio Conferences and the emergence of the concept of sustainable development.[39] In a few instances, the General Assembly addressed concrete substantive issues, as for instance in the area of the protection of marine resources and the marine environment. Its resolution that called for a global moratorium of large-scale pelagic driftnet fishing on the high seas by the end of 1992[40] was widely complied with by states, including the principal large-scale driftnet fishing nations Japan, Korea and Taiwan.[41] Although failing to adopt a similar morato-

---

35 According to Article 18 (2) and (3) of the Charter of the United Nations, recommendations of the General Assembly can be adopted by simple majority or two-thirds majority for important questions which are enumerated or determined by the Assembly.

36 UNGA Res. 34/188 (1979).

37 UNGA Res. 37/7 (1982).

38 UNGA Res. 42/187 (1987).

39 U. Beyerlin, Umweltvölkerrecht, 2000, 14.

40 UN Doc. A/RES/46/215 (1991), para. 3.

41 See the consecutive reports of the UN Secretary-General on this issue, e.g. Report of the Secretary-General, Large-Scale Pelagic Driftnet Fishing and its Impact on the Living Marine Resources of the World's Oceans and Seas, UN Doc A/47/487 (1992), Report of the Secretary-General, Large-Scale Pelagic Driftnet Fishing, Unauthorized Fishing in Zones of National Jurisdiction and Fisheries By-catch and Discards, and Other Developments, UN Doc. A/53/33 (1998); for an analysis G.J. Hewison, "The Legally Binding Nature of the Moratorium on Large-Scale High Seas Driftnet Fishing", Journal of Maritime Law and Commerce 25 (1994), 557-579; D.R. Rothwell, "The General Assembly Ban on Driftnet Fishing" in: D. Shelton (ed.), Commitment and Compliance: The Role of Non-Binding Norms in the International System, 2000, 121-146 (Table 5.1.).

rium with respect to trawling on the bottom of the high seas, the United Nations General Assembly's call upon regional fisheries organisations and their members to address the issue in 2006[42] had led a number of organisations to regulate such fishing practices by 2008.[43]

UNGA resolutions have also been influential on legal development. This is for example illustrated by the resolution on permanent sovereignty over natural resources as an international legal right.[44] In combination with the principle that states have the responsibility not to cause damage to the environment of other states and areas beyond national jurisdiction, as reflected in principle 2 of the Rio Declaration, the principle of sovereignty over natural resources represents one of the cornerstones of international environmental law today.[45] It has since been expressed in a number of important environmental treaties.[46] Acting through resolutions, the United Nations General Assembly has also established a number of institutions such as the United Nations Programme for Environment[47] and the Commission on Sustainable Development[48] that have played –often acting through nonbinding instruments – an important role in fostering the concept of sustainable development.

b) United Nations Environment Programme

The United Nations Environment Programme, established in 1972 by a General Assembly resolution, is the only UN body that is specifically mandated to focus on environmental issues.[49] UNEP is not an international organisation but a subsidiary organ of the United Nations General Assembly. UNEP's main decision-making body, the Governing

---

42 UN Doc. A/Res. 61/105 (2006), para. 83.

43 UN Doc. A/RES/63/112 (2008), para. 105.

44 UNGA Res. 1803 (XVII) (1962).

45 Compare Principle 21 of the Rio Declaration.

46 See e.g. Article 192 United Nations Convention on the Law of the Sea, 1833 UNTS 3, 21 ILM 1261 (1982); Article 15 (1) Convention on Biological Diversity, 1760 UNTS 79, 31 ILM 818 (1992); Preamble of the United Nations Framework Convention on Climate Change, 1771 UNTS 107, U.N. Doc. A/AC.237/18 (Part II)/Add.1, 31 ILM 849 (1992).

47 UN Doc. A/RES/27/2997 (1972).

48 UN Doc. A/RES/47/191 (1992).

49 Compare UN Doc. A/RES/27/2997 (1972), section I, para. 2.

Council, renamed United Nations Environment Assembly of UNEP in 2013, cannot adopt binding instruments, but can recommend (nonbinding) policies to states.[50]

Since its establishment, UNEP has made important contributions to the development and implementation of international and national environmental law and policies[51] by producing and promoting an extensive array of guidelines and codes of ethics on substantive environmental issues.[52] The so-called Goals and Principles of Environmental Impact Assessment[53] for instance have been widely endorsed by states at the national level and prepared the ground for binding treaty law such as the UN ECE Convention on Environmental Impact Assessment.[54] A direct influence of UNEP Guidelines on later Conventions is also notable in the case of the Cairo Guidelines and Principles for the Environmentally

---

50 UN Doc. A/RES/27/2997 (1972), Section I, para. 2 (a). According to rule 48 of the rules of procedure, the Governing Council can decide by majority vote. However, in practice it strives to reach consensus.

51 P. Sands, Principles of International Environmental Law, 2003, 83.

52 These include, in chronological order, the 1978 draft Principles of Conduct in the Field of the Environment for the Guidance of States in the Conservation and Harmonious Utilization of Natural Resources Shared by Two or More States, ILM 17 (1978), 1091; the Guidelines Concerning the Environment Related to Offshore Mining and Drilling Within the Limits of National Jurisdiction, UNEP GC Dec. 10/14/(VI) (1982); the 1985 Montreal Guidelines for the Protection of the Marine Environmental against Pollution from Land-based Sources, UNEP GC Dec. 13/18(II) 1985; the 1987 Cairo Guidelines and Principles for the Environmentally Sound Management of Hazardous Wastes, UNEP GC Dec. 14/30 (1987); the UNEP Goals and Principles of Environmental Impact Assessment adopted in 1987, UNEP GC Dec. 14/25 (1987); the 1987 London Guidelines for the Exchange of Information on Chemicals in International Trade, UNEP GC Dec. 14/27 (1987) and amended by UNEP GC Dec. 15/30; the UNEP Guidelines on Compliance with and Enforcement of Multilateral Environmental Agreements, UNEP GCSS.VII/4 (1992); the UNEP Guidelines for the development of national legislation on access to information, public participation and access to justice in environmental matters, UNEP/GCSS.XI/L.5 (2010); the Guidelines for the development of national legislation on liability, response action and compensation for damage caused by activities dangerous to the environment, UNEP/GCSS.XI/L.5 (2010).

53 UNEP Goals and Principles of Environmental Impact Assessment, UNEP GC Dec. 14/25 (1987).

54 U. Beyerlin, Umweltvölkerrecht, 2000, 13; P. Sands, Principles of International Environmental Law, 2003, 802.

Sound Management of Hazardous Wastes which prepared the ground for the 1989 Basel Convention for the Control of Transboundary Movement of Hazardous Waste. The impact of the UNEP London Guidelines for the Exchange of Information on Chemicals in International Trade on the Convention on the Prior Informed Consent Procedure for Certain Hazardous Chemicals and Pesticides in International Trade (PIC Convention)[55] will be the subject of one of the case studies below.

One of the main objectives of UNEP Guidelines is to provide concrete guidance for national environmental legislation, in particular in developing countries which often have not achieved the regulatory density of most developed countries. In 2010, UNEP's Governing Council at its 11th Special Session in Bali, Indonesia, adopted two sets of Guidelines as guidance for the development of national legislation in issue areas where there have been important legal developments in most developed countries in recent years: the UNEP Guidelines for the development of national legislation on access to information, public participation and access to justice in environmental matters[56] and the UNEP Guidelines for the development of national legislation on liability, response action and compensation for damage caused by activities dangerous to the environment.[57] The process and the discussions leading up to their adoption show how serious states take these instruments even though they are entirely voluntary. The adoption of both sets of Guidelines was preceded by controversial discussions at the 25th Governing Council in 2009 over the inclusiveness of the norm making process. After further rounds of negotiations in 2009, states controversially discussed whether they could be adopted or should be simply taken note of or adopted. Brazil, fearing the potential political and legal effects of the clearly non-

---

[55] Convention on the Prior Informed Consent Procedure for Certain Hazardous Chemicals and Pesticides in International Trade (10 September 1998), 38 ILM 1 (1999).

[56] UNEP Guidelines for the development of national legislation on access to information, public participation and access to justice in environmental matters, UNEP/GCSS.XI/L.5 (2010).

[57] UNEP Guidelines for the development of national legislation on liability, response action and compensation for damage caused by activities dangerous to the environment, UNEP/GCSS.XI/L.5 (2010).

binding Guidelines for other international legal developments and negotiations, strongly tried to prevent their "adoption".[58]

c) Food and Agriculture Organization

The Food and Agriculture Organization (FAO) has become one of the most active organisations in matters of environmental protection. With its objective being "to promote the conservation of natural resources and the adoption of improved methods of agricultural production"[59], it has a clear environmental mandate. The FAO may initiate and approve treaties on food and agriculture,[60] and its plenary body, the FAO Conference, may make recommendations to governments, if necessary by a two-thirds majority vote.[61]

To pursue its objectives in the absence of the competence to adopt binding instruments, the FAO organs have in addition to the initiation of a number of important treaties[62] issued a number of nonbinding recommendations related to the protection of resources and food security. The FAO Code of Conduct on the Distribution and Use of Pesticides[63] and the FAO Code of Conduct for Responsible Fisheries[64] – both sub-

---

58 This information is based on the author's personal experience as a member of the German delegation to the 11th Special Session of the UNEP Governing Council from 24-26 February 2010 in Bali, Indonesia.

59 Constitution of the Food and Agriculture Organization, Article I (2) (c).

60 Constitution of the Food and Agriculture Organization, Article XIV.

61 According to Article IV para. 3 of the Constitution of the Food and Agriculture Organization, this requires a two-thirds majority of the votes cast.

62 International Plant Protection Convention, 6 December 1951, 150 UNTS 67; Convention on Fishing and Conservation of the Living Resources of the High Seas, 29 April 1958, 17 UST 138; 559 UNTS 285; the Convention on the Prior Informed Consent Procedure for Certain Hazardous Chemicals and Pesticides in International Trade, 10 September 1998, 38 ILM 1 (1999); the International Treaty on Plant Genetic Resources for Food and Agriculture, 3 November 2001, FAO Res. 3/2001, available at http://www.planttreaty.org/texts_en.htm.

63 FAO International Code of Conduct on the Distribution and Use of Pesticides, originally adopted by FAO Conference Res. 10/85 of 28 November 1985, see Report of the Conference of FAO on its Twenty-third Session of 9 – 28 November 1985, Annex 1, available at www.fao.org.

64 FAO Code of Conduct for Responsible Fisheries, Rome, FAO, 1995, compare for the text the Report of the Conference of FAO, Twenty-Eighth Ses-

ject of two case studies below[65] – are but two particularly remarkable examples of a number of similar nonbinding initiatives.[66] In addition to codes of conduct, the FAO also develops more concrete international plans of action, for example in the area of forestry[67] and fisheries.[68]

As will be seen in greater detail in the case studies, the FAO employs a number of non-confrontational compliance-enhancing mechanisms to promote the implementation of its nonbinding recommendations. This regularly includes periodic reporting procedures but also sometimes, as in the case of the Code of Conduct for Plant Germplasm Collection and Transfer, even direct complaint procedures for private actors.

d) World Health Organization

The World Health Organization (WHO), established in 1946 to ensure "the attainment of all peoples of the highest possible level of health",[69] has the competence to adopt Conventions[70] and Agreements, regulations,[71] and to make recommendations.[72] In adopting a number of non-

---

sion, 20-31 October 1995, Annex 1, the text is also available at: ftp://ftp.fao.org/docrep/fao/005/v9878e/v9878e00.pdfs.

[65] See in this Part further below, at B.I. and B.II.

[66] Other FAO Codes of Conduct besides the Pesticide Code and the Fisheries Code include the International Code of Conduct for Plant Germplasm Collecting and Transfer, 1993, Report of the Conference of FAO, Twenty-Seventh Session, Appendix E, available at http://www.fao.org; already developed but not yet adopted is the Draft FAO Code of Conduct on Biotechnology as it relates to Genetic Resources for Food and Agriculture, the draft is available at ftp://ftp.fao.org/ag/cgrfa/cgrfa9/r9w18ae.pdf.

[67] The 1985 Tropical Forestry Action Plan developed by the Food and Agriculture Organization was a nonbinding global strategy on forest management and protection in which 74 countries had participated already by 1991.

[68] These are the International Plan of Action for Reducing Incidental Catch of Seabirds in Longline Fisheries, the IPOA for Conservation and Management Sharks and the IPOA for the Management of Fishing Capacity, all adopted in 1999, and the International Plan of Action on Illegal, Unreported and Unregulated Fishing adopted in 2001; all available at http://www.fao.org/fishery/ccrf/2,3/en.

[69] Constitution of the World Health Organization, Article 1.

[70] Constitution of the World Health Organization, Article 19.

[71] Constitution of the World Health Organization, Article 21.

[72] Constitution of the World Health Organization, Article 23.

binding recommendations on health issues with environmental implications, for example air quality and drinking water quality,[73] the WHO has indirectly contributed to environmental standard-setting in these fields. It has also greatly contributed to technical standardisation through the establishment of the Codex Alimentarius Commission in cooperation with the FAO.[74]

e) International Maritime Organization

Mandated "to encourage and facilitate the general adoption of the highest practicable standards in matters concerning the maritime safety, efficiency of navigation and prevention and control of marine pollution from ships",[75] the International Maritime Organization (IMO)[76] has an environmental mandate. It has pursued this mandate primarily through the promotion and further development of important Conventions.[77] But it is also authorised to consider and make recommendations to member states,[78] and has thus adopted a great number of codes of conduct, guidelines and recommendations to prevent and control pollution from ships.[79] The resolutions are usually adopted by consensus of all

---

[73] 1993 WHO Guidelines for Drinking Water Quality; 1999 WHO Air Quality Guidelines; consider also the more recent 2003 WHO Guidelines for safe recreational water environments; all available at www.who.int.

[74] On the Codex Alimentarius Commission, see further below in the section on technical standards at A.VI.1.

[75] Convention on the International Maritime Organization (hereinafter IMO Convention), 289 UNTS 3, Article 1 (a).

[76] The original name was "Inter-Governmental Maritime Consultative Organization".

[77] The most important ones are the International Convention for the Safety of Life at Sea (hereinafter SOLAS), 1 November 1974, 1184 UNTS 278; the Convention on the Prevention of Marine Pollution by Dumping of Wastes and Other Matters (hereinafter London Convention), 29 December 1972, 1046 UNTS 120, 11 ILM 1294 (1972); the International Convention for the Prevention of Pollution from Ships, 2 November 1973, 1340 UNTS 184, 2 ILM 1319 (1973), as modified by the Protocol Relating to the 1973 Convention for the Prevention of Pollution from Ships (17 February 1978, entry into force 2 October 1983), 17 ILM 546 (1978) (hereinafter MARPOL).

[78] IMO Convention, Article 2 (a).

[79] E.g. the 1997 IMO Guidelines to Assist Flag States in the Implementation of IMO Instruments, IMO Assembly Res. A.847(29); the 2002 Revised GE-

IMO Members.[80] Although of nonbinding nature, these recommendations are in practice widely accepted and implemented by states and have contributed to the harmonisation of the formerly widely disparate domestic regulations.[81]

Nonbinding codes and guidelines of the IMO generally serve one of two purposes. They either address issues not (yet) covered by treaty law and are subsequently made legally binding for the respective states through incorporation into an existing treaty or by establishing a new treaty;[82] or they serve to supplement existing treaty law in order to facilitate implementation, as is often the case with respect to highly technical issues requiring flexible norm changes.

Both of these functions are clearly visible in how the IMO has addressed safety and pollution management by ships. In response to growing concern over poor management and safety of ships, the IMO adopted nonbinding guidelines in 1989.[83] After some experience with these guidelines, the International Management Code for the Safe Operation of Ships and for Pollution Prevention (ISM Code) was adopted in 1993.[84] The objectives of the ISM Code are *inter alia* to ensure safety at sea and to prevent pollution arising as a result of inadequate safety and risk management.[85] One of its central requirements is the establishment of safety and management systems for every ship which in-

---

SAMP Hazard Evaluation Procedure for Chemical Substances Carried by Ships, adopted by a IMO/FAO/UNESCO-IOC/WMO/WHO/IAEA/UN/UNEP Joint Group of Experts on the Scientific Aspects of Marine Environmental Protection, GESAMP Reports and Studies, No. 64.

[80] IMO, Implications of the United Nations Convention on the law of the sea for the International Maritime Organization, IMO Doc. LEG/MISC/3/Rev.1 (6 January 2003), p. 5.

[81] J.D. Aston, Sekundärgesetzgebung internationaler Organisationen zwischen mitgliedstaatlicher Souveränität und Gemeinschaftsdisziplin, 2005, 155.

[82] The International Code for the Construction and Equipment of Ships carrying Dangerous Chemicals in Bulk (IBC Code) adopted by the IMO Marine Safety Committee in 1983 through resolution MSC.4(48) became legally binding for chemical tankers constructed on or after 1 July 1986 under chapter VII of the International Convention for the Safety of Life at Sea (SOLAS).

[83] IMO Resolution A.647 (16) of 19 October 1989.

[84] IMO Resolution A.741 (18) of 4 November 1993, Annex (hereinafter ISM Code).

[85] ISM Code, para.1.2.1.

cludes a safety and environmental policy.[86] Interestingly, these safety and management systems must be designed not only to ensure compliance with all mandatory regulations but also to take into account the codes, guidelines and standards recommended by the IMO.[87] After five years of operation, the ISM Code became legally binding in 1998 through the entry into force of an amendment to the International Convention for the Safety of Life at Sea (SOLAS) which had been adopted in 1994. The amendment introduced a new chapter to SOLAS which includes a reference to the ISM Code.[88] In addition, the IMO issues (nonbinding) guidelines on the implementation of these requirements by administrations.[89] The Maritime Safety Committee and the Marine Environment Protection Committee of the IMO are mandated by the Assembly to amend these Guidelines if necessary.[90] This adds additional flexibility by facilitating updates.

An example of a code prescribing highly detailed and technical requirements which later became legally binding through reference is the International Maritime Dangerous Goods Code (IMDG Code) adopted in 1965.[91] This code seeks to prevent pollution and accidents by establishing minimum international norms of safety for the shipping of dangerous goods. It establishes a classification of such goods and, focussing on the particular dangers of each substance, outlines requirements for the sending, transport and packaging of these substances. The specificity and technical nature of these recommendations likens them to technical standards. Just as the ISM Code was later incorporated into binding treaty law, the IMDG Code became legally binding for ratifying Parties in 2004 through the entry into force of an amendment[92] to SO-

86 ISM Code, para. 1.4.

87 ISM Code, para. 1.2.3.

88 SOLAS, Chapter IX.

89 Guidelines on the Implementation of the International Safety management (ISM) Code by Administrations, IMO Resolution A.788(19) (1995), revised through the Revised Guidelines on the Implementation of the International Safety management (ISM) Code by Administrations, IMO Resolution A.913(22) (2001), IMO Doc. A 22/Res.913 (2002), Annex.

90 IMO Assembly Resolution A.913(22) (2001), IMO Doc. A 22/Res.913 (2002), para. 4.

91 IMO Assembly Resolution A81(IV) (1965), revoked by IMO Assembly Resolution A.716(17) (1991).

92 IMO Marine Safety Committee resolution MSC.123(75) (2002).

LAS which introduced a mandatory reference to the IMDG Code. Annex III of MARPOL 73/78 also contains a reference to the IMDG Code which makes the code binding for ratifying Parties. According to Regulation 3 of Part A of Chapter VII, the carriage of dangerous goods in packaged form must now be in compliance with the IMDG Code.[93] It has to be noted however that the IMDG Code had already been widely incorporated into the domestic law of Parties before the amendment.[94]

In some cases, nonbinding instruments of the IMO prepare the ground for new conventions. The IMO considers the introduction of alien invasive species through ballast water to be one of the four greatest threats to the marine environment.[95] The adoption of the Guidelines for the Control and Management of Ships' Ballast Water to Minimize the Transfer of Harmful Aquatic Organisms and Pathogens in 1997[96] answered calls of Agenda 21 to better control ballast water in order to avoid contamination by alien invasive species.[97] The guidelines provided the basis for the development of the International Convention for the Control and Management of Ships' Ballast Water and Sediments.[98] Some of the most important obligations of the Convention include recommendations from the Guidelines, such as the requirement to exchange ballast water in the open sea and to discharge ballast water to onshore reception and treatment facilities. But the Convention also goes beyond the Guidelines in providing for certification requirements

---

93 For instance, Chapter VII, Part A, Regulation 3 of SOLAS now provides that "[T]he carriage of dangerous goods in packaged form shall be in compliance with the relevant provisions of the IMDG Code".

94 F.L. Kirgis, "Specialized law-making processes" in: C.C. Joyner (ed.), The United Nations and International Law, 1997, 65-94 (70-73); J.E. Alvarez, International Organizations as Law-makers, 2005, 221.

95 The other three are marine pollution, destruction of habitat and overexploitation of marine resources.

96 IMO Assembly Resolution A.868(20) (1997), Annex.

97 Agenda 21, UN Doc A/Conf.151/26 (1992), in section 17.30 (a) (iv) demands that "States, acting ... within the framework of IMO ... should assess the need for additional measures to address degradation of the marine environment ...[F]rom shipping by ... [C]onsidering the adoption of appropriate rules on ballast water discharge to prevent the spread of non-indigenous organisms".

98 International Convention for the Control and Management of Ships' Ballast Water and Sediments, adopted on 13 February 2004, not in force as of March 2010).

and the possibility of inspection of ships, including empowering coast guards to delay ballast water discharge if there are concerns.[99] Moreover, as in the other examples mentioned, nonbinding guidelines continue to play a supplementary role under the new Convention. Under the Annex to this Convention, governments must for example take into account the relevant Guidelines developed by the IMO when approving the ballast water management plan of ships.[100] The guidelines are developed by the Marine Environment Protection Committee and adopted in the form of nonbinding resolutions.[101]

The role of the IMO in setting legally binding and nonbinding norms for marine environmental protection is greatly enhanced through references in the Convention on the Law of the Sea[102] to "international rules and standards".[103] Additionally reinforced and safeguarded through the availability of dispute settlement under UNCLOS, the prescription of norms of the IMO is a prime example of the successful interplay of norm setting activities by international institutions (and national legislatures) with an international treaty and an international judiciary.[104]

---

99 International Convention for the Control and Management of Ships' Ballast Water and Sediments, Arts 7, 9.

100 International Convention for the Control and Management of Ships' Ballast Water and Sediments, Annex, Regulation 5.1.

101 IMO Guidelines for Approval of Ballast Water Management Systems (G8), IMO Doc. MEPC.174(58) (2008).

102 United Nations Convention on the Law of the Sea 1982 (10 December 1982), 1833 UNTS 3, 21 ILM 1261.

103 E.g. UNCLOS, in Articles 211, 218, 220; see on this interplay and its legitimacy R. Wolfrum, "Legitimacy of International Law and the Exercise of Administrative Functions: The Example of the International Seabed Authority, the International Maritime Organization (IMO) and International Fisheries Organizations" in: A. von Bogdandy/R. Wolfrum/J. von Bernstorff/P. Dann/M. Goldmann (eds.), The Exercise of Public Authority by International Institutions: Advancing International Institutional Law 2010, 917-940; for details see also the assessment in Part 2, at A.I.2., further below.

104 R. Wolfrum, "Legitimacy of International Law and the Exercise of Administrative Functions: The Example of the International Seabed Authority, the International Maritime Organization (IMO) and International Fisheries Organizations" in: A. von Bogdandy/R. Wolfrum/J. von Bernstorff/P. Dann/M. Goldmann (eds.), The Exercise of Public Authority by International Institutions: Advancing International Institutional Law 2010, 917-940 (936).

f) International Civil Aviation Organization

The International Civil Aviation Organization (ICAO) was founded by Article 43 of the Convention on International Civil Aviation (Chicago Convention)[105] with a view to promoting efficient and economical air transport,[106] but also more generally the development of "all aspects of international civil aeronautics".[107] Although environmental issues were not specifically foreseen in the Chicago Convention, rising concern over noise and – later on – about aircraft engine emissions led the ICAO Assembly to take on environmental issues as early as 1968 (noise pollution) and 1971 (engine emissions).

The main instrument used by ICAO is the adoption of so-called "standards and recommended practices" (SARPs) by the ICAO Council.[108] SARPs are developed and recommended to the Council by the Council's Committee on Aviation Environmental Protection consisting of states, intergovernmental and non-governmental organisations representing industry and environmental interests. SARPs on aircraft engine emissions and aircraft noise have first been adopted in 1971 (noise pollution) and 1981 (engine emissions) and since then been continuously revised and tightened. Generally speaking, they include specific emission standards for various polluting substances such as oxides of nitrogen or carbon monoxide as well as noise levels, which provide the certification standard for national authorities. In addition to precise standards, ICAO also produces a number of recommended practices for state authorities. The SARPs, i.e. both standards and the recommended practices, are included in Annex 16 to the Chicago Convention.[109]

In contrast to Multilateral Environmental Agreements, the inclusion in an Annex does not necessarily render these SARPs legally binding. The standards and practices are included in the Annex to the International Convention on Civil Aviation "for convenience",[110] but the various

---

[105] Convention on International Civil Aviation (hereinafter Chicago Convention), 7 December 1944, 15 UNTS 295.

[106] Chicago Convention, Article 44 (d) and (h).

[107] Chicago Convention, Article 44 (i).

[108] The Council is composed of 36 representatives of members elected by the ICAO Assembly (Article 50 Chicago Convention). The Council can take decisions by majority vote if necessary, Article 54 (l) and 52 Chicago Convention.

[109] Chicago Convention, Annex 16, Volume I (on noise reduction) and Volume II (on aircraft engine emissions).

[110] Chicago Convention, Article 54 (l).

Annexes have not been ratified with the Chicago Convention, nor does the Chicago Convention clearly demand the ratification of these Annexes or clearly gives the ICAO Council the power to enact legally binding norms. The Chicago Convention leaves the legal status of SARPs in "purposeful ambiguity".[111] It only establishes a rather softly-worded obligation by stipulating that "each contracting State undertakes to collaborate in securing the highest practicable degree of uniformity in regulations, standards, procedures and organization" ... and "to this end the International Civil Aviation Organization shall adopt and amend ... international standards and recommended practices and procedures..." [112] In referring to the "highest practicable degree", the wording of the provision does not clearly render the standards and recommended practices legally binding. States only have to notify the Council if they cannot comply with a newly adopted standard.[113] A similar obligation for recommended practices is however missing.

In addition to SARPs, ICAO produces a sizeable amount of (nonbinding) guidance material, including for instance the Airport Air Quality Guidance Manual issued in 2007,[114] the Guidance on the Balanced Approach to Aircraft Noise Management[115] or an ICAO Circular on operational opportunities to minimize fuel use and reduce emissions.[116] A number of resolutions calling on states to adopt various environmental measures, recently in particular on climate change related measures,[117] complete the toolbox of recommendatory instruments of the organisation.

ICAO has recently been focussing on the reduction of greenhouse gas emissions. Its responsibility for addressing the reduction of greenhouse gases is recognized by the Kyoto Protocol. According to Article 2 (2) of the Kyoto Protocol, Annex I (developed) states have the responsibility

---

[111] J.E. Alvarez, International Organizations as Law-makers, 2005, 223.

[112] Chicago Convention, Article 37.

[113] Chicago Convention, Article 38.

[114] Airport Air Quality Guidance Manual, 2007, available at http://www.icao.int/icaonet/dcs/9889/9889_en.pdf.

[115] ICAO, Guidance on the Balanced Approach to Aircraft Noise Management (2004, revised 2007).

[116] ICAO Circular 303-AN/176 (2004).

[117] E.g. ICAO Assembly Res. A36-22, Appendix L urges States to adopt measures relating to emissions related charges and taxes, emissions trading, carbon offsets and clean development mechanism.

for limiting or reducing greenhouse gas emissions from aviation bunker fuels by cooperating through ICAO.[118] In line with this obligation of states, ICAO is currently working on developing specific policies, including guidance on an emissions trading system for aircraft. A document providing a template and guidance on voluntary measures and a so-called "draft guidance" on the use of emissions trading has already been issued.[119]

g) International Atomic Energy Agency

The International Atomic Energy Agency (IAEA) has the mandate to develop the peaceful uses of nuclear energy.[120] One of the main activities of the IAEA besides its role as the nuclear inspectorate under the Treaty on Non-Proliferation of Nuclear Weapons[121] and the facilitation of treaty-making[122] is to establish standards for the safety of nuclear technology and related health concerns.[123] Of course, the safety of nuclear technology has environmental implications. To achieve its standard-setting task, the plenary of the IAEA, the General Conference,[124] dis-

---

[118] Article 2 (2) of the Kyoto Protocol to the United Nations Framework Convention on Climate Change, 10 December 1997, UN Doc FCCC/CP/1997/7/Add.1, 37 ILM 22 (1998).

[119] ICAO, Draft Guidance on the use of Emissions Trading for Aviation, ICAO Doc. 9885; the template and guidance on Voluntary Measures is available at http://www.icao.int/env/measures.htm.

[120] Constitution of the International Atomic Energy Agency, Article II.

[121] Article III of the Treaty on the Non-proliferation of Nuclear Weapons, 1 July 1968, 729 UNTS 161; 7 ILM 8809 (1968). 35 ILM 1439 (1996).

[122] The IAEA has initiated, sponsored and provides secretariat functions for a number of treaties with relevance for environmental protection such as the Vienna Convention on Civil Liability for Nuclear Damage, 21 May 1963, 1063 UNTS 265, 727 ILM 2 (1963); the 1980 Convention on the Physical Protection of Nuclear Material, 18 ILM 1419; the 1994 Convention on Nuclear Safety, 17 June 1994, 1963 UNTS 293, 33 ILM 1514 (1994).

[123] Article III.A.6 of the IAEA Statute authorises the Agency to adopt standards of safety for the purposes of protecting health and minimizing danger to life and property.

[124] The principal organs of the IAEA are the General Conference of all IAEA member states, the Board of Governors and the secretariat headed by a Director General.

poses of the competence to make recommendations to member states, if necessary acting through majority vote.[125]

In accordance with this mandate, the IAEA has developed and adopted a number of advisory recommendations aimed at protecting people and the environment from harmful radiation. One can generally distinguish between safety standards and codes of conduct. The safety standards prescribe detailed norms covering nuclear safety, radiation protection, safe waste management, the transport of radioactive materials, the safety of nuclear fuel cycle facilities and quality assurance.[126] The codes of conduct, namely the Code of Conduct on the Safety and Security of Radioactive Sources[127] and the Code of Practice on the International Transboundary Movement of Radioactive Waste[128], provide more general advice on policies that states should adopt. These include, *inter alia*, checklists on necessary legislation, the designation of regulatory bodies with specific regulatory competencies in this area as well as prior notification and consent mechanisms for exports and imports of radioactive sources[129] and waste.[130] In the case of the Code of Conduct on the Safety and Security of Radioactive Sources, the IAEA additionally issued a further guiding document on how best to comply with the import-export related provisions of the code of conduct.[131]

---

125 IAEA Statute, Article V.D.

126 E.g. Regulatory Control of Radioactive Discharges into the Environment (2000), Safety Guide No. WS-2-G.3; Requirements for Near Surface Disposal of Radioactive Waste (1990), WS-R-1; for details, see http://www.iaea.org/Publications/Standards/index.html.

127 Compare IAEA Code of Conduct on the Safety and Security of Radioactive Sources, IAEA GC(47)/RES/7 (19 September 2003), Part B. The code is published as IAEA/CODEOC/2004, available at: http://www-pub.iaea.org/MTCD/publications/PDF/Code-2004_web.pdf.

128 IAEA, Code of Practice on the International Transboundary Movement of Radioactive Waste, Requirements, TS-R-1, IAEA GC(XXXIV)/920 (1990), Annex 1, printed in 30 ILM (1991), available also at: http://www.iaea.org/Publications/Documents/Infcircs/Others/inf386.shtml; for an analysis see P. Sands, Principles of International Environmental Law, 2003, 697-698.

129 Code of Conduct on the Safety and Security of Radioactive Sources, paras 23-29.

130 Code of Practice on the International Transboundary Movement of Radioactive Waste, paras 5 et seq.

131 IAEA Guidance on the Import and Export of Radioactive Sources (2005), IAEA/CODEOC/IMP-EXP/2005.

Also of interest are the further activities of the IAEA to promote observance with these instruments. In the case of the Code of Conduct on the Safety and Security of Radioactive Sources, states are asked to write to the Director General expressing their support for the code and the additional guidance and their willingness to work towards following the guidance as a sign of their "political commitment".[132] The Director-General then publishes a regularly updated list of states which have notified the Director General accordingly. According to the list, 94 states have done so.[133] States are also asked to designate national contact points for the prior notification and consent mechanism. Again, a list of states that have done so is published; this also shows compliance by an overwhelming majority of states.[134] Finally, ICAO also urges each state to respond to a self-assessment questionnaire regarding their efforts to follow the guidance, and to make available an update on the responses if they change. A list of states that have sent back the questionnaire is also published.[135]

These subtle pressure tools are employed despite the fact that the recommendations are clearly nonbinding. Nothing in the Statute bestows binding force upon them, nor does it authorize the IAEA to ask for responses by states.[136] Although not legally bound, states nevertheless widely comply with the IAEA recommendations,[137] or strive to work towards their implementation. Commentators explain their effectiveness with the high and widely accepted level of expertise that the IAEA

---

[132] IAEA GC (47)/RES/7.B (2003), para. 4 and 5; IAEA GC (48)/RES/10.D (2004), para. 8.

[133] The list of States that have a made a political commitment with regard to the Code of Conduct on the Safety and Security of Radioactive Sources and the Supplementary Guidance on the Import and Export of Radioactive Sources is available at http://www.iaea.org/Publications/Documents/Treaties/codeconduct_status.pdf.

[134] See http://www.iaea.org/Publications/Documents/Treaties/codeconduct_status.pdf.

[135] See http://www.iaea.org/Publications/Documents/Treaties/codeconduct_status.pdf.

[136] P. Birnie/A. Boyle, International law and the environment, 2002, 457.

[137] J.E. Alvarez, International Organizations as Law-makers, 2005, 231; B. Kellman, "Protection of Nuclear Materials" in: D. Shelton (ed.), Commitment and compliance: the role of non-binding norms in the international legal system, 2000, 486-505 (495).

and its instruments represent.[138] Experts from the field stress that nonbinding recommendations fill in the gaps of treaty law, because they provide a much-needed method of harmonizing state practices in a highly technical field without the need to resort to lengthy and cumbersome treaty making or treaty amendment.[139] The nonbinding recommendations furnish the necessary details and guidance without intruding on sovereignty in such a sensitive area for states.[140]

h) Organisation for Economic Co-operation and Development

A good example of an organisation that shapes international and national environmental policy and law through nonbinding instruments is the Organisation for Economic Co-operation and Development (OECD). In contrast to most other international organisations, the main decision-making body of the OECD, the OECD Council, has the competence to adopt decisions binding upon OECD member states in addition to nonbinding recommendations.[141] Nevertheless, the OECD Council makes only infrequent use of this possibility with respect to environmental matters.[142]

(1) Environmental recommendations addressed to states

While only mandated to promote policies to achieve economic growth, economic development and the expansion of world trade,[143] the OECD Council started to address environmental issues as early as the 1970s. Among the reasons for the turn to environmental issues were the differences among environmental standards of OECD member states which

---

138 P. Birnie/A. Boyle, International law and the environment, 2002, 456.

139 B. Kellman, "Protection of Nuclear Materials" in: D. Shelton (ed.), Commitment and compliance: the role of non-binding norms in the international legal system, 2000, 486-505 (487).

140 B. Kellman, "Protection of Nuclear Materials" in: D. Shelton (ed.), Commitment and compliance: the role of non-binding norms in the international legal system, 2000, 486-505 (495).

141 Article 5 a) of the Convention on the Organisation for Economic Co-operation and Development (14 December 1960), 888 UNTS 179 (hereinafter OECD Convention).

142 A.C. Kiss/D. Shelton, International environmental law, 2004, 92.

143 OECD Convention, Article 1.

were assumed to have negative economic, political and trade implications, as well as the perception that the capacities of some member states to address environmental issues were too limited.[144]

The environmental issues thus addressed are manifold. A number of recommendations concern procedures for environmental protection, including transboundary environmental protection[145], the polluter-pays principle,[146] environmental impact assessment,[147] the use of economic instruments in environmental policy[148], integrated pollution prevention and control[149], pollutant release and transfer registers,[150] "green" public procurement[151] and environmental information.[152] In addition to these mostly procedural means, the OECD Council has also addressed a number of substantive issues, sometimes acting through binding decisions but mostly by means of nonbinding recommendations. These include recommendations on air and water quality,[153] environment and

---

144 P. Sands, Principles of International Environmental Law, 2003, 103.

145 The most important ones of these Recommendations were the OECD Recommendation on Principles Concerning Transfrontier Pollution, C(74)224 (1974); the OECD Recommendation on Equal Right of Access in Relation to Transfrontier Pollution, C(76)55 (1976); the OECD Recommendation on Implementation of a Regime of Equal Right of Access and Non-Discrimination in Relation to Transfrontier Pollution, C(77)28 (1977); the OECD Recommendation on Strengthening International Co-operation on Environmental Protection in Transfrontier Regions, Recommendations C(78)77 (1978).

146 OECD Recommendation on the Implementation of the Polluter-Pays Principle, C(74)223 (1974); OECD Recommendation on the Application of the Polluter-Pays Principle to Accidental Pollution, C(89)88 (1989).

147 E.g. the OECD Recommendation on the Assessment of Projects with Significant Impacts on the Environment, C(79)116 (1979).

148 OECD Recommendation on the Use of Economic Instruments in Environmental Policy, C(90)177 (1991).

149 OECD Recommendation on Integrated Pollution Prevention and Control, C(90)164 (1990).

150 OECD Recommendation C(96)41 (1996).

151 OECD Recommendation C(2002) 3 (2002).

152 OECD Recommendation on Reporting on the State of the Environment, C(79)114 (1979), OECD; Recommendation on Environmental Indicators and Information, C(90)165; OECD Recommendation on Environmental Information C(98)67 (1998).

153 E.g. the OECD Recommendation on Control of Air Pollution from Fossil Fuel Combustion, C(85)101 (1985); OECD Recommendation on the Con-

tourism,[154] energy[155], noise[156], the transboundary movement of waste[157] and chemicals.[158]

The OECD is generally quite successful in influencing national legal systems and policies through nonbinding instruments. Its recommendations in the field of the environment have had a significant influence on the national legal orders of OECD member states.[159] They have also shaped environmental standards in other regions and at the global level.[160] Salient examples of this influence include the implementation of the polluter-pays principle in OECD states and procedures allowing for access for claimants of neighbouring states to national remedies, procedures and information in cases of transboundary pollution as recommended in the above mentioned OECD Recommendations.[161]

---

trol of Eutrophication of Waters, C(74)221 (1974); Recommendation on Control of Water management Policies and Instruments, C(85)101 (1985).

154 OECD Recommendation on Environment and Tourism, OECD C(79)115 (1979).

155 E.g. OECD Recommendation on Coal and the Environment, C(79)117 (1979); Recommendation on Environmentally Favourable Energy Options and their Implementation, C(85)102 (1985).

156 E.g. OECD Recommendation on Noise Prevention and Abatement, C(74)217 (1974); OECD Recommendation on Strengthening Noise Abatement Policies, C (85)103 (1985).

157 OECD Recommendation on Transfrontier Movements of Hazardous Waste C(83)180 (1984); OECD Resolution on International Cooperation Concerning Transfrontier Movements of Hazardous Waste, C(85)100 (1985); OECD Recommendation on Exports of Hazardous Wastes from the OECD Area, C(86)64 (1986); OECD Recommendation on Transfrontier Movements of Hazardous Wastes, C(88)90 (1988); OECD Recommendation on the Reduction of Transfrontier Movements of Wastes, C(90)178 (1991).

158 E.g. OECD Recommendation on Measures to Reduce All Man-Made Emissions of Mercury to the Environment, C(73)172 (Final) (1973); OECD Recommendation on Information Exchange Related to Export of Banned or Severely Restricted Chemicals, C(84)37 (Final) (1984).

159 U. Beyerlin, Umweltvölkerrecht, 2000, 79.

160 P. Sands, Principles of International Environmental Law, 2003, 103.

161 See on this work P. Birnie/A. Boyle, International law and the environment, 2002, chapters 3 and 5.

(2) OECD Recommendation on Common Approaches on the Environment and Officially Supported Export Credits

The OECD Recommendation on Common Approaches on the Environment and Officially Supported Export Credits[162] is a nonbinding instrument which seeks to establish environmental minimum requirements for the export credit policies of member states.

The loans, insurances and guarantees provided by export credit agencies to domestic companies investing abroad constitute one of the largest sources of global public financing. They often support large infrastructure projects with significant environmental impact. Traditionally, export credit agencies have not taken the environmental impact of financed projects into consideration; both environmental regulation and enforcement were left to host governments. In 1998, the OECD responded to pressure from the United States and NGOs[163] and developed – through its Trade Committee's Working Party on Export Credits and Credit Guarantees – the draft "Recommendation on Common Approaches on Environment and Officially Supported Export Credits"[164]. The Recommendation was not adopted immediately, but was adopted unanimously by the OECD Council in 2003 and revised in 2007.[165] The Recommendation distinguishes between various categories of projects according to their environmental sensitivity. For the most sensitive projects, the Recommendation calls for the preparation of an

---

[162] OECD, Revised Council Recommendation on Common Approaches on the Environment and Officially Supported Export Credits, TAD/ECG(2007)9, available at http://web domino1.oecd.org/olis/2007doc.nsf/Linkto/tad-ecg(2007)9.

[163] The Export-Import Bank of the United States had as the first Export Credit Agency developed and applied minimum environmental standards, but found itself at a competitive disadvantage to Export Credit Agencies of other industrialized states which had not adopted such policies. The United States hence started lobbying OECD members that common standards be established for all OECD member states. The efforts of the United States were supported by a global NGO campaign led by ECA Watch.

[164] Draft OECD Recommendation on Common Approaches on Environment and Officially Supported Export Credits and Credit Guarantees, OECD Doc. TD/ECG(2000)11/REV6 (December 14, 2001).

[165] OECD, Revised Council Recommendation on Common Approaches on the Environment and Officially Supported Export Credits adopted by the OECD Council of 12 June 2007, TAD/ECG(2007)9, available at http://www.olis.oecd.org/olis/2007doc.nsf/linkto/tad-ecg(2007)9.

Environmental Impact Statement before the final commitments, for transparency during the review process and for notification of and consultation with affected groups. member states are to submit annual progress reports to the Working Party on Export Credits.

Although the draft recommendation at first failed to be adopted by the OECD Council, its standards were nevertheless soon adopted throughout the OECD even without formal adoption. Most OECD member states have adopted the Common Approaches domestically as internal agency policies rather than national legislative acts.[166] As illustrated by the recent decision of Austria, Germany and Switzerland to suspend support for the controversial Ilisu Dam project in Turkey,[167] states also take decisions to enforce the standards. Some states, including the United States, the United Kingdom and Canada and Japan, even go beyond the minimal standards of the Recommendation in their implementing practice.[168]

(3) OECD Guidelines addressed to private actors

A particularly interesting example of a nonbinding initiative of the OECD is one of its oldest. With the OECD Guidelines for Multinational Enterprises,[169] the OECD aims to create rules and standards addressed directly to multinational corporations on a variety of issues. Since 1991, the Guidelines also include an environmental chapter. The OECD Guidelines are particularly interesting not only for directly addressing private actors but also for their novel compliance mechanism. For this reason, the OECD Guidelines will be analysed in more detail in one of the case studies below.[170]

---

166 For details on the implementation in Germany, see the discussion on national implementation through distributive administration in Part 2, at B.II.2.c)(2)., further below.

167 Compare the Article in the Financial Times of 23 December 2008, available at http://www.ft.com/cms/s/0/f08aa2d2-d091-11dd-ae00-000077b07658.html.

168 J. Salzman, "Decentralized Administrative Law in the Organization for Economic Cooperation and Development", Law and Contemporary Problems 68 (2005), 189-224 (210).

169 OECD, Guidelines for Multinational Enterprises, 15 I.L.M. 969 (1976); the Guidelines were updated in 2011 and are in this version available at http://www.oecd.org/dataoecd/43/29/48004323.pdf.

170 See in this Part 1, at B.III., further below.

2. *Multilateral Environmental Agreements*

International environmental law is increasingly developed, monitored and – sometimes – enforced by the treaty bodies of Multilateral Environmental Agreements (MEAs). This specialisation and sectoralisation has also brought about a fragmented landscape of dynamically evolving independent treaty bodies acting separately from each other. MEAs such as the Framework Convention on Climate Change and the Kyoto Protocol have created a complex web of institutional bodies that hardly differ functionally from formal international organisations.[171] They usually comprise a Conference of the Parties which acts as the plenary and main decision making body, a secretariat and possibly a number of further treaty bodies established to implement the parent Convention and its Protocols.

In the attempt to overcome or at least alleviate the strict consent principle of traditional international law doctrine which is often at odds with the exigencies of environmental law for flexible and adaptive regulation, some Conferences of the Parties are authorised to take decisions which are legally binding upon Parties, for instance on adjusting Annexes to the Protocols of the treaties which are directly binding for Parties. In some cases, for example under the Montreal Protocol on Substances that Deplete the Ozone Layer (Montreal Protocol)[172] or the Convention on International Trade in Endangered Species of Wild Fauna and Flora (CITES),[173] such decisions can even be taken by a two-thirds majority if consensus fails.

However, nonbinding instruments also play an important role under MEAs. Even under those MEAs where COPs can take binding decisions for member states, the respective treaty bodies often resort to explicitly nonbinding recommendations. The OSPAR Commission established under the Convention for the Protection of the Marine Envi-

[171] It is increasingly argued that institutionalization, autonomous will formation and the ability of Conferences of the Parties to enter into agreements in practice has evolved to a point where they can no longer be distinguished from traditional international organisations. In this sense compare M. Schroeder, Die Koordinierung der internationalen Bemühungen zum Schutz der Umwelt, 2005, 169 et seq.; different U. Beyerlin, Umweltvölkerrecht, 2000, 80.

[172] Montreal Protocol on Substances that Deplete the Ozone Layer (16 September 1987), 1522 UNTS 3, 26 ILM 1550 (1987), Article 2 (9) (c).

[173] Convention on International Trade in Endangered Species of Wild Fauna and Flora (3 March 1973), 993 UNTS 243, Article XV.

ronment of the North-East Atlantic (OSPAR Convention)[174] for instance has the power to adopt legally binding decisions as well as recommendations, both by a three-quarters majority.[175] Nevertheless the OSPAR Commission uses non-binding recommendations alongside binding decisions.

This suggests that nonbinding instruments are not simply used due to lack of competence, but that there is a need for nonbinding cooperation even when comparatively flexible binding instruments are also available. COPs resort to nonbinding instruments to supplement treaty norms. The nonbinding instruments serve as supplements to the treaty system, and are deliberately used as another option instead of changing an Annex or striving for an amendment to the treaty which would require lengthy and burdensome ratification processes.

As will be seen, nonbinding recommendations, often adopted as guidelines or codes of conduct, are particularly often employed within those MEAs which do not comprise a flexible law making possibility beside amendment procedures.[176] For other MEAs such as the Montreal Protocol or CITES that can be characterised by a high level of legalisation and comprise over some kind of flexible law making, nonbinding instruments are less prominent but nevertheless still used.[177] Overall, however, the need for nonbinding instruments appears to be lower when a treaty is already highly specified through Protocols, and where treaty bodies under these Protocols have the power to take binding de-

---

[174] Convention for the Protection of the Marine Environment of the North-East Atlantic (22 September 1992), 2354 UNTS 67; 32 ILM 1069 (1993).

[175] OSPAR Convention, Articles 10 (3) and 13 (1), (2).

[176] Compare only the Guidelines adopted by the Conference of the Contracting Parties of the Convention on Wetlands of International Importance especially as Waterfowl Habitat (22 February 1971) (hereinafter Ramsar Convention), 996 UNTS 245; 11 ILM 963 (1972); the Guidelines are available at http://www.ramsar.org/key_guidelines_index.htm.

[177] For the Montreal Protocol see the next section; for CITES see e.g. the recommendations of the Standing Committee of CITES recommending to states the suspension of trade with certain parties as a consequence of findings of non-compliance by these parties, compare e.g. CITES Notification 2009/003 of 3 February 2009, available at http://www.cites.org/eng/notif/2009/E003.pdf (for a similar assessment H.E. Ott, Umweltregime im Völkerrecht: eine Untersuchung zu neuen Formen internationaler institutionalisierter Kooperation am Beispiel der Verträge zum Schutz der Ozosnschicht und zur Kontrolle grenzüberschreitender Abfallverbringungen, 1998, 208).

cisions. Conferences of the Parties that do not have an explicit competence to adopt binding decisions, such as the ones under the Convention on Biological Diversity[178] and the United Nations Framework Convention on Climate Change (UNFCCC)[179] produce a great number of nonbinding recommendatory instruments.[180]

a) Vienna Convention and Montreal Protocol on Substances that Deplete the Ozone Layer

The Montreal Protocol is an example of a treaty that is defined by highly complex and detailed rules. It sets out a flexible law-making mechanism according to which the Meeting of the Parties (MOP) can adopt binding changes to the Annexes of the Protocol even with a two-thirds majority, and without the possibility of opting-out.[181] In light of these characteristics, the need for the MOP to resort to nonbinding instruments in addition to its binding decision making appears comparatively weak.

Nevertheless, even within this treaty regime, nonbinding instruments play a role. Nonbinding instruments were often employed in the earlier days of the Montreal Protocol when new scientific findings constantly required reconsideration of the rules of the Protocol. Scientists revealed threats by substances that had so far not been regulated or at least not with the necessary stringency. While the Protocol allowed making adjustments and amendments to the Protocol and the Annexes, binding adjustments sometimes did not find the immediate support of all State Parties, or some needed longer transitional periods.

---

178 Convention on Biological Diversity (5 June 1992), 1760 UNTS 79; 31 ILM 818 (1992), for more details on the Convention on Biological Diversity and the use of nonbinding instruments in its context see further below in this section.

179 Compare United Nations Framework Convention on Climate Change (9 May 1992), 1771 UNTS 107; 31 ILM 849 (1992), Article 7 (2) (g).

180 See e.g. the COP-Decision 2/CP.13 entitled "Reducing emissions from deforestation in developing countries: approaches to stimulate action", para. 1. The Conference of the Parties "[I]nvites Parties to further strengthen and support ongoing efforts to reduce emissions from deforestation and forest degradation on a voluntary basis." It contains a list of measures in an Annex.

181 Montreal Protocol, Article 2.9.

In this situation, nonbinding resolutions and declarations served as a means to build consensus by all states, to exert political pressure on slow states and as a basis for certain groups of states to move ahead faster. The Helsinki Declaration of 1989 adopted at the first Conference of the Parties in 1989 is an example of a nonbinding resolution being employed to build consensus and keep the issue of more stringent measures on the agenda. By means of this Declaration, which was included in the final report, all Governments declared that they agreed on phasing out the production and consumption of chlorofluorocarbons controlled by the Montreal Protocol as soon as possible but not later than the year 2000, and for that purpose to tighten the timetable agreed upon in the Montreal Protocol.[182] In a similar example, all State Parties expressed their intent to reduce emissions of methyl bromide after becoming aware that the substance is more dangerous than previously thought, and as substitutes became available.[183] The issue of the phasing out of methyl bromide and other substances was further addressed by a number of declarations by groups of states, using declarations as tools to build political pressure and as a basis for cooperation beyond the legal requirements. A number of states most of which are developed, repeatedly declared that they would phase out methyl bromide faster than required, and that they recommend to others to do the same.[184] Similarly, a number of states declared at COP 5 that they would limit the use of hydrofluorocarbons – so far used as a substitute for other ozone-depleting substances – to absolutely necessary applications and to phase out the consumption as soon as possible but not later than the year 2015.[185] Another resolution expressed the consensus of State Parties to refrain from authorising the use of halons, although the authorisation of such use was not prohibited through the respective Annex A of the Montreal Protocol.[186]

With the inclusion of a number of additional substances and more stringent timetables into the Montreal Protocol over time, nonbinding resolutions and declarations are now used less frequently. In some in-

---

182 UNEP/OzL.Pro.1/5, Appendix I.

183 UNEP/OzL.Pro.4/15, Annex XV.

184 See for example the Declarations on Methyl Bromide at the Fifth Meeting and Seventh Meeting of the Parties to the Montreal Protocol, UNEP/OzL. Pro.5/12, Annex XV, UNEP/OzL.Pro.7/12, Annex X.

185 See in particular UNEP/OzL.Pro.5/12, Annex VI.

186 UNEP/OzL.Pro.2/3, Annex VII.

stances, however, states still resort to nonbinding instruments to address specific issues insufficiently addressed by the Protocol. A decision of the MOP on the problem of illegal trade in ozone-depleting substances outlines that in order to address this issue more effectively, State Parties "may wish to consider implementing domestically on a voluntary basis the following measures ...".[187] More recently, Declarations adopted by all Governments have been used to show the commitment of states to implementing the Montreal Protocol,[188] or, as in the 2008 Doha Declaration, to show their commitment to undertake specific efforts such as the replenishment of the Multilateral Fund or the efforts to destroy remaining banks of ozone-depleting substances.[189] Participants in negotiations on the Montreal Protocol report that these nonbinding instruments regularly form part of the package deals over which states bargain at international conferences.[190]

b) Convention on Biological Diversity

Most of the products of the treaty bodies of the Convention on Biological Diversity are nonbinding. The COP of the Convention on Biological Diversity does not have the competence to adopt legally binding decisions.[191] Instead of opting for a treaty amendment, the COP therefore regularly takes legally nonbinding decisions to further the implementation of the provisions of the Convention on Biological Diversity and thereby fulfil its mandate.[192]

---

[187] UNEP/OzL.Pro.19/7, Decision XIX/12.

[188] Montreal Declaration, UNEP/OzL.Pro.19/7, Annex IV.

[189] Doha Declaration, UNEP/OzL.Conv.8/7-UNEP/OzL.Pro.20/9, Annex IV.

[190] H.E. Ott, Umweltregime im Völkerrecht: eine Untersuchung zu neuen Formen internationaler institutionalisierter Kooperation am Beispiel der Verträge zum Schutz der Ozonschicht und zur Kontrolle grenzüberschreitender Abfallverbringungen, 1998, 213.

[191] Convention on Biological Diversity (5 June 1992), 1760 UNTS 79; 31 ILM 818 (1992), Article 23 (4) c) – f).

[192] Article 23 (4) (i) of the Convention on Biological Diversity requires the Conference of the Parties to "consider and undertake any additional action that may be required for the achievement of the purposes of this Convention in light of the experiences gained in its operation".

For example, a number of guidelines have recently been adopted through COP decisions with a view to furthering the implementation of in-situ conservation regulated in Article 8 of the Convention on Biological Diversity. This concerns in particular the issue of indigenous peoples' participation and protection, which is recognized as significant for any in-situ conservation strategy in Article 8 (j) of the Convention on Biological Diversity. Guidelines on environmental impact assessment requirements have already been adopted.[193] A code of ethical conduct to ensure respect for the cultural and intellectual heritage of indigenous and local communities relevant for the conservation and sustainable use of biological diversity is being development at the time of writing.[194]

Another example is that of the so-called Bonn Guidelines on Access to Genetic Resources and Fair and Equitable Sharing of the Benefits Arising out of Their Utilization.[195] With the Bonn Guidelines, the COP aimed to concretise the broadly framed provisions on access and benefit sharing of the Convention[196] and to outline possible elements of state measures and bilateral agreements on access and benefit sharing. The COP thus did not attempt to settle the complex issue, but instead resorted to an expressly nonbinding instrument as a first step in an evolutionary process towards an international agreement.[197] The nonbinding

---

193 E.g. the so-called Akwé: Kon Voluntary Guidelines for the Conduct of Cultural, Environmental and Social Impact Assessments regarding Developments Proposed to Take Place on, or which are likely to Impact on, Sacred Sites and on Lands and Waters Traditionally Occupied or Used by Indigenous and Local Communities, included in COP Decision VII/16, para. F, Annex, UN Doc. UNEP/CBD/COP/7/21.

194 The 9th COP in 2008 agreed on "draft elements of a code of ethical conduct to ensure respect for the cultural and intellectual heritage of indigenous and local communities relevant for the conservation and sustainable use of biological diversity", compare COP 9 Decision IX/13 "Article 8(j) and related provisions", UN Doc. UNEP/CBD/COP/DEC/IX/13.

195 Bonn Guidelines on Access to Genetic Resources and Fair and Equitable Sharing of the Benefits Arising out of Their Utilization, Part A, CBD Dec. VI/24, 2002, UN Doc. UNEP/CBD/COP/6/20, available at www.biodiv.org.

196 Convention on Biological Diversity, Articles 8(j), 10 (c), 15, 16 and 19.

197 An international agreement is envisaged for the 10th Conference of the Parties in Nagoya in 2010. For the past difficulties and opinions see Compilation of Submissions Provided by Parties, Governments, International Organizations, Indigenous and Local Communities, and Relevant Stakeholders in

instrument was perceived as being limited in its capacity to provide sufficient legal security on the matter,[198] and eventually Parties to the Convention on Biologial Diversity adopted the legally binding Nagoya Protocol on Access to Genetic Resources and the Fair and Equitable Sharing of Benefits Arising from their Utilization to the Convention on Biological Diversity in 2010.[199]

Some of recommendations to states are called action plans and address specific substantive issues. One example is the action plan on capacity-building for access to genetic resources and benefit-sharing[200] adopted by the Conference of the Parties to the Convention on Biological Diversity.[201]

V. Operational Procedures and Safeguard Policies

In contrast to recommendations addressed to states or private actors, the operational policies of financial institutions such as the World Bank or the International Monetary Fund constitute norms developed by the organisation which are formally addressed to other institutional organs of the same institution. Despite this "internal" character, however, these policies constitute *de facto* policy prescriptions for developing countries, because loan policies are conditioned upon compliance with these policies. The operational policies of the World Bank or the Operational Strategy and Programs of the Global Environment Facility (GEF) based on guidance by Conferences of the Parties of MEAs constitute distinct instruments formally nonbinding for states but which neverthe-

---

Preparation for the Third Meeting of the Ad Hoc Open-Ended Working Group on Access and Benefit-Sharing, Note by the Executive Secretary, 2004, UN Doc. UNEP/CBD/WG-ABS/3/INF/1 of 8 December 2004, para. 5, available at www.biodiv.org.

[198] On the merits of a legally nonbinding versus a legally binding approach in this issue area see e.g. M. Dross/F. Wolff, "Do We Need a New Access and Benefit sharing Instrument?", Yearbook of International Environmental Law 15 (2004), 95-118 (100).

[199] The Nagoya Protocol is available at http://www.cbd.int/abs/doc/protocol/nagoya-protocol-en.pdf.

[200] UNEP/CBD/COP/DEC/VII/19, F, Annex, available at www.cbd.int/decisions.

[201] Convention on Biological Diversity, 5 June 1992, 1760 UNTS I-30619.

less exercise great practical and legal constraints on borrowing countries.

1. World Bank

The environmental activities of the World Bank[202] are not foreseen in the respective Articles of Agreement of its various institutions.[203] None of the provisions deals specifically with environmental protection.[204] Rather, the World Bank was established financially to support reconstruction and development projects and promote structural reforms in less developed countries by means of capital loans. Irrespective of the narrow development mandate, the World Bank has however become increasingly responsive to environmental concerns and the environmental side effects of projects, and today incorporates a number of environmental and sustainable development considerations into its work. Although representing a "mission creep" which raises questions of legitimacy,[205] the broadening of the mandate can nevertheless be justified. For a justification, one can either point to subsequent practice that has – by being widely accepted and acquiesced by states – modified the mandate,[206] or to a modern interpretation of the development objectives of the bank which cannot be considered independently from a modern development conception based on concept of "sustainable development".[207]

---

[202] The term World Bank refers to the International Bank for Reconstruction and Development (IBRD) and the International Development Association (IDA). The World Bank Group additionally comprises the International Finance Corporation (IFC), the International Centre for the Settlement of Investment Disputes (ICSID) and the Multilateral Investment Guarantee Agency (MIGA).

[203] World Bank Articles of Agreement of 27 December 1945, 2 UNTS 143 (as amended).

[204] According to Article I (i) of the Articles of Agreement of the World Bank, the purposes of the Bank are "... to assist in the reconstruction and development of territories of members ...".

[205] J.E. Alvarez, International Organizations as Law-makers, 2005, 239.

[206] G. Handl, Multilateral Development Banking: Environmental Principles and Concepts Reflecting General International Law and Public Policy, 2001, 25.

[207] C. Holstein, Der Umweltschutz in der Tätigkeit der Weltbankgruppe: Instrumente, rechtliches Mandat und Bedeutung für das internationale Umweltrecht, 2001, 89 et seq.

Since the 1980s, environmental policies have been introduced into the loan policies of the Bank via a number of detailed Operational Procedures and Bank Policies (originally named Operational Directives) on issues such as environmental impact assessment, pest management, the safety of dams, and the rights of indigenous people.[208] Similar procedures and safeguards have also been established by regional development banks such as the Inter-American Development Bank[209] and the Asian Development Bank.[210] These procedures and policies are developed as internal staff rules by the World Bank management.[211] Many of the requirements in operational procedures are of an essentially procedural nature. The Operational Procedure 4.01 on Environmental Assessment (1999) for example establishes the need for environmental assessment for each project, and the OP 4.10 on Indigenous Peoples (2005) requires that procedures are in place through which the needs and complaints of indigenous peoples are taken into account. The Op-

---

[208] The Operational Procedures with environmental implications include the OP 4.00 Piloting the Use of Borrower Systems to Address Environmental and Social Safeguard Issues in Bank-Supported Projects (2005), OP 4.01 Environmental Assessment (1999); OP 4.02 Environmental Action Plans (2000), OP 4.04 Natural Habitats (2001), OP 4.07 Water Resources Management (2000), OP 4.09 Pest Management (1998), OP 4.10 Indigenous Peoples (2005); OP 4.36 Forests (2002), OP 4.37 Safety of Dams (2001), OP 7.50 Projects of International Waterways (2001), OP 10.04 Economic Evaluation of Investment Operations (1994). The operational manual of the bank which includes all current Operational Procedures, Bank Procedures and interim instructions to bank staff is available at http://go.worldbank.org/DZDZ9038D0. For a detailed analysis see G. Handl, Multilateral Development Banking: Environmental Principles and Concepts Reflecting General International Law and Public Policy, 2001; C. Holstein, Der Umweltschutz in der Tätigkeit der Weltbankgruppe: Instrumente, rechtliches Mandat und Bedeutung für das internationale Umweltrecht, 2001, 59 et seq.

[209] Environment and Safeguard Policy of the Inter-American Development Bank, see http://www.iadb.org.

[210] Safeguard Policy Statement of the Asian Development Bank, information available at http://www.adb.org/Environment/default.asp; on common rules and procedures in multilateral development banking G. Handl, Multilateral Development Banking: Environmental Principles and Concepts Reflecting General International Law and Public Policy, 2001.

[211] For details on the development process, see L. Boisson de Chazournes, "Policy Guidance and Compliance: The World Bank Operational Standards" in: D. Shelton (ed.), Commitment and Compliance: The Role of Non-Binding Norms in the International Legal System, 2000, 281-302 (284).

erational Procedures 7.50 on projects of International Waterways are also interesting; these require consultation with riparian states which can lead, in the case of protest, to additional independent expert assessments. But borrowers must also comply with some substantive standards. The Operational Procedures 4.09 on Pest Management (2004) for example aim to reduce the negative environmental impact of chemicals, pesticides and other substances by establishing substantive criteria for pesticide selection and use. These include the criterion that pesticides must have minimal effects on non-target species and the environment, and that pesticides must be packed and stored in accordance with the FAO Guidelines for Packaging and Storage of Pesticides.

Although constituting internal staff guidelines, the Operational Procedures are incorporated into loan agreements and therefore bind the borrowing state. States that economically depend on the loans of the bank for their development often do not really have a choice whether or not to accept these conditions. This factual pressure has the effect of making the borrowing state contractually bound to observe the World Bank's internal policies.

The Operational Policies have also served as a model for the OECD Recommendation on Common Approaches on the Environment and Officially Supported Export Credits and for national legislation regarding environmental impact assessment, and have influenced the policies of other banks as well as other important instruments and declarations.[212]

In response to criticism directed at the lack of transparency and accountability, the World Bank has been a forerunner of creating improved accountability mechanisms in the form of review mechanisms. The World Bank Inspection Panel creates a procedure through which affected individuals or groups can challenge projects financed by the Bank if they are in violation of the Bank's own policies. Similar procedures have also been established by other development banks, namely

---

212 Such as the Rio Declaration on Environment and Development (13 June 1992), UN Doc. A/CONF.151/26 (vol. I); 31 ILM 874 (1992); see on this J.E. Alvarez, International Organizations as Law-makers, 2005, 239; L. Boisson de Chazournes, "Policy Guidance and Compliance: The World Bank Operational Standards" in: D. Shelton (ed.), Commitment and Compliance: The Role of Non-Binding Norms in the International Legal System, 2000, 281-302 (299)

the Inter-American Development Bank[213] and the Asian Development Bank.[214] The procedures serve to enforce compliance of the institutions with their own policies and raise the institutions' public accountability.

2. *Global Environment Facility*

The Global Environment Facility, founded in 1991,[215] today is the largest provider of funds for projects aimed at improving and protecting the environment. The institution provides grants in six focus areas, namely biodiversity protection, climate change, international waters, land degradation, ozone layer protection and the spread of persistent organic pollutants. It is also the designated funding institution under several Multilateral Environmental Agreements,[216] and in this function funds initiatives that help developing countries meet their obligations under these Conventions. The highest governing body is the GEF Assembly with representatives from 177 member countries. Responsible for the development of guiding policies but also for the approval of GEF projects is the GEF Council. Comprised of 32 member states among which 16 are developing countries, 14 developed economies and 2 economies in transition, the GEF Council is more open to developing country participation than the World Bank. This is also somewhat reflected in the fact that the GEF Council decides by consensus and if consensus fails, by a double majority vote consisting of 60 per cent of the members present and 60 percent of the votes representing donors. It is however noteworthy that the GEF, like the World Bank and other similarly functioning funding mechanisms, does not operate under a system of

---

[213] The Inter-American Development Bank has established the Independent Consultation and Investigation Mechanism, see for details http://www.iadb.org/.

[214] The Asian Development Bank has established the Compliance Review Panel, for details available at http://compliance.adb.org/.

[215] Instrument Establishing the Global Environment Facility (16 March 1994) 33 ILM 1273 (1994).

[216] GEF is the designated funding mechanism for the Convention on Biological Diversity, the UNFCCC, the Stockholm Convention on Persistent Organic Pollutants (22 May 2001), 40 ILM 532 (2001), and the United Nations Convention to Combat Desertification in Countries Experiencing Serious Drought and/or Desertification, Particularly in Africa (17 Juni 1994), 1954 UNTS 3; 33 ILM 1328 (1994).

one-state-one-vote. With the outsourcing of funding from MEAs, the balanced representation in MEAs is consequently weakened.[217]

Proposed projects are evaluated and approved by the Council on the basis of various formally nonbinding guiding instruments adopted by the Council. The GEF funds projects on the basis of a set of focal area strategies as well as Strategic Programs which are revised at each replenishment cycle. The Strategic Programs define eligibility criteria for projects, including the objectives and expected outcomes. More detailed procedures outlining the project cycle are provided by the Operational Manual that is developed by the GEF secretariat.[218]

The Operational Strategy and the Operational Program, similarly to the World Bank Operational Procedures, establish a number of criteria of eligibility of projects. They are to a large extent based on the guidance issued by the Conferences of the Parties of the MEAs for which the GEF functions as the funding mechanism.[219] As a result, one can say that the Conferences of the Parties in conjunction with the GEF Council determine the funding policies of the GEF. This means that in similarity to the World Bank Operational Procedures, instruments outlining internal policies have external effects on those countries in need of the funds despite the fact that the institutions do not have the competence to adopt decisions that are legally binding for states. An indirect legal effect of these instruments is furthermore that developing countries can only benefit from technical and financial assistance under MEAs through compliance with these criteria. Since COPs of MEAs together with the GEF Council can therefore condition financial assistance and loans through decisions taken by a two-third majority if consensus fails, the consensual underpinnings are eroded through the use of these instruments.[220]

---

[217] E. Hey, "International Institutions" in: D. Bodansky/J. Brunnée/E. Hey (eds.), Oxford Handbook of International Environmental Law, 2007, 749-769 (763).

[218] GEF, Focal Area and Strategic Programming for GEF-4, document available at http://www.gefweb.org/uploadedFiles/Focal%20Area%20 Strategies_10.04.07.pdf.

[219] Compare e.g. the guidance to the GEF issued by the 9th Conference of the Parties to the Convention on Biological Diversity in 2008, UNEP/CBD/COP/DEC/IX/31 (9 October 2008).

[220] This problem is identified and analysed by E. Hey, "International Institutions" in: D. Bodansky/J. Brunnée/E. Hey (eds.), Oxford Handbook of International Environmental Law, 2007, 749-769 (756 et seq.).

VI. Technical standards

Technical standards are here understood as documents which provide, for common and repeated use, guidelines or characteristics for products or production-related processes and methods.[221] Technical standards can thus be understood, in line with the definitions in the Annex of the WTO Agreement on Technical Barriers to Trade (TBT Agreement),[222] not only as standards that define the specific characteristics of a product, but also as norms prescribing requirements on how to handle, produce, transport, package, label or deal in other specific ways with specific products.[223] In contrast to other more normative policy-oriented requirements, technical standards are specific descriptions which usually define requirements or characteristics of products with high specificity but are not designed for general application. It must however not be overlooked that technical standards have important policy implications, and therefore the distinction between policy-oriented and technical norms is not always clear. Harmonisation of product characteristics or production methods almost always includes a policy objective. For example, the adoption of a technical standard outlining pesticide residue limits for foods has obviously implications for environmental protection. It would mean that those states or businesses that formerly had lower limits by adopting the standard would shift towards a stricter policy on pesticide use for food production.

Institutions which develop and issue technical standards therefore indirectly take part in policy making exercises that have important regulatory implications for states and even directly for private actors. Their effect is reinforced when they are acknowledged in treaty law, as is the case for technical standards of the Codex Alimentarius Commission

[221] This definition is oriented at the definition in the Agreement on Technical Barriers to Trade [in Annex 1A of the Agreement Establishing the World Trade Organization, 1867 UNTS 154; 33 ILM 1144 (1994)]. According to Annex I, No. 2 of the TBT Agreement, a "standard" is a "[D]ocument approved by a recognized body that provides, for common and repeated use, rules, guidelines or characteristics for products or related processes and production methods, with which compliance is not mandatory. It may also include or deal exclusively with terminology, symbols, packaging, marking or labelling requirements as they apply to a product, process or production method."

[222] *Ibid.*

[223] Examples are in particular the ISO 14000 standards issued by the International Organization for Standardization, available at www.iso.org (1 March 2010); for some details see in this section further below at 5.c).

and the International Organization for Standardization under the WTO Agreements on Technical Barriers to Trade (TBT Agreement) and that on Sanitary and Phytosanitary Measures (SPS Agreement).[224]

1. Codex Alimentarius Commission

In cooperation with the FAO, the WHO established the Codex Alimentarius Commission in 1963 with a view to preparing and promoting food standards that protect the health of consumers and to facilitate international food trade.[225] Over 180 states are Members of the Commission. After acceptance by governments, the standards are published in the Codex Alimentarius. The Codex Alimentarius mainly establishes food standards that describe characteristics of products, such as the maximum residue limits of pesticides or of drugs in foods or general standards for food additives, contaminants and toxins in foods. But the Codex Alimentarius also issues codes of practice which define the production, processing, transport and storage practices for foods. Moreover, the Commission also adopts so-called Codex Alimentarius Principles that set out policies in key areas. These include a risk analysis of foods derived from modern biotechnology and guidelines for the interpretation of these principles and the general standards.

The standards need not be ratified but nevertheless must be formally accepted by states. Still, from a legal perspective, any state remains free to withdraw any time, since even formal acceptance of the standard cannot entail an international law obligation.[226] While not legally binding, the standards are widely accepted by corporations involved in food trade which adapt to the pressure of the markets of food trade. With the standards being as widely accepted as they are, non-compliance would be a serious obstacle to trade in the respective foodstuff.[227] The wide acceptance of these standards and further pressure to comply also derives

---

224 For details, see the analysis in Part 2, at A.I.2.c), further below.

225 Statutes of the Codex Alimentarius Commission, Article 1(a). The Statutes are included in the Codex Alimentarius Commission Procedural Manual (18th ed. 2008), available at http://www.codexalimentarius.net/web/procedural_manual.jsp.

226 J.E. Alvarez, International Organizations as Law-makers, 2005, 222.

227 D.M. Leive, International regulatory regimes: case studies in health, meteorology, and food, 1976, 547; J.E. Alvarez, International Organizations as Law-makers, 2005, 223.

from the recognition of the Codex Alimentarius as an international standard in international trade agreements such as the WTO Agreement on the Application of Sanitary and Phytosanitary Measures[228] and the TBT-Agreement[229] as well as the North Atlantic Free Trade Agreement.[230] The consequential influence of the Commission has triggered discussions on the legitimacy of the Commission and possible means for improvement.[231]

2. *International Organization for Standardization*

One of the most important organisations concerned with technical standardisation is the International Organization for Standardization (ISO). The ISO is an organisation of 148 national standard-setting bodies which are in turn constituted of private bodies (in most developed countries), governmental agencies (in most developing countries) or of mixed private-public bodies.[232] It is thus not an international organisation but a hybrid private-public organisation. Its standards are developed in technical committees constituted by expert delegates from the national standard-setting bodies.

Of particular relevance for the context of this study – norms prescribing certain desirable behaviour – are the environmental management standards of the ISO. These are procedural standards such as the ISO 14000 series and the ISO 26000 series which provide rules on how to integrate environmental considerations into the planning, management, operation and the controlling processes of an enterprise. These management standards differ from those technical standards only defining

---

228 Agreement on Sanitary and Phytosanitary Measures [in Annex 1A of the Agreement Establishing the World Trade Organization, 1867 UNTS 154; 33 ILM 1144 (1994)], Articles 2.2 and 3.1.

229 TBT Agreement, Article 2.6.

230 Compare e.g. Articles 905 and 915 of the North Atlantic Free Trade Agreement (17 December 1992), 32 ILM 289, 605 (1993).

231 See e.g. A. Herwig, "The Contribution of Global Administrative Law to Enhancing the Legitimacy of the Codex Alimentarius Commission", in: O. Dilling/M. Herberg/G. Winter (eds.), Transnational administrative rule making: performance, legal effects and legitimacy, 2011, 171 – 212.

232 For details on the structure of the ISO see e.g. G.G. Sander, "International Organization for Standardization (ISO)", in: R. Wolfrum (ed.), Max Planck Encyclopedia of Public International Law, 2010, para. 5 et seq., online at: www.mpepil.com.

certain characteristics of a product. They are still considered technical standards for the purposes of this overview, since they directly affect production processes and methods which are inextricably linked to characteristics of goods. The Agreement on Technical Barriers to Trade of the WTO recognises this relationship by broadly defining a "standard" to include "rules, guidelines or characteristics for products or related processes and production methods...".[233] This wording also implies that "standards" may be more than those rules prescribing characteristics of products, and may also include other rules and guidelines that do not directly define characteristics of products.

B. Illustrative case studies

In line with the objective of this study to analyse the role of institutions, the following case studies focus on international organisations which not only develop and issue nonbinding instruments, but also those which attempt to establish mechanisms to promote their implementation.

Apart from these similarities, the case studies concern three very different issue areas which each have distinct challenges, and they represent three completely different approaches to how to address these problems and how to promote the norms adopted. In the case of fisheries, the FAO attempts to supplement existing treaty instruments by issuing a code of conduct and related instruments which entail substantive norms and are mainly aimed at improving the legal and policy framework of states. In the case of the voluntary system on prior informed consent, treaty law was non-existent at the moment of its adoption. The voluntary system set up by a cooperative initiative of the FAO and UNEP paved the way for an important treaty in the field of the control of hazardous substances. And the international institutions in contrast to the fisheries case not only promoted the system, but also managed the multi-level system at the international level. The third case of the OECD Guidelines shows an attempt of an international organisation directly to address multinational corporations through international norms in an area not covered by international law. The implementation system used represents an example for the tendency to establish com-

---

[233] TBT Agreement, Annex 1, No. 2.

plaint mechanisms for private actors outside of domestic and international judicial systems on the basis of international procedures.

I. Fisheries Regulation: the FAO Code of Conduct for Responsible Fisheries[234]

1. *The institutional framework*

The institutional framework of the Food and Agriculture Organization (FAO)[235] which adopted the Code of Conduct for Responsible Fisheries in 1995[236] is that of a typical international organisation. It comprises a political plenary body, the FAO Conference, which is constituted by representatives of all 189 FAO member states and the European Union, a smaller executive organ, the FAO Council, as well as an administrative organ, the secretariat. Much of the substantive work is undertaken by committees established as subsidiary bodies to the Council. The FAO Conference can make recommendations to states by a two-third

---

[234] The following case study on the FAO and its Code of Conduct for Responsible Fisheries is partly based on my previously published article entitled "Legal Challenges of Nonbinding Instruments: The Case of the FAO Code of Conduct for Responsible Fisheries", in: A. von Bogdandy, R. Wolfrum, J. von Bernstorff, P. Dann, M. Goldmann (eds.), The exercise of public authority by international institutions: advancing international institutional law 2010, 511-540.

[235] On the FAO K. Mechlem, "Food and Agriculture Organization of the United Nations (FAO)", in: R. Wolfrum (ed.), Max Planck Encyclopedia of Public International Law 2010, online at: www.mpepil.com; a detailed account of the institutional and legal framework of the FAO is also given by G.G.R. Blom, "Institutional and legal aspects of the Food and Agriculture Organization of the United Nations (FAO)", Revue de Droit International 74 (1996), 227-332.

[236] FAO Code of Conduct for Responsible Fisheries (31 October 1995), FAO Doc. 95/20/Rev/1, available at: ftp://ftp.fao.org/docrep/fao/005/v9878e/9878e00.pdf; generally on the CCRF W. Edeson, "Closing the Gap: The Role of 'Soft' International Instruments to Control Fishing", Australian Yearbook of International Law 20 (1999), 83-104; G. Moore, "The Code of Conduct for Responsible Fisheries" in: E. Hey (ed.), Developments in International Fisheries Law, 1999, 85-105.

majority vote,[237] and otherwise all organs, including the Council and its committees, can take decisions by majority vote.[238]

Mandated *inter alia* with the conservation of natural resources,[239] one of the main fields of activity of the FAO is fisheries policy. The main body responsible for fisheries policy is the Committee on Fisheries (COFI), a sub-committee to the FAO Council.[240] The COFI meets every two years and is open to any FAO Member.[241] In the last meeting period between 2007 and 2009, 131 FAO Members were members of the Committee. Important substantive work is conducted by two sub-committees established by COFI, namely the Sub-Committee on Fish Trade and the Sub-Committee on Aquaculture.[242] Also open to all members, the meetings of the sub-committees have a smaller number of participants (usually between 40 and 60 government representatives). As is the case in other international organisations, the government representatives attending the meetings of COFI and the sub-committees are not high level diplomats, but most often government officials from specialised state ministries. In the case of COFI these are usually the agricultural ministries or ministries specifically responsible for fisheries issues.[243] In addition to state representatives, approximately 30 environmental, labour and industry NGOs and a great number of the most important international organisations, including the World Bank, the World Trade Organization (WTO), the IMO and numerous regional

237 Article 4 (2) of the Constitution of the Food and Agriculture Organization (16 October 1945) (hereinafter FAO Constitution), CTS 1945/32; 40 AJIL Supp. 76, available at http://www.fao.org/DOCREP/003/X8700E/x8700e01.htm#P8_10.

238 FAO General Rules of Procedure, Article XII; Rules of Procedure of the Council of FAO, Rule IV (1); e.g. Rules of Procedure of the Committee on Fisheries, Rule V (2). All legal texts are available at http://www.fao.org/Legal/index_en.htm.

239 Compare FAO Constitution, Article I (2) c).

240 FAO Constitution, Article V para 6.

241 Rule III of the Rules of Procedure of the Committee on Fisheries (COFI), available at: http://www.fao.org/Legal/index_en.htm.

242 These are the Sub-Committee on Fish Trade and the Sub-Committee on Aquaculture. The power to establish sub-committees derives from Rule XXX para. 10 of the General Rules of the Organization, available at: http://www.fao.org/Legal/index_en.htm.

243 In the case of Germany, this is the Ministry of Food, Agriculture and Consumer Protection.

fisheries organisations participate in the meetings of COFI as observers.[244] This makes COFI the main international forum for fisheries issues.

The only body of the FAO which is exclusively composed of international civil servants, and is formally independent of governments,[245] is the FAO secretariat. The Fisheries and Aquaculture Department of the secretariat is responsible for all activities related to the Code of Conduct for Responsible Fisheries (CCRF), and its work is guided by the CCRF and COFI. Composed of 74 professional staff at the headquarters alone, the Department disposes of considerable human resources. In addition to preparing meetings of COFI, it carries out important functions in the follow-up procedures and the coordination with other international organisations. Oversight of the secretariat and the Secretary-General by the governing bodies appears to be rather weak.[246] The Committee on Fisheries has some control over the Fisheries and Aquaculture Department through its budgetary competence. Apart from budgetary control, review and oversight of specific activities of the secretariat is mainly provided for by an internal reporting mechanism. The secretariat reports on its activities in relation to the CCRF in biannual progress reports to the COFI.[247] The COFI in turn reports to the FAO Council.[248]

2. *The regulatory challenge*

After decades of ever more extensive fishing activities in response to an increasing demand from a growing world population, fish stocks around the world are in danger of being overexploited or depleted. The FAO estimates that more than half of the marine fish stocks are fully

---

244 Rule III of the Rules of Procedure of the Committee for Fisheries, available at http://www.fao.org/Legal/index_en.htm.

245 FAO Constitution, Article VIII (2).

246 This conclusion was drawn by an expert evaluation, compare FAO, The Challenge of Renewal, Independent External Evaluation of the Food and Agriculture Organization, FAO: Working Draft (2007) Box 4.3. For an in-depth assessment of oversight mechanisms and control by states compare Part 3 on legitimacy, further below.

247 This reporting mechanism is foreseen by Article 4.2 CCRF.

248 Rules of Procedure of the Committee on Fisheries, Rule VI.

exploited and 28 percent are either overexploited or depleted.[249] Any solution to this state of affairs faces complex regulatory challenges. It requires cooperation across jurisdictional zones by a multitude of different actors with various economic and social interests in a subject area marked by fierce economic competition. Free riding must be prevented through monitoring and enforcement at sea. Further, it is now understood that long-term sustainable use largely depends on the protection of the living and non-living environment of the resource, i.e. a consequent ecosystem approach is required. In addition, the uncertainty over reproduction levels and impact of environmental degradation makes a precautionary approach to fisheries management indispensable for successful regulation. The high level of uncertainty with regards to numbers, problems and possible responses requires highly flexible and adaptable regulation.

Treaty instruments so far only inadequately reflect these requirements.[250] The framework of the United Nations Convention on the Law of the Sea (UNCLOS)[251] establishes very general duties to protect, to take measures or to cooperate. [252] Apart from these general obligations state parties to the convention enjoy great regulatory autonomy with regards to the substantive regulation and management of fisheries.[253] The treaty establishes separate zones of jurisdiction by attributing to coastal states the rights and responsibilities over fisheries resources in the territorial sea and the exclusive economic zones, subject to general duties to protect the marine environment and to avoid overexploitation.[254] In addition, fishing on the high seas is only subject to a very rudimentary legal framework. Articles 117 to 119 UNCLOS merely establish a duty to cooperate and outline some of the main features to

249 FAO, The State of World Fisheries and Aquaculture 2008 (Rome 2009), Part I, page 30, available at: ftp://ftp.fao.org/docrep/fao/011/i0250e/i0250e.pdf.

250 For an overview of existing legal frameworks aiming at the protection of marine environment see E.J. Techera, Marine environmental governance: from international law to local practice, 2012.

251 United Nations Convention on the Law of the Sea 1982 (10 December 1982), 1833 UNTS 3, 21 ILM 1261.

252 E.g. in Articles 64, 118 and 197 UNCLOS.

253 For details on the emergence and shortcomings of the fisheries regime under UNCLOS see R. Wolfrum, Die Internationalisierung staatsfreier Räume: die Entwicklung einer internationalen Verwaltung für Antarktis, Weltraum, Hohe See und Meeresboden, 1984, 188 et seq.

254 Article 61 (2) UNCLOS.

guide such cooperation. UNCLOS thus hardly restricts the principle of the freedom to fish on the high seas, eschews meaningful internationalisation and control and basically leaves resource allocation to the competition between fleets.[255] The provisions on fisheries are based on the maximum sustainable yield concept which does not conform to economic or ecological principles, but rather aims at maximum food production.[256] In addition, UNCLOS in general only insufficiently takes into account modern concepts of fisheries protection, as it only contains minimal ecosystem considerations[257] and lacks norms on precaution. This increases the danger of overexploitation.[258]

The main strategy employed internationally to address these regulatory lacunae is cooperation of states in regional fisheries organisations. In particular fisheries management organisations, with a mandate to prescribe binding conservation and management measures to be implemented by states, have become the centre of attention. To improve persisting enforcement problems and thereby strengthen regional fisheries organisations, two multilateral treaties were developed in the 1990s. One is the Fish Stocks Agreement; an implementation agreement to Article 64 UNCLOS with a specific focus on highly migratory fish stocks and those straddling the border of the EEZ and the high seas. It incorporates the precautionary[259] and ecosystem[260] approaches and thereby

---

255 R. Wolfrum, Die Internationalisierung staatsfreier Räume: die Entwicklung einer internationalen Verwaltung für Antarktis, Weltraum, Hohe See und Meeresboden, 1984, 210.

256 R. Wolfrum, Die Internationalisierung staatsfreier Räume: die Entwicklung einer internationalen Verwaltung für Antarktis, Weltraum, Hohe See und Meeresboden, 1984, 205.

257 According to Arts 61 (4) and 119 (1) b) UNCLOS, states have to take into consideration only species "associated or dependent upon harvested species" when setting conservation and management measures for harvested species, while non-target species or the ecosystem as such do not play a role.

258 This is also noted by E.J. Techera, Marine environmental governance: from international law to local practice, 2012, at 68.

259 The precautionary approach is a general principle of the Fish Stocks Agreement, compare Article 5 (c), but also Article 6 as well as Annex II which contains a set of guidelines on how to implement the approach.

260 In particular Articles 5 (d) and (j), 10 (d) Fish Stocks Agreement; for details, see G. Vigneron, "Compliance and International Environmental Agreements: A Case Study of the 1995 United Nations Straddling Fish Stocks Agreement", Georgetown International Environmental Law Review 10 (1998), 581-624 (588-590).

remedies some of the lacunae of UNCLOS.[261] Most remarkable however are the enforcement mechanisms. The Fish Stocks Agreement establishes a duty to cooperate with regional fisheries organisations and to comply with their conservation and management measures.[262] Only those states that are members or that comply with the conservation measures have access to the resources of the area covered by the respective regional fisheries agreement.[263] The Fish Stocks Agreement further provides for the possibility of boarding a foreign ship in order to verify whether it complies with the respective conservation and management measures. If there are "clear grounds for believing that a vessel has committed a serious violation", the inspectors may take the vessel to the nearest port for further inquiries.[264] In only addressing stocks which straddle the boundaries of the exclusive economic zones and the high seas, the Fish Stocks Agreement however remains limited in its scope *ratione materiae*. And at least so far, it has not achieved the ratification numbers necessary for the mechanism to function effectively. Even though recently ratification numbers are rising at a faster rate, many important fishing nations such as China or Chile are still absent.[265]

The other multilateral treaty is the FAO Compliance Agreement.[266] It was specifically developed to improve enforcement of management

---

[261] For a detailed analysis compare e.g. C.J. Carr, "Recent developments in compliance and enforcement for international fisheries", Ecology law quarterly 24 (1997), 847-860; D.A. Freestone/Z. Makuch, "The new international environmental law of fisheries: the 1995 United Nations Straddling Stocks Agreement", Yearbook of international environmental law 7 (1996), 3-51; G. Vigneron, "Compliance and International Environmental Agreements: A Case Study of the 1995 United Nations Straddling Fish Stocks Agreement", Georgetown International Environmental Law Review 10 (1998), 581-624.

[262] Fish Stocks Agreement, Articles 8 (1)-(3) and (5).

[263] Fish Stocks Agreement, Article 8 (4).

[264] Fish Stocks Agreement, Article 21 (8), a "serious violation" is defined in Article 21 (11) Fish Stocks Agreement.

[265] The Fish Stocks Agreement has been ratified by 77 States as of 1 March 2010. Important fishing nations such as China and Taiwan, Peru, Chile, Indonesia, Thailand, the Philippines, Malaysia, Mexico, Vietnam and Argentina are still missing.

[266] Agreement to Promote Compliance with international conservation and management measures by fishing vessels on the high seas (24 November 1993), 1860 UNTS 148; 33 ILM 968 (hereinafter Compliance Agreement).

measures on the high seas.[267] With this objective, it strengthens regional fisheries organisations in two ways. First, it imposes on states the obligation that vessels that fly their flag do not fish in a way that undermines the conservation and management measures of regional organisations.[268] It does so more specifically by requiring states to establish a system of authorisations to fish, and only to grant such authorisation if they are able adequately to enforce their international obligation not to undermine conservation measures vis-à-vis ships flying their flag.[269] The Agreement also restricts the capacity of states to authorise fishing vessels that have recently reflagged to avoid the consequences of a previous non-compliance.[270] Secondly, the Compliance Agreement obliges states to keep detailed records of their authorisations and to report the data to the global registry of the FAO.[271] Although such measures are theoretically useful to address non-compliance of non-participants of regional organisations and the problem of enforcement on the high seas, the agreement has only limited effect for global fisheries. This is due to its limited scope deriving from its specific focus on the High Seas,[272] which prevents it from addressing bad management or enforcement problems in the EEZ and the territorial sea. And it also has not yet attained the ratification numbers necessary to avoid free riders and make the enforcement mechanisms function well.[273]

3. The nonbinding response: the Code of Conduct for Responsible Fisheries

In addition to additional treaty making, the FAO resorted to a voluntary instrument – the Code of Conduct for Responsible Fisheries (CCRF) – to improve the state of global fisheries resources and address

---

[267] For details see e.g. C.J. Carr, "Recent developments in compliance and enforcement for international fisheries", Ecology law quarterly 24 (1997), 847-860 (851 et seq.)

[268] FAO Compliance Agreement, Article III (1) (a).

[269] FAO Compliance Agreement, Article III (2) and (3).

[270] FAO Compliance Agreement, Article III (5).

[271] FAO Compliance Agreement, Arts IV and VI.

[272] FAO Compliance Agreement, Article II (1).

[273] As of May 2009, 38 States and the European Union had ratified the Compliance Agreement, compare http://www.fao.org/Legal/treaties/012s-e.htm.

the remaining gaps in fisheries law. The Conference of the Food and Agriculture Organization adopted the CCRF unanimously in 1995.

Closer observation reveals that the CCRF is not the only – albeit it is the main – document of a rather complex and continuously growing array of different nonbinding instruments produced by the FAO through various organs and bodies. Three types of instruments produced in the context of the CCRF can be distinguished, namely:

(1) the main instrument entitled CCRF adopted by the FAO Conference;

(2) the International Plans of Action (IPOAs) which are usually adopted by the Committee of Fisheries and a Strategy for Improved Information, also adopted by the Committee of Fisheries;

(3) Technical Guidelines and related supplements that are elaborated under the auspices of the FAO secretariat.

These will now be considered in turn.

a) The CCRF

The CCRF is expressly voluntary, as are the further instruments adopted under its framework.[274] However, some parts of the CCRF are based on UNCLOS or may have achieved binding international law status between some of the states. This is the case in particular for those states that have ratified the FAO Compliance Agreement, which forms "... an integral part of the code".[275]

The CCRF fills some of the gaps left by the limited scope of other fisheries instruments.[276] The norms of the CCRF and related instruments are addressed to all states, not just members of the FAO, as well as fishing entities,[277] governmental and non-governmental organisations and –

274 CCRF, Article 1.1; IPOA-IUU para 4; Technical Guidelines usually include a phrase that they have "no formal legal status", e.g. FAO Technical Guidelines on Aquaculture Development, 2007.

275 CCRF, Article 1.1.

276 The potential of the CCRF to complement more limited fisheries treaties is emphasized by W. Edeson, "Closing the Gap: The Role of 'Soft' International Instruments to Control Fishing", Australian Yearbook of International Law 20 (1999), 83-104 (90).

277 The term "fishing entities" should be understood as a reference to Taiwan, province of China, which is not recognised as a member state of the United Nations or the FAO.

by contrast to other soft and hard law instruments – all persons involved in some way or another in the conservation, management or development of fisheries.[278] Similarly wide is the scope of territorial application and the substantive scope *ratione materiae*. The CCRF comprises all activities related to fisheries ranging from conservation and management through to trade in fish products and aquaculture.[279] The territorial scope of the CCRF is defined as "global".[280]

The CCRF establishes the only framework for fisheries governance that integrates all actors involved in such activities worldwide. Being nonbinding, the norms of the CCRF can easily link the activities of a large variety of state and non-state actors even across sectoral boundaries. Furthermore, the main instrument of the CCRF represents a remarkably innovative and complete statement of principles for fisheries and is as such unequalled in international governance and law.[281] Two of the central elements of the concept of sustainable development, namely the principle of sustainable use and the principle of the integration of environmental considerations and development needs,[282] are specified in the context of fisheries.[283] A related principle that is manifest throughout the CCRF and implementing instruments is the precautionary principle[284] and its ecosystem orientation.[285]

However, the main achievement of the CCRF and implementing instruments is the translation and concretisation of general principles and concepts of international law, such as the precautionary principle, into fisheries-specific rules and proposals for action.[286] Taken together, all of the different nonbinding instruments provide for a rather complete system of norms that can be directly implemented without the need for much further consideration or concretisation. It therefore serves as a

---

278 CCRF, Article 1.2.

279 CCRF, Article 1.3.

280 CCRF, Article 1.2.

281 G. Moore, "The Code of Conduct for Responsible Fisheries" in: E. Hey (ed.), Developments in International Fisheries Law, 1999, 85-105 (96).

282 P. Sands, Principles of International Environmental Law, 2003, 253.

283 CCRF, Articles 2(a) and 6(1).

284 CCRF, Article 6(5).

285 The ecosystem approach is manifest in Articles 6(1), (2), (3) and (8) CCRF.

286 G. Moore, "The Code of Conduct for Responsible Fisheries" in: E. Hey (ed.), Developments in International Fisheries Law, 1999, 85-105 (98).

"toolbox" of ideas and valuable strategies for responsible states.[287] The different sets of norms amount to a cascade of norms, ranging from more general and rarely altered ones developed at the highest political level and the more specific action plans adopted by COFI through to specific and highly flexible norms developed and administered by the experts of the FAO secretariat.

In the cascade of norms, the thematic sections of the Code constitute a first level of concretisation. The provisions in these Articles outline what actions should be taken by states and private actors in order to implement the principles in a range of issue areas from fisheries management and operations to aquaculture development, research, coastal management and trade.[288] For example, the thematic section on fisheries management translates the general precautionary principle[289] into factors that states need to take into account in fisheries management. These include environmental and social conditions and non-target fisheries as well as natural phenomena.[290]

b) International Plans of Action

A higher degree of specificity is achieved by the International Plans of Action (IPOAs). These are collections of norms that address specific problems such as the decline of sharks or illegal, unregulated or unreported fishing.[291] As the IPOAs are adopted within the framework of the CCRF, they share its wide scope with regards to addressees and territorial application. But IPOAs also contain norms prescribing in great

---

[287] W. Edeson, "The International Plan of Action on Illegal Unreported and Unregulated Fishing: The Legal Context of a Non-Legally Binding Instrument", The International Journal of Marine and Coastal Law 16 (2001), 603-623 (623); J.K. Ferrell, "Controlling flags of convenience: one measure to stop overfishing of collapsing fish stocks", Lewis & Clark Law School Environmental Law 35 (2005), 323-390 (330).

[288] CCRF, Articles 7-12.

[289] CCRF, Article 6(5).

[290] CCRF, Articles 7(5.2) and (5.5).

[291] So far, four International Plans of Action have been adopted, the IPOA for Reducing Incidental Catch of Seabirds in Longline Fisheries (IPOA-Seabirds), the IPOA for Conservation and Management Sharks (IPOA-Sharks), the IPOA for the Management of Fishing Capacity (IPOA-Capacity), all adopted in 1999, and the IPOA-IUU, adopted in 2001. All IPOAs are available at: www.fao.org/fi.

detail the requirements for national law and policy. For example, the IPOA on Illegal, Unreported and Unregulated Fishing specifies in detail the kind of information vessel monitoring systems or authorisations to fish should contain.[292] Sometimes timetables for the adoption of national plans of action are included.[293] The recent Strategy for Improving Information on Status and Trends of Capture Fisheries[294] aims to concretise and implement the CCRF chapter on research[295] by calling on states to establish data collection systems at the national and global level.

c) FAO Technical Guidelines for Responsible Fisheries

A further concretisation of both CCRF and IPOAs is achieved by the Technical Guidelines and supplementary documents. The Technical Guidelines are texts usually containing general explanations of certain provisions of the CCRF.[296] Most importantly, they include suggestions and recommendations on how objectives outlined in the code can be achieved, for example through the provision of best practices elaborated by leading experts in a particular field. The recently developed Technical Guidelines on the implementation of the ecosystem approach to fisheries management, for example, provides best practices for decision-makers on how best to integrate ecosystem considerations into fisheries management systems by making use of the scientific method of ecosystem modelling.[297]

---

[292] IPOA-IUU, paras 42-49.

[293] IPOA-IUU, para. 25.

[294] FAO, Strategy for Improving Information on Status and Trends of Capture Fisheries (2003), available at: http://www.fao.org/DOCREP/006/Y4859T/Y4859T00.htm.

[295] CCRF, Article 12.

[296] All Technical Guidelines and accompanying supplements are available at http://www.fao.org/fishery/ccrf/4/en.

[297] FAO Technical Guidelines for Responsible Fisheries on Fisheries Management, No.4 Suppl.2 Add.1 (2008), available at ftp://ftp.fao.org/docrep/fao/011/i0151e/i0151e00.pdf.

Annexes to the Technical Guidelines often include guidance on specific technical subjects, as for example on how to mark fishing gear.[298] Recently, the FAO secretariat has even started to develop supplements to Technical Guidelines – so-called "companion documents" – which reach an even higher degree of specificity.[299] Finally, the Guidelines often include references to or include as annexes very specific guiding nonbinding instruments of other international organisations such as the IMO.

4. Norm development

a) Development of the CCRF and International Plans of Action

The subject matter of the CCRF falls within the general objectives outlined in the FAO Constitution, which comprise the promotion of the conservation of natural resources and improvement of the processing, marketing and distribution of food and agricultural products.[300] However, the FAO Constitution does not specifically authorise the FAO Conference to adopt a code of conduct.[301] It is also difficult to base the adoption of the FAO codes of conduct on the general competence of the FAO Conference to issue recommendations to states or other organisations,[302] because neither of the respective provisions provides for the possibility to address directly private actors and non-members or "fishing entities".[303] It was however implicitly accepted by all FAO

---

[298] For instance, the FAO Technical Guidelines for Responsible Fisheries on Fishing Operations contain an Annex III which outlines a "Proposed System for the Marking of Fishing Gear."

[299] E.g. the document on "Compliance to FAO Technical Guidelines for Responsible Fisheries: Health management for responsible movement of live aquatic animals" as announced in FAO Technical Guidelines for Responsible Fisheries No. 5 Aquaculture Development, Suppl. 2.

[300] Article I(2)c) and d) FAO Constitution.

[301] This is also stressed by R. Wolfrum, "Introduction" in: R. Wolfrum/ V. Röben (eds.), Developments of International Law in Treaty Making, 2005, 1-13 (6).

[302] FAO Constitution, Article IV(3), (4).

[303] For the categorization of codes of conduct as recommendations G.G.R. Blom, "Institutional and legal aspects of the Food and Agriculture Organization of the United Nations (FAO)", Revue de Droit International 74 (1996), 227-332 (261).

Members that the FAO is competent to adopt such codes of conduct. Similarly, specific procedural requirements for the development and adoption of the CCRF and the other instruments are lacking. In the absence of any pre-existing procedural rules, the organs and sub-entities of the FAO have used their broad unspecific mandates to develop the various instruments in ad hoc procedures.

The development of the CCRF was initiated by COFI. After the idea was endorsed by two important intergovernmental conferences, the CCRF was developed by consultations at the FAO open to expert delegates from Members, non-members, intergovernmental and non-governmental organisations. There was apparently also considerable input from the secretariat.[304] The resulting draft was further reviewed by COFI and by the FAO Council which both established open-ended expert working groups for that purpose. The text of the CCRF was first endorsed by the FAO Council before its final unanimous approval by the FAO Conference in 1995 by means of Resolution 4/95.[305] Even if the drafting was heavily influenced by formally independent experts as well as the FAO secretariat and NGOs, all important decisions in the elaboration processes of the CCRF were taken by higher political bodies. This indicates – as was confirmed by participants – that the technical specialist input remained secondary, and that instead political debate and bargaining was determinative for the code's contents.[306]

Political control by governments is less pronounced in the development of the International Plans of Action. Here, the experts' drafts underwent an elaboration process involving few political decisions. Again initiated by COFI, these instruments are usually elaborated in expert consultations organised by the secretariat in cooperation with particularly active states, discussed and negotiated at so-called Technical Consultations at the FAO headquarters, and finally adopted by COFI.

---

304 For detailed information on the process of developing the CCRF, compare Annex 1 of the text of the CCRF, FAO Doc. 95/20/Rev/1.

305 FAO Conference Res. 4/95.

306 W. Edeson, "The Role of Technical Bodies" in: R. Wolfrum/V. Röben (eds.), Developments of International Law in Treaty Making, 2005, 63-91 (82 and 90).

b) Development of Technical Guidelines

In Resolution 4/95 of the FAO Conference by which member states adopted the CCRF,[307] the FAO secretariat is provided with a wide mandate for the elaboration of "Technical Guidelines for Responsible Fisheries". The Technical Guidelines need not necessarily be adopted by a political body of the organisation in every case. Acting on this broad mandate, the FAO secretariat's Fisheries Department is almost constantly engaged in the development of guidelines and supplements to the CCRF.[308] It does so with considerable autonomy from any interference of the governing bodies, sometimes with the participation of other international or non-governmental organisations.[309] Individual governments often initiate and lead such processes, but sometimes the Fisheries Department relies mainly on its own expertise.[310] The secretariat even develops Technical Guidelines on issues that are not explicitly mentioned in the CCRF, but which should, according to the secretariat and experts, be dealt with in order to implement the code's objectives.[311] The independence of the secretariat from the political level enables it to act swiftly on new developments and to adapt the norms of the CCRF to technological or scientific developments, adding flexibility to the overall mechanism.[312]

Even when COFI is involved in the elaboration of these guidelines, majority voting as foreseen in the Rules of Procedure of the FAO can be

[307] FAO Conference Res. 4/95 of 31 October 1995, para. 5, empowers the FAO "... to elaborate, as appropriate, technical guidelines in support of implementation of the Code."

[308] The Fisheries Department of the FAO secretariat has developed and published about 20 Technical Guidelines for Responsible Fisheries as of 2010; all Guidelines are available at http://www.fao.org/fishery/ccrf/4/en.

[309] Thus, the Technical Guidelines on Marine Protected Areas are being developed by the FAO with the World Bank and the NGO International Union for the Conservation of Nature (IUCN).

[310] This was the case for the development of the Technical Guidelines on Aquaculture.

[311] For example, the CCRF does not address movement of live aquatic animals, but the FAO secretariat has developed the FAO Technical Guidelines on Aquaculture Development, Suppl. 2 on "Health Management For Responsible Movement of Live Aquatic Animals" (2007).

[312] W. Edeson, "Closing the Gap: The Role of 'Soft' International Instruments to Control Fishing", Australian Yearbook of International Law 20 (1999), 83-104 (85).

resorted to in practice. In one interesting recent example, COFI initiated the development of technical guidelines regarding marine protected areas even against the expressly stated will of a member state.[313] This not only suggests that states take this activity seriously even though the matter concerns only the elaboration of voluntary technical guidelines supplementing a nonbinding instrument. The incident also illustrates the readiness of COFI to take majority decisions at this lower level of regulatory activity, thereby underscoring a departure from consensual intergovernmental processes.

5. Norm adaptation

The CCRF can be revised any time through the bodies of the FAO, taking into account new development in fisheries and COFI reports.[314] No specific procedure other than the general rules of procedure of the bodies of the FAO or the revision process is provided for. This bestows flexibility not only in terms of amending the code, but also with respect to the manner of a possible revision. The breadth of approach of the CCRF and the possibility of adopting further instruments under its framework appear to have made any revision unnecessary so far. Thus, new developments can easily be addressed through the development of specific International Plans of Action and Technical Guidelines. These implementing instruments can be developed and adopted at a lower political level or – in the case of Technical Guidelines – are not even necessarily subject to approval by any political body. In some cases, these instruments address subjects which have not even been mentioned in the CCRF itself.[315]

---

[313] FAO, Report of the Twenty-Sixth Session of the Committee on Fisheries, 7-11 March 2005, para. 103, available at: ftp://ftp.fao.org/docrep/fao/008/a0008e/a0008e00.pdf.

[314] CCRF, Article 4.3.

[315] Take for example the development of the FAO Technical Guidelines on "Aquaculture development. 2. Health management for responsible movement of live aquatic animals", 2007, available at ftp://ftp.fao.org/docrep/fao/010/a1108e/a1108e00.pdf.

6. Norm implementation and follow-up

In addition to norm development, the FAO also engages in less visible but equally significant compliance management. Its main elements are a reporting mechanism as well as implementation assistance. Both are important features of a non-confrontational managerial strategy known from compliance mechanisms in multilateral environmental agreements and highlighted by scholars for its compliance-inducing effects.[316]

The FAO Conference with the adoption of the CCRF has authorised the FAO secretariat to give advice to developing countries and establish an Interregional Assistance Program.[317] The CCRF and the implementing instruments serve as a basis for the formulation and design of capacity building projects and for mechanisms of legal, financial and technical assistance. More concretely, the FAO provides the institutional platform, executive know-how and funding to help local communities and developing states with implementation of the code. For example, the advisory service of the Fisheries Department assists governments in the formulation and revision of fisheries legislation[318] and multilateral fisheries agreements such as the Convention on the Sustainable Management of Lack Tanganyika, which is based on the CCRF.[319] By means of the Global Partnership for Responsible Fisheries ("FishCode") and a corresponding financing institution ("FishCode trust fund") which draws on external donations as well as regular program resources of the FAO, the FAO further funds and manages capacity building projects designed to help states, but also communities, fishermen and fish workers to shift to responsible fisheries.

---

316 A. Chayes/A. Handler Chayes, The New Sovereignty: Compliance with International Regulatory Agreements, 1995, 154 and 197; R. Wolfrum, "Means of Ensuring Compliance with and Enforcement of International Environmental Law", Recueil des cours de l'Académie de Droit International de La Haye 272 (1998), 13-154 (110).

317 FAO Conference Res. 4/95, para. 4.

318 For example, the FAO secretariat has assisted in the revision of domestic legislation of a number of developing countries, including Angola, Namibia, Malaysia, The Maldives, Vietnam, Barbados, Antigua and Barbuda, compare for details www.fao.org/fi.

319 Convention on the Sustainable Management of Lack Tanganyika (12 June 2003), available at http://faolex.fao.org/faolex.

The secretariat is also charged with the monitoring of implementation, and must report accordingly to the COFI.[320] The reporting mechanism is based on voluntary questionnaires developed by the secretariat with the approval by COFI.[321] The questionnaires are continuously revised by the secretariat, which is in this regard acting on the basis of specific suggestions from COFI.[322] They are sent out to states as well as organisations, including regional fisheries organisations and NGOs. The results provide the input for the progress report on implementation presented by the secretariat to COFI biannually.[323] Details of these compliance-enhancing tools will be discussed in the analysis on the potential effectiveness of nonbinding instruments in Part 2.[324]

7. Preliminary assessment and outlook

One is left to wonder whether a voluntary instrument of an organisation without any enforcement capabilities can effectively address the regulatory challenges outlined above.

Although nonbinding, the CCRF has not been ineffective. To the contrary, when considering the limited possibilities available to the FAO, it has been increasingly effective as a tool to change policy and legal framework of relevant actors at the international, regional and domestic level.[325] Norms of responsible fisheries, including in particular norms that take into account ecosystem and precautionary approaches, are now widely established among most relevant actors, including govern-

---

320 CCRF, Article 4(2) and FAO Conference Res. 4/95, para. 6; references to reporting to and of FAO are equally included in all of the IPOAs, see IPOA Seabirds, para. 24, IPOA-Sharks, para. 31; IPOA-Capacity, para. 44, IPOA-IUU, para. 87.

321 FAO Council, Report of its Hundred and Twelfth Session, 1997, CL 112/REP, para. 29; Report of the Twenty-Second Session of the Committee on Fisheries, 1997, FIPL/R562 (En), para. 29.

322 See e.g. the revision in 2001 which was based on an improved format suggested by COFI at its 23rd session in 1999.

323 FAO, Committee on Fisheries, Report of the Twenty-Second Session, 17-20 March 1997, para. 29.

324 See on this Part 2, at B.I.

325 Compare for a detailed assessment of the impact on policy and legal frameworks in various regions G. Hosch/G. Ferraro/P. Failler, "The 1995 FAO Code of Conduct for Responsible Fisheries: Adopting, implementing or scoring results?", Marine Policy 35 (2011), 189-200.

ments, international institutions, non-governmental organisations and industry.[326]

More specific progress has also been made for example with respect to monitoring, control and surveillance systems. The CCRF has contributed to improvements in food safety and quality assurance systems. States and regional organisations have generally taken a number of legal and policy measures to implement the 2005 International Plan of Action to Prevent, Deter and Eliminate Illegal, Unreported and Unregulated Fishing.[327] More generally, as discussed above, the CCRF serves as a point of reference and a framework for discourse which programs, coordinates and integrates fisheries activities of most actors at various levels of governance and in science. This is underlined by citation analysis of the FAO which demonstrates that since its adoption in 1995, the CCRF and related instruments are increasingly cited in published work in the fields of law, management and science. According to this analysis, the CCRF today belongs to the category of highly cited publications. This result indicates that the CCRF is reaching its stakeholders, but also that there appears to be significant under-representation of research from developing countries.[328]

The capacity of the CCRF has however been limited in terms of changing actual fishing practices on the ground. While contributing to the establishment of norms, implementation of and compliance with such norms remains insufficient in most countries and regions.[329] The core problems that the CCRF aimed to address are far from being solved 15 years after its adoption. Fish stocks continue to deteriorate in most parts of the world. One of the reasons is that the conservation and management measures such as the establishment of quotas and authorisation systems cannot be and are not set through the nonbinding code

---

[326] *Ibid.*, at 199.

[327] FAO, Committee on Fisheries, Analysis of the implementation and impact of the FAO Code of Conduct for Responsible Fisheries Since 1995, COFI/2009/Inf.10 (September 2008), Para. 4.

[328] J. Parker/D. Doulman/J. Collins, "Citation analysis for the 1995 FAO Code of Conduct for Responsible Fisheries", Marine Policy 34 (2010), 139–144.

[329] This is empirically underscored by the detailed studies of G. Hosch/G. Ferraro/P. Failler, "The 1995 FAO Code of Conduct for Responsible Fisheries: Adopting, implementing or scoring results?", Marine Policy 35 (2011), 189-200, at 199; T.J. Pitcher/D. Kalikoski/G. Pramod (eds.), Evaluations of Compliance with the FAO (UN) Code of Conduct for Responsible Fisheries, 2006, available at http://www.fisheries.ubc.ca/archive/publications/reports/14-2.pdf.

of conduct, but must be established through regional organisations and states. And more importantly, the CCRF cannot overcome the enforcement problem in international fisheries governance. Enforcement still remains decentralised and must ultimately be undertaken by states. States remain highly reluctant to enact and enforce meaningful conservation and fisheries management measures where these are politically sensitive and may hurt the short term interests of fisheries industries. As the FAO reports indicate, only limited progress has been made in the implementation of sound management practices.[330] Considerable implementation problems continue with respect to a lack of fisheries management plans, the implementation of the ecosystem and precautionary approaches, subsidies leading to overcapacity and the lack of financial and human resources.[331] Moreover, the CCRF's impact has also been limited where enforcement has trade implications as in the case of port and flag state controls.[332] Thirteen years after the adoption of the CCRF, overexploitation of stocks is still increasing dramatically.[333]

These findings are supported by a studies which examined the compliance of states with the central provisions on fisheries management of the CCRF.[334] In one study undertaken at the University of British Co-

330 FAO, Committee on Fisheries, Analysis of the implementation and impact of the FAO Code of Conduct for Responsible Fisheries Since 1995, UN Doc. COFI/2009/2, para. 11 et seq.

331 FAO, Committee on Fisheries, Progress in the implementation of the Code of Conduct for Responsible Fisheries, related international plans of action and strategy (2009), COFI/2009/2, para. 11 and 39 et seq.; FAO, Committee on Fisheries Progress in the implementation of the Code of Conduct for Responsible Fisheries, related international plans of action and strategy (2007), COFI/2007/2, paras 35-38; FAO, Committee on Fisheries, COFI/2005/2, Progress in the implementation of the 1995 Code of Conduct for Responsible Fisheries, related International Plans of Action (2005), paras. 33-36.

332 FAO, Committee on Fisheries, Analysis of the implementation and impact of the FAO Code of Conduct for Responsible Fisheries Since 1995, UN Doc. COFI/2009/Inf.10 (September 2008), para. 9.

333 FAO, The State of World Fisheries and Aquaculture 2008 (Rome 2009), Part I, page 30, available at: ftp://ftp.fao.org/docrep/fao/011/i0250e/i0250e.pdf.

334 T.J. Pitcher/D. Kalikoski/G. Pramod (eds.), Evaluations of Compliance with the FAO (UN) Code of Conduct for Responsible Fisheries, 2006, available at http://www.fisheries.ubc.ca/archive/publications/reports/14-2.pdf; similarly G. Hosch/G. Ferraro/P. Failler, "The 1995 FAO Code of Conduct for Responsible Fisheries: Adopting, implementing or scoring results?", Marine Policy 35 (2011), 189-200, at 199.

lumbia, the fishing practices of 53 countries responsible for 96 per cent of the global catch were examined.[335] The researchers found that even though many fisheries laws adopted after the adoption of the CCRF reflect its principles and recommendations, serious implementation and enforcement problems persist in practice. The study in particular points to poor compliance with the fisheries management provisions of the CCRF among those countries.[336] Some of the key provisions of the CCRF are currently not implemented. The research shows that more than 90 per cent of the countries examined failed to reduce their own excess fishing capacity, that over 80 per cent of the countries had unsatisfactory scores when it came to irresponsible practices such as catching juvenile fish, and that only very few countries had methods to ensure that fish and shellfish would not be fatally trapped in lost fishing gear and traps. Only Norway, the U.S., Canada, Australia, Iceland and Namibia received overall compliance scores of 60 per cent, but at the same time 28 countries that are responsible for 40 per cent of the global catch basically failed in all relevant aspects.[337]

Generally speaking, developing countries thus tend to have greater compliance problems. This is also reflected in FAO reports, which found that, besides lack of political will and economic factors, the predominant constraints for more rapid progress are insufficient resources and institutional incapacity, as well as awareness deficits in developing countries.[338] In the case of shrimp aquaculture, for instance, studies suggest that only a few countries have so far implemented the strategies of the CCRF.[339] But the study led by the University of British Columbia

---

335 T.J. Pitcher/D. Kalikoski/G. Pramod (eds.), Evaluations of Compliance with the FAO (UN) Code of Conduct for Responsible Fisheries, 2006, available at http://www.fisheries.ubc.ca/archive/publications/reports/14-2.pdf.

336 T.J. Pitcher/D. Kalikoski/G. Pramod/K. Short, "Not honouring the Code", Nature 457 (2009), 658-659.

337 *Ibid.*

338 FAO, The State of World Fisheries and Aquaculture 2008 (Rome 2009), Part I, page 9 et seq., available at: ftp://ftp.fao.org/docrep/fao/011/i0250e/i0250e.pdf.

339 S.M.N. Alam/C. Kwei Lin/A. Yakupitiyage/H. Demaine/M.J. Phillips, "Compliance of Bangladesh shrimp culture with FAO code of conduct for responsible fisheries a development challenge", Ocean and Coastal Management 48 (2005), 177-188 (186); D. Barnhizer, "Waking from Sustainability's "Impossible Dream": The Decisionmaking Realities of Business and Government", Georgetown International Environmental Law Review 18 (2006), 595-690 (677).

also mentions that "disappointing scores from some European Union nations, with the resources and know-how to implement the code, reinforce a low priority given to improving fisheries management."[340] Assuming that this is correct, implementation problems are thus not limited to capacity but to a lack of political will.

The implementation and enforcement of the norms established through the CCRF and widely accepted in discourse may thus require not only a nonbinding instrument but international treaty norms which follow the example of the Fish Stocks Agreement and that of some progressive regional fisheries management organisations. Similarly, researchers stress that the voluntary code of conduct was required in 1995 to establish the framework and forge consensus on norm change, but that international binding law may now be necessary for stock conservation and enforcement.[341]

II. Regulating international trade in chemicals and pesticides: the introduction of Prior Informed Consent through nonbinding instruments

1. The institutional framework

The voluntary PIC system is to a large extent the product of the close cooperation of two international institutions, the FAO and UNEP. The institutional framework of the FAO has been described in detail above. In the case of pesticide regulation, the FAO Committee of Agriculture, a subsidiary body of the FAO Council, plays the central role in the development of norms and in the oversight of the secretariat. As also mentioned already, UNEP is – in contrast to the FAO – not an international organisation. It is not based on an international treaty but was established by the United Nations General Assembly through Resolution 2997 in 1972.[342] UNEP does not have the competence to take binding decisions, but its main governing body, the Governing Council, which , which was renamed from Governing Council into the United Nations Environment Assembly of UNEP in 2013, may however make recom-

340 T.J. Pitcher/D. Kalikoski/G. Pramod/K. Short, "Not honouring the Code", Nature 457 (2009), 658-659 (658).

341 T.J. Pitcher/D. Kalikoski/G. Pramod/K. Short, "Not honouring the Code", Nature 457 (2009), 658-659 (659).

342 UN Doc. A/RES/27/2997 (1972).

mendations to states.[343] Even though it could take these decisions by majority vote,[344] the United Nations Environment Assembly, just as the Governing Council previously, usually strives for consensus when adopting recommendations for states in order to enhance their authority. The United Nations Environment Assembly reports to the General Assembly through the UN Economic and Social Council. UNEP has its own secretariat, which is headed by an Executive Director.

For the introduction of the voluntary PIC system, the secretariats of the FAO and UNEP joined forces and managed the system through a joint FAO/UNEP secretariat. Each secretariat remained however largely responsible for their specific field of expertise, i.e. UNEP for chemicals and the FAO secretariat for pesticides. Most importantly, the secretariats of the FAO and UNEP established a joint FAO/UNEP Group of Experts which played the central role in running the voluntary PIC system. This will be explained in detail below following an introduction to the regulatory challenge and the way it was addressed through the PIC system.

2. *The regulatory challenge*

The use of chemicals, pesticides and other substances is a complex issue that illustrates the challenges of sustainable development particularly well. On the one hand, the use of chemicals and pesticides is central for economic development, the fight against diseases such as malaria and yellow fever, particularly in developing countries,[345] and has greatly contributed to food production much needed to fight hunger and starvation in the developing world. The downside is also well-known. Improper pesticide and chemical usage carries high environmental and health risks.[346] Pesticides are usually produced in large quantities and directly (and often manually) introduced into the natural environment

---

343 UN Doc. A/RES/27/2997 (1972), Sec. I, para. 2 (a).

344 UNEP, Rules of Procedure of the Governing Council, Rule 48, RoPs available at http://www.unep.org/download_file.multilingual.asp?FileID=11.

345 J. Ross, "Legally Binding Prior Informed Consent", Colorado Journal of International Environmental Law and Policy 10 (1999), 499-529 (501 et seq.).

346 The World Health Organization estimated in 1990 that there ware a minimum of 3 million acute severe cases of pesticide poisoning, 20000 unintentional deaths and 25 occupational poisonings each year, compare World Health Statistics Quarterly 43 (1990), 139-144.

at large quantities. Coupled with their high persistence levels, this often results in long-term and transboundary pollution, particularly of water supplies. This leads to the accumulation of toxic substances in the food chain and has negative effects on fauna and flora. Misuse of chemicals and pesticides has often led directly to the destruction of wild animal populations.[347] While beneficial for agriculture, the increased use of artificial pesticides has also brought about an increasing dependency on pesticides to fight pest outbreaks.[348] Residues from pesticides pose direct risks to human food consumers, especially where foods are sold without proper handling or washing. Populations in developing countries are often particularly vulnerable. This at least partly derives from their greater dependency on a healthy environment for subsistence.[349] In addition, poor social and economic conditions contribute to misuse.[350] Although exact measurement is hardly possible, studies suggest that by the 1990s, people in developing countries suffered approximately three-quarters of the reported poisonings from pesticides even though these countries accounted for only one-fifth of global pesticide usage.[351]

---

[347] J. Ross, "Legally Binding Prior Informed Consent", Colorado Journal of International Environmental Law and Policy 10 (1999), 499-529 (504); P. Sands, Principles of International Environmental Law, 2003, 625.

[348] For a more detailed analysis of these problems A. Pearse, Seeds of Plenty, Seeds of Want: Social and Economic Implications of the Green Revolution, 1980; B. Dinham, "The Success of a Voluntary Code in Reducing Pesticide Hazards in Developing Countries", Green Globe Yearbook of International Co-operation on Environment and Development 3 (1996), 29-36 (29).

[349] B. Dinham, "The Success of a Voluntary Code in Reducing Pesticide Hazards in Developing Countries", Green Globe Yearbook of International Co-operation on Environment and Development 3 (1996), 29-36 (30).

[350] See for details of the negative effects B. Dinham, "The Success of a Voluntary Code in Reducing Pesticide Hazards in Developing Countries", Green Globe Yearbook of International Co-operation on Environment and Development 3 (1996), 29-36 (29); R.L. Paarlberg, "Managing Pesticide Use in Developing Countries" in: P.M. Haas/R.O. Keohane/M.A. Levy (eds.), Institutions for the Earth: Sources of Effective International Environmental Protection, 1993, 309-350 (309 et seq.); J. Ross, "Legally Binding Prior Informed Consent", Colorado Journal of International Environmental Law and Policy 10 (1999), 499-529 (503 ff.).

[351] These are figures of a study prepared by the U.S. Agency for International Development of 1990, cited by R.L. Paarlberg, "Managing Pesticide Use in Developing Countries" in: P.M. Haas/R.O. Keohane/M.A. Levy (eds.), Insti-

So why does this issue require international regulatory efforts and cooperation? Indeed, if all states adequately regulated and controlled pesticide and chemical use, the issue could be left to domestic law. However, the regulatory capabilities of individual states differ hugely. Many developing countries lack the legal, administrative and enforcement capacities to control and handle these substances. With uncontrolled international trade, trade substances that are banned or severely regulated in industrialised countries but nevertheless produced in these countries can be exported to developing countries where they are often used with little knowledge of the associated risks.[352] International cooperation and assistance is therefore a matter of solidarity and equity, but due to the transboundary nature of the issue also in the immediate interest of developed countries. The natural movements of pesticides across the globe as well as globalising markets of products and foodstuffs facilitate the spread of poisonous residues across boundaries and their eventual accumulation in all of the various natural environmental media.[353] This may contribute to the situation whereby exporting substances to states with poor pesticide management ultimately results in harmful effects even in (the exporting) states which themselves have stringent regulations on pesticide use.[354]

3. The nonbinding response: the voluntary PIC procedure

In the context of chemical and pesticide usage, PIC basically requires that chemicals or pesticides which are banned or severely restricted in

---

tutions for the Earth: Sources of Effective International Environmental Protection, 1993, 309-350 (310) see also J. Ross, "Legally Binding Prior Informed Consent", Colorado Journal of International Environmental Law and Policy 10 (1999), 499-529 (501).

[352] On this problem generally C. Mehri, "Prior Informed Consent: An Emerging Compromise for Hazardous Exports", Cornell International Law Journal 21 (1988), 365-389 (366).

[353] P.-T. Stoll, "Hazardous Substances and Technologies" in: F.L. Morrison/R. Wolfrum (eds.), International, Regional and National Environmental Law, 2000, 437-451 (437).

[354] V.P. Nanda/B.C. Bailey, "Nature and scope of the problem" in: G. Handl/R. E. Lutz (eds.), Transferring Hazardous Technologies and Substances, 1989, 3-39 (42); J. Ross, "Legally Binding Prior Informed Consent", Colorado Journal of International Environmental Law and Policy 10 (1999), 499-529 (506).

an exporting state should not be imported without the consent of the importing country.[355] More concretely, the PIC mechanism establishes a complex multi-level procedure. On the national (exporting country) level, a network of so-called designated national authorities (DNA) were asked to notify the secretariat run jointly by the FAO and UNEP[356] whenever pesticides or chemicals were either banned or severely restricted in a country.[357] The international instruments not only defined in detail which measures qualified as a ban or a severe restriction,[358] but also required the information to be submitted in a certain format.[359] Countries were asked to provide information on the identity of the substance, the reasons for the control action, possible alternative measures and any additional information available to the authorities.[360] The international institutions would then notify the designated national authorities of other countries of these measures.[361] On the basis of the information received, the designated national authorities had to evaluate whether and under what conditions to allow future imports. A summary of the resulting decision, the so-called "importing country response", was to be sent to the secretariat. The international instruments prescribe specific time frames (90 days) as well as the format for the response in a clear effort to streamline the procedure through interna-

---

355 Amended London Guidelines, Part I (g) and (h); FAO Pesticide Code, Article 2.

356 Although cooperating closely with regards to secretariat functions, the UNEP and FAO secretariats each retained the lead in their respective speciality, i.e. UNEP for chemicals other than pesticides, and FAO for pesticides.

357 FAO Pesticide Code (1989), Articles 2, 9.1., 9.5. and 9.7.; UNEP London Guidelines (1989), paras 1 (h) and 6.

358 See the definitions in the FAO Pesticide Code (1989), Article 2; UNEP London Guidelines (1989), para. 1 (b), (c).

359 Thus, the 'FAO Guidelines on the operation of prior informed consent', adopted with the amendments to the FAO Pesticide Code through FAO Conference Res. 6/89, contain a detailed format for the notification in the Annex; the Guidelines can be found in the Report of the FAO Conference on its Twenty-fifth session, Appendix E, available at http://www.fao.org/docrep/x5588e/x5588e0l.htm; see also UNEP Guidelines, Annex I, 'Form for notification of control action'.

360 FAO Pesticide Code (1989), Article 9 (1), (2); UNEP London Guidelines (1989), para. 6.

361 FAO Pesticide Code (1989), Article 9 (1); UNEP London Guidelines (1989), para. 7.4.

tional procedural norms. In addition, exporting countries would have to make sure that importing countries were notified of any specific export that would occur or was in the process of occurring.[362]

If properly implemented, the international prior informed consent (PIC) mechanism can address some of the problems outlined above. PIC offers an *ex ante* procedural safeguard allowing states to take an informed decision on whether to allow the import in question, based on the information provided by the exporter and the international institution. If no response occurred, the substance was not to be exported for lack of consent.[363] In the case of an import ban, the instruments required the exporting country's national agency to enforce the prohibition vis-à-vis its industry by taking the relevant control actions, and possibly to prohibit the further production of the substance in question.[364] In case of a permit requirement, the exporting country had to ensure that the exporter obtained a proper permit. In both cases, this mechanism effectively shifted the enforcement burden from importing countries with low regulatory capacity to exporting countries with higher regulatory capacity.[365] Furthermore, the PIC clause, by requiring the provision of information on the substance and its regulation in the exporting state, enables the importing state to take an informed decision when accepting the import, and to take the necessarily regulatory and administrative measures based on the information provided before harm is done.[366]

---

362 FAO Pesticide Code (1989), Article 9 (3)-(5); UNEP London Guidelines (1989), para. 8.

363 FAO Guidelines on PIC, Step 6; UNEP London Guidelines, para. 7.3.

364 FAO Guidelines on PIC, Step 8; UNEP London Guidelines, para. 7.4.

365 D.G. Victor, "'Learning by Doing' in the Nonbinding International Regime to Manage Trade in Hazardous Chemicals and Pesticides" in: D.G. Victor/K. Raustiala/E.B. Skolnikoff (eds.), The Implementation and Effectiveness of International Environmental Commitments: Theory and Practice, 1998, 221-281 (237).

366 C. Redgwell, "Regulating Trade in Dangerous Substances: Prior Informed Consent under the 1998 Rotterdam Convention" in: A. Kiss/D. Shelton/K. Ishibashi (eds.), Economic Globalization and Compliance with International Environmental Agreements, 2003, 75-88 (75); for the various regulatory approaches and advantages of PIC in the environmental field, compare C. Mehri, "Prior Informed Consent: An Emerging Compromise for Hazardous Exports", Cornell International Law Journal 21 (1988), 365-389; generally on PIC in this

4. Norm development process

a) Mandate

As already seen in the context of the FAO Code of Conduct for Responsible Fisheries, the adoption of a code of conduct addressed not only to states but also to private actors is not foreseen in the FAO Constitution. It only contains a possibility for the FAO Conference to make recommendations to members, but not to address "all public and private entities engaged in or associated with the distribution and use of pesticides"[367] as does the FAO Pesticide Code. In contrast, the UNEP London Guidelines were only addressed to governments, and thus it remained within the mandate of UNEP to recommend policies in order to promote international cooperation.[368]

The founding documents of both institutions contain a general substantive mandate which provides the basis for these voluntary instruments, and the PIC clause in particular. Even though not equipped with a general mandate to protect health or the environment or to regulate trade, the references in the FAO Constitution to the "conservation of natural resources"[369] and the "improvement of administration relating to nutrition, food and agriculture"[370] are providing sufficiently close links to the objectives of PIC in order to give the FAO a general mandate for the matter. A substantive mandate for UNEP's Governing Council could be seen in its general mandate to "promote international cooperation in the environment field and to recommend, as appropriate, policies to this end".[371] The aim of the PIC procedure is the enhanced protection of health and the environment at the global level through ex-

---

context P. Birnie/A. Boyle, International law and the environment, 2002, 431-432.

367 See Article 1.1 of the FAO International Code of Conduct on the Distribution and Use of Pesticides (Revised Version), adopted through FAO Council Resolution 1/123 (2002), available at http://www.fao.org/DOCREP/005/Y4544E/y4544e00.htm.

368 UN Doc. A/RES/27/2997 (1972), para. 2 (a).

369 FAO Constitution, Article I.2. (c).

370 FAO Constitution, Article I.2. (b).

371 UN Doc. A/RES/27/2997 (1972), Article 2(a).

change of information and regulation of trade of hazardous substances.[372]

b) Norm elaboration and adoption

The PIC clause was first introduced in 1989 through amendments to the International Code of Conduct on the Distribution and Use of Pesticides (FAO Pesticide Code")[373] and the London Guidelines for the Exchange of Information on Chemicals in International Trade[374] of UNEP (London Guidelines).[375] Following an initiative of the FAO Director General, the FAO Pesticide Code had been adopted by the FAO Conference in 1985 after a consultation process which included the FAO Committee on Agriculture and other interested international organisations.[376] The London Guidelines were adopted by UNEP's Governing Council in 1987.[377]

The PIC clause was incorporated into the Pesticide Code and the London Guidelines through a resolution of the FAO Conference and a decision of UNEP's Governing Council in 1989.[378] These decisions were preceded by a political struggle between industry and many developed states on the one hand and developing countries supported by NGOs on the other. Pressure from NGOs and developing countries could not

---

372 Compare the "Introduction to the Guidelines" included in UNEP GC Decision 15/30 (1989) and the Report of the Conference of FAO, Twenty-Fifth Session of 11-29 November 1989, paragraph 112.

373 International Code of Conduct on the Distribution and Use of Pesticides, originally adopted by FAO Conf. Res. 10/85 of 28 November 1985; compare Report of the Conference of FAO on its Twenty-third Session of 9-28 November 1985, Annex 1, available at www.fao.org.

374 UNEP London Guidelines for the Exchange of Information on Chemicals in International Trade, originally adopted by UNEP Governing Council, Decision UNEP/GC/DEC/14/27 of 17 June 1987.

375 UNEP was established as a UN programme through UN Doc. A/RES/27/2997 (1972).

376 FAO Conference Res. 10/85 of 28 November 1985.

377 UNEP GC Dec. 14/27 of 17 June 1987.

378 UNEP GC Dec. 15/30 (25 May 1989), available at http://www.chem.unep.ch/ethics/english/longuien.htm; FAO Conference Res. 6/89 (29 November 1989).

at first convince governments of many industrialised states[379] to regulate against the interests of their chemical and agro-chemical industries, at least not on a decisive scale, until the mid-1980s.[380] As the political momentum towards PIC gained headway in the United Nations General Assembly and UNEP's Governing Council,[381] many industrialised states turned to the OECD to develop their own instrument on the issue, arguably in an attempt to forestall the development of more restrictive norms in the UN organisations.[382] It is interesting to see that they employed exactly the same strategy as in the case of corporate responsibility codes, to be discussed below, where OECD states successfully established an alternative to UN norms.[383] In the case of PIC, the "preemptive consensus"[384] established at the OECD became manifest in the OECD Guiding Principles.[385] In this document, OECD member states explicitly rejected any export control but insisted on one-time-only export notification. According to some commentators, these principles were then actively used as arguments against the introduction of PIC at

---

379 This foremost included the United States, the United Kingdom, Germany and Japan. Compare for the history R.L. Paarlberg, "Managing Pesticide Use in Developing Countries" in: P.M. Haas/R.O. Keohane/M.A. Levy (eds.), Institutions for the Earth: Sources of Effective International Environmental Protection, 1993, 309-350 (320); M. Pallemaerts, "Developments in International Pesticide Regulation", Environmental Policy and Law 18 (1988), 62-69 (66).

380 An exception were the United States of America, since they had implemented a notification system for exports, compare D.G. Victor, "'Learning by Doing' in the Nonbinding International Regime to Manage Trade in Hazardous Chemicals and Pesticides" in: D.G. Victor/K. Raustiala/E.B. Skolnikoff (eds.), The Implementation and Effectiveness of International Environmental Commitments: Theory and Practice, 1998, 221-281 (220 and 228).

381 UNEP Dec. 85(V), UNEP Report of the Governing Council on the work of its fifth session, UN Doc. A/32/25 (1977); UNGA Res. A/37/51 (1982), paras. 112-113.

382 M. Pallemaerts, "Developments in International Pesticide Regulation", Environmental Policy and Law 18 (1988), 62-69 (65).

383 See the analysis of the OECD Guidelines in this Part at B.III, further below.

384 R.L. Paarlberg, "Managing Pesticide Use in Developing Countries" in: P.M. Haas/R.O. Keohane/M.A. Levy (eds.), Institutions for the Earth: Sources of Effective International Environmental Protection, 1993, 309-350 (322).

385 OECD 'Guiding Principles on Information Exchange Related to Export of Banned or Severely Restricted Chemicals', OECD Doc. C(84)/37(final) (1984).

several levels, including in the European Community, UNEP, and the FAO.[386]

The later inclusion of PIC in the two nonbinding instruments was largely due to the efforts of a non-governmental organisation: the Pesticide Action Network (PAN). Building on its enormous expertise in the field, this NGO was able to convince a majority of developing countries of the G 77 in the UNEP Governing Council of the importance of PIC and related action.[387] Using its observer position in the decisive meeting of the Governing Council, it distributed an amendment proposal for the inclusion of PIC which was taken up by Senegal during negotiations.[388] As a majority in the Governing Council in favour of PIC emerged, the UNEP Governing Council reached the compromise that the London Guidelines should be adopted immediately without PIC, but that PIC should be included in two years time.[389] The adoption of PIC thus highlights not only the influence of NGOs, but also that the possibility of majority voting may lead to compromises above the lowest common denominator even if member states do not actually decide by majority. The political momentum of the developments at UNEP as well as continuing political pressure exerted by developing countries and NGOs forced a parallel development within the FAO.[390]

While NGOs were thus highly influential with respect to political agenda setting, they have not been included in the further elaboration process. Once the main governing bodies of UNEP and the FAO had

---

386 R.L. Paarlberg, "Managing Pesticide Use in Developing Countries" in: P.M. Haas/R.O. Keohane/M.A. Levy (eds.), Institutions for the Earth: Sources of Effective International Environmental Protection, 1993, 309-350 (322); M. Pallemaerts, "Developments in International Pesticide Regulation", Environmental Policy and Law 18 (1988), 62-69 (65).

387 For a historical account and assessment of the negotiations see R.L. Paarlberg, "Managing Pesticide Use in Developing Countries" in: P.M. Haas/R.O. Keohane/M.A. Levy (eds.), Institutions for the Earth: Sources of Effective International Environmental Protection, 1993, 309-350 (323); M. Pallemaerts, "Developments in International Pesticide Regulation", Environmental Policy and Law 18 (1988), 62-69 (66).

388 R.L. Paarlberg, "Managing Pesticide Use in Developing Countries" in: P.M. Haas/R.O. Keohane/M.A. Levy (eds.), Institutions for the Earth: Sources of Effective International Environmental Protection, 1993, 309-350 (323).

389 UNEP GC/14/27 (1987), Annex I, sixth preambular paragraph, available at www.unep.org.

390 FAO Conference Res. 5/87, para. 1, available at www.fao.org.

agreed on the incorporation of PIC into the instruments, both organisations charged expert working groups with the task of developing drafts for amendments of the London Guidelines and the Pesticide Code.[391] This elaboration process did not build on predetermined procedural rules.

5. Norm adaptation

The incorporation of the PIC procedure into the UNEP London Guidelines and the FAO Pesticide Code by means of resolutions of the main political organs demonstrates how nonbinding instruments provide tools which can be flexibly and speedily adapted to meet newly arising challenges. The FAO Pesticide Code was again subject to revision in 2002 to reflect changes in management practices and international conventions relating to pesticide use, including the adoption of the Convention on the Prior Informed Consent Procedure for Certain Hazardous Chemicals and Pesticides in International Trade (PIC Convention)[392] which made the PIC clause in the code superfluous.[393] With this revision, the FAO Council followed Article 12.10 of the Pesticide Code which stresses that the "[g]overning Bodies of FAO should periodically review the relevance and effectiveness of the Code. The Code should be considered a dynamic text which must be brought up to date as required, taking into account technical, economic and social progress."[394]

---

[391] FAO Conference Resolution 5/87, para. 2; UNEP Governing Council Dec. 14/27 of 17 June 1987, preambular para. 12.

[392] Convention on the Prior Informed Consent Procedure for Certain Hazardous Chemicals and Pesticides in International Trade (10 September 1998), 38 ILM 1 (1999); in detail on the regulatory objectives and main provisions as well as the limitations of this Convention N.S. Zahedi, "Implementing the Rotterdam Convention: the challenges of transforming aspirational goals into effective controls on hazardous pesticide exports to developing countries", Georgetown international environmental law review 11 (1999), 707-739 (714 et seq.).

[393] FAO Code of Conduct on the Distribution and Use of Pesticides (revised version 2002), adopted by FAO Council Resolution 1/123 (2002).

[394] FAO Pesticide Code, Article 12.10.

6. *Norm implementation and follow-up*

a) Overview

The international institutions had several key functions in the above described procedure. As indicated, a FAO/UNEP secretariat served as the information hub between the designated national authorities of the exporting and importing states.[395] The implementation also required a number of administrative decisions at the international level. The FAO and UNEP had to ensure that the nature of the control action in the exporting state met the definitions of a ban or a severe restriction. Any substance that was subject to a relevant control action and thus fulfilled the conditions for PIC to apply was listed.[396] Those control actions which were in effect before the establishment of PIC and thus were not subject to the notification requirement had to be compiled in an initial PIC list. Finally, the procedure recommended that other substances which were not subject to any ban or severe restriction but which could nevertheless pose acute hazards due to the specific conditions of use in some (developing) countries would also be listed.[397] The decision whether a particular substance posed such a hazard required an assessment and a decision that was entirely independent from regulatory decisions on the national level. Once a chemical or pesticide was included in the list, so-called decision guidance documents were developed in order to enable the designated national authorities to make an informed decision.[398]

b) International administration by the FAO/UNEP Joint Group of Experts

With the exception of information dissemination and the drafting of the decision guidance documents, the above mentioned management and administrative activities were undertaken in practice by the so-called Joint Group of Experts established jointly by UNEP and the FAO. The

---

395 FAO Guidelines on PIC, Step 5; UNEP London Guidelines, para. 7.2. (b).

396 FAO Pesticide Code (1989), Article 9.8.; UNEP London Guidelines (1989), para. 7.2. (a).

397 FAO Pesticide Code (1989), Guidelines on PIC, step 4; UNEP London Guidelines, para. 1 (h) and Annex II.

398 FAO Pesticide Code (1989), Article 9.8.1. and FAO Guidelines on the operation of PIC, step 4; UNEP London Guidelines (1989), para. 7.2. (b).

FAO/UNEP Joint Group of Experts was made up of ten experts; five chosen by each organisation, with due consideration given to equal representation of developing and developed countries. Representatives of other international organisations, namely the World Health Organization, the OECD, the International Labour Organization and the European Economic Community participated as observers in the meetings of the group. NGOs could participate in the meetings, but were excluded when particularly sensitive matters were discussed. Their access was formalised by access rules adopted by the FAO/UNEP Joint Meetings, which limited the number of observers to two from industry and two from public interest groups.[399] In practice, however, most groups interested in the work of the expert group were represented, including industry associations and a number of NGOs.[400]

The Joint Group of Experts constituted a standing body which took decisions of an administrative nature with implications for the PIC procedure and participating countries as well as industry. The experts administered the central PIC list by assessing which control actions met the definitions of the instruments, by determining which substances were to be included in the initial PIC list and by deciding which substances had to be included as a result of particular conditions of use that turned these substances into hazards. One would expect that such significant decisions were taken on the basis of pre-determined clear procedural and substantive rules delegating such decision making power to the expert group and legitimising its activities.

First of all, the expert group did not act with a clear mandate. None of the founding documents of the FAO and UNEP foresee the possibility

---

[399] Reported by D.G. Victor, "'Learning by Doing' in the Nonbinding International Regime to Manage Trade in Hazardous Chemicals and Pesticides" in: D.G. Victor/K. Raustiala/E.B. Skolnikoff (eds.), The Implementation and Effectiveness of International Environmental Commitments: Theory and Practice, 1998, 221-281 (240).

[400] Pesticide producers were represented by the Groupement international des associations nationals de fabricants de produits agrochimiques (GIFAP, today Global Crop Protection Federation – GCPF); chemical producers were represented by the Chemical Manufacturers Association (CMA) and European Chemical Industry Council (CEFIC); the participating NGOs were the World Wide Fund for Nature (WWF), the International Organization of Consumer Unions (IOCU, today Consumers International – CI), and the Pesticide Action Network (PAN).

of joint working groups.[401] And neither instrument explicitly provides a legal basis for the reliance on an expert group to establish a PIC list. The resolution of the FAO Conference which amended the FAO Pesticide Code and the London Guidelines only mandated both institutions to engage in unspecified cooperation in the operation and implementation of the PIC procedure.[402] Neither the FAO Pesticide Code nor the UNEP London Guidelines however envisaged the establishment of such a group, nor did they authorise their respective secretariats to do so. The UNEP London Guidelines only mention that an "informal consultative process"[403] may be used to assist UNEP in identifying substances for the PIC list, and the Guidelines to the FAO Pesticide Code mention that the selection of pesticides would be done in cooperation with UNEP.[404] The rudimentary procedural rules on the composition of expert groups which can be found in Annex 2 of the UNEP London Guidelines only applied to the particular problem of acutely hazardous substances and can therefore hardly be seen as a mandate.[405]

The lack of a mandate was mirrored in the complete lack of procedural rules regarding the composition and decision-making of the expert group.[406] The Joint Expert Group even established procedural law on its own when adopting rules for observer admittance.[407]

---

[401] The absence of a mandate for joint working groups is also emphasised by J.E. Alvarez, International Organizations as Law-makers, 2005, 232.

[402] Para. 5.2 of the amended UNEP London Guidelines (1989) states that "UNEP should share with FAO the operational responsibility for the implementation of the PIC procedure...", and para. 5.3 that "UNEP should collaborate with FAO in reviewing the implementation for the PIC procedure." See also FAO Conference Resolution 6/89, para. 3 and the 'Guidelines on the operation of prior informed consent (PIC)', included in the Report of the Conference of FAO, Twenty-fifth Session of 11-29 November 1989, Annex to Appendix E.

[403] UNEP London Guidelines, para. 7.2. a).

[404] UNEP London Guidelines, para. 7.2 a); FAO Guidelines to the FAO Pesticide Code, Step 4.

[405] UNEP London Guidelines, Annex II, paras 2 and 3.

[406] Specifics on the establishment and the running of a joint secretariat can only be found in a Memorandum of Understanding between the Executive Director of UNEP and the Director-General of the FAO. The Memorandum of Understanding is mentioned in the Report of the Conference of FAO, Twenty-sixth Session of 9-27 November 1991, para. 125. The Memorandum of Understanding was formalized by letter exchange in November 1992, see D.G. Victor, "'Learning by Doing' in the Nonbinding International Regime to Manage

Even where procedural rules existed in the FAO Pesticide Code and the London Guidelines as well as the FAO Conference's Guidelines on the operation of prior informed consent,[408] the decision-making practice of the Joint Expert Group deviated from them. With one of its first decisions, the FAO/UNEP Joint Expert Group clarified that in addition to "banned" or "severely restricted" chemicals and pesticides, others which were rejected for registration or voluntarily withdrawn from registration by a manufacturer would be included in the PIC procedure as provided for by the FAO Conference Report but not by the FAO Pesticide Code and the UNEP London Guidelines.[409] In other words, the expert group broadened the criteria set out by the two instruments on the basis of a Conference report which was never put to a vote. In addition, the Joint Expert Group also developed more detailed definitions than those included in the voluntary instruments, and subsequently undertook a revision of the reporting requirements, so that they would be in harmony with the new criteria.[410]

In deciding on the initial PIC list and on updates to the PIC list,[411] the expert group decided autonomously to deviate from existing procedures. When the procedures that were suggested in Annex II of the London Guidelines to set up an initial PIC list proved to be useless in

---

Trade in Hazardous Chemicals and Pesticides" in: D.G. Victor/K. Raustiala/E. B. Skolnikoff (eds.), The Implementation and Effectiveness of International Environmental Commitments: Theory and Practice, 1998, 221-281 (footnote 46).

407 D.G. Victor, "'Learning by Doing' in the Nonbinding International Regime to Manage Trade in Hazardous Chemicals and Pesticides" in: D.G. Victor/K. Raustiala/E.B. Skolnikoff (eds.), The Implementation and Effectiveness of International Environmental Commitments: Theory and Practice, 1998, 221-281 (240).

408 'Guidelines on the operation of prior informed consent (PIC'), printed in the Report of the Conference of FAO, Twenty-fifth Session of 11-29 November 1989, Annex to Appendix E.

409 Report of the Conference of FAO on its Twenty-fifth Session (11-29 November 1989), para. 117.

410 D.G. Victor, "'Learning by Doing' in the Nonbinding International Regime to Manage Trade in Hazardous Chemicals and Pesticides" in: D.G. Victor/K. Raustiala/E.B. Skolnikoff (eds.), The Implementation and Effectiveness of International Environmental Commitments: Theory and Practice, 1998, 221-281 (242).

411 By the end of 1997, i.e. shortly before the adoption of the Rotterdam Convention, the PIC list included twenty-two of the most dangerous chemicals.

practice,[412] the Joint Expert Group accepted a list that had been adopted by the European Community.[413] In other words, an EU regulation was adopted at the international level simply because it was a reliable list, and despite the fact that many of the substances of greater importance for developing countries were not included. The Joint Expert Group also established a procedure for the removal of a substance from the list.[414]

Furthermore, the Expert Group had to set priorities regarding the choice of substances which would have to be included in the PIC list. This was necessary because the inclusion of all substances subject to relevant control actions by far exceeded the capacity of the system. Each substance entering the system had to be accompanied by decision guidance documents, and the preparation of these documents was time-consuming and resource-intensive.[415] Therefore, the Joint Expert Group adopted priorities for the inclusion of substances, even though this was not foreseen in either the FAO Pesticide Code or the UNEP London Guidelines. The priorities were mainly based on the width of usage and production of a certain substance, and largely built on the expert sense of the members of the group. In some cases, the group assessed certain control actions on a case by case basis. In that context, the Joint Expert Group even introduced new information tools not foreseen in the FAO Pesticide Code or the UNEP London Guidelines, such as the "information data sheets" for substances that remained outside the PIC process

---

412 D.G. Victor, "'Learning by Doing' in the Nonbinding International Regime to Manage Trade in Hazardous Chemicals and Pesticides" in: D.G. Victor/K. Raustiala/E.B. Skolnikoff (eds.), The Implementation and Effectiveness of International Environmental Commitments: Theory and Practice, 1998, 221-281 (241).

413 D.G. Victor, "'Learning by Doing' in the Nonbinding International Regime to Manage Trade in Hazardous Chemicals and Pesticides" in: D.G. Victor/K. Raustiala/E.B. Skolnikoff (eds.), The Implementation and Effectiveness of International Environmental Commitments: Theory and Practice, 1998, 221-281 (243).

414 D.G. Victor, "'Learning by Doing' in the Nonbinding International Regime to Manage Trade in Hazardous Chemicals and Pesticides" in: D.G. Victor/K. Raustiala/E.B. Skolnikoff (eds.), The Implementation and Effectiveness of International Environmental Commitments: Theory and Practice, 1998, 221-281 (243).

415 Over 1000 control actions existed at the time of the launch of the PIC procedure.

but for which information dissemination nevertheless seemed necessary.[416]

A particularly salient example of the autonomous decision-making by the Joint Expert Group is the listing of substances which pose hazards due to specific conditions of use in developing countries. Again, the implementation by the Joint Expert Group deviated from the procedures suggested in the FAO Pesticide Code and the UNEP London Guidelines on this problem. First, the Joint Expert Group did not rely on WHO experts as was stipulated in the procedures.[417] Second, the Joint Expert Group used a list of substances provided by environmental groups as a starting point instead of the list that was suggested in the instruments (a WHO list).[418] Most of the substances suggested by NGOs entered the PIC procedure. Besides providing another example for the autonomous decision-making of the Joint Expert Group, this is another striking incident of NGO influence made possible by their observer status, now in the Joint Group of Experts.[419] In other instances, however, NGO proposals for the inclusion of non-regulated substances were refused by the Joint Expert Group. This clearly shows how the Joint Expert Group took decisions on concrete contested issues.

Overall, the practice confirms that the Joint Expert Group applied its general mandate in a very flexible manner that deviated from the procedures outlined in the main instruments. Given the amount of administrative decision-making, it is striking that none of these operational de-

---

[416] See for example the case of the pesticide bifenthrin which was "severely restricted" in the Netherlands but only due to the specific circumstances which applied in the Netherlands, compare D.G. Victor, "'Learning by Doing' in the Nonbinding International Regime to Manage Trade in Hazardous Chemicals and Pesticides" in: D.G. Victor/K. Raustiala/E.B. Skolnikoff (eds.), The Implementation and Effectiveness of International Environmental Commitments: Theory and Practice, 1998, 221-281 (242).

[417] FAO Pesticide Code (1989), Guidelines on PIC, step 4; UNEP London Guidelines, Annex II, para. 3.

[418] D.G. Victor, "'Learning by Doing' in the Nonbinding International Regime to Manage Trade in Hazardous Chemicals and Pesticides" in: D.G. Victor/K. Raustiala/E.B. Skolnikoff (eds.), The Implementation and Effectiveness of International Environmental Commitments: Theory and Practice, 1998, 221-281 (246).

[419] R.L. Paarlberg, "Managing Pesticide Use in Developing Countries" in: P.M. Haas/R.O. Keohane/M.A. Levy (eds.), Institutions for the Earth: Sources of Effective International Environmental Protection, 1993, 309-350 (fn. 55).

cisions were made by a governing body.[420] While this contradicts ideals of delegation of power from member states, the expert group was able to react flexibly to newly emerging necessities and new scientific findings and managed to establish or adapt existing procedures to problems arising in practice.

7. Preliminary assessment and outlook

Considering the initial opposition from many industrialised states and agrochemical industry, the successful introduction of a widely accepted PIC system to regulate international trade in pesticides within less than a decade is striking, in particular if one takes into account that the system was of voluntary nature. Already in 1991, i.e. two years after the inclusion of PIC provisions in the two nonbinding instruments, 109 states had designated national authorities for the implementation of the procedure, and this number rose to 154 in 1997.[421] In this respect, compliance with PIC has been nearly perfect.[422]

Crucial for the success of the voluntary PIC system was its institutionalisation. The FAO and UNEP not only facilitated the adoption of the scheme, but also established follow-up mechanisms and institutions for the implementation of the PIC system at the international level. The listing of PIC pesticides and chemicals through the Joint Expert Com-

---

[420] D.G. Victor, "'Learning by Doing' in the Nonbinding International Regime to Manage Trade in Hazardous Chemicals and Pesticides" in: D.G. Victor/K. Raustiala/E.B. Skolnikoff (eds.), The Implementation and Effectiveness of International Environmental Commitments: Theory and Practice, 1998, 221-281 (239).

[421] M. Pallemaerts, "International Transfer of Restricted or Prohibited Substances, Regulation of Chemicals", Yearbook of International Environmental Law 2 (1991), 170-175 (170); M.A. Mekouar, "Pesticides and Chemicals: The Requirement of Prior Informed Consent" in: D. Shelton (ed.), Commitment and Compliance: The Role of Non-Binding Norms in the International Legal System, 2000, 146-163 (157).

[422] D.G. Victor, "'Learning by Doing' in the Nonbinding International Regime to Manage Trade in Hazardous Chemicals and Pesticides" in: D.G. Victor/K. Raustiala/E.B. Skolnikoff (eds.), The Implementation and Effectiveness of International Environmental Commitments: Theory and Practice, 1998, 221-281 (258).

mittee, for example, is being regarded by commentators as a central element for the success of the initiative.[423]

A further important aspect seems to have been the flexibility of the nonbinding system. The nonbinding nature of the PIC system allowed for a comparatively rapid adoption of the norms and establishment of institutions. The subject matter could be addressed much faster than through a multilateral treaty.[424] The nonbinding approach proved to be sufficiently flexible with respect to adjustment and implementation to enable the kind of "learning by doing" which was so important in the start-up phase of PIC,[425] and which is hardly conceivable under binding rules.[426] The voluntary instruments provided a framework in which all relevant states could move forward in a "moving consensus" towards stricter standards without strict amendment procedures.[427]

At the initial stage, a number of important aspects of the untested system were not yet clear. As could be seen from the practice of the expert group, important decisions of how to set up the system and how to implement the procedure could be flexibly decided upon as the system emerged. Facilitated by the nonbinding nature of the scheme, decisions

---

423 M.A. Mekouar, "Pesticides and Chemicals: The Requirement of Prior Informed Consent" in: D. Shelton (ed.), Commitment and Compliance: The Role of Non-Binding Norms in the International Legal System, 2000, 146-163 (159).

424 R.E. Lutz/G.D. Aron, "Codes of Conduct and Other International Instruments" in: G. Handl/R.E. Lutz (eds.), Transferring Hazardous Technologies and Substances: The International Legal Challenge, 1989, 129-151 (157).

425 J. Ross, "Legally Binding Prior Informed Consent", Colorado Journal of International Environmental Law and Policy 10 (1999), 499-529 (515); D.G. Victor, "'Learning by Doing' in the Nonbinding International Regime to Manage Trade in Hazardous Chemicals and Pesticides" in: D.G. Victor/K. Raustiala/E.B. Skolnikoff (eds.), The Implementation and Effectiveness of International Environmental Commitments: Theory and Practice, 1998, 221-281 (257); D.G. Victor, "The Use and Effectiveness of Nonbinding Instruments in the Management of Complex International Environmental Problems", American Society International Law Proceedings 91 (1997), 241-250 (246).

426 D.G. Victor, "'Learning by Doing' in the Nonbinding International Regime to Manage Trade in Hazardous Chemicals and Pesticides" in: D.G. Victor/K. Raustiala/E.B. Skolnikoff (eds.), The Implementation and Effectiveness of International Environmental Commitments: Theory and Practice, 1998, 221-281 (264).

427 R.L. Paarlberg, "Managing Pesticide Use in Developing Countries" in: P.M. Haas/R.O. Keohane/M.A. Levy (eds.), Institutions for the Earth: Sources of Effective International Environmental Protection, 1993, 309-350 (345).

of the expert group could be taken at a level below the governing body.[428] This independence of the expert body is hardly imaginable under a binding instrument. In fact, one of the main differences between the voluntary PIC system and that of the later PIC Convention which was developed on the basis of the nonbinding system at a later stage is that under the PIC Convention listing proposals of the expert group must be approved by the Conference of the Parties.[429] This does not mean, however, that the accountability of the expert group to political bodies should not have been clearer for the sake of its legitimacy.[430] The issue hints at the tension between the need for flexibility and bases for legitimacy which will be discussed in more detail in Part 3.[431]

The relative success of the voluntary PIC system does not mean however that the environmental issue had been resolved. Numerous problems still persist today, in particular in many of the developing country regions such as Africa and Latin America.[432] A PIC system does not automatically develop the regulatory capacities of developing coun-

---

[428] D.G. Victor, "'Learning by Doing' in the Nonbinding International Regime to Manage Trade in Hazardous Chemicals and Pesticides" in: D.G. Victor/K. Raustiala/E.B. Skolnikoff (eds.), The Implementation and Effectiveness of International Environmental Commitments: Theory and Practice, 1998, 221-281 (239).

[429] Rotterdam Convention, Articles 7.3 and 9.3.

[430] Less concerned with the vague mandate for the Joint Expert Group, but rather citing it as a positive effect of informal procedure D.G. Victor, "'Learning by Doing' in the Nonbinding International Regime to Manage Trade in Hazardous Chemicals and Pesticides" in: D.G. Victor/K. Raustiala/E.B. Skolnikoff (eds.), The Implementation and Effectiveness of International Environmental Commitments: Theory and Practice, 1998, 221-281 (257).

[431] See on this in detail the legitimacy analysis in Part 3, further below.

[432] B. Dinham, "The Success of a Voluntary Code in Reducing Pesticide Hazards in Developing Countries", Green Globe Yearbook of International Co-operation on Environment and Development 3 (1996), 29-36 (31); R.L. Paarlberg, "Managing Pesticide Use in Developing Countries" in: P.M. Haas/R.O. Keohane/M.A. Levy (eds.), Institutions for the Earth: Sources of Effective International Environmental Protection, 1993, 309-350 (333); in particular for Asia and the Pacific, however, the technical and infrastructural capabilities were "clearly increased", as stated in the FAO "Analysis of Government Responses to the First Questionnaire on the International Code of Conduct on the Distribution and Use of Pesticides", 1993, in the introduction.

tries.[433] And the inclusion of substantive standards in the PIC process does not necessarily translate into export restraints.[434] Even though the overwhelming majority of countries found the procedures "useful",[435] a third of the developing countries did not report to FAO on national actions, and even 56 per cent of developed countries reported that they had not or had only partly been able to use the procedures for advising their pesticide exporters and industry of the decisions of the importing states.[436] In other words, although the voluntary system has fulfilled a number of important functions by providing information through guidance documents, and by stimulating learning and legislative processes in many developing countries,[437] it was slow to resolve any of the underlying regulatory issues.

Eventually, states and international organisations therefore pushed for a binding Convention.[438] The PIC procedure as developed by UNEP and the FAO was adopted without major changes as central part of the PIC

---

433 This is acknowledged but falsely attributed to the voluntary character of the system by J. Ross, "Legally Binding Prior Informed Consent", Colorado Journal of International Environmental Law and Policy 10 (1999), 499-529 (517).

434 R.L. Paarlberg, "Managing Pesticide Use in Developing Countries" in: P.M. Haas/R.O. Keohane/M.A. Levy (eds.), Institutions for the Earth: Sources of Effective International Environmental Protection, 1993, 309-350 (342).

435 This was stated in the responses of 94 percent of the countries. Compare FAO, 'Analysis of government responses to the second questionnaire on the state of implementation of the International Code of Conduct on the Distribution and Use of Pesticides' (1996).

436 FAO, 'Analysis of government responses to the second questionnaire on the state of implementation of the International Code of Conduct on the Distribution and Use of Pesticides' (1996).

437 B. Dinham, "The Success of a Voluntary Code in Reducing Pesticide Hazards in Developing Countries", Green Globe Yearbook of International Co-operation on Environment and Development 3 (1996), 29-36 (33).

438 FAO, 'Analysis of government responses to the second questionnaire on the state of implementation of the International Code of Conduct on the Distribution and Use of Pesticides' (1996); N.S. Zahedi, "Implementing the Rotterdam Convention: the challenges of transforming aspirational goals into effective controls on hazardous pesticide exports to developing countries", Georgetown international environmental law review 11 (1999), 707-739 (709).

Convention. Signed in 1998, the PIC Convention entered into force in 2004 and has been ratified by 127 states as of December 2008.[439]

III. Establishing norms of corporate social responsibility: the case of the OECD Guidelines for Multinational Enterprises

1. The institutional framework

The Organisation for Economic Co-operation and Development (OECD) was founded as an international organisation with legal personality[440] in 1961 as the successor to the Organisation of European Economic Cooperation.[441] By 2011, thirty-four states, all of which can be politically classified as market democracies, have ratified the OECD Convention.[442] In addition, the OECD has working relationships with 70 non-member states from Africa, Asia, Europe, Latin America and the Middle East, a fact that contributes to the organisation's self-proclaimed "global reach".[443]

The main organ of the OECD is the OECD Council. Composed of all Members, it meets in sessions of Ministers or of Permanent Representatives.[444] Decisions are taken by consensus and are not effective for abstaining Members;[445] a fact that indicates the strong intergovernmental

---

[439] See for a specific analysis of this shift to a binding instrument Part 2, at A. I.1.b) (2), further below.

[440] Compare in particular OECD Convention, Article 19.

[441] OECD Convention, Article 15.

[442] So-called "enhanced engagement" is envisaged for relationships with Brazil, China, India, South Africa and Indonesia, not excluding the possibility of future membership depending on "the willingness, preparedness and ability of these countries to adopt OECD practices, policies and standards". For these developments, compare the OECD Council Resolution on Enlargement and Enhanced Engagement, section I, para. 1, available at http://www.oecd.org.

[443] These countries participate in a variety of ways in the activities of the OECD, often as observers in the various committees. For details see the general information at www.oecd.org. According to the information on this website, 24 non-member states participate in at least one OECD Committee, and 50 non-member states are engaged in at least one working party, scheme or programme.

[444] OECD Convention, Article 7.

[445] Article 6 of the OECD Convention stipulates in paragraphs 1 and 2 that decisions and recommendations require the agreement of all members. How-

character of the OECD. As in other international organisations, numerous committees and a secretariat support the work of the Council. A noteworthy element of the OECD is the institutionalised participation of industry and trade unions through Advisory Bodies. These Advisory Bodies enjoy consultative status in the OECD. The Business Industry Advisory Committee to the OECD (BIAC) represents the major industrial and employers' organisations of all member states.[446] The Trade Union Advisory Committee to the OECD (TUAC) coordinates the views of the trade union movement in the industrialised countries and represents roughly 56 national trade union centres in the 30 OECD industrialised countries.[447] More recently, OECD Watch – a network of NGOs from around the world[448] – has been granted certain consultative rights in the Investment Committee.[449] Access to the Investment Committee provides industry, trade unions and NGOs with an avenue of participation at the international level.

2. *The regulatory challenge*

International standards of corporate responsibility seek to address challenges arising in the context of globalisation and the ensuing proliferation of powerful private actors such as multinational corporations (MNCs).[450] The process of globalisation weakens the regulatory capa-

---

ever, abstentions do not invalidate such acts, but lead to the inapplicability of the act on the abstaining Member.

446 The BIAC roughly represents eight million companies, compare www.biac.org.

447 The TUAC thus represents roughly 66 million workers, see www.tuac.org.

448 See for details www.oecdwatch.org.

449 See the OECD Council Decision C(2000)96 (27 June 2000), Part II, available at http://www.olis.oecd.org/olis/2000doc.nsf/LinkTo/c(2000)96-final.

450 For in-depth analyses of the issue of multinational enterprises and their corporate responsibility according to international law see K. Weilert, "Transnationale Unternehmen im rechtsfreien Raum? Geltung und Reichweite völkerrechtlicher Standards", Zeitschrift für ausländisches öffentliches Recht und Völkerrecht 69 (2009), 883-917; E. Morgera, Corporate accountability in international environmental law, 2009; J.A. Zerk, Multinationals and Corporate Social Responsibility: Limitations and Opportunities in International Law, 2008; P. Muchlinski, Multinational enterprises and the law, 1995; K. Nowrot, Normative Ordnungsstruktur und private Wirkungsmacht: Konsequenzen der Be-

bility of most states with respect to the protection of collective goods and regulation of economic processes.[451] Technological advancement in transportation and communication systems and the liberalisation of the movement of products and means of production increases the ability of transnational business actors to avoid territorially limited national regulation. Multinational corporations (MNCs)[452] are able to take advantage of favourable business conditions in those states with lower social and environmental standards and thereby avoid costly environmental and social regulation.[453] For a number of reasons, governments may not be able to impose stricter standards even if they wish to do so. One such reason is that governments in particular in the developing world compete for foreign investment, and might stand to lose foreign investment if higher standards or better enforcement threatens to trigger the moving of an MNC to another country.[454] Specifically in the context of foreign direct investment, the investor-state relationship is often unbalanced due to power asymmetries arising in particular from the need for

teiligung transnationaler Unternehmen an den Rechtssetzungsprozessen im internationalen Wirtschaftssystem, 2006.

451 H. Keller, "Codes of Conduct and their Implementation: the Question of Legitimacy" in: R. Wolfrum/V. Röben (eds.), Legitimacy in International Law, 2008, 219-298 (232).

452 Defining this kind of enterprise is difficult. According to the OECD Guidelines for Multinational Enterprises, Chapter I, para. 3, "multinational enterprises usually comprise companies or other entities established in more than one country and so linked that they may co-ordinate their operations in various ways. While one or more of these entities may be able to exercise a significant influence over the activities of others, their degree of autonomy within the enterprise may vary widely from one multinational enterprise to another. Ownership may be private, state or mixed."; for definitional problems see also P. Muchlinski, Multinational enterprises and the law, 1995, 12-15.

453 Regulatory avoidance by private actors is mentioned by B. Simma/A. Heinemann, "Codes of Conduct" in: W. Korff et al. (eds.), Handbuch der Wirtschaftsethik, Band 2: Ethik wirtschaftlicher Ordnungen, 1999, 403-418 (404-405).

454 S.D. Murphy, "Taking multinational corporate codes of conduct to the next level", Columbia journal of transnational law 43 (2005), 389-433 (399); E. Westfield, "Globalization, governance, and multinational enterprise responsibility: corporate codes of conduct in the 21st century", Virginia journal of international law 42 (2002), 1075-1108 (1077).

investment by the territorially bound state.[455] This often impedes balanced regulation taking into account investor and public interests.[456] Furthermore, in particular so-called least developed countries often struggle adequately to draft, administer and enforce laws, especially where the issue area involves a high level of complexity as in environmental protection.[457] Furthermore, in authoritarian non-democratic states, the interests of the government may not coincide with the interests of its people in a high level of social and environmental protection.[458] Finally, even if willing to do so, many multinational corporations also have difficulties in meeting the home state level of social and environmental protection in countries with less stringent standards. It is often difficult for companies to compete on the basis of their home standards, or to control a lack of due diligence on the part of their foreign subsidiaries.[459]

Harmonised international rules could establish a level playing field and thereby address the challenges resulting from disparate standards or gaps in national law. The search for an international solution to the problem of the regulation of multinational corporations began in the 1970s and continues today.[460] Basic standards of conduct of transna-

---

455 A.A. Fatouros, "On the implementation of international codes of conduct: an analysis of future experience", American University Law Review 30 (1980-1981), 941-972 (951); E. Westfield, "Globalization, governance, and multinational enterprise responsibility: corporate codes of conduct in the 21st century", Virginia journal of international law 42 (2002), 1075-1108.

456 C.-T. Ebenroth, Code of Conduct – Ansätze zur vertraglichen Gestaltung internationaler Investitionen, 1987, 32.

457 S.D. Murphy, "Taking multinational corporate codes of conduct to the next level", Columbia journal of transnational law 43 (2005), 389-433 (398).

458 S.D. Murphy, "Taking multinational corporate codes of conduct to the next level", Columbia journal of transnational law 43 (2005), 389-433 (399).

459 For a very instructive and insightful description of these challenges see M. Herberg, "Private Authority, global governance, and the law: The case of environmental self-regulation in multinational enterprises" in: G. Winter (ed.), Multilevel Governance of Global Environmental Change: Perspectives from Science, Sociology and the Law, 2006, 149-178 (152-153).

460 On these developments in detail R. Grosse, "Codes of Conduct for Multinational Enterprises", Journal of World Trade Law 16 (1982), 414-433; S.D. Murphy, "Taking multinational corporate codes of conduct to the next level", Columbia journal of transnational law 43 (2005), 389-433; J.A. Zerk, Multinationals and Corporate Social Responsibility: Limitations and Opportunities in International Law, 2008, 244-262.

tional corporations relating to labour concerns were stipulated in the Tripartite Declaration of Principles concerning Multinational Enterprises and Social Policy elaborated by Governments, employer and employee associations under the auspices of the International Labour Organization (ILO) in 1977 ('ILO Tripartite Declaration').[461]

Discussions on a comprehensive UN Code of Conduct for Transnational Corporations started at the Second Session of the UN Commission on Transnational Corporations in 1976.[462] The Commission had been set up by the UN Economic and Social Council as part of the endeavour, led mainly by developing countries, to establish a New International Economic Order. From the beginning, the binding or non-binding nature of the code was an issue of disagreement that was never finally settled.[463] By 1990, however, states at UNCTAD had accepted the possibility that the code would remain a voluntary instrument.464 With respect to the substance of the code, negotiators could not agree on a number of issues related to the minimum standards for investors and the rights of host states.[465] Developed states feared that the UN efforts would affect freedom of investment and the protection of proprietary rights and therefore be hostile to business interests and free markets.[466] Arguably with the intent to prevent more stringent action by

---

461 ILO Tripartite Declaration Concerning Multinational Enterprises and Social Policy, 17 ILM (1978) 422, adopted by the Governing Body of the ILO at its 204th Session (November 1977), amended at its 279th (November 2000) and 295th Session (March 2006), available at http://www.ilo.org.

462 On the UN Draft Code of Conduct on Transnational Corporations in comparison with the OECD Guidelines and other similar instruments see E. Morgera, Corporate accountability in international environmental law, 2009, 78 et seq.

463 S.J. Rubin, "Transnational Corporations and International Codes of Conduct: a study of the relationship between international legal cooperation and economic development", American University Law Review 30 (1980-1981), 907-921 (916).

464 See the Draft Code of Conduct on Transnational Corporations, UN Doc. E/1990/94 (1990), Annex.

465 J.A. Zerk, Multinationals and Corporate Social Responsibility: Limitations and Opportunities in International Law, 2008.

466 J.E. Alvarez, International Organizations as Law-makers, 2005, 255; P.T. Muchlinski, "Global Bukowina' Examined: Viewing the Multinational Enterprise as a Transnational Law-making Community" in: G. Teubner (ed.), Global Law without a state, 1996, 79-108 (95).

the UN, the Reagan Administration of the United States of America supported alternative forums such as the OECD that were seen as more favourable to business interests.[467]

The swift adoption of a voluntary instrument in the OECD – the OECD Guidelines for Multinational Enterprises – has at least partly contributed to the failure of global and more far-reaching approaches at the UN level. The Draft UN Code of Conduct on Transnational Corporations ('UN Draft Code')[468] was never adopted. The negotiations were finally abandoned in 1992, when the focus shifted to the Uruguay Round of multilateral trade negotiations of the GATT which led to the establishment of the World Trade Organization.

3. The nonbinding response: the OECD Guidelines for Multinational Enterprises

The OECD member states promulgated the Guidelines for Multinational Enterprises[469] as an Annex to the Declaration on International Investment and Multinational Enterprises in June 1976.[470] In addition to the Guidelines, the Declaration also contained two further instruments, one on national treatment and one on incentives and disincentives for investment.

---

[467] J.E. Alvarez, International Organizations as Law-makers, 2005, 255; D.J. Plaine, "The OECD Guidelines for Multinational Enterprises", International Lawyer 11 (1977), 339-346 (340).

[468] UN Doc. E/1990/94 (12 June 1990).

[469] OECD Guidelines for Multinational Enterprises, 15 ILM. 969 (1976); the latest updated version of 2011 is available at http://www.oecd.org/ dataoecd/43/29/48004323.pdf; on the OECD Guidelines in the context of environmental protection E. Morgera, "An Environmental Outlook on the OECD Guidelines for Multinational Enterprises: Comparative Advantages and Legitimacy in relation to Other International Initiatives", Georgetown International Environmental Law Review 18 (2006), 751-777.

[470] Declaration on International Investment and Multinational Enterprises, OECD Doc. C(76)99/Final (21 June 1976), as amended by OECD Doc. C(79)102/Final, OECD Doc. C/M(84)7 Part II(Final) item 99 and 100, OECD Doc. C/M(91)12/FINAL, Item 111.II c), OECD Doc. C/M(2000)17; the latest version of the Declaration is available as part of a booklet issued by the OECD on the OECD Guidelines for Multinational Enterprises, at http://www.oecd.org/dataoecd/43/29/48004323.pdf.

In contrast to negotiations at UNCTAD, Member states of the OECD managed quickly to agree on the formulation of comparatively business-friendly guidelines.[471] Although not all OECD member states were at the time exporters of investment like the United States,[472] they represented a group with a relatively homogenous interest structure compared to the UN.[473] OECD member states largely shared the view that even though standards for multinational corporations might be necessary to respond to public concerns and thereby improve investment climate, the overriding goals were the liberalisation of investment and the creation of greater stability through the harmonisation of social standards.[474] They could thus agree more easily than UN member states on the adoption of a less stringent and nonbinding instrument. While a nonbinding instrument was the preferred option, some also hoped that this would be a first step towards binding norms.[475] More controversial issues such as the definition of multinational enterprises were left out.[476]

---

[471] I. Seidl-Hohenveldern, "International Economic 'Soft Law'", Recueil des Cours de l'Académie de Droit International 163 (1979), 169-246 (197).

[472] For example, Germany had a balance between incoming and outgoing investment, the US was and still is the home base for the greatest number of multinationals, while smaller countries were primarily importers of investment, compare R. Blanpain, The OECD Guidelines for Multinational Enterprises and Labour Relations 1976-1979, 1979, 48.

[473] E. Morgera, Corporate accountability in international environmental law, 2009, 102; B. Simma/A. Heinemann, "Codes of Conduct" in: W. Korff et al. (eds.), Handbuch der Wirtschaftsethik, Band 2: Ethik wirtschaftlicher Ordnungen, 1999, 403-418 (407).

[474] T.W. Vogelaar, "The OECD Guidelines: Their Philosophy, History, Negotiation, Form, Legal Nature, Follow-Up Procedures and Review" in: N. Horn (ed.), Legal Problems of Codes of Conduct for Multinational Enterprises, 1980, 127-140 (128); J.A. Zerk, Multinationals and Corporate Social Responsibility: Limitations and Opportunities in International Law, 2008, 248.

[475] R. Blanpain, The OECD Guidelines for Multinational Enterprises and Labour Relations 1976-1979, 1979, 50.

[476] For example, no attempt was made to reach agreement on the complex issue of a definition for multinational enterprises, and more technical questions were left to technical committees. See for an overview of the negotiations R. Blanpain, The OECD Guidelines for Multinational Enterprises and Labour Relations 1976-1979, 1979, 48 et seq.

a) General scope

The OECD Guidelines constitute a collection of principles and standards of good practice recommended by governments to multinational corporations.[477] Originally restricted to those corporations operating in the territories of the member states, the revision in the year 2000 gave the Guidelines a potentially global scope by removing the strict territorial reference. Enterprises from "adhering governments" are now encouraged to observe the Guidelines "wherever they operate".[478] The adhering governments commit themselves to promote the Guidelines and encourage their use.[479] As of 2011, 43 governments had become "adhering governments" to the OECD Guidelines, comprising the 30 Members of the OECD[480] as well as 13 non-Members.[481]

The notion of "governments adhering to the Guidelines"[482] is a noteworthy expansion of the membership concept. Unlike traditional treaty law, with the Guidelines states do not have to become a member of an organisation or a party to a treaty in order to participate. It suffices to simply declare the adherence to a particular instrument. A number of developing countries that are not members of the OECD have thus recently joined the ranks of the "adhering governments" of the Guidelines, including Egypt as the first African state. By becoming "adhering governments", states can flexibly choose to cooperate on a specific issue without agreeing to any legally binding treaty. At the same time, "adherence" requires a clear public act that signals that a state intends to participate as an "adherent".[483] "Adherence" therefore creates a situation akin to membership without a legally binding basis.

---

[477] OECD Declaration on Investment and Multinational Enterprises, para. I, in conjunction with OECD Guidelines, Chapter I, para. 1 and 2.

[478] OECD Guidelines, Chapter I, para. 3.

[479] OECD Guidelines, Chapter I, para. 11.

[480] The European Community is formally not a Member of the OECD, but the European Commission is given a special status that goes beyond being a mere observer.

[481] Argentina, Brazil, Chile, Egypt, Estonia, Israel, Latvia, Lithuania, Morocco, Peru, Romania, Slovenia and Columbia.

[482] OECD Guidelines, e.g. in Chapter I, paras 3, 9 and 11.

[483] Thus, for instance, Columbia officially "became the 43rd adherent" in 2011, compare http://www.oecd.org/document/40/0,3746,en_2649_34889_49258792_1_1_1_1,00.html.

As flexible and advantageous as this construction may be in terms of facilitating cooperation, the attempt to combine traditional institutional structures with new flexible approaches to membership leads to the somewhat odd result that adherence and membership rights may fall apart. Only OECD member states but not all adhering governments have the right to vote in the Investment Committee, the decision-making body of the OECD responsible for decisions concerning review and implementation of the OECD Guidelines. At the same time, a decision of the OECD Council can of course not bind non-Members. The Council decisions which require adhering states to establish National Contact Points and to adhere to a set of procedural rules outlining implementation issues thus are not binding for all adhering governments. A Council decision that is directed at all adhering states therefore represents a combination of a binding decision addressed to OECD member states and a set of communications addressed to non-Member adhering states. Such communications to non-Members are foreseen by Article 9 of the OECD Convention. The Rules of Procedure of the OECD in Rule 18 (a) (iv) clarify that such communications may also be contained in decisions of the Council.[484]

b) Nonbinding character

The Guidelines are not legally binding for corporations. Their voluntary nature is explicitly highlighted at the outset of the instrument where it provides that "observance of the Guidelines by enterprises is voluntary and not legally enforceable".[485] As indicated already, the OECD Guidelines not only establish standards and principles for corporations, but also outline the actions that must be taken by governments.[486] The instrument does not however specify whether the guide-

---

[484] The view sometimes (perhaps accidentally) voiced by Commentators that the decisions of the Council are binding on adhering states are therefore not accurate in this generality. Inaccurate insofar A. Böhmer, "The Revised 2000 OECD Guidelines for Multinational Enterprises – Challenges and Prospects after 4 Years of Implementation", Policy Papers on Transnational Economic Law 3 (2004), (4).

[485] OECD Guidelines, Chapter I, para. 1.

[486] The provisions entail the commitment to treat enterprises equitably and in accordance with international law, not to use the Guidelines for protectionist purposes and to cooperate with other states in case of conflicting requirements by several adhering countries, see OECD Guidelines, Chapter I, paras 7-8.

lines are voluntary and legally nonbinding for governments. The language employed is also ambiguous ("governments will implement them").[487] The nonbinding nature of the norms for governments however derives from the formally nonbinding nature of the Declaration on International Investment and Multinational Enterprises of which the OECD Guidelines form an integral part.

c) Substantive norms

The substantive norms of the OECD Guidelines comprise general principles as well as issue-specific standards of behaviour that cover a wide range of issues, including access to information about the activities of enterprises,[488] the protection of human rights[489] and labour rights[490], protection of the environment[491] as well as consumer interests,[492] the fight against corruption[493] and competition-distorting practices,[494] as well as norms on scientific and technological advancement[495] and taxation.[496]

Sustainable development has become one of the leading principles of the Guidelines.[497] Of relevance, also from an environmental perspective, may also be the emphasis of the policy that enterprises should "respect the internationally recognised human rights of those affected by their activities".[498] Other "general policies" mentioned in Chapter II of the OECD Guidelines underline the attempt to emphasise both the benefits and dangers of the activities of multinational corporations in a balanced

---

487 OECD Guidelines, Chapter I, para. 11.

488 Chapter III on "Disclosure".

489 Chapter IV on "Human Rights".

490 Chapter V on "Employment and Industrial Relations".

491 Chapter VI on "Environment".

492 Chapter VIII on "Consumer Interests".

493 Chapter VII on "Combating Bribery, Bribe Solicitation and Extortion".

494 Chapter X on "Competition".

495 Chapter IX on "Science and Technology".

496 Chapter XI on "Taxation".

497 OECD Guidelines, Chapter II, para. 1.

498 OECD Guidelines, Chapter II, para. 2.

way.[499] Thus the Guidelines on the one hand emphasise the positive effects of international investment but on the other hand stress the need to achieve transparency and to prevent the abuse of economic power.

It is stressed throughout the OECD Guidelines that they prescribe standards and principles consistent with applicable laws,[500] and that the overriding responsibility of corporations remains to obey national laws.[501] They are thus not intended as substitutes for domestic law, and do not override such laws. However, corporations should "seek ways to honour such principles and standards to the fullest extent which does not place them in violation of domestic law".[502]

The revision of the Guidelines in 2000 and the update in 2011 resulted in a much greater emphasis on existing international law and international standards. National jurisdiction and law ceases to be the main point of reference.[503] Thus, the Guidelines not only stress the responsibility of governments to treat and regulate multinational corporations in accordance with international law.[504] They also go beyond existing international law which does not apply directly to corporations in stipulating that corporations should respect the human rights of those affected by their activities "consistent with the host government's international obligations and commitments".[505] Moreover, the chapter on the environment starts with the general provision that "enterprises should ... in consideration of relevant international agreements, principles, objectives and standards, take due account of the need to protect the envi-

---

499 Compare OECD Guidelines, Chapter I; A.A. Fatouros, "Les Principes directeurs de l'OECD à l'intention des entreprises multinationales: perspectives actuelles et possibilités futures" in: C. Reymond (ed.), Études de droit international en l'Honneur de Pierre Lalive, 1993, 231-240 (234).

500 OECD Guidelines, Chapter I, paras 1 and 8; Chapter V on "Employment" in the chapeau; Chapter VI on "Environment", chapeau; Chapter X on "Competition", chapeau; Chapter XI on "Taxation", chapeau.

501 OECD Guidelines, Chapter I, para. 2.

502 OECD Guidelines, Chapter I, para. 2.

503 J.A. Zerk, Multinationals and Corporate Social Responsibility: Limitations and Opportunities in International Law, 2008, 250; J. Murray, "A new phase in the regulation of multinational enterprises: the role of the OECD", The Industrial law journal 30 (2001), 255-270 (264).

504 OECD Guidelines, Chapter I, paras 8 and 9.

505 OECD Guidelines, Chapter II, para. 2; see in particular Chapter IV on "Human Rights".

ronment ..."[506] In other words, international human rights and international environmental obligations directly inform the responsibilities of corporations.

Most of the substantive chapters of the OECD Guidelines reflect and incorporate existing treaty law and international nonbinding instruments.[507] As clarified in the Commentary on the Chapter on the Environment, the text of the environmental chapter reflects to principles of the Rio Declaration on Environment and Development, Agenda 21, and takes into account the Convention on Access to Information, Public Participation in Decision Making and Access to Justice in Environmental Matters ('Aarhus Convention')[508] as well as the ISO standard on Environmental Management.[509] And the preface of the Guidelines mentions the 1948 Universal Declaration of Human Rights, the ILO Declaration on Fundamental Principles and Rights at Work and the Copenhagen Declaration for Social Development in addition to the Rio Declaration and Agenda 21 as the guiding "international legal and policy framework in which business is conducted."[510] In other words, the OECD Guidelines – just like other nonbinding instruments analysed in this study – flexibly integrate international standards into one instrument, irrespective of their legal nature or general acceptance. Given that the OECD Guidelines are addressed to corporations but also contain recommendations for states, these references have two effects. First, the referenced rules of treaties become recommendations addressed to all adhering states irrespective of ratification of the specific treaty by an adhering state. And secondly, the rules contained in treaty and other nonbinding instruments which are so far only applicable to states are now directly addressed to private actors, irrespective of whether their home state has ratified or adopted the respective instrument.

---

506 OECD Guidelines, Chapter VI, chapeau.

507 The general reference to international agreements, principles, objectives and standards in the chapter V on the environment also includes a reference to other nonbinding instruments. Otherwise the inclusion of "standards" and "objectives" in addition to the terms "international agreements" and "principles" would not make any sense.

508 Convention on Access to Information, Public Participation in Decision-Making and Access to Justice in Environmental Matters (28 June 1998), 2161 UNTS 447, 38 ILM 517 (1999); for a similar assessment, see Commentary on the OECD Guidelines for Multinational Enterprises, para. 30.

509 OECD Guidelines, Commentary on the Environment, para. 60.

510 OECD Guidelines, Preface, para. 8.

A closer look at the environmental chapter shows that the standards mostly remain procedural rather than substantive as well as vague rather than concrete. The overall emphasis on procedural rules such as the need to establish environmental management systems or to adhere to environmental impact assessment reflects modern conceptions of self-regulatory approaches which are sometimes referred to as next generation environmental law or so-called reflexive law.[511] They stress information, management and learning over strict command-and-control regulation. This is seen by some as a factor that enhances acceptance by business actors.[512]

However, one must not overlook that an enterprise can implement procedures such as those on environmental impact assessment but still continue business as usual, since it is not forced to take a particular decision in accordance with its assessments. For example, even if an environmental impact assessment is undertaken as recommended by the OECD Guidelines, an enterprise could still go ahead with the project since it must only "prepare an appropriate impact assessment".[513] This possibility for an enterprise to comply with the Guidelines and still continue business as usual is also facilitated by the soft language of the OECD Guidelines. To take up the example of environmental impact assessment again, such an assessment is only necessary in cases "where the activities ... are subject to a decision of a competent authority",[514] i.e. where some national law is already in place. Similarly, environmental management systems should only comprise targets for performance "where appropriate".[515] Information dissemination and reporting on environmental performance is conditioned by the possibility of taking into account concerns about, *inter alia*, cost and business confidentiality.[516] Precautionary measures must be taken only in a "cost-effective"

---

[511] R.B. Stewart, "A new generation of environmental regulation?", Capital University Law Review 29 (2001), 21-182; D.C. Esty, "Next generation environmental law", Capital University Law Review 29 (2001), 183-204.

[512] This is seen as one of the main achievements of the OECD Guidelines by C. Wilkie, "Enhancing Global Governance: Corporate Social Responsibility and the International Trade and Investment Framework" in: J.J. Kirton/M.J. Trebilcock (eds.), Hard Choices, Soft Law: Voluntary Standards in Global Trade, Environment and Social Governance, 2004, 288-322 (299).

[513] OECD Guidelines, Chapter VI, para. 3.

[514] OECD Guidelines, Chapter VI, para. 3.

[515] OECD Guidelines, Chapter VI, para. 1 b).

[516] OECD Guidelines, Chapter VI, para. 2.

way.[517] Finally, the OECD Guidelines do not provide for independent monitoring, the essence of an effective self-regulatory system. Where corporations are asked to " seek to improve corporate environmental performance" – a rare but softly worded substantive norm – the norms only apply to the level of the supply chain "where appropriate",[518] thus leaving much discretion to corporations in this relevant application to the supply chain. In sum, the OECD Guidelines therefore often do not go much further than promoting self-regulation.

However, with the revisions in 2000 and 2011, more and more substantive norms have been incorporated, indicating a gradual development and concretization of the Guidelines in terms of substance. The update in 2011 has for instance incorporated the need that corporate environmental performance should aim at developing products and services that reduce greenhouse gas emissions. And while earlier versions of the norms simply referred to the need to improve environmental performance over the longer term, the 2011 update explicitly mentions strategies for emission reduction, efficient resource utilisation and recycling, and strategies on biodiversity protection.[519]

4. Norm development process

As mentioned, the OECD Guidelines were adopted after a comparatively short negotiation phase in 1976.[520] Lengthy negotiation processes were avoided, and the participation of stakeholders was limited to comments from the Advisory Bodies representing business and trade union interests, BIAC and TUAC.[521]

Having been adopted as one part of an intergovernmental declaration instead of a resolution of an OECD body, the OECD Guidelines are technically speaking of intergovernmental character and do not consti-

---

[517] OECD Guidelines, Chapter VI, para. 4.

[518] OECD Guidelines, Chapter VI, para. 6 (chapeau).

[519] OECD Guidelines, Chapter VI, para. 6b) and 6d).

[520] The complete process from the adoption of the mandate for the Committee on International Investment and Multinational Enterprises on 21 January 1975 to the promulgation of the Guidelines on 21 June 1976 took only one year and five months. For details on the process R. Blanpain, The OECD Guidelines for Multinational Enterprises and Labour Relations 1976-1979, 1979, 48 et seq.

[521] R. Blanpain, The OECD Guidelines for Multinational Enterprises and Labour Relations 1976-1979, 1979, 50.

tute secondary law of the OECD. This technique of "outsourcing" the adoption of the Guidelines enabled OECD states to respond to new regulatory tasks and political circumstances.[522] None of the instruments available to the OECD Council provided for the possibility of directly addressing private actors. The form of the Declaration also avoided the risk of non-adoption due to a veto or of an abstention of one state that existed for decisions or recommendation, which according to the Convention have to be made by "mutual agreement of all the Members".[523]

After it had been adopted, however, the Declaration was immediately brought to the attention of the OECD Council, which was "taking note of the Declaration."[524] It then by means of a decision took on the task of implementing the Declaration.[525] By endorsing the Declaration and by taking on its implementation, the OECD Council integrated the OECD Guidelines into the OECD legal structure.[526]

The legal basis for these activities of the Council can be found in the provisions of the OECD Convention that authorise the Council to

[522] T.W. Vogelaar, "The OECD Guidelines: Their Philosophy, History, Negotiation, Form, Legal Nature, Follow-Up Procedures and Review" in: N. Horn (ed.), Legal Problems of Codes of Conduct for Multinational Enterprises, 1980, 127-140 (133).

[523] OECD Convention, Article 6 para. 1.

[524] OECD, Decision of the Council on Inter-governmental Consultation Procedures on the Guidelines for Multinational Enterprises of 21 June 1976, C(1976)117, in the preamble. The decision is available as a reprint in N. Horn (ed.), Legal Problems of Codes of Conduct for Multinational Enterprises, 1980, 462.

[525] Compare the Decision of the Council on Inter-governmental Consultation Procedures on the Guidelines for Multinational Enterprises of 21 June 1976, C(1976)117, paras 1-4; for the most recent version of the Council Decision by which the Council endorses the revised Guidelines and establishes the implementation procedures, see OECD Council Decision C(2000)96/FINAL (27 June 2000). With the latest review the Committee on International Investment and Multinational Enterprises was renamed to Investment Committee (IC).

[526] T.W. Vogelaar, "The OECD Guidelines: Their Philosophy, History, Negotiation, Form, Legal Nature, Follow-Up Procedures and Review" in: N. Horn (ed.), Legal Problems of Codes of Conduct for Multinational Enterprises, 1980, 127-140 (133).

adopt binding decisions[527] and communications to non-members and organisations.[528] According to Article 1 of the OECD Convention, any action of the Council must serve the aims of the Convention. These are the promotion of policies designed to achieve economic growth, employment, a rising standard of living, financial stability, sound economic expansion of Members and non-Members, as well as expansion of world trade on a multilateral, non-discriminatory basis. According to Article 5 (a) of the OECD Convention, the Council can take decisions that are binding on Members in order to achieve the aims of the Organisation.

At first sight, these mainly economic objectives of the OECD hardly suffice as an authorisation for the adoption of environmental policies such as those contained in Chapter 5 of the Guidelines. The pursuit of environmental policies can only be justified if economic and environmental goals are considered as intertwined objectives. The revisions of the OECD Guidelines in 1991 to include environmental objectives and above all in 2000 to include sustainable development as a guiding principle clearly reflect a re-orientation of the OECD in this sense. The integrationist core of the concept of sustainable development serves as a leitmotif which links the economic and social objectives of the OECD to environmental protection. This is illustrated by the discussions at the OECD Council Meeting in 1998 where Ministers agreed to interpret "sustainable" to include social, environmental and economic considerations and where they stressed the objective of the OECD to exploit its multi-disciplinary expertise in order to "pursue the integration of economic, environmental and social policies to enhance welfare".[529] Nevertheless, the turn to environmental policies effectively stretches the limits of the mandate of the OECD.

---

[527] According to Article 5 a) of the OECD Convention, the Council can take decisions that are binding on Members in order to achieve the aims of the organisation.

[528] Article 12 of the OECD Convention enables the organisation to address communications to non-member states and organisations.

[529] See Communiqué from the OECD Ministerial Council Meeting of 27-28 April 1998, available at http://www.g7.utoronto.ca/oecd/index.html.

5. Norm adaptation

The OECD Guidelines have been reviewed and revised periodically.[530] The five revisions undertaken in 1979, 1984, 1991, 2000 and 2010/2011 have each led to changes in the content and procedural regime of the OECD Guidelines. For instance, the chapter on environmental protection was included in 1991. The 2000 review is of particular interest not only for the substantial substantive amendments such as the inclusion of the concept of sustainable development, a strengthened implementation mechanism and a stronger emphasis of international norms, but also for the way it was conducted. The OECD used to have rather opaque decision-making processes and lacked any participatory element in decision-making.[531] This remains true in many respects. For example, all meetings of OECD bodies are held in private[532] and the minutes or reports of meetings of Committees are not publicly available. In contrast, the revision in 2000 as well as the update in 2011 was accompanied by a full-fledged public consultation process. It drew heavily on the various stakeholders represented through BIAC, TUAC and OECD Watch as well as the expertise of the OECD secretariat. As these actors gained influence in the actual drafting process, the influence of governments became less dominant.

Following the initiation of the review process by the Ministerial Meeting of the OECD Council in 1998,[533] the Committee on International Investment and Multinational Enterprises (CIME) launched a review in the course of which it undertook extensive consultations with Member and non-member states, as well as non-state actors such as the business community, labour and NGO representatives.[534] The CIME also col-

---

530 Such a review is foreseen in the Declaration on International Investment and Multinational Enterprises, section VI.

531 J. Salzman, "Decentralized Administrative Law in the Organization for Economic Cooperation and Development", Law and Contemporary Problems 68 (2005), 189-224 (194-195).

532 See Rule 5 of the Rules of Procedure of the OECD, available at www.oecd.org.

533 See OECD Council Meeting at Ministerial Level of 27-28 April 1998, Communiqué, para. 26, available at http://www.g7.utoronto.ca/oecd/oecd98.htm.

534 These notice-and-comment procedures were often internet-based, and personal consultations were held at conferences, as for example in Budapest in 1998 and at the OECD in 1999.

laborated with other OECD Committees and sub-committees[535] as well as experts from the secretariat.[536] The main decisions in this process were taken in an informal consultation convened by the Chair of the Working Party to CIME. Later known as the "Hague Process", it started with a brainstorming meeting of TUAC, BIAC, selected NGOs and experts from the ILO during which members spoke in their personal capacity. In several of these meetings, participants discussed drafts prepared by the Chair and the OECD secretariat so far unknown to governments. Only the last of these meetings included a few selected government representatives. The draft produced in these meetings then served as the basis for a public notice-and-comment procedure.[537] This was posted on the internet and all interested actors could post comments which could be viewed by the public. The draft was then amended by the secretariat in order to reflect the comments received.[538] One can assume that the secretariat influenced the process through a selection process of which comments were taken into account and which were not. OECD Member states then adopted a new declaration on the

[535] For an account of the (then ongoing) review process, see the paper by the OECD Directorate for Financial, Fiscal and Enterprise Affairs, 'Review of the OECD Guidelines for Multinational Enterprises: Framework for the Review' (1999), available at www.oecd.org; for a detailed analysis of the process see J. Salzman, "Decentralized Administrative Law in the Organization for Economic Cooperation and Development", Law and Contemporary Problems 68 (2005), 189-224 (216).

[536] Collaboration with the secretariat included the actual drafting of norms and legal advice, as evidenced by the acknowledgments in the booklet on the revised Guidelines published by the OECD which explicitly name the individuals of the secretariat, see OECD, 'The OECD Guidelines for Multinational Enterprises: Revision 2000', available at http://www.oecd.org.

[537] J. Salzman, "Decentralized Administrative Law in the Organization for Economic Cooperation and Development", Law and Contemporary Problems 68 (2005), 189-224 (216).

[538] The details of this process are only known to insiders. My description draws on the excellent account of J. Salzman, "Decentralized Administrative Law in the Organization for Economic Cooperation and Development", Law and Contemporary Problems 68 (2005), 189-224 (216-217).

basis of the final draft,[539] and the OECD Council adopted a new decision on implementation.[540]

As remarkable as this process was in terms of the involvement of stakeholders, civil society and the secretariat, it is important to note that the process was not framed by procedural rules. Rather, the CIME and its Working Party, acting on the basis of a very broad mandate, could develop the procedure without specific guidance by the higher political body on an *ad hoc* basis. In particular the Chairperson of the Working Party to CIME could structure and influence the process with substantial discretion. Neither the contributions of the OECD secretariat and the participation of the Advisory Bodies and NGOs, nor the public notice-and-comment procedure were pre-determined. In addition, the influence of governments in the actual drafting was slim. Governments mainly exercised their control through the act of adoption.

While remaining inclusive with respect to stakeholders, the updating process of the Guidelines in 2010/2011 was conducted in a more government-driven process that even included terms of reference adopted by all adhering governments.[541] It is remarkable that these terms of reference also include modalities on the consultative process of the update itself, i.e. some form of procedural rules on process that was still lacking in the previous revision.[542]

6. Norm implementation and follow-up

After governments had adopted the OECD Guidelines, the OECD Council – by means of a binding decision[543] – established a multi-level institutional structure with a view to supporting the implementation of the instrument as well as procedural rules to guide these implementation activities. At the national level, states are required to set up National Contact Points (NCPs) charged with promotional activities, the handling of inquiries and facilitating dialogue between business and

---

539 OECD Declaration for International Investment and Multinational Enterprises, OECD Doc. C/M(2000)17.

540 OECD Doc. C(2000)96, available at http://www.olis.oecd.org/olis/2000doc.nsf/LinkTo/c(2000)96-final.

541 The terms of reference are available at http://www.oecd.org/dataoecd/61/41/45124171.pdf.

542 *Ibid.*, Part IV.

543 OECD Convention, Article 5 (a).

other actors.[544] The NCPs can be composed of a senior government official or a government office headed by a senior official, or alternatively can be organised as a cooperative body of government officials from one or several national agencies, possibly including representatives from non-governmental bodies.[545] At the international level, the Investment Committee (IC),[546] is mandated to be the main international forum for dialogue between member states, non-member adhering states, international organisations and the OECD Advisory Bodies and NGOs. It is open to all OECD member states,[547] and deals with all matters that relate to the Guidelines, as well as maintaining the dialogue with the National Contact Points.[548] Non-Member adhering governments are not given participatory rights in the IC, but since 2004, non-members fully participate in the Working Party to the IC in questions related to the Guidelines.[549] Representatives from non-members that are associated

---

[544] Implementation Procedures of the OECD Guidelines for Multinational Enterprises [hereinafter "Implementation Procedures"], available at http://www.oecd.org/dataoecd/43/29/48004323.pdf, para. 1.

[545] Implementation Procedures, Procedural Guidance, available at http://www.oecd.org/dataoecd/43/29/48004323.pdf, Part I, section A., para. 2. In practice, 21 NCPs are located in one government department; seven NCPs are composed of representatives from more than one government department. Nine NCPs are tripartite, involving the government (many of these also involve multiple government departments), representatives of the business and industry sector and representations of trade unions. Two NCPs are organized quadripartite, i.e. they additionally involve NGO representatives, compare OECD Guidelines for MNEs: 2006 Annual Meeting of the NCPs, Report of the Chair, available at http://www.oecd.org/dataoecd/23/33/37439881.pdf.

[546] The OECD Council decided in 2004 to merge the Committee on Capital Movements and Invisible Transactions and the Committee on International Investment and Multinational Enterprises and establish a new Investment Committee. See OECD Council Resolution C(2004)3 and CORR1, Articles 1 and 2, available at http://webnet3.oecd.org/OECDgroups.

[547] The representatives are usually senior officials of national ministries or central banks, compare OECD Investment Committee, Promoting Investment for Growth and Sustainable Development Worldwide, booklet available at http://www.oecd.org/dataoecd/63/10/35250560.pdf, p. 5.

[548] See OECD Council Decision C(2000)96/FINAL, Part II, available at http://www.olis.oecd.org/olis/2000doc.nsf/LinkTo/c(2000)96-final.

[549] Decision of the Investment Committee of 20 September 2004, DAF/INV(2004)1 and DAF/INV/M(2004)1, para.3, available at http://webnet3.oecd.org/OECDgroups. The importance of non-member cooperation is equally highlighted by the IC through the establishment of the Advisory

with the OECD and from other international institutions and organisations such as the International Monetary Fund (IMF), the U.N. Conference on Trade and Development (UNCTAD), the World Bank and the World Trade Organization (WTO) participate as observers. [550]

In its decision, the Council has also adopted detailed so-called Procedural Guidance. This guidance document not only covers the composition of NCPs, but also establishes general criteria which should guide the work of the NCPs and the IC. For the NCPs, these are visibility, accessibility, transparency and accountability;[551] timeliness and efficiency are mentioned for both the NCPs and the IC.[552] The activities of the NCPs are also subjected to oversight by the Investment Committee. NCPs must report annually to the IC on their activities[553], and the IC includes this information in its periodical reports to the OECD Council.[554]

Most importantly however, the Procedural Guidance establishes a complex system of mediation and review: the procedure for the "implementation in specific instances".[555] This procedure is designed to resolve allegations of misconduct by an enterprise. It has two phases.

The first consultation and mediation phase is conducted by the National Contact Points (NCPs).[556] It can be triggered by the business community, worker organisations, non-governmental organisations or "other interested parties concerned".[557] In response to their application, the NCP first makes an initial assessment as to whether the issues raised merit further examination. If this is found to be the case, the NCP consults with the parties involved. In this consultation and mediation

---

Group on Cooperation with Non-Members on Investment Matters in 2001; see for the mandate Decision of the Investment Committee, DAF/INV/M(2004), available at http://webnet3.oecd.org/OECD groups.

550 OECD Council Res. on the Terms of Reference of the Investment Committee, C(2004)3 and CORR1 (22 April 2004).

551 For details on these criteria, compare the Investment Committee's Commentary on the Implementation Procedures of the OECD Guidelines for Multinational Enterprises, attached to the procedures, para. 8.

552 Procedural Guidance, Part I. C. and Part II. 4.

553 Procedural Guidance, Part I, D.1.

554 Implementation procedures, Part II, para. 7.

555 Procedural Guidance, Part I.C.

556 Procedural Guidance, Part C.

557 Procedural Guidance, Part C (chapeau).

phase, the NCP may seek guidance authorities, experts, worker organisations or business community as well as non-governmental organisations, consult with other NCPs and ask for guidance from the Investment Committee on the interpretation of the Guidelines. With their agreement, the NCP assists the parties in resolving the issue in a conciliatory manner. In case the parties do not reach an agreement or if one party is unwilling to participate in the procedures, the NCP will – as a third step – issue a statement and make – "as appropriate"[558] and therefore at its own discretion – recommendations on how the Guidelines should be understood.

In the second phase of the procedure, the activities of the NCPs can be reviewed by the Investment Committee (IC) at the international level. Adhering governments, the European Commission, an advisory body or OECD Watch, an NGO specialised in working on OECD issues and tracking OECD activities – can ask the IC to review whether a particular NCP has properly applied the procedural rules issued by the Council and whether it has correctly interpreted the OECD Guidelines. In its review, the IC considers whether an NCP has correctly handled the procedure, and if necessary issues a clarification on whether an NCP "has correctly interpreted the Guidelines in specific instances".[559] In other words, the Investment Committee oversees the activities of the NCP in what resembles an appeal on legal grounds in judicial systems. The individual enterprise whose interests are directly involved – even though it cannot initiate a review – is granted a procedural right to be heard, either in writing or orally, during the clarification procedure before the IC.[560]

7. Preliminary assessment and outlook

The issue of corporate environmental and social responsibility which emerged in the 1970s has gained renewed political momentum in the new century.[561] In addition to the growing number of private initia-

---

558 Procedural Guidance, Part C. 3 c.

559 Procedural Guidance, Part C. 2 c; see also the Commentary on the Procedural Guidance, para. 23.

560 Implementation procedures, Part II, para. 4.

561 Morgera speaks of a development towards accountability, compare E. Morgera, "From Stockholm to Johannesburg: From Corporate Responsibility to Corporate Accountability for the Global Protection of the Environment", Review of European Community and International Environmental Law

tives,[562] international efforts have also intensified. This is illustrated by the fundamental revision of the OECD Guidelines in 2000 and 2010/11, but also by a number of initiatives at the UN level. The draft UN Norms on the Responsibilities of Transnational Corporations and Other Business Enterprises with Regard to Human Rights adopted by the UN Commission on Human Rights' Sub-commission on the Promotion and Protection of Human Rights[563] in 2003 which contain a number of comparatively stringent environmental and social norms is a further example. So far the UN Norms have, however, not been adopted by the Human Rights Council. The Global Compact[564] established in 2000 and the UNEP Principles for Responsible Investment[565] adopted under the auspices of the UN on the initiative of the Secretary-

---

13 (2004), 214-222; E. Morgera, Corporate accountability in international environmental law, 2009.

[562] Compare e.g. the numerous codes of conduct and standards issued by the International Chamber of Commerce (ICC) such as the ICC Business Charter for Sustainable Development of 1991, available at http://www.iccwbo.org ; the standards of the International Standardization Organization (e.g. ISO 14000), of the Forest Stewardship Council and the Marine Stewardship Council, and numerous corporate codes of conduct. An overview and assessment is undertaken by J. Friedrich, "Codes of Conduct", in: R. Wolfrum (ed.), Max Planck Encyclopedia of Public International Law, 2010, online at: www.mpepil.com.

[563] UN Doc. E/CN.4/Sub.2/2003/12/Rev.2 (2003); on this significant development e.g. N. Rosemann, "Business Human Rights Obligations – The Norms of the Responsibility of Transnational Corporations and Other Business Enterprises with Regard to Human Rights", Nordic Journal of Human Rights 23 (2005), 47-62; K. Nowrot, "Die UN-Norms on the Responsibility of Transnational Corporations and Other Business Enterprises with Regard to Human Rights – Gelungener Beitrag zur transnationalen Rechtsverwirklichung oder das Ende des Global Compact?", Beiträge zum Transnationalen Wirtschaftsrecht 21 (2003), 1-27; D. Weissbrodt/M. Kruger, "Norms on the Responsibilities of Transnational Corporations and Other Business Enterprises with Regard to Human Rights", American Journal of International Law 97 (2003), 901-922.

[564] Global Compact of the United Nations, information available at http://www.unglobalcompact.org; for further analysis of the Global Compact compare K. Nowrot, "The New Governance Structure of the Global Compact – Transforming a "Learning Network" into a Federalized and Parliamentarized Transnational Regulatory Regime", Beiträge zum Transnationalen Wirtschaftsrecht 47 (2005), 1-38 (6 et seq.); A. Blüthner, "Ein Globalisierungspakt über Werte und Effizienz" in: S. Hobe (ed.), Kooperation oder Konkurrenz internationaler Organisationen, 2001, 155.

[565] Compare http://www.unpri.org/.

General also establish principles addressed to private actors. In comparison to the OECD Guidelines, governmental input into these standard-setting and implementation activities is only peripheral.[566] The initiatives more resemble process-oriented self-regulatory learning networks of private business actors and NGOs that concentrate on developing best practices rather than establishing public minimum standards.[567] All of these efforts are in line with calls for the active promotion of corporate responsibility in two of the main policy instruments of the sustainable development agenda, namely the Agenda 21[568] and the Johannesburg Plan of Implementation.[569]

All these international "voluntary" instruments on corporate responsibility are evidence for an emerging and increasingly converging web of international environmental norms for multinational corporations.[570] These norms obviously represent a new regulatory approach to sustainable development and environmental problems previously unknown in

---

566 The UN Global Compact is controlled by the Global Compact Board composed of 20 elected representatives from Business, Trade Unions, NGOs and the UN (2 representatives), details available at www.unglobalcompact.org.

567 A. Blüthner, "Ein Globalisierungspakt über Werte und Effizienz" in: S. Hobe (ed.), Kooperation oder Konkurrenz internationaler Organisationen, 2001, 155 (75).

568 Agenda 21 states in section 30.10.:

"Business and industry, including transnational corporations, should be encouraged:

a. To report annually on their environmental records, as well as on their use of energy and natural resources;

b. To adopt and report on the implementation of codes of conduct promoting the best environmental practice, such as the Business Charter on Sustainable Development of the International Chamber of Commerce (ICC) and the chemical industry's responsible care initiative."

569 The Plan of Implementation of the World Summit on Sustainable Development (Johannesburg Plan of Implementation), UN Doc. A/CONF.199/20, Resolution 2, Annex, states in para. 49 that action at all levels is required to "[A]ctively promote corporate responsibility and accountability, based on the Rio principles, including through the full development and effective implementation of intergovernmental agreements and measures, international initiatives and public private partnerships and appropriate national regulations, and support continuous improvement in corporate practices in all countries".

570 E. Morgera, Corporate accountability in international environmental law, 2009, 113

international law.[571] With the objective to supplement national and international law, they directly address private actors and attempt to provide an internationally defined framework for their activities which does not only rely on domestic regulation. Such an approach deviates from traditional treaty-making on the one hand, and from soft law approaches traditionally outlining recommendations to states on the other. In contrast to other instruments presented in this study which also contain norms addressed to private actors, the OECD Guidelines stand for a new approach to implementation. Instead of primarily building on legislative and administrative implementation by states, the OECD itself sets up a complex procedure to promote implementation of the Guidelines. It directly implicates private actors by allowing any interested private actors to bring complaints to specific bodies supervised by an international institution and acting under international procedures. Even if not ideal implementation tools, these implementation procedures have certainly enhanced the impact and role of the Guidelines.[572] And even though the implementation mechanism remains imperfect, such weaknesses with respect to implementation can arguably be advantageous for maintaining a flexible and legitimate process accepted by relevant actors. Linkages between legal binding instruments using nonbinding norms as references may become possible not despite, but because of their particular weakness as compared treaty law.[573]

[571] As a matter of doctrine, international treaty law cannot directly obligate private actors which are not considered subjects of international law.

[572] E. Assadourian, "The State of Corporate Responsibility and the Environment", Georgetown International Environmental Law Review 18 (2006), 571-594 (588); E. Morgera, "An Environmental Outlook on the OECD Guidelines for Multinational Enterprises: Comparative Advantages and Legitimacy in relation to Other International Initiatives", Georgetown International Environmental Law Review 18 (2006), 751-777 (775).

[573] S.F. Vendzules, "The Struggle for Legitimacy in Environmental Standards Systems: The OECD Guidelines for Multinational Enterprises", 21 Colorado Journal of International Environmental Law & Policy 21 (2010), 451–489 (488).

Irrespective of the overall effectiveness of the OECD Guidelines,[574] it is remarkable that the OECD, as well as the other organisations such as the ILO or the UN, attempt to fill gaps in national and international law by directly striving to influence private actors. This approach and the institutional set-up of the OECD Guidelines hint at some of the legitimacy challenges that may arise in the context of nonbinding instruments, and which will be discussed later on in this study. The direct addressing of private entities to some extent breaks down the traditional separation of the national and the international level, and calls into question traditional bases for legitimacy that are associated with that separation. This is not to say that states do not play an important role, but rather that the importance of the state as the addressee of norms, and as the polity in which decisions are taken as to the means and ways of implementation, is diminished.

C. Characteristics of nonbinding instruments and parameters for analysis

I. Nonbinding status

This study stresses "binding" and "nonbinding" as a parameter for distinguishing between international instruments. From a traditional perspective, this may be seen as trite. Nonbinding instruments are not intended to create international legal rights and obligations between parties, and that should matter. However, whether or not the concept of "bindingness" should continue to play a decisive doctrinal role in distinguishing between instruments is sometimes and increasingly thrown into doubt.[575] I believe, as explained already in the Introduction of this

574 For a study on implementation and effectiveness see C.N. Franciose, "A critical assessment of the United States' implementation of the OECD guidelines for multinational enterprises", Boston College international and comparative law review 30 (2007), 223-236; for a critical view of their effectiveness that stresses the need for the internalization of moral norms which cannot be achieved by codes imposed from the outside, consider J. Dine, "Multinational enterprises: international codes and the challenge of sustainable development", Non-State actors and international law 1 (2001), 81-106.

575 For example by M. Goldmann, "Inside Relative Normativity: From Sources to Standard Instruments for the Exercise of International Public Authority" in: A. von Bogdandy/R. Wolfrum/J. von Bernstorff/P. Dann/M. Gold-

study, that upholding that distinction is to the contrary necessary and should be stressed.[576] The reason is not that I presume legally binding norms to be superiour to nonbinding ones, or that I assume that enforcement would make a difference. But neglecting the distinction would not only run counter to actual state practice but also risk upsetting the common understandings and legal procedures associated with "bindingness". It would furthermore be of little help for dealing with nonbinding instruments. Nothing would be actually gained from giving up that distinction, and states would most likely become weary of using nonbinding instruments in known forms but instead most likely "escape" into even more and other informal ways and means. Finally, the distinction matters for implementation and compliance.[577]

The nonbinding status of acts of international organisations derives from the constitutive treaty outlining whether the specific body has the power to adopt binding instruments or not. In all cases where the instrument is not adopted as an act of an organisation or treaty body, the question whether an instrument is binding or not depends on the "manifest intent" of states.[578] The intention of states in turn generally can be deduced from the language employed and the circumstances of the conclusion of the document.[579] The wording is the decisive key to the subjective intentions of the Parties.[580] The avoidance of treaty lan-

---

mann (eds.), The Exercise of Public Authority by International Institutions: Advancing International Institutional Law, 2010, 661-711 (709-710).

576 Similarly J. Klabbers, "Goldmann Variations" in: A. von Bogdandy/ R. Wolfrum/J. von Bernstorff/P. Dann/M. Goldmann (eds.), The Exercise of Public Authority by International Institutions, 2010, 713-725 (722).

577 Clearly in that way and in favour of the distintion e.g. D. Bodansky, The art and craft of international environmental law, 2010, 102; see on the connection of status and compliance also the analysis in Part 2, at B.I.1., further below.

578 "Manifest intent" as the decisive element of distinction is for example stressed by O. Schachter, "International Law in Theory and Practice: General Course in Public International Law", Recueil des cours de l'Académie de Droit International de La Haye 178 (1982), 9-396 (123 et seq.).

579 The International Court of Justice in the Aegan Sea Continental Shelf Case (Greece v. Turkey), ICJ Rep. 3 (1978), at para. 96 affirmed that in order to determine the binding nature of an agreement it "must have regard above all to its actual terms and to the particular circumstances in which it was drawn up".

580 In the Case Concerning Maritime Delimitation and Territorial Questions between Qatar and Bahrain (Jurisdiction and Admissibility), ICJ Rep. 6 (1994), para. 41, the ICJ stressed that the words of the Agreement are decisive as the

guage by using "should" instead of "shall" strongly indicates that states did not intend to be legally bound.[581] The concrete assessment can at times be difficult however.[582]

While often the lack of intention on the part of states to be legally bound can only be deduced implicitly from the language and circumstances of adoption,[583] many nonbinding instruments are more explicit. All the instruments discussed in the case studies expressly include a statement to clarify their "non-legally binding" or "voluntary" nature.[584] Other examples are the Bonn Guidelines on Access to Genetic Resources and the Fair and Equitable Sharing of Benefits Arising Out of Their Utilisation adopted by the Conference of the Parties to the Convention on Biological Diversity in 2002[585] or the FAO International Code of Conduct for Plant Germplasm Collecting and Transfer.[586] With

---

expression of the intention of states, "[w]hatever may have been the motives of each of the Parties".

[581] However, some declarations are generally held in non-mandatory language ("should"), they sometimes also resort to mandatory language to bestow principles with particular authority. The Declaration of Principles on the Sea-Bed and Ocean Floor provides that "[t]he area shall not be subject to appropriation ..., and no State shall claim or exercise sovereignty or sovereign rights over any part thereof", compare UNGA Res. 2749 (XXV) of 17 December 1970, para. 2. Similarly, the influential Principle 4 of the nonbinding Rio Declaration provides that "[i]n order to achieve sustainable development, environmental protection shall constitute an integral part of the development process and cannot be considered in isolation from it", compare Rio Declaration on Environment and Development (13 June 1992), UN Doc. A/CONF.151/26 (vol. I); 31 ILM 874 (1992), principle 4.

[582] See on this issue C. Chinkin, "Normative Development in the International Legal System" in: D. Shelton (ed.), Commitment and Compliance: the Role of Non-binding Norms in the International Legal System, 2000, 21-42 (38-39).

[583] P. Sands, Principles of International Environmental Law, 2003, 126.

[584] OECD Guidelines, Chapter I, para. 1.; FAO Pesticide Code, para. 1.1; FAO CCRF, Article 1.1.

[585] The Bonn Guidelines at para. 7 provide explicitly that the "[T]he present Guidelines are voluntary". The Bonn Guidelines were adopted as Convention on Biological Diversity COP Decision VI/24, available at http://www.cbd.int/doc/publications/cbd-bonn-gdls-en.pdf.

[586] Article 3.1. of the FAO International Code of Conduct for Plant Germplasm Collecting and Transfer provides that "the Code is voluntary". The Code

similar care to avoid binding obligations, states included a disclaimer in the founding document of the recently adopted Strategic Approach to International Chemical Management (SAICM). In this so-called Dubai Declaration, states "acknowledge that as a new voluntary initiative in the field of international management of chemicals, the Strategic Approach is not a legally binding instrument".[587] The newly negotiated so-called Non-legally binding instrument on all types of forests[588] emphasizes its nonbinding nature already in its title. This emphasis on the nonbinding nature of these instruments illustrates that states in practice take great care to distinguish between a legally binding instrument and one which is non-legally binding. And as mentioned, the distinction remains highly relevant for states since it determines whether a particular instrument must be ratified and implemented into national law. Given the significance of this distinction in state practice, this study therefore holds on to this traditional distinction without necessarily implying that nonbinding instruments do not have substantial legal effects or some steering capacity.

II. Norm characteristics

The overview at the beginning of this part has indicated that nonbinding instruments prescribe – similar to treaty law – a large variety of different types of provisions. A closer look reveals further differences and commonalities of the characteristics of the norms of different instruments which may have implications for the implementation and compliance and therefore for their regulatory and steering capacity.

---

has been adopted by the FAO Conference at its 27th session in November 1993, available at http://www.fao.org/ag/AGp/AGPS/pgr/icc/icce.htms.

[587] The Dubai Declaration was adopted at the first session of the International Conference on Chemicals Management held in Dubai, 2-4 February 2006. All SAICM instruments so far adopted are available at http://www.saicm.org/documents/saicm%20texts/SAICM_publication_ENG.pdf.

[588] Non-legally binding instrument on all types of forests, adopted by UN Doc. A/RES/62/98 (17 December 2007), Annex, available at http://www.un.org/esa/forests/about-history.html; details and an assessment of this instrument are provided by K. Kunzmann, "The Non-legally Binding Instrument on Sustainable Management of All Types of Forests – Towards a Legal Regime for Sustainable Forest Management?", German Law Journal 9 (2008), 981-1005.

First, one can distinguish norms with respect to their specificity. The spectrum of norms in the instruments analysed ranges from general aspirational objectives that share similarities with similar provisions in framework conventions to precise and concrete rules and standards that can be measured and allow for monitoring of compliance (e.g. Technical Guidelines under the framework of the FAO CCRF, IMO codes, World Bank Operational Standards). This means that one must discard the assumption that nonbinding instruments are typically unspecific and vague. Nonbinding does not always mean soft and imprecise norms, but rather there is often a correlation between nonbinding norms and precise substance.[589] Nonbinding instruments such as codes of conduct or guidelines often contain surprisingly detailed measures, best practices in addition to general rules which could all be contained in treaty law.

Just as has been pointed out for binding obligations of international law,[590] one can further distinguish between norms prescribing a certain conduct, norms prescribing a specific result and norms which prescribe general objectives to be reached through an evolutionary process at some time in the future. Nonbinding norms often postulate long-term objectives, such as for example the effective conservation and management of the living aquatic resources in case of the FAO Code of Conduct for Responsible Fisheres.[591] Prescriptions of results refer to a certain more concrete result to be achieved by states.[592] To again take the

---

[589] H. Neuhold, "The Inadequacy of Law-Making by International Treaties: "Soft Law" as an Alternative" in: V. Röben (ed.), Developments of International Law in Treaty Making, 2005, 39-52 (51); P.-M. Dupuy, "Soft law and the international law of the environment", Michigan Journal of International Law 12 (1990–1991), 420-435 (429); H. Hillgenberg, "A Fresh Look at Soft Law", European Journal of International Law 10 (1999), 499-515 (501) A.E. Boyle, "Some Reflections on the Relationship of Treaties and Soft Law", International and Comparative Law Quarterly 48 (1999), 901-913 (903).

[590] See for this distinction of rules in international law and its relevance in particular for implementation R. Wolfrum, "Obligation of Result Versus Obligation of Conduct: Some Thoughts About the Implementation of International Obligations", in: M.H. Arsanjani, J.K. Cogan, R.D. Sloane and S. Wiessner (eds.), Looking to the Future: Essays on International Law in Honor of W. Michael Reisman, 2010, Chapter 20.

[591] CCRF, Art. 6.1.

[592] R. Wolfrum, "Obligation of Result Versus Obligation of Conduct: Some Thoughts About the Implementation of International Obligations", in: M.H. Arsanjani, J.K. Cogan, R.D. Sloane and S. Wiessner (eds.), Looking to the Fu-

FAO Code of Conduct for Responsible Fisheries as an example, it stipulates in Art. 7.2.2.d) that "biodiversity of aquatic habitats and ecosystems is conserved". Nonbinding instruments rarely however stipulate concrete results such as emission limitations or quotas. Exceptions exist however, as for instance the Copenhagen Accord where each State commits to a specific emission reduction. Whether concerned with prescriptions of result or objectives, the means of achieving the envisaged objectives and results are left to the addressees, i.e. to states and private actors. Increasingly, however, one can also find prescriptions of conduct or actions, i.e. norms that recommend a particular action,[593] in nonbinding instruments in the environmental field.[594] These types of norms are much less permissive as to which actions should be taken by the addressees. They generally intrude much more directly with domestic legal systems if states decide or are pressured to comply.

A further distinction relates to procedural versus substantive norms. Nonbinding instruments often entail and emphasise procedural rather

---

ture: Essays on International Law in Honor of W. Michael Reisman, 2010, Chapter 20.

[593] Similar, albeit for treaty law, R. Wolfrum, "Obligation of Result Versus Obligation of Conduct: Some Thoughts About the Implementation of International Obligations", in: M.H. Arsanjani, J.K. Cogan, R.D. Sloane and S. Wiessner (eds.), Looking to the Future: Essays on International Law in Honor of W. Michael Reisman, 2010, Chapter 20.

[594] Many of the examples mentioned in the overview in this Part, at I, contain such prescriptions of actions; see e.g. table of activities and concrete measures contained in the Global Plan of Action of the Strategic Approach to Chemicals Management, adopted in 2006 through the International Conference on Chemicals Management (SAICM), available at http://www.saicm.org/documents/saicm texts/SAICM_publication_ENG.pdf; the concrete legislation proposals of the UNEP Guidelines for the development of national legislation on liability, response action and compensation for damage caused by activities dangerous to the environment, UNEP/GCSS.XI/L.5 (2010); the prescriptions for forest policies contained in the Non-legally binding instrument on all types of forests, adopted by UN Doc. A/RES/62/98 (17 December 2007), Annex, available at http://www.un.org/esa/forests/about-history.html; the voluntary PIC procedure in the FAO Pesticide Code of Conduct discussed in II.2.; the Technical Guidelines and International Plans of Action adopted under the framework of the FAO Code of Conduct for Responsible Fisheries as discussed in this Part 1, at B. I., but also many of the norms of the Code themselves which makes it possible to assess compliance with the CCRF, a fact that is stressed by T.J. Pitcher/D. Kalikoski/G. Pramod (eds.), Evaluations of Compliance with the FAO (UN) Code of Conduct for Responsible Fisheries, 2006.

than substantive standards. Procedural environmental rules prescribe for example that states or non-state actors should conduct environmental impact assessments (as in the OECD Guidelines), comply with prior informed consent procedures (voluntary PIC system), or establish authorisation and monitoring systems (FAO CCRF). It appears that these types of norms have generally made an impact on the addressees. Substantive rules of behaviour on the other hand prescribing the end of certain damaging behaviour, such as the requirement to manage fisheries responsibly in accordance with reproduction numbers[595] or the 2 degree Celsius objective of the Copenhagen Accord[596] are less often found.

Some characteristics of the norms of nonbinding instruments can be found in all instruments analysed. First, unlike most treaty law, they rarely contain specific objectives such as timetables or targets, as for example specific fishery quotas or timetables for reducing fishing capacities or subsidies (e.g. FAO CCRF). Only some examples exist where declarations contain timetables and targets,[597] but these do not usually define individual responsibilities of states. A rare exception was the Copenhagen Accord adopted at the 15th Conference of the Parties to the UNFCCC in 2009. It contains an Appendix I where states should fill in their national economy-wide emission targets for 2020 as well as an Appendix II where developing country Parties should report their nationally appropriate mitigation actions to be taken.[598] Most developed countries followed suit and reported concrete targets,[599] which if com-

---

595 FAO CCRF, Article 6.3.

596 Copenhagen Accord, adopted by the UNFCCC Conference of the Parties at its fifteenth session, Decision 2/CP.15 (2009), available at http://unfccc.int/resource/docs/2009/cop15/eng/11a01.pdf.

597 E.g. the UNGA Resolution 46/215 of 20 December 1991 provided a deadline (31 December 1992) for the cessation of driftnet fishing. One remarkable example is also that of the Declarations of the International North Sea Conferences. The London Declaration adopted by the International North Sea Conferences for instance outlines the objective to phase out dumping of wastes in the North Sea by 1989 and a 50 percent reduction of phosphorus, nitrogen and hazardous substances by 1995, but does not specify the respective individual burden of each participating state.

598 Copenhagen Accord, Appendix I and II.

599 Compare the list of responses of developed country Parties with concrete timelines and quantified emission targets according to Appendix I of the Copenhagen Accord, available at http://unfccc.int/home/items/5264.php; develop-

bined are however unlikely to be sufficiently ambitious to reach the 2 degree that was collectively set in the Copenhagen Accord.[600]

Second, a characteristic that appears to be shared by all analysed instruments prescribing norms is that they tend to promulgate comparatively new and innovative norms in areas where binding rules are inexistent or insufficiently developed. Their underlying purpose is mostly to promote change rather than preserve the status quo.[601] Innovative approaches are often first adopted through nonbinding norms, in particular when compared with existing legal instruments.[602] The nonbinding nature apparently often facilitates agreement on progressive norms.[603] All of the analysed instruments refer to progressive concepts of international environmental law. Nonbinding instruments in the environmental sectors typically include and further develop the concept of sustainable development, reflect the precautionary principle, the ecosystem approach or the principle of common but differentiated responsibilities. Procedural principles such as the need for environmental impact assessments or prior informed consent are also typical elements.

III. Origin and norm development

This parameter refers to both norm development process and the adoption of instruments. First of all, nonbinding instruments can be distinguished based on the public or private nature of their origin. The public category on which this study focuses comprises instruments that are adopted by states in intergovernmental fora or in institutions which are

ing country Parties states reported their nationally appropriate mitigation actions, see http://unfccc.int/home/items/5265.php.

600 Copenhagen Accord, op. para. 1 ("We agree that deep cuts in global emissions are required according to science, and as documented by the IPCC Fourth Assessment Report with a view to reduce global emissions so as to hold the increase in global temperature below 2 degrees Celsius").

601 A.A. Fatouros, "On the implementation of international codes of conduct: an analysis of future experience", American University Law Review 30 (1980-1981), 941-972 (944).

602 See for this argument with respect to the CCRF, W. Edeson, "Closing the Gap: The Role of 'Soft' International Instruments to Control Fishing", Australian Yearbook of International Law 20 (1999), 83-104 (102).

603 I. Seidl-Hohenveldern, "International Economic 'Soft Law'", Recueil des Cours de l'Académie de Droit International 163 (1979), 169-246 (210).

founded and formally controlled by states. The category also comprises instruments adopted by states in Conferences of the Parties of treaty regimes such as the Convention on Biological Diversity[604] or the UN Framework Convention on Climate Change,[605] instruments adopted in organs of international organisations as well as intergovernmental declarations and initiatives which are not adopted by a particular institutional body even though states may use an international institution as the negotiating forum. The formal control by state actors over the adoption of an instrument – either directly through vote or through governing bodies and committees of international organisations (e.g. COFI, Investment Committee) – gives rise to a distinct claim to legitimacy and authority.[606].

The elaboration of nonbinding instruments often takes less time than negotiating a treaty instrument. However, even nonbinding instruments may take years of complex negotiations. These are often reminiscent of negotiations for treaty instruments. Observers to negotiations on nonbinding instruments frequently confirm that states negotiate these instruments with the same care as if they were negotiating a treaty.[607] Regarding norm development, one can differentiate between processes under strict state control and those which provide for access by non-

---

604 For example, the 9th Conference of the Parties of the Convention on Biological Diversity in 2008 in Bonn established a working group to develop a code of ethical conduct to promote respect for the heritage of indigenous and local communities, as one element in the implementation of Article 8 (j) of the Convention on Biological Diversity, compare UNEP/CBD/IX/13 (2008). The Working Group in 2009 recommended recommend the 'The Tkariwaié:ri Code of Ethical Conduct on respect for the cultural and intellectual heritage of indigenous and local communities relevant for the conservation and sustainable use of biological diversity', for possible adoption at the 10th COP in 2010, compare UNEP/CBD/WG8J/6/4 (2009).

605 E.g. the Copenhagen Accord, UNFCCC COP Decision CP2/15 (2009).

606 For a similar argument see S.R. Ratner, "Business" in: D. Bodansky/J. Brunnée/E. Hey (eds.), The Oxford Handbook of International Environmental Law, 2007, 807-828 (820); for further discussion of the issue of legitimacy see Part 3, in this study.

607 This is for instance reported by one of the advisors to the German Government to the negotiations of the Non-Legally Binding Instrument on all Types of Forests in the United Nations Forum on Forests, K. Kunzmann, "The Non-legally Binding Instrument on Sustainable Management of All Types of Forests – Towards a Legal Regime for Sustainable Forest Management?", German Law Journal 9 (2008), 981-1005 (985).

state actors. As already mentioned, the elaboration processes of nonbinding instruments are generally conducive to the influence of non-state actors. Generally speaking, expert input to these processes is also high.

Nonbinding instruments obviously differ greatly from treaty instruments because of their lack of ratification requirements. Nonbinding instruments have the obvious advantage that, once adopted, they can immediately serve as the basis of cooperation. The risk of not reaching a meaningful number of ratifications is avoided. Furthermore, in contrast to most binding instruments and decisions,[608] nonbinding instruments could formally often be adopted by majority. Organs of international organisations that do not have the authority to issue binding norms can often decide by majority, whereas this is rarely the case for decisions with binding effects. Nonbinding "flagship" instruments such as the FAO CCRF or the OECD Guidelines are however most often adopted by consensus, presumably in order to secure their authority and legitimacy. These instruments can thus claim full governmental support. As indicated by the above analysis, nonbinding instruments developed by subsidiary bodies or the secretariats are however more prone to majority decision making.[609]

IV. Non-state actor involvement

As with treaty law instruments, nonbinding instruments differ significantly in the extent to which non-state actors are included in norm development and implementation. One can however observe that the nonbinding nature of an instrument often seems to facilitate the influence of non-state actors such as international organisations, expert groups and secretariats. Nonbinding instruments can also generally be characterised as particularly conducive to the inclusion of private actors

[608] There is a tendency however to flexibilise decision-making in international treaty regimes by allowing for voting by consensus, opt-out procedures or decision-making by two-thirds majority if consensus fails, compare J. Brunnée, "COPing with Consent: Law-Making under Multilateral Environmental Agreements", Leiden Journal of International Law 15 (2002), 1-52.

[609] Take for example the adoption of International Plans of Action by the FAO Committee on Fisheries which is not composed of all states, or of the Technical Guidelines adopted by the secretariat with or without the initiative of COFI. See for details already the case studies in this Part, at B.

such as non-governmental organisations (NGOs).[610] They often provide one of the main opportunities how NGOs can get involved and push their agenda. In fact, many nonbinding instruments would not exist if not for NGOs, and they have played an important role in the development and growing importance of nonbinding instruments.[611] They deliberately integrate non-state actors such as industries, trade unions or NGOs, encouraging them to become active in the formulation and specifically in the implementation phase.[612]

Although generalisations are difficult to make, states appear to be less wary of these influences when the instrument in question is nonbinding. Important differences exist with respect to the degree and means of integration of these actors from one institution and instrument to another. Generally speaking, a corporatist approach traditionally focussing on industry and trade union associations can be distinguished from an approach more focussed on the regulated admittance of specific organisations as observers with a lower degree of institutionalisation. The OECD is an example of a corporatist approach. Here non-state actors participate in the institutional mechanisms through representative bodies such as the Business and Industry Advisory Committee, the Trade Union Advisory Committee and, most recently, the NGO OECD Watch. In contrast, UN organisations allow particular non-governmental organisations which meet certain criteria[613] to participate in their

---

[610] J.A. Fuentes Véliz, "L'evolution du role des organisations non gouvernementales dans le droit de l'environnement", Revue Européenne de droit de l'Environnement 4 (2007), 401-430 (421 et seq.).

[611] J.A. Fuentes Véliz, "L'evolution du role des organisations non gouvernementales dans le droit de l'environnement", Revue Européenne de droit de l'Environnement 4 (2007), 401-430 (421)

[612] This is also observed H. Keller, "Codes of Conduct and their Implementation: the Question of Legitimacy" in: R. Wolfrum/V. Röben (eds.), Legitimacy in International Law, 2008, 219-298 (249); D. Shelton, "Law, Non-Law and the Problem of 'Soft Law'" in: D. Shelton (ed.), Commitment and Compliance: The Role of Non-Binding Norms in the International Legal System, 2000, 1-18 (13).

[613] Compare UN ECOSOC Res. 1296 (XLIV) (1968) and UN ECOSOC Res. 1996/31 (1996). The general criteria are often refined by the respective organization, as for example by the FAO through the "FAO Policy Concerning Relations with International Non-Governmental Organizations"; Report of the ninth session of the Conference, para. 497 and Resolution 44/57 entitled "Granting of Observer Status (in respect of international governmental and non-governmental organizations)"; all documents available at www.fao.org.

meetings as observers. The PIC example has shown how environmental NGOs can use this right to drive environmental norm setting. By pushing for the adoption of nonbinding instruments, NGOs often successfully place issues on the international agenda, and eventually may in this manner even trigger customary or treaty law development.[614]

V. Addressees

In contrast to treaty law or customary law rules, nonbinding instruments are not only addressed to states, but also establish norms directed at non-state actors. This can be seen as one defining characteristic of codes of conduct which distinguishes them from other forms of nonbinding instruments.[615] For instruments prescribing rules of corporate social responsibility such as the OECD Guidelines for Multinational Enterprises, private corporations are actually the main addressees. States are in these cases only incidentally addressed for implementation and promotion purposes.[616] Furthermore, nonbinding instruments go beyond the contracting parties of the constitutive treaty of an organisation or treaty regime and also address non-contracting parties, nonmember states[617] or other actors such as organisations or NGOs. Finally, some nonbinding instruments are not at all directed at actors external to the institutions, but are directed at organs or the staff of institutions. Although formally without any binding effect for states, these

614 Compare on this issue in more detail C. Chinkin, "Normative Development in the International Legal System" in: D. Shelton (ed.), Commitment and Compliance: the Role of Non-binding Norms in the International Legal System, 2000, 21-42 (31 et seq.).

615 Compare for this also J. Friedrich, "Codes of Conduct", in: R. Wolfrum (ed.), Max Planck Encyclopedia of Public International Law, 2010, online at: www.mpepil.com.

616 A.A. Fatouros, "On the implementation of international codes of conduct: an analysis of future experience", American University Law Review 30 (1980-1981), 941-972 (947).

617 As illustrated by the case of the OECD Guidelines, the outreach to nonmembers effectively establishes differentiated obligations and rights for adhering governments. Only member states of the OECD are bound by the OECD Decision to establish NCPs, and only members have voting rights in the Investment Committee while adhering governments can participate as observers.

instruments may have implications for states if they establish conditions for loan policies of international institutions.[618]

VI. Adaptability

While it lies in the nature of nonbinding instruments that they can be easily adapted, which is one of their advantages, one must acknowledge that treaty law as well is developing towards greater adaptability and flexibility. Of course, changes and revisions of nonbinding instruments can be easily achieved through a decision of the political institutional body or by participating governments. In contrast, amendments to treaty law usually only become binding for those states which ratify or otherwise accept the respective instrument.[619] Furthermore, adopting changes to nonbinding instruments is generally easier than for treaty instruments, since the voting procedures of most political bodies of international organisations provide for majority decision-making in the case of nonbinding recommendations.[620] In particular when one compares nonbinding instruments to treaty instruments that make use of innovative flexible techniques such as the Framework/Protocol approach, contracting-in or opting-out procedures and in particular flexible amendment procedures which allow for majority decision-making,[621] even though in most cases states must ratify the amendments. But these flexibilities in modern environmental treaty law diminish this advantage

---

[618] On these instruments, see already above in this Part 1, at A.IV.

[619] Article 40 (4) of the Vienna Convention on the Law of Treaties (23 May 1969), UN Doc. A/Conf.39/27; 1155 UNTS 331; 8 ILM 679 (1969). A notable exception is Article 2 (9) of the Montreal Protocol on Substances that Deplete the Ozone Layer (16 September 1987), 26 ILM 1550 (1987), which provides that amendments to Annexes become binding by decision of the Meeting of the Parties.

[620] E.g. FAO Rules of Procedure, Rule XII. See already the overview of the various organisations provided in this Part 1, at A.IV.

[621] Amendments to the Vienna Convention for the Protection of the Ozone Layer (22 March 1985), 1513 UNTS 323; 26 ILM 1529 (1987), can be taken by a three-fourth majority [Article 9 (3)]; an amendment to the Montreal Protocol by a two-thirds majority [e.g. Article 2 (9)]; an amendment to the Kyoto Protocol to the United Nations Framework Convention on Climate Change (11 December 1997), UN Doc FCCC/CP/1997/7/Add.1; 37 ILM 22 (1998) by three-fourth majority if consensus fails [Article 20 (3)].

of nonbinding instruments. Moreover, in practice, changes to the main nonbinding instruments are often taken by consensus or unanimously in order to gain much needed legitimacy and authoritativeness.[622]

VII. Follow-up and compliance mechanisms

The parameter that most obviously correlates with the effectiveness of an instrument is the existence and design of follow-up and compliance mechanisms. In particular international organisations and treaty institutions often use their institutional and financial potentials as well as membership obligations to enhance the effectiveness of their instruments. Voluntary or even compulsory reporting of States or the participating private entities has been included in numerous codes in recent years, irrespective of the non-binding nature of the instruments. International organisations however avoid individual declarations of non-compliance or contentious procedures. The spectrum of co-operative implementation mechanisms often includes facilitative means such as financial, technological and institutional assistance, in particular to developing countries. It is this institutionalisation of a reiterated discourse and of compliance-enhancing means which distinguishes the one-time adoption of a nonbinding declaration from instruments with institutional underpinnings and support. At the same time, the existence of such mechanisms can somewhat level out differences between treaty law and nonbinding instruments. In much of international environmental law, international enforcement and contentious dispute settlement is the exception rather than the rule. Thus, one can ascertain a tendency of convergence of nonbinding instruments and treaty law when nonbinding instruments are promoted and supported by cooperative compliance mechanisms and follow-up processes with strong institutional underpinnings.[623] Differences between treaty law and non-

[622] Legitimacy hereby derives from the fact that the decision is carried by the consensus of all states and therefore can be seen as an expression of the community interest expressed in the convergence of opinions, compare R. Wolfrum, "Konsens im Völkerrecht" in: H. Hattenhauer (ed.), Mehrheitsprinzip, Konsens und Verfassung, 1986, 79-91 (87).

[623] A. Peters/I. Pagotto, "Soft Law as a New Mode of Governance: A Legal Perspective", New Modes of Governance Project 04/D11 (2006), (27).

binding instruments are in this way reduced, although they continue to exist.[624]

Even if generally cooperative and non-contentious, the design of compliance mechanisms of nonbinding instruments may however differ greatly from one instrument to another, with consequences for their effectiveness. One can generally distinguish two main approaches. In the more traditional approach, implementation of an instrument by states appears as the main objective of the procedure. This approach strives to impact national legislative and administrative processes and employs mechanisms such as state reporting, capacity building and international cooperation and information (e.g. FAO CCRF). The OECD Guidelines exemplify another approach whereby states attempt directly to steer private actors' behaviour by means of international norms (OECD Guidelines, UN Global Compact). Follow-up and implementation mechanisms consequently attempt directly to influence private actors through international procedures. This entails some decision-making of public agencies vis-à-vis individuals (e.g. the OECD specific instances procedure).[625]

---

[624] For details on this assessment, see Part 2, at B.I. and III.

[625] For a detailed assessment of these types of procedures, see Part 2, at C.I.1.

Part 2

The functions and limits of nonbinding instruments

A. The international level

One can ascertain the impact of nonbinding instruments on the international level in two ways: their influence on and relationship with formal international environmental law on the one hand, and their role in shaping and linking the activities of international institutions on the other. Both have important implications for the potential of nonbinding instruments to induce changes in the behaviour of their addressees, at least to the extent that one can assume that most states generally tend to follow their international obligations[1] and that state and private actors are today increasingly influenced by international institutions.[2]

I. The interplay of nonbinding instruments with international law

1. The precursory function of nonbinding instruments

The role of nonbinding instruments as precursors to customary or treaty law is widely acknowledged but rarely analysed in detail. This section explores which characteristics of nonbinding instruments facilitate or hamper the emergence of customary law – which may also be of interest for negotiators of such instruments. With respect to treaty law, actors that support the environmental objectives of these instruments often tend to favour a shift to binding rules as soon as politically feasi-

---

1 This is famously expressed by Louis Henkin who observed that "almost all nations observe almost all principles of international law and almost all of their obligations almost all the time," compare L. Henkin, How nations behave: law and foreign policy, 1979, 47.

2 See e.g. the contributions in J.E. Alvarez, "International Organizations: Then and Now", American Journal of International Law 100 (2006), 324-347; A. von Bogdandy/R. Wolfrum/J. von Bernstorff/P. Dann/M. Goldmann (eds.), The Exercise of Public Authority by International Institutions: Advancing International Institutional Law, 2010.

ble. A closer look at the example of the negotiations of the PIC Convention will show that it may not be that simple, and that a shift from a non-binding instrument to binding law carries distinct advantages and disadvantages.

a) Role in the development of customary international law

It is widely acknowledged that nonbinding instruments may contribute to the emergence of new customary international law.[3] The following analysis will not fully reopen the long-standing debate, but merely confirm this possibility. In addition, a closer look at the process of customary law formation in the context of nonbinding instruments can provide us with a more differentiated view of this issue. As will be seen, the characteristics of a particular instrument – as identified in Part 1 of this study – can be linked to its potential to contribute to the emergence of customary law rules.

(1) Nonbinding instruments and customary law

In traditional international law doctrine, customary law can be described as "usages generally accepted as expressing principles of law".[4] It is considered to emerge as a result of uniform state practice which is

---

[3] A.E. Boyle, "Some Reflections on the Relationship of Treaties and Soft Law", International and Comparative Law Quarterly 48 (1999), 901-913 (904); H. Hillgenberg, "A Fresh Look at Soft Law", European Journal of International Law 10 (1999), 499-515 (514); P.-M. Dupuy, "Soft law and the international law of the environment", Michigan Journal of International Law 12 (1990 – 1991), 420-435 (432); H. Neuhold, "The Inadequacy of Law-Making by International Treaties: "Soft Law" as an Alternative" in: V. Röben (ed.), Developments of International Law in Treaty Making, 2005, 39-52 (52); T. Gruchalla-Wesierski, "A Framework for Understanding 'Soft Law'", McGill Law Journal 30 (1984-1985), 37-88 (54); A. Peters/I. Pagotto, "Soft Law as a New Mode of Governance: A Legal Perspective", New Modes of Governance Project 04/D11 (2006), 1 et seq. (23); B. Simma/A. Heinemann, "Codes of Conduct" in: W. Korff et al. (eds.), Handbuch der Wirtschaftsethik, Band 2: Ethik wirtschaftlicher Ordnungen, 1999, 403-418 (415).

[4] Permanent Court of International Justice, The Case of the S.S. "Lotus", Judgement of 7 September 1927, PCIJ Series A – No. 10, at 18.

based on the *opinio juris sive necessitates* – "the belief that this practice is rendered obligatory by the existence of a rule of law requiring it."[5]

The adoption of nonbinding instruments can under certain circumstances reflect an existing or emerging *opinio juris* of states. This has been widely discussed and accepted with respect to resolutions of the United Nations General Assembly (UNGA). UNGA resolutions and the attitude of states towards their adoption are accepted as possibly containing evidence of the existing or emerging *opinio juris* of states,[6] in particular if adopted by consensus.[7] The International Court of Justice hinted at this possibility in the *Nicaragua Case* when explaining that "opinio juris may, though with all due caution, be deduced from, *inter alia*, the attitude of the Parties and the attitude of States towards certain UNGA resolutions".[8] It further confirmed and elaborated on this possibility in its Advisory Opinion on the *Legality of the Threat or Use of Nuclear Weapons*, stating that resolutions of the UNGA "can, in certain circumstances, provide evidence important for establishing the existence of a rule or the emergence of an opinio juris".[9] In the recent judgment

---

[5] International Court of Justice (ICJ), North Sea Continental Shelf Cases (Federal Republic of Germany/Denmark; Federal Republic of Germany/ Netherlands), Judgement of 20 February 1969, ICJ Reports 1969, p. 3, para. 77; on the formation of customary law in the context of environmental law see P.-M. Dupuy, "Formation of customary international law and general principles" in: J. Brunnée/D. Bodansky/E. Hey (eds.), Oxford Handbook of International Environmental Law, 2007, 449-466; generally on customary law formation G. Dahm/J. Delbrück/R. Wolfrum, Völkerrecht, Band I/1, Die Grundlagen. Die Völkerrechtssubjekte, 1989, 56 et seq.

[6] A.A. D'Amato, The concept of custom in international law, 1971, 86; A. Verdross, Die Quellen des universellen Völkerrechts, 1973, 116; G.J.H. van Hoof, Rethinking the sources of international law, 1983, 182; J.A. Frowein, "Der Beitrag der internationalen Organisationen zur Entwicklung des Völkerrechts", Zeitschrift für ausländisches öffentliches Recht und Völkerrecht 36 (1976), 147-167 (150); A. Verdross/B. Simma, Universelles Völkerrecht: Theorie und Praxis, 1984, 359.

[7] Generally on this discussion T. Treves, "Customary International Law", in: R. Wolfrum (ed.), Max Planck Encyclopedia of Public International Law, 2010, para. 44, online at: www.mpepil.com.

[8] ICJ, Military and Paramilitary Activities in und against Nicaragua (Nicaragua v. United States of America), Judgment of 27 June 1986, ICJ Reports 1986, p. 14, para. 188.

[9] ICJ, Legality of the Threat or Use of Nuclear Weapons, Advisory Opinion of 8 July 1996 , I.C.J Reports (1996), p. 226, para. 70.

on the *Case Concerning Armed Activities in the Territory of the Congo (Democratic Republic of the Congo v. Uganda)*, the ICJ as in other cases relied on UNGA resolutions to support its finding that a certain rule was part of customary international law, and looked to the resolutions to assess whether the customary law rule applied to the case at hand.[10]

Even though UNGA resolutions can be particularly influential due to the particularly authoritative nature of the General Assembly,[11] the underlying argument has much wider application. When certain conditions relating to the authority of the adopting body and the conditions of adoption are met, any nonbinding instrument adopted in multilateral fora by state representatives which outlines a sufficiently concrete norm may also contribute to the formation of customary law rules.[12] Due to

---

[10] ICJ, Case Concerning Armed Activities in the Territory of the Congo (Democratic Republic of the Congo v. Uganda), Judgment of 19 December 2005, ICJ Reports 2005, p. 27, para. 244; compare also Iran-United States Claims Tribunal which in the case Sedco Inc. v. Nat'l Iranian Oil Co., 25 ILM 629 (1986), at para. 33 stated that "United Nations General Assembly resolutions are not as such binding upon States and generally are not evidence of customary law. Nevertheless, it is generally accepted that such resolutions in certain specified circumstances may be regarded as evidence of customary international law and can contribute – among other factors – to the creation of such law".

[11] For instance, a customary law rule has arguably arisen from the call for a global moratorium on large-scale pelagic driftnet fishing in the high seas by the United Nations General Assembly, UN Doc. A/RES/46/215 (1991), para. 3, which was adopted by consensus and received support from a very large number of states. Compare for this argument also P. Sands, Principles of International Environmental Law, 2003, 589.

[12] P.-M. Dupuy, "Formation of customary international law and general principles" in: J. Brunnée/D. Bodansky/E. Hey (eds.), Oxford Handbook of International Environmental Law, 2007, 449-466 (459); in this sense already R. Higgins, "The Role of Resolutions of International Organizations in the Process of Creating Norms in the International System" in: W.E. Butler (ed.), International Law and the International System, 1987, 21-30 (28); O. Schachter, "International Law in Theory and Practice: General Course in Public International Law", Recueil des cours de l'Académie de Droit International de La Haye 178 (1982), 9-396 (129); T. Gruchalla-Wesierski, "A Framework for Understanding 'Soft Law'", McGill Law Journal 30 (1984-1985), 37-88 (54). This possibility is also accepted by most authors dealing with codes of conduct and nonbinding instruments other than General Assembly Resolutions, compare B. Simma/A. Heinemann, "Codes of Conduct" in: W. Korff et al. (eds.), Hand-

the authority of large-scale UN conferences, the legally nonbinding 1972 Stockholm Declaration and the 1992 Rio Declaration have arguably contributed to the emergence or at least the development of customary international environmental law,[13] even though the influence of these principles on international environmental law can hardly be assessed in terms of binding customary law status but goes much further.[14] Furthermore, nonbinding recommendations such as the OECD Principles Concerning Transfrontier Pollution adopted in 1974[15], the

buch der Wirtschaftsethik, Band 2: Ethik wirtschaftlicher Ordnungen, 1999, 403-418 (415); H. Hohmann, Präventive Rechtspflichten und -prinzipien des modernen Umweltvölkerrechts, 1992, 232; P. Muchlinski, "Human Rights, social responsibility and the regulation of international business: The development of international standards by intergovernmental organisations", Non-State Actors and International Law 3 (2003), 123-152 (128); H.W. Baade, "The Legal Effects of Codes of Conduct for Multinational Enterprises", German Yearbook of International Law 22 (1979), 11-52 (23); I. Brownlie, "Legal effects of codes of conduct for MNEs: Commentary" in: N. Horn (ed.), Legal Problems of Codes of Conduct for Multinational Enterprises, 1980, 39-43 (42); U. Dieckert, Die Bedeutung unverbindlicher Entschließungen internationaler Organisationen für das innerstaatliche Recht der Bundesrepublik Deutschland, 1993, 250; J.E. Alvarez, International Organizations as Law-makers, 2005, 146-147.

13 Principle 21 of the Stockholm Declaration, outlining that the sovereign rights of states to exploit their resources must be balanced with the general obligation of states to avoid transboundary harm, can be considered customary law today, compare e.g. ICJ, Legality of the Threat or Use of Nuclear Weapons, Advisory Opinion of 8 July 1996, I.C.J Reports (1996), p. 226, para. 29; A.C. Kiss/D. Shelton, International environmental law, 2004, 190. The legal status of other principles such as the precautionary principle, the principle of common but differentiated responsibilities or the principle of cooperation is still disputed, compare for an overview of the debate J. Brunnée, "The Stockholm Declaration and the Structure and Processes of International Environmental Law" in: A. Chircop (ed.), The future of ocean regime building: essays in tribute to Douglas M. Johnston, 2009, 41-62; U. Beyerlin, "Different Types of Norms in International Environmental Law: Policies, Principles and Rules" in: J. Brunnée (ed.), Oxford Handbook of International Environmental Law, 2007, 425-448.

14 On this J. Brunnée, "The Stockholm Declaration and the Structure and Processes of International Environmental Law" in: A. Chircop (ed.), The future of ocean regime building: essays in tribute to Douglas M. Johnston, 2009, 41-62.

15 OECD Principles Concerning Transfrontier Pollution, OECD C(88)84(Final) (1974), Principle 9.

1987 UNEP Principles on Shared Resources[16] together with important treaty law provisions[17] played an important role in the recognition of the customary legal duty to notify and inform other affected states in emergency situations that endanger shared natural resources.[18]

This role of nonbinding instruments should however be approached with caution. One should not be too quick to infer *opinio juris* simply from the act of adoption of a clearly nonbinding instrument.[19] Most of these instruments are adopted by states with the knowledge that they are nonbinding.[20] States are aware which language indicates nonbinding status, and whether the respective international body has the competence to enact and interpret binding law or not. Most notably, general provisions in nonbinding instruments or in their adopting decisions which emphasise their overall voluntary or nonbinding character[21] make it difficult to qualify the act of adoption and subsequent supportive statements of states as an expression of *opinio juris.*[22] In fact, these provisions clearly communicate quite the opposite, namely that the addressees must not consider themselves obliged in any legally binding

---

[16] UNEP Principles of Conduct in the Field of the Environment for the Guidance of States in the Conservation and Harmonious Utilization of Natural Resources Shared by Two or More States, 17 ILM 1097 (1978), Principle 9.

[17] Most notably Article 198 of the United Nations Convention on the Law of the Sea (UNCLOS), 1833 UNTS 3, 21 ILM 1261 (1982).

[18] A.C. Kiss/D. Shelton, International environmental law, 2004, 191-194.

[19] G. Dahm/J. Delbrück/R. Wolfrum, Völkerrecht, Band I/1, Die Grundlagen. Die Völkerrechtssubjekte, 1989, 72; I. MacGibbon, "Means for the Identification of International Law: General Assembly Resolutions: Custom, Practice and Mistaken Identity" in: B. Cheng (ed.), International law: teaching and practice, 1982, 10-26 (23); K. Döhring, Völkerrecht, 2003, § 4, mn 308; P. Malanczuk, Akehurst's modern introduction to international law, 1997, 53.

[20] This is stressed by K. Döhring, Völkerrecht, 2003, § 4, mn 308; S.M. Schwebel, The legal effect of resolutions and codes of conduct of the United Nations, 1986, 11.

[21] Compare e.g. OECD Guidelines, Chapter I, para. 1; FAO Code of Conduct on the Distribution and Use of Pesticides, Article 1.1; FAO Code of Conduct for Responsible Fisheries, Article 1.1.

[22] B.-O. Bryde, Internationale Verhaltensregeln für Private – Völkerrechtliche und verfassungsrechtliche Aspekte, 1981, 24; R.E. Lutz/G.D. Aron, "Codes of Conduct and Other International Instruments" in: G. Handl/R.E. Lutz (eds.), Transferring Hazardous Technologies and Substances: The International Legal Challenge, 1989, 129-151 (155).

way. Any positive response to such an instrument cannot simply be interpreted to negate this clear statement of non-bindingness. As long as the voluntary character of the respective instrument remains the undisputed common denominator, the formation of a customary law rule cannot be logically inferred. However, this remark of caution does not deny the possibility that even these nonbinding instruments may shape practice which eventually is accompanied by an emerging *opinio juris* of states. *Opinio juris* must not necessarily be present at the time of adoption of a certain instrument or a policy, but may emerge with continuous repetition and practice in accordance with a certain norm over time.[23] But the simple reference to these instruments that stress the voluntary nature of their norms does not suffice.

Further, caution is also warranted with respect to the existence of the element of state practice. *Opinio juris* alone does not suffice. As stated by the ICJ in the Nicaragua case in the context of UNGA resolutions and customary law rules, "[the] mere fact that States declare their recognition of certain rules is not sufficient... to consider these as being customary international law, and as applicable as such to those States. ... The Court must satisfy itself that the existence of the rule in the *opinio juris* of States is confirmed by state practice."[24] This is not the place to revisit the argument that opinio juris may be the sole decisive element for the formation of customary international law.[25] The element of state practice secures that the established concept of customary law does not become a mere tool for wishful policy arguments without any basis in the real world, and thus safeguards the normative force of the concept of customary law.

---

[23] Compare already A. Verdross, "Entstehungsweisen und Geltungsgrund des universellen völkerrechtlichen Gewohnheitsrechts", Zeitschrift für ausländisches öffentliches Recht und Völkerrecht 29 (1969), 635-653 (639 et seq.); for the possibility that codes of conduct may culminate in hard rules in spite of disclaimers regarding their legal nature B.-O. Bryde, Internationale Verhaltensregeln für Private – Völkerrechtliche und verfassungsrechtliche Aspekte, 1981, 25.

[24] ICJ, Military and Paramilitary Activities in und against Nicaragua (Nicaragua v. United States of America), Judgment of 27 June 1986, ICJ Reports 1986, p. 14, para. 184.

[25] See for such as view e.g. B. Cheng, "Custom: The Future of General State Practice in a Divided World" in: R.J. Macdonald (ed.), The Structure and Process of International Law: Essays in Legal Philosophy Doctrine and Theory, 1983, 513-554 (533).

A different question is what should count as state practice. In times of increasing cooperation through international organisations and multilateral fora, one means for identifying state practice may be to look not only at the sum of individual acts of states, but also at collective statements of states in these fora.[26] The collective acts and actions of states are *inter alia* evidenced in acts of bodies of international organisations such as resolutions and declarations.[27] However, while the responses of states to international norms can be considered as relevant indicators, simple voting for an international instrument and its adoption should not be considered sufficient to actually ascertain a customary law rule. Confirming the existence of state practice requires further evidence. As clarified by Rosalyn Higgins, who was one of the first to point out the role of nonbinding resolutions for customary law development, "resolutions cannot be a substitute for ascertaining custom".[28] Further practice of states must be examined in order to ensure that what states say in international fora coincides with what they actually do.[29] In other words, voting and debating cannot be equated with actual change in behaviour, and thus they alone cannot fulfil the requirement of state practice.[30] In particular if one accepts that acts of states in international organisations such as voting on nonbinding instruments may possibly reflect an emerging *opinio juris*, care must be taken to uphold the requirement of state practice in order not to erode the concept of custom-

---

[26] The significance of collective acts was first pointed out by R. Higgins, The development of international law through the political organs of the United Nations, 1963, 2 and 117; Dissenting Opinion of Judge Tanaka to ICJ, South West Africa (Libera v. South Africa), Judgment of 18 July 1966, ICJ Reports 1966, p. 248, at p. 290; J.E. Alvarez, International Organizations as Lawmakers, 2005, 146-148.

[27] R. Higgins, The development of international law through the political organs of the United Nations, 1963, 2.

[28] R. Higgins, "The Role of Resolutions of International Organizations in the Process of Creating Norms in the International System" in: W.E. Butler (ed.), International Law and the International System, 1987, 21-30 (27).

[29] P.-M. Dupuy, "Formation of customary international law and general principles" in: J. Brunnée/D. Bodansky/E. Hey (eds.), Oxford Handbook of International Environmental Law, 2007, 449-466 (459); A. Verdross/B. Simma, Universelles Völkerrecht: Theorie und Praxis, 1984, 637.

[30] A.A. D'Amato, The concept of custom in international law, 1971, 88; I. MacGibbon, "Means for the Identification of International Law: General Assembly Resolutions: Custom, Practice and Mistaken Identity" in: B. Cheng (ed.), International law: teaching and practice, 1982, 10-26 (20).

ary law and thus authority of international law.[31] Otherwise one would also "count the articulation of a rule twice",[32] namely both as opinio juris and state practice.

With these reservations in mind, it can be concluded that nonbinding instruments may play an important role in customary law formation, in particular in times of increased activities of international institutions. As aptly described by Jonathan Charney, multilateral fora of international institutions adopting formally nonbinding instruments contributes to a more formalised process of customary law formation.[33] Representatives of states and other actors come together, debate problems of common concern and express their shared understandings with the adoption of proposals, resolutions or codes of conduct, frequently in nonbinding form. The products of these discussions are communicated to all states and organisations which may positively respond to the proposals and recommendations either by implementing a particular norm or by other forms of support. A rule or principle which receives a positive response can eventually enter the realm of international law irrespective of the technical legal status of the original instrument.[34] This does not mean that these institutions thereby take on legislative functions. State practice and *opinio juris* remain necessary elements which must be clearly discernable. But through the adoption of these instruments, international institutions formalise and make more transparent the processes that may eventually result in the emergence of a new customary law rule.[35] The actual codification of norms in this way facili-

---

[31] B. Simma/P. Alston, "The Sources of Human Rights Law: Custom, Jus Cogens, and General Principles", Australian Yearbook of International Law 12 (1992), 82-108 (96-100); with respect to GA resolutions in this sense already C. Tomuschat, "Die Charta der wirtschaftlichen Rechte und Pflichten der Staaten: Zur Gestaltungskraft von Deklarationen der UN-Generalversammlung", Zeitschrift für ausländisches öffentliches Recht und Völkerrecht 36 (1976), 444-491 (468).

[32] B. Simma/P. Alston, "The Sources of Human Rights Law: Custom, Jus Cogens, and General Principles", Australian Yearbook of International Law 12 (1992), 82-108 (96).

[33] J.I. Charney, "Universal International Law", American Journal of International Law 87 (1993), 529-551 (543 et seq.)

[34] J.I. Charney, "Universal International Law", American Journal of International Law 87 (1993), 529-551 (545).

[35] J.I. Charney, "Universal International Law", American Journal of International Law 87 (1993), 529-551 (547).

tates the identification of relevant practice by providing a point of comparison that is needed to identify those types of conduct which support a particular rule from those that are irrelevant.[36] Cross-references to norms in (binding and nonbinding) instruments of other organisations[37] further enhance the formation of a common understanding that contributes to this process.[38]

(2) Distinguishing between nonbinding instruments

It is necessary further to distinguish among nonbinding instruments according to their characteristics along the lines of the parameters identified in Part 1 above.[39] As stated by the ICJ in its Advisory Opinion on the *Legality of the Threat or Use of Nuclear Weapons*, the relevant circumstances on which the role of a UNGA resolution depends are "its content and the conditions of its adoption".[40] One must thus for instance distinguish those instruments expressing an opinion on international law from the more common ones that express the will of states to introduce new norms and possibly change the status quo.[41] The potential of an instrument will depend on its authority, which in turn depends on whether it is promulgated by a widely accepted and legitimated forum and whether it is widely supported by states.[42] An instrument adopted by consensus by the United Nations General Assembly, as for example the moratorium on large-scale pelagic drift net fishing in

---

[36] U. Fastenrath, "Relative Normativity in International Law", European Journal of International Law 4 (1993), 305-340 (319); T. Gruchalla-Wesierski, "A Framework for Understanding 'Soft Law'", McGill Law Journal 30 (1984-1985), 37-88 (60-61).

[37] Consider for details the discussion at A.II, further below in this Part.

[38] P.-M. Dupuy, "Soft law and the international law of the environment", Michigan Journal of International Law 12 (1990 – 1991), 420-435 (424-428).

[39] See Part 1, at C, further above.

[40] ICJ, Legality of the Threat or Use of Nuclear Weapons, Advisory Opinion of 8 July 1996, ICJ Reports 1996, p. 226, para. 70.

[41] Similarly T. Treves, "Customary International Law", in: R. Wolfrum (ed.), Max Planck Encyclopedia of Public International Law, 2010, para. 44, online at: www.mpepil.com.

[42] U. Fastenrath, "Relative Normativity in International Law", European Journal of International Law 4 (1993), 305-340 (319).

the high seas[43] or the FAO CCRF of the Conference of the FAO, certainly indicate greater authority than resolutions adopted by majorities, and those nonbinding instruments adopted by lower level bodies as for instance the numerous International Plans of Action and Guidelines adopted by the Committee on Fisheries of the FAO. It also depends on the extent to which this forum communicates that the norm under consideration is considered to be part of international law development.[44] Furthermore, as mentioned above, nonbinding instruments that expressly stress their nonbinding nature should be distinguished from those that do not, and in particular from those purporting to codify international law. And finally, for conforming practice to ensue, institutionalisation of discourse over compliance with a particular instrument can make a difference.[45]

Another characteristic identified in Part 1 that matters in this regard is whether norms is to whom a particular instruments and its norms are addressed. When particular norms of a nonbinding instrument are addressed to non-state actors such as multinational corporations, customary law formation is rendered particularly difficult. While it is theoretically possible for such norms eventually to become customary law that directly binds multinational corporations, such a progressive step in legal doctrine is more likely to be part of future treaty law developments.[46] As long as international law doctrine does not accept international legal obligations for multinational corporations, these norms cannot materialise into customary law obligations even if widely supported and complied with by these actors.[47] And even if one accepted a

---

43 UN Doc. A/RES/46/215 (1991), para. 3.

44 J.I. Charney, "Universal International Law", American Journal of International Law 87 (1993), 529-551 (544).

45 See for this feature of nonbinding instruments and its potential the analysis in this Part 2, at B.I.4.and 5. The importance of follow-up mechanisms for customary law formation was already stressed early in the "soft law" debate by H.W. Baade, "The Legal Effects of Codes of Conduct for Multinational Enterprises", German Yearbook of International Law 22 (1979), 11-52 (21 and 23); B.-O. Bryde, Internationale Verhaltensregeln für Private – Völkerrechtliche und verfassungsrechtliche Aspekte, 1981, 24.

46 J.A. Zerk, Multinationals and Corporate Social Responsibility: Limitations and Opportunities in International Law, 2008, 305.

47 In this sense also C. Chinkin, "Normative Development in the International Legal System" in: D. Shelton (ed.), Commitment and Compliance: the Role of Non-binding Norms in the International Legal System, 2000, 21-42 (30); H.W. Baade, "The Legal Effects of Codes of Conduct for Multinational

limited international legal status of multinational corporations with the effect that they can be directly obliged by international law, such a status would not necessarily extend to the ability of these actors to create customary international law.

Norms of nonbinding instruments addressed to non-state actors could however indirectly influence the content of customary law, namely insofar as they may clarify how states must treat non-state actors.[48] A distinction must therefore be drawn between the emergence of customary law rules obliging private actors and the contribution of nonbinding instruments addressed to private actors to customary law obligations of states. Although for the most part directed at multinational corporations, the OECD Guidelines for instance also entail commitments for governments relating to the treatment of multinational corporations. In particular, they should encourage corporations to comply with the norms in the Guidelines.[49] This could give rise to the customary law rule that states must ensure a certain minimum environmental and social accountability of multinational corporations. The content of these minimum requirements would then be defined by those norms of the OECD Guidelines outlining the responsibilities of multinational corporations. Several of the principles of corporate environmental responsibility such as the precautionary principle, the requirement to conduct an environmental impact assessment and the notification requirements in the case of environmental damage expressed in the OECD Guidelines or in the Draft UN Norms on the Responsibilities of Transnational Corporations and Other Business Enterprises with Regard to Human Rights could thus eventually become part of international cus-

---

Enterprises", German Yearbook of International Law 22 (1979), 11-52 (9); R.E. Lutz/G.D. Aron, "Codes of Conduct and Other International Instruments" in: G. Handl/R.E. Lutz (eds.), Transferring Hazardous Technologies and Substances: The International Legal Challenge, 1989, 129-151 (155); I. Seidl-Hohenveldern, "International Economic 'Soft Law'", Recueil des Cours de l'Académie de Droit International 163 (1979), 169-246 (212).

[48] B.-O. Bryde, Internationale Verhaltensregeln für Private – Völkerrechtliche und verfassungsrechtliche Aspekte, 1981, 24.

[49] Take for example the provision in the OECD Guidelines, Chapter I, para. 2, which stipulates that "governments adhering to the Guidelines encourage the enterprises ... to observe the Guidelines wherever they operate", and the substantive requirement in the OECD Guidelines, Chapter I, para. 8, to treat enterprises equitably.

tomary law obliging states.[50] So far, however, there is still insufficient state practice to confirm the existence of such rules, also because the number of developing states that participate in the process of the OECD Guidelines is still very limited, and nothing indicates that OECD Members feel legally obliged to act.[51]

Although nonbinding instruments may contribute to the development of customary law, this discussion indicates that much depends on the particular characteristics of a specific instrument. Given the above mentioned limitations, the contribution of many of these instruments to customary law development does not appear as their principal function, and hardly captures their specific potential.

b) Role in the development of general principles

Nonbinding international instruments can also contribute to the emergence of 'general principles recognised by civilized nations' referenced as a source of international law in Article 38 (1) (c) of the Statute of the International Court of Justice (ICJ). Their influence in this regard can be perceived in two ways, depending on how one approaches the concept of 'general principles'.

According to the traditional understanding which is based upon the drafting history of the ICJ Statute, 'general principles' are those recognised in the domestic legal orders of states.[52] If general principles are understood in this sense, the impact of international nonbinding instruments on such principles will consequently depend on the extent to which principles adopted in such instruments are implemented in domestic legal systems. This rather indirect influence depends on the availability of compliance mechanisms that enhance the implementation and adoption of the relevant norms in the various domestic legal systems.

---

[50] J.A. Zerk, Multinationals and Corporate Social Responsibility: Limitations and Opportunities in International Law, 2008, 276 et seq.

[51] J.A. Zerk, Multinationals and Corporate Social Responsibility: Limitations and Opportunities in International Law, 2008, 300 and 304.

[52] G.J.H. van Hoof, Rethinking the sources of international law, 1983, 136-139; see also the references in B. Vitanyi, "Les positions doctrinales concernant le sens de la notion de 'principes généraux de droit reconnus par les nations civilisée'", Revue générale de droit international public 86 (1982), 48 (96-102).

But it can also be argued that it is possible to identify 'general principles' directly at the international level.[53] The emergence of representative international fora articulating international principles, as for example through UNGA resolutions, and the broad acceptance of such principles by states also constitutes an appropriate mechanism for the objective validation of general principles. Given the possibility to objectively identify the validation of a rule through the consensus of states at the international level, there is no further need to restrict the concept of 'general principles' to those accepted at the domestic level.[54] International nonbinding instruments such as resolutions adopted by UN organs can – alongside judicial decisions and treaty law – play a significant role in the making and identification of general principles.[55]

However, as for customary law, one must remain careful not to give up any objective element in ascertaining general principles. A complete lack of practice and thus of objective element would rid international law of one its fundamental pillars.[56] Thus, the doctrinal proposal of some authors to rely exclusively on *opinio juris* for the ascertainment of general principles – as opposed to concrete rules – goes too far.[57] As international law in the form of principles that shape rules could thus evolve in the absence of any state practice, this conception would erode

53 Compare for this conception C.M. Bassiouni, "A Functional Approach to 'General Principles of International Law'", Michigan journal of international law 11 (1990), 768-818 (772); B. Simma/P. Alston, "The Sources of Human Rights Law: Custom, Jus Cogens, and General Principles", Australian Yearbook of International Law 12 (1992), 82-108 (102); W. Weiß, "Allgemeine Rechtsgrundsätze des Völkerrechts", Archiv des Völkerrechts 39 (2001), 394-431 (409); S. Kadelbach/T. Kleinlein, "Überstaatliches Verfassungsrecht. Zur Konstitutionalisierung im Völkerrecht", Archiv des Völkerrechts 44 (2006), 235-266 (255); N. Petersen, Demokratie als teleologisches Prinzip: Zur Legitimität von Staatsgewalt im Völkerrecht, 2009, 67 et seq.

54 *Ibid.*

55 Similarly W. Weiß, "Allgemeine Rechtsgrundsätze des Völkerrechts", Archiv des Völkerrechts 39 (2001), 394-431 (410); B. Simma/P. Alston, "The Sources of Human Rights Law: Custom, Jus Cogens, and General Principles", Australian Yearbook of International Law 12 (1992), 82-108 (102).

56 G. Dahm/J. Delbrück/R. Wolfrum, Völkerrecht, Band I/1, Die Grundlagen. Die Völkerrechtssubjekte, 1989, 66.

57 This is proposed by S. Kadelbach/T. Kleinlein, "Überstaatliches Verfassungsrecht. Zur Konstitutionalisierung im Völkerrecht", Archiv des Völkerrechts 44 (2006), 235-266 (261); N. Petersen, Demokratie als teleologisches Prinzip: Zur Legitimität von Staatsgewalt im Völkerrecht, 2009, 72.

the objective element of practice that has been a fundamental doctrinal basis of international law which prevents that international law becomes too easily subject to influences of ideology. Even if state practice is not necessary to the same extent as in customary law processes, some form of objective recognition in legal practice is necessary to fulfil the requirements expressed in Article 38 (1) (c) ICJ Statute. General principles could for instance not be ascertained in the face of widespread opposing practice by states.[58] Provided that some form of objective acceptance by states is ascertainable, nonbinding instruments can play a role in identifying general principles at the international level.

c) Role in the development of treaty law

The adoption of nonbinding instruments is often a first step in the development of treaty law.[59] A nonbinding instrument such as a resolution of an international organisation or treaty body frequently stands at the beginning of this process. Although nonbinding, resolutions or declarations shape and restrain future negotiations on the issue by defining politically the principles, objectives and negotiations.[60]

Nonbinding instruments are often adopted as a first step in treaty law development for a number of reasons. First of all, nonbinding instruments often allow states to go beyond the status quo and adopt com-

---

[58] G. Dahm/J. Delbrück/R. Wolfrum, Völkerrecht, Band I/1, Die Grundlagen. Die Völkerrechtssubjekte, 1989, 66.

[59] A.E. Boyle, "Some Reflections on the Relationship of Treaties and Soft Law", International and Comparative Law Quarterly 48 (1999), 901-913 (904-905); P. Muchlinski, "Human Rights, social responsibility and the regulation of international business: The development of international standards by intergovernmental organisations", Non-State Actors and International Law 3 (2003), 123-152 (128); A. Peters/I. Pagotto, "Soft Law as a New Mode of Governance: A Legal Perspective", New Modes of Governance Project 04/D11 (2006), (23); B. Simma, "Remarks" in: G.F. Handl, "A Hard Look at Soft Law", American Society of International Law Proceedings 82 (1988), 371-393 (379) (pointing out that in the process of development of hard human rights law, a soft law stage is rather the rule than the exception).

[60] R. Wolfrum, "Vorbereitende Willensbildung und Entscheidungsprozeß beim Abschluß multilateraler völkerrechtlicher Verträge" in: J. Ipsen/E. Schmidt-Jortzig (eds.), Recht-Staat-Gemeinwohl: Festschrift für Dietrich Rauschning, 2001, 407-418 (412).

paratively progressive norms.[61] It is often easier to reach political consensus when the norms are nonbinding.[62] Negotiation periods are shorter, and governments need not be concerned with domestic legal and political obstacles. Without ratification requirements, the instrument can be adopted and applied immediately. Governments can also postpone questions regarding their capacity to comply, and need not fear legal consequences in the meantime.

Second and perhaps most important is their usefulness to initiate a broad discourse among all relevant actors. Nonbinding instruments more easily allow the integration of non-state actors and reluctant states into a continuous discourse on the creation and implementation of substantive norms at a point in time when the issue is still contested.[63] Although not legally binding, these instruments generally frame the expectations of the actors in the international legal discourse by legitimising certain behaviour and delegitimising alternative norms and points of view. More specifically, they set the agenda for issues dealt with at the international level and pre-define possible approaches and principles and future legal rules.[64] By generating and defining political and legal discourse, nonbinding instruments thus serve as catalysts of lawmaking processes which pave the way for the adoption of international treaties.

Finally, as will be seen in further detail below, nonbinding instruments also influence and shape behaviour in the desired way. When states begin acting in accordance with a particular instrument or once they realise that a proposed system is feasible, moving towards hard law usually requires only little additional political efforts. The networks of government officials that frequently emerge around nonbinding instru-

---

61 E. Louka, International Environmental Law: Fairness, Effectiveness and World Order, 2006, 60.

62 A.C. Kiss/D. Shelton, International environmental law, 2004, 89; J.B. Skjaerseth/O.S. Stokke/J. Wettestad, "Soft Law, Hard Law, and Effective Implementation of International Environmental Norms", Global Environmental Politics 6 (2006), 104-120 (115).

63 A.C. Kiss/D. Shelton, International environmental law, 2004, 89.

64 R. Wolfrum, "Vorbereitende Willensbildung und Entscheidungsprozeß beim Abschluß multilateraler völkerrechtlicher Verträge" in: J. Ipsen/E. Schmidt-Jortzig (eds.), Recht-Staat-Gemeinwohl: Festschrift für Dietrich Rauschning, 2001, 407-418 (412); R.E. Lutz/G.D. Aron, "Codes of Conduct and Other International Instruments" in: G. Handl/R.E. Lutz (eds.), Transferring Hazardous Technologies and Substances: The International Legal Challenge, 1989, 129-151 (157).

ments and the institutionalised discourse that is associated with these networks facilitate such a process: cooperation under the nonbinding instrument will already have created common understandings and policies on the respective issue.[65]

In similarity to processes of customary law formation, these considerations suggest that the precursory function is enhanced through the institutionalisation of discourse within multilateral fora and committees as well as follow-up procedures that help to shape behaviour and initiate learning processes by the participating actors.

(1) Examples from practice

Practice overwhelmingly confirms the relevance of nonbinding norms for treaty development.[66] Prominent examples include the United Nations Declaration on Outer Space[67] which was the forerunner to the Treaty on Outer Space;[68] and the United Nations Resolution on the Seabed as the Common Heritage of Mankind,[69] a code of conduct which preceded in particular Part XI of the United Nations Conven-

65 K. Raustiala, "The Architecture of International Cooperation: Transgovernmental Networks and the Future of International Law", Virginia Journal of International Law 43 (2002-2003), 1-92 (85-86).

66 For this assessment, compare e.g. E. Brown Weiss, "Introduction" in: E. Brown Weiss (ed.), International Compliance with Nonbinding Accords, 1997, 1-20 (5).

67 Declaration of Legal Principles Governing Activities of States in Exploration and Use of Outer Space (13 December 1963), UN Doc. A/RES/5515 (1963).

68 Treaty on Principles Governing Activities of States in Exploration and Use of Outer Space, Including the Moon and other Celestial Bodies (27 January 1967), UNGA Res. 2222(XXI), 610 UNTS 205.

69 Declaration of Principles Governing Sea-Bed and Ocean Floor, and Subsoil Thereof, Beyond Limits of National Jurisdiction, UN G.A. Res. 2749 (XXV) of 17 December 1970, U.N. Doc. A/8028 (1971), 10 ILM 220 (1971), at 24.

tion on the Law of the Sea[70] by defining that the seabed, the sub-soil and their resources belong to the common heritage of mankind.[71]

International environmental law appears as an area of law where this role of nonbinding instruments is particularly significant. The impact of nonbinding declarations such as the Stockholm Declaration on the Human Environment of 1972 on the principles and processes of international environmental law as well as treaty law can hardly be overestimated.[72] International organisations such as the Food and Agriculture Organization (FAO) or the International Maritime Organization (IMO) successfully adopt international nonbinding instruments and then strive to develop a binding treaty on that basis. Thus, the International Treaty on Plant Genetic Resources for Food and Agriculture, which entered into force in 2004 and has 120 state parties as of April 2009,[73] emerged from the FAO International Undertaking on Plant Genetic Resources adopted by the FAO in 1983.[74] A similar process occurred in the development of the Basel Convention on the Control of Transboundary Movements of Hazardous Wastes and Their Disposal (Basel Convention)[75] which is based on the UNEP Cairo Guidelines and Principles for the Environmentally Sound Management of Hazardous Wastes.[76] Also, the concept of environmental impact assessment in a

---

70 United Nations Convention on the Law of the Sea (10 December 1982), 1833 UNTS 3; 21 ILM 1261 (1982), 21 ILM 1261 (1982).

71 On these developments R. Wolfrum, Die Internationalisierung staatsfreier Räume: die Entwicklung einer internationalen Verwaltung für Antarktis, Weltraum, Hohe See und Meeresboden, 1984, 331 et seq.

72 Instead of many J. Brunnée, "The Stockholm Declaration and the Structure and Processes of International Environmental Law " in: A. Chircop (ed.), The future of ocean regime building: essays in tribute to Douglas M. Johnston, 2009, 41-62; specifically for the impact on marine law see the contributions in M.H. Nordquist (ed.), The Stockholm declaration and law of the marine environment, 2003.

73 International Treaty on Plant Genetic Resources for Food and Agriculture. The treaty is deposited at the FAO and available at http://www.fao.org/legal/treaties/033t-e.htm.

74 International Undertaking on Plant Genetic Resources, adopted by FAO Conference Res. 8/83 (1983).

75 Basel Convention on the Control of Transboundary Movements of Hazardous Wastes and Their Disposal, 1673 UNTS 126, 28 ILM 657 (1989).

76 Cairo Guidelines and Principles for the Environmental Sound Management of Hazardous Waste, UN Doc. UNEP WG 122/3 (1985).

transboundary context which was developed from the UNEP Guidelines on Environmental Impact Assessment and other nonbinding instruments in 1987[77] became the basis for the development of the Convention on Environmental Impact Assessment in a Transboundary Context (Espoo Convention) which was adopted in 1991 and entered into force in 1997.[78]

Nonbinding instruments of corporate responsibility such as the OECD Guidelines for Multinational Enterprises which reflect an emerging consensus on principles and standards for multinational corporations are increasingly seen as useful models for future treaty law norms on this issue.[79] The FAO Code of Conduct for Responsible Fisheries (CCRF), analysed in Part 1,[80] also appears to influence treaty making. For example, the riparian states of Lake Tanganyika agreed to develop sustainable fisheries management policies based on the objectives of the CCRF in the Convention on the Sustainable Management of Lake Tanganyika.[81]

A similar development can be observed with respect to the FAO International Plan of Action against Illegal, Unreported and Unregulated (IUU) Fishing. Enforcement through control measures by port states is one of the central means of combating illegal and unreported fishing. Already in 2005, the FAO Committee on Fisheries had acknowledged "that there was a need to strengthen port State measures ... given that the lack of agreed, binding measures provided a loophole". Accord-

---

77 UNEP Goals and Principles of Environmental Impact , UNEP/GC14/25 (1987).

78 UNECE Convention on Environmental Impact Assessment in a Transboundary Context, 30 ILM 800 (1991).

79 United Nations General Assembly, Human Rights Council, Report of the Special Representative of the Secretary-General (SRSG) on the issue of human rights and transnational corporations and other business enterprises, "Business and Human Rights: Mapping International Standards of Responsibility and Accountability for Corporate Acts", U.N. Doc. A/HRC/4/035 of 9 February 2007, para. 49; J.A. Zerk, Multinationals and Corporate Social Responsibility: Limitations and Opportunities in International Law, 2008, 298 and 305; more hesitant K. Weilert, "Transnationale Unternehmen im rechtsfreien Raum? Geltung und Reichweite völkerrechtlicher Standards", Zeitschrift für ausländisches öffentliches Recht und Völkerrecht 69 (2009), 883-917 (908).

80 Compare the case study in Part 1, at B.I, further above.

81 Article 7 para.2 b) of Convention on the Sustainable Management of Lake Tanganyika of 12 June 2003, available at http://faolex.fao.org/faolex/index.htm.

ingly, in order to establish binding minimum standards for port state measures, the 131 states represented at the Committee on Fisheries of the FAO in 2007 started the process of developing a new legally binding international agreement to be developed on the basis of the International Plan of Action on Illegal, Unreported and Unregulated Fishing which was developed under the framework of the FAO Code of Conduct for Responsible Fisheries as well as the FAO Model Scheme on Port State Measures to Combat IUU Fishing.[82] In 2009, the FAO Conference adopted the Agreement on Port State Measures to Prevent, Deter and Eliminate Illegal, Unreported and Unregulated Fishing which *inter alia* establishes detailed rules on the possibility of inspections of vessels, the denial of port entry or denial of landing facilities in cases of IUU fishing.[83] The shift to a binding agreement indicates that states seem to find some additional value in the development of a legally binding treaty when they need to avoid loopholes. In other words, binding instruments seem to be considered necessary in order that state actors can be reassured that all port states will indeed apply the measures. This can be taken as a sign for one central limitation of nonbinding instruments, namely that they are not able to provide for legal certainty and reliability.

One of the most striking examples of the relevance of a nonbinding precursor for a treaty is the development of the PIC Convention.[84] The next section will take a closer look at this development in order to provide a better picture of the significance of nonbinding instruments for international law-making.

(2) Case study: from the voluntary PIC procedure to the PIC Convention

(i) The shift to binding treaty law

Having established the voluntary PIC system through nonbinding instruments in 1989, the FAO Council and the UNEP Governing Coun-

---

[82] FAO, Report of the twenty-seventh session of the Committee on Fisheries, Rome, 5–9 March 2007, FAO Fisheries Report No. 830, FIEL/R830, para. 68.

[83] Agreement on Port State Measures to Prevent, Deter and Eliminate Illegal, Unreported and Unregulated Fishing, approved by FAO Members on 22 November 2009 at the Thirty-sixth Session of the FAO Conference.

[84] See above in this Part, at A.I.1c).

cil in two separate but coordinated decisions mandated their secretariats in 1994/1995 jointly to develop a draft of a binding convention.[85] After two years of negotiations, the Convention on the Prior Informed Consent Procedure for Certain Hazardous Chemicals and Pesticides in International Trade (PIC Convention)[86] was signed in 1998. It entered into force in 2004 and has 131 Parties as of April 2010.[87]

Ever since the voluntary procedure of Prior Informed Consent had been established, the need for binding norms was debated. Developing states generally favoured such a transformation into binding law, and large exporters of pesticides such as the United States, Germany, the United Kingdom and Japan opposed it.[88] Believing that a binding convention would be more effective than the voluntary system, a coalition of some European Union and developing countries eventually succeeded when UNEP's Governing Council decided to establish a working group to explore the development of a legally binding instrument in 1991.[89] The issue received a further major political push through the Rio Conference in 1992, where it was included as an objective in chapter 19 of Agenda 21.[90]

---

85 FAO Council Decision CL 107/11 (November 1994); UNEP Governing Council Decision 18/12, (May 1995).

86 Rotterdam Convention on the Prior Informed Consent Procedure for Certain Hazardous Chemicals and Pesticides in International Trade (10 September 1998), 38 ILM 1 (1999), also available at http://www.pic.int/en/ConventionText/ONU-GB.pdf.

87 General on the PIC Convention see C. Redgwell, "Regulating Trade in Dangerous Substances: Prior Informed Consent under the 1998 Rotterdam Convention" in: A. Kiss/D. Shelton/K. Ishibashi (eds.), Economic Globalization and Compliance with International Environmental Agreements, 2003, 75-88; N.S. Zahedi, "Implementing the Rotterdam Convention: the challenges of transforming aspirational goals into effective controls on hazardous pesticide exports to developing countries", Georgetown international environmental law review 11 (1999), 707-739; R.W. Emory, "Probing the protections in the Rotterdam Convention on Prior Informed Consent", Colorado journal of international environmental law and policy (2001), 47-69.

88 R.L. Paarlberg, "Managing Pesticide Use in Developing Countries" in: P. M. Haas/R.O. Keohane/M.A. Levy (eds.), Institutions for the Earth: Sources of Effective International Environmental Protection, 1993, 309-350 (344 et seq.).

89 UNEP Governing Council Decision 16/35 (1991).

90 Agenda 21, as one of the central instruments adopted by the UNCED Conference at Rio de Janeiro, in chapter 19 para. 19.38 outlines the objective of "achieving full participation in and implementation of the PIC procedure, in-

This negotiation history illustrates the attitude of states towards binding norms. States which were supportive of strong international regulation advocated legally binding norms, and those with an interest in a weak regime opposed such a move. Other negotiations on the question of the legal nature of a particular instrument point in a similar direction. For instance, in the negotiations on an instrument of Access and Benefit Sharing (ABS) under the Convention for Biological Diversity, the issue of legal bindingness was a major negotiating issue. Developing states with an interest in a strong ABS regime strongly advocated a legally binding Protocol on Access and Benefit Sharing under the Convention on Biological Diversity.[91] A binding Protocol was eventually adopted in 2010: the Nagoya Protocol on Access to Genetic Resources and the Fair and Equitable Sharing of Benefits Arising from their Utilization to the Convention on Biological Diversity.[92] State actors and other actors thus clearly distinguish between binding and nonbinding rules even in cases where international enforcement of binding rules would be unlikely. And, even though states may – empirically speaking – actually comply as much with nonbinding norms as with binding ones,[93] actors apparently have the perception that binding instruments give rise to more effective regimes.

The development of the PIC Convention is also a clear example of how nonbinding instruments can be a necessary first step in the development

---

cluding possible mandatory applications through legally binding instruments contained in the Amended London Guidelines and in the FAO International Code of Conduct, taking into account the experience gained within the PIC procedure."

91 At the 9th meeting of the ABS Working Group in 2010 in Cali, Columbia, "developing countries from the Latin America and the Caribbean Group (GRULAC), Asia-Pacific Group, African Group and the Like-Minded Megadiverse Countries (LMMC) noted that the nature of the draft protocol is not up for negotiation," compare Environmental Negotiation Bulletin, Vol. 9 No. 503 (2010), p. 3, available at http://www.iisd.ca/download/pdf/enb09503e.pdf.

92 Nagoya Protocol on Access to Genetic Resources and the Fair and Equitable Sharing of Benefits Arising from their Utilization to the Convention on Biological Diversity, adopted through Decision X/1 at the 10th Conference of the Parties to the Convention on Biological Diversity. The text of the Nagoya Protocol is available at http://www.cbd.int/abs/doc/protocol/nagoya-protocol-en.pdf.

93 E. Brown Weiss, "Introduction" in: E. Brown Weiss (ed.), International Compliance with Nonbinding Accords, 1997, 1-20 (1).

of an international treaty. It is highly unlikely that the Convention would have been adopted without the prior existence of the voluntary PIC system.[94] The existence and functioning of the procedure may have convinced laggard states that a regime based on binding norms was conceivable and feasible. The voluntary norms also reduced the perception of complexity of the new legal regime for states, since state actors had already experience with the system.[95] The voluntary norms thus contributed to efficient negotiations (which only lasted two years) and possibly to the wide support the Convention enjoys today.

Given that with over 150 states participating,[96] more states embraced the voluntary PIC system than have ratified the PIC Convention today, one may ask if there was any added value of developing a binding Convention.[97] Irrespective of substantive changes, the shift by itself may be an achievement in terms of stability and legal certainty. The shift to binding norms can be understood as an expression of the need of states eventually to base a complex regime on clearly binding forms of law.[98] While states may not comply with a voluntary system without risking

---

94 M.A. Mekouar, "Pesticides and Chemicals: The Requirement of Prior Informed Consent" in: D. Shelton (ed.), Commitment and Compliance: The Role of Non-Binding Norms in the International Legal System, 2000, 146-163 (163).

95 K. Kummer, "Prior Informed Consent for Chemicals in International Trade: The 1998 Rotterdam Convention", Review of European Community and International Environmental Law 8 (1999), 323-330 (329).

96 N.S. Zahedi, "Implementing the Rotterdam Convention: the challenges of transforming aspirational goals into effective controls on hazardous pesticide exports to developing countries", Georgetown international environmental law review 11 (1999), 707-739 (709).

97 Scholars often stress the effectiveness of the voluntary system, compare M.A. Mekouar, "Pesticides and Chemicals: The Requirement of Prior Informed Consent" in: D. Shelton (ed.), Commitment and Compliance: The Role of Non-Binding Norms in the International Legal System, 2000, 146-163 (163); D.G. Victor, "'Learning by Doing' in the Nonbinding International Regime to Manage Trade in Hazardous Chemicals and Pesticides" in: D.G. Victor/K. Raustiala/E.B. Skolnikoff (eds.), The Implementation and Effectiveness of International Environmental Commitments: Theory and Practice, 1998, 221-281 (250); more critical of both the voluntary approach but also of the PIC Convention N.S. Zahedi, "Implementing the Rotterdam Convention: the challenges of transforming aspirational goals into effective controls on hazardous pesticide exports to developing countries", Georgetown international environmental law review 11 (1999), 707-739.

98 J.E. Alvarez, International Organizations as Law-makers, 2005, 232.

great reputational damage, this may not be the case with binding norms, because such a breach would contradict the legitimate expectations of the other actors in a much greater way. Only binding forms of law may therefore be able to provide the degree of trust which is the source of legal certainty and stability in international law and which ultimately secures long-term cooperation.

(ii) Comparing the voluntary and binding PIC systems

A comparison of the PIC Convention and voluntary PIC system shows the great extent to which the voluntary PIC system influenced the legally binding PIC regime.

The PIC procedure in the PIC Convention is almost entirely modelled on the voluntary PIC system as it was foreseen in the voluntary instruments and as it had been further developed in the first years in practice.[99] Its development shows that nonbinding instruments can contribute much more to treaty law development than merely shaping general discourse. Rather, the voluntary PIC system provided a complex and detailed model which only had to be copied into treaty form.[100]

One of the main contributions and advantages of the voluntary system was that it enabled a learning process. As has been outlined above,[101] the voluntary PIC system was subject to numerous changes and adjustments that could – mainly due to the nonbinding nature of the instruments – easily be implemented by the expert group on PIC on the basis

---

[99] J. Ross, "Legally Binding Prior Informed Consent", Colorado Journal of International Environmental Law and Policy 10 (1999), 499-529 (519); N.S. Zahedi, "Implementing the Rotterdam Convention: the challenges of transforming aspirational goals into effective controls on hazardous pesticide exports to developing countries", Georgetown international environmental law review 11 (1999), 707-739 (714). One of the few changes was the expansion of the definitional scope with respect to hazardous pesticides formulations. Whereas the FAO Pesticide Code and the Guidelines only covered "acutely hazardous pesticide formulations", Article 2 (a) and (c) PIC Convention now include the somewhat broader category of "severely hazardous pesticide formulations". Compare K. Kummer, "Prior Informed Consent for Chemicals in International Trade: The 1998 Rotterdam Convention", Review of European Community and International Environmental Law 8 (1999), 323-330 (325).

[100] Step 4 of the FAO Guidelines on the operation of prior informed consent (PIC), Report of the 25th session of the FAO Conference, Appendix E, available at http://www.fao.org/docrep/x5588e/x5588e0l.htm.

[101] Compare the case study in Part 1, at B.II., further above.

of practical experiences and feedback from states and organisations. The voluntary PIC system had therefore already undergone a learning period before discussions on the PIC Convention started. The Convention then codified the practice as it had emerged under the voluntary system.[102] For example, as outlined above,[103] the FAO/UNEP Joint Expert Group had enlarged the definitional scope of the PIC system to also consider substances which were not subject to any regulatory action because they were withdrawn from the domestic market. It had appeared under the voluntary system that manufacturers in particular in the US voluntarily took substances from the market as a result of negotiations between government agencies and manufacturers. The U.S. Environment Protection Agency, for instance, initiated voluntary cancellations or voluntary restrictions through negotiations with companies rather than formally regulating these substances.[104] Without any formal regulatory action by the EPA, the substances were not "banned" or "severely restricted" and consequently could not enter the PIC system. Companies could thus still export these substances without informing the importing state. The practice of the UNEP/FAO Joint Group nevertheless to include such substances is now codified in the PIC Convention. Its wide definition of "final regulatory action" ensures that the terms "banned" and "severely restricted" encompass the voluntary withdrawal of a substance from the market or from the domestic authorization process.

The adoption of a treaty – even if it includes flexible amendment procedures – usually renders it more difficult to pursue this kind of flexible learning processes. Under the PIC Convention, for instance, any decision of the expert committee on the listing of a chemical in Annex III requires a consensus decision of the Conference of the Parties.[105] The

---

[102] For a similar assessment K. Kummer, "Prior Informed Consent for Chemicals in International Trade: The 1998 Rotterdam Convention", Review of European Community and International Environmental Law 8 (1999), 323-330 (329); M.A. Mekouar, "Pesticides and Chemicals: The Requirement of Prior Informed Consent" in: D. Shelton (ed.), Commitment and Compliance: The Role of Non-Binding Norms in the International Legal System, 2000, 146-163 (163).

[103] See the case study in Part 1, at B.II., further above.

[104] N.S. Zahedi, "Implementing the Rotterdam Convention: the challenges of transforming aspirational goals into effective controls on hazardous pesticide exports to developing countries", Georgetown international environmental law review 11 (1999), 707-739 (717 et seq.).

[105] PIC Convention, Article 22 (5) b).

kind of flexible adjustments that were undertaken by the experts under the voluntary PIC system are thus no longer possible.

The need for these learning processes increases with the complexity of the issues and uncertainty regarding the underlying processes. As environmental problems are typically defined by particular complexity and uncertainty, nonbinding instruments have much to contribute as learning frameworks. Institutionalisation of follow-up procedures and the establishment of expert bodies in the case of the voluntary PIC system helped to ensure that the instrument could play this role.

Successful learning processes that are stimulated by a nonbinding instrument require that actors carefully consider when to shift to a binding form of cooperation. There may be a considerable trade-off between flexibility and learning on the one hand and the value of having binding legal rules on the other. Considering that the PIC Convention basically copied the nonbinding system but did not substantively move the issue forward, it could have been more effective to postpone this codification. Indeed, some commentators considered the development of the PIC Convention premature because not enough had been learned with respect to the implementation of the voluntary PIC procedure.[106] The voluntary PIC system had just started to function when important actors turned to new negotiations instead of learning from the implementation of the existing rules. Codification in a treaty also led to the loss of information and experience. According to Victor's study of lists of participants, the shift to negotiations for a binding convention led to a shift of responsibility within governments away from operational ministries on agriculture to foreign ministries. This meant that many of those government officials who had participated in the implementation of the voluntary system did not participate in the negotiations of the PIC Convention.[107] In addition, the shift to negotiations for binding rules added considerable costs.[108]

---

[106] J. Ross, "Legally Binding Prior Informed Consent", Colorado Journal of International Environmental Law and Policy 10 (1999), 499-529 (524); D.G. Victor, "'Learning by Doing' in the Nonbinding International Regime to Manage Trade in Hazardous Chemicals and Pesticides" in: D.G. Victor/K. Raustiala/E.B. Skolnikoff (eds.), The Implementation and Effectiveness of International Environmental Commitments: Theory and Practice, 1998, 221-281 (264).

[107] D.G. Victor, "'Learning by Doing' in the Nonbinding International Regime to Manage Trade in Hazardous Chemicals and Pesticides" in: D.G. Victor/K. Raustiala/E.B. Skolnikoff (eds.), The Implementation and Effectiveness

Apart from making the voluntary PIC system binding, the PIC Convention does not make any substantive changes the previous PIC system. Unlike the Basel Convention, the new treaty for example does not prohibit certain particularly dangerous exports – they are merely subjected to a PIC procedure.[109] Although also modelled upon a voluntary PIC system, the shift to a legally binding regime in the case of the Basel Convention also led to a substantive widening of its scope. The difference between the two conventions can at least in part be explained by the high-profile character of the negotiations of the Basel Convention which were accompanied by scandals of illicit waste disposal practices.[110] This shows that the relative few changes undertaken in the establishment of the PIC Convention should not be generalised in the sense that a shift to binding rules may not also stimulate further substantive development.[111] Nonbinding instruments may help to forge consensus on a specific issue while leaving more meaningful cooperation to a legally binding regime once actors accept the necessity or once public pressure rises. In the case of the PIC Convention, proposals for an expansion failed. For example, proposals of a group of states led by the European Union favouring a broader framework convention for

---

of International Environmental Commitments: Theory and Practice, 1998, 221-281 (260).

[108] D.G. Victor, "'Learning by Doing' in the Nonbinding International Regime to Manage Trade in Hazardous Chemicals and Pesticides" in: D.G. Victor/K. Raustiala/E.B. Skolnikoff (eds.), The Implementation and Effectiveness of International Environmental Commitments: Theory and Practice, 1998, 221-281 (259).

[109] This is also criticised and suggested as a possible improvement for the PIC Convention by N.S. Zahedi, "Implementing the Rotterdam Convention: the challenges of transforming aspirational goals into effective controls on hazardous pesticide exports to developing countries", Georgetown international environmental law review 11 (1999), 707-739 (737).

[110] C. Redgwell, "Regulating Trade in Dangerous Substances: Prior Informed Consent under the 1998 Rotterdam Convention" in: A. Kiss/D. Shelton/K. Ishibashi (eds.), Economic Globalization and Compliance with International Environmental Agreements, 2003, 75-88 (79).

[111] K. Kummer, "Prior Informed Consent for Chemicals in International Trade: The 1998 Rotterdam Convention", Review of European Community and International Environmental Law 8 (1999), 323-330 (330).

chemicals were not supported by another group of industrialised states led by the United States of America.[112]

There is one noteworthy procedural difference between the voluntary procedure and the binding Convention. It concerns the decision on which substances are subjected to the prior informed consent procedure. While the voluntary system allowed the joint expert group to make the listing decision, under the PIC Convention the decision must be taken by consensus of state parties represented in the Conference of the Parties.[113] The expert body of the PIC Convention – the Chemical Review Committee – only makes proposals for listings or removals from the list and prepares the respective decision guidance documents.[114] In contrast to the voluntary system where the decisions of the UNEP/FAO Joint Expert Group were not subject to approval by a political body,[115] the PIC Convention now in fact gives a veto to any state on such listings.

This politicisation of the science input may be part of the price to be paid for a binding Convention, because states tend to seek greater political control over decisions that will be legally binding for them. Any single state can now block the procedure, which increases the danger of lobbying by a single producer for a veto.[116] The PIC Convention is thus blocked politically in its central mechanism on a number of dangerous substances. For example, the unanimous decision of the experts to recommend the inclusion of chrysotile asbestos in the respective Annex III which was already made in 2005[117] could not be agreed upon in the three following meetings of the Conferences of the Parties in 2005, 2006 and 2008. A number of asbestos producing countries did not question the soundness of the scientific assessment, but nevertheless opposed the

---

[112] K. Kummer, "Prior Informed Consent for Chemicals in International Trade: The 1998 Rotterdam Convention", Review of European Community and International Environmental Law 8 (1999), 323-330 (325).

[113] PIC Convention, Article 22 (5) (b).

[114] PIC Convention, Article 5-9.

[115] See the analysis in Part 2, at B.II., further above.

[116] N.S. Zahedi, "Implementing the Rotterdam Convention: the challenges of transforming aspirational goals into effective controls on hazardous pesticide exports to developing countries", Georgetown international environmental law review 11 (1999), 707-739 (727).

[117] Report of the Chemical Review Committee on the work of its first Meeting, UNEP/FAO/RC/CRC.1/28 (2005), Annex I.B.

listing.[118] Similar problems exist with regard to other substances. Only one chemical of lesser economic importance – tributyl tin compounds – was listed in three meetings of the Conference of the Parties despite numerous recommendations by the Review Committee to list.[119] These difficulties have thrown serious doubts on the effectiveness of the linkage to consensus decision of the Conference of the Parties for so-called "live chemicals", i.e. chemicals of continuing economic importance.[120]

2. *The supplementary function of nonbinding instruments*

Nonbinding instruments often function as supplements to treaty norms. Where treaty law is insufficiently precise, nonbinding instruments play a supplementary role by providing concretisations, guidance for interpretation, or by serving as standards that gain legal significance through references in treaties.

a) Interpretation and concretisation of treaty law through nonbinding instruments

Nonbinding instruments frequently serve as tools to concretise and clarify international legal principles and rules.[121] Regarding treaty law, this may happen in several ways.

First, nonbinding instruments provide definitions or criteria to concretise, clarify or delimit general or vague terms of a rule of international

---

[118] Environmental negotiation bulletin, Summary report of the 4th Conference of the Parties of the PIC Convention in 2008, available at http://www.iisd.ca/download/pdf/enb15168e.pdf.

[119] Endosulfin has been recommended for listing by the Chemicals Review Committee in 2006, but was not included by the Conference of the Parties neither in its meeting in 2006 nor in 2008.

[120] International Institute for Sustainable Development, Environmental negotiation bulletin, Summary report of the 4th Conference of the Parties of the PIC Convention in 2008, available at http://www.iisd.ca/download/pdf/enb15168e.pdf.

[121] J.E. Alvarez, "International Organizations: Then and Now", American Journal of International Law 100 (2006), 324-347 (329); U. Fastenrath, "Relative Normativity in International Law", European Journal of International Law 4 (1993), 305-340 (315).

law or of an international treaty.[122] Nonbinding instruments may contribute to the concretisation of norms in disputed areas of international law, for instance of rules of international law on the protection of foreign investment.[123] Regarding treaty law, it is in this context pointed out by some authors that treaties often rely on linguistic conventions and definitions for the determination of their substantive scope. Common understandings on such linguistic conventions are often developed and defined through nonbinding instruments.[124] The UNEP Cairo Guidelines and Principles for the Environmentally Sound Management of Hazardous Wastes,[125] for instance, could be seen as providing a common understanding of what constitutes "environmentally sound management" according to Article 4 of the Basel Convention on the Control of Transboundary Movements of Hazardous Wastes and Their Disposal.[126]

Second, nonbinding instruments are today indispensable insofar as they are used to provide concrete rules and standards for implementation of a particular obligation. Nonbinding instruments establish guidance, best practices or due diligence requirements for the application and implementation of the norms of a particular treaty. By some referred to as "ecostandards"[127], such instruments are often adopted by Conferences

---

[122] U. Fastenrath, "Relative Normativity in International Law", European Journal of International Law 4 (1993), 305-340 (314); A.E. Boyle, "Some Reflections on the Relationship of Treaties and Soft Law", International and Comparative Law Quarterly 48 (1999), 901-913 (905); A. Peters/I. Pagotto, "Soft Law as a New Mode of Governance: A Legal Perspective", New Modes of Governance Project 04/D11 (2006), (23); H.W. Baade, "The Legal Effects of Codes of Conduct for Multinational Enterprises", German Yearbook of International Law 22 (1979), 11-52 (48).

[123] See B.-O. Bryde, Internationale Verhaltensregeln für Private – Völkerrechtliche und verfassungsrechtliche Aspekte, 1981, 25-31; H.W. Baade, "The Legal Effects of Codes of Conduct for Multinational Enterprises", German Yearbook of International Law 22 (1979), 11-52 (39).

[124] U. Fastenrath, "Relative Normativity in International Law", European Journal of International Law 4 (1993), 305-340 (314-315).

[125] UNEP Cairo Guidelines and Principles for the Environmentally Sound Management of Hazardous Wastes, UNEP/GC.14/17 (1987).

[126] A.E. Boyle, "Some Reflections on the Relationship of Treaties and Soft Law", International and Comparative Law Quarterly 48 (1999), 901-913.

[127] P. Contini/P.H. Sand, "Methods to expedite environment protection: international ecostandards", American Journal of International Law 66 (1972), 37-59.

of Parties to assist and guide parties in the implementation of treaty obligations.[128] For instance, the Basel Convention regularly adopts a number of technical guidelines which assemble best practices on how to deal with particular parts of wastes that fall under the scope of the Convention:[129] the 8th Conference of the Parties to the Basel Convention in 2007 adopted a number of Technical Guidelines on persistent organic pollutants which define best practices for the environmentally sound management of such wastes as required by Article 4 of the Basel Convention.[130] The Specific Guidelines for Assessment of Carbon Dioxide Streams for Disposal into Sub-seabed Geological Formations accepted by the 2nd Meeting of the Parties in 2007[131] to the 1996 London Protocol to the Convention on the Prevention of Marine Pollution by Dumping of Wastes and Other Matter[132] provide the technical details designed to guide parties in their carbon capture and storage activities which are now allowed following the amendment to Annex 1 of the London Protocol in 2006.[133]

Third, norms prescribed by nonbinding instruments which have not yet become customary law rules may nevertheless guide the interpretation

---

128 E.g. the Bonn Guidelines on access to genetic resources and the fair and equitable sharing of the benefits arising from their utilization adopted by the Conference of the Parties to the Convention on Biological Diversity, Decision VI/24 (2002); on this role see A.E. Boyle, "Some Reflections on the Relationship of Treaties and Soft Law", International and Comparative Law Quarterly 48 (1999), 901-913 (905).

129 Compare Basel Convention, Article 1 (a) in conjunction with Annex 1 and 3.

130 Conference of the Parties to the Basel Convention, Decision VIII/16, UNEP/CHW.8/16 (5 January 2007).

131 Twenty-Ninth Consultative Meeting of Contracting Parties to the Convention on the Prevention of Marine Pollution by Dumping of Wastes and Other Matter 1972 & Second Meeting of Contracting Parties to the 1996 Protocol to the Convention on the Prevention of Marine Pollution by Dumping of Wastes and Other Matter 1972, LC 29/17 (14 December 2007), para. 4.3. and Annex 4 (LC 29/4).

132 1996 Protocol to the Convention on the Prevention of Marine Pollution by Dumping of Wastes and Other Matter, 36 ILM 1 (1997).

133 Meeting of the Parties to the London Protocol Resolution LP.1(1) (2006), para. 102. On this development J. Friedrich, "Carbon capture and storage: a new challenge for international environmental law", Zeitschrift für ausländisches öffentliches Recht und Völkerrecht 67 (2007), 211-227.

of treaty law provisions by providing authoritative interpretations.[134] United Nations General Assembly (UNGA) resolutions have for example been used by the International Court of Justice (ICJ) for the authoritative interpretation of the UN Charter in question regarding decolonisation policies.[135] There is no apparent reason why one should accept such a role for resolutions of the UNGA but not for those of other multilateral fora, provided that these other nonbinding instruments also emanate from a body with particular authority for the issue in question.

One possible mode of impact on the interpretation of treaty law is a reference to nonbinding instruments in preambles of treaties.[136] Nonbinding instruments can be used for interpretation by virtue of the preamble being recognised as a means to determine the context of the terms of a treaty in accordance with Article 31 (2) of the Vienna Convention on the Law of Treaties.[137] This can be achieved through a direct reference to nonbinding instruments, as exemplified by the references to specific ministerial declarations and resolutions of the GA which can be found for instance in the preamble of the Framework Convention on Climate Change,[138] the UN Convention on the Law of the Sea[139] or the

---

[134] This has for example been accepted in the context of the interpretation of a treaty provision referencing "new accepted trends of the Third Conference on the Law of the Sea" by Judge Jiménez de Aréchaga in his Dissenting Opinion, Continental Shelf (Tunisia/Libyan Arab Jamahiriya), ICJ Reports 1982, 100, para. 33; generally in support of this role of nonbinding norms A.E. Boyle, "Some Reflections on the Relationship of Treaties and Soft Law", International and Comparative Law Quarterly 48 (1999), 901-913 (905); J.E. Alvarez, International Organizations as Law-makers, 2005, 249; A. Peters/I. Pagotto, "Soft Law as a New Mode of Governance: A Legal Perspective", New Modes of Governance Project 04/D11 (2006), (23).

[135] ICJ, Western Sahara, Advisory Opinion of 16 October 1975, ICJ Reports 1975, p. 12, paras 53 et seq.

[136] References to the Stockholm Declaration or the Rio Declaration are frequent, take for instance the reference to the Stockholm Declaration in the Framework Convention on Climate Change; see also the reference to the UNGA Resolution 2749 (XXV) of 17 December 1970 f.

[137] Vienna Convention on the Law of Treaties (23 May 1969), 1155 UNTS 331, 8 ILM 679 (1969), Article 31 (2).

[138] Compare e.g. the numerous references in the preamble of the Framework Convention on Climate Change to various GA resolutions (e.g. GA resolutions 44/228 of 22 December 1989 on the UN Conference on Environment and Development, 43/53 of 6 December 1988, 46/169 of 19 December 1991 on the protection of the global climate, 44/206 of 22 December 1989 on the possible ad-

Fish Stocks Agreement.[140] Another possibility which opens the door for a broader impact of various nonbinding instruments is that of general references that indicate objectives of the treaty. The WTO Agreement for example refers to sustainable development as one of its objectives in the preamble. Nonbinding instruments which constitute widely accepted international norms that concretise these objectives may then become influential in the interpretation of norms of the respective treaty, even when they are nonbinding. The potential role of nonbinding instruments in the interpretation of exception clauses in WTO law will be discussed in more detail in the next section.

b) Example: interpretation of Article XX GATT

Nonbinding instruments may play a role in the interpretation of Article XX of the General Agreement on Tariffs and Trade (GATT), Article XIV of the General Agreement on Trade in Services (GATS) or similar provisions in regional trade agreements such as Article 2101 of the North American Free Trade Agreement. These exception clauses are of central importance for international and national environmental law. They define the extent to which WTO Members are allowed to use trade-related measures for environmental protection even though such measures contravene their obligations under WTO law.

Trade-related environmental measures are not only prescribed through multilateral environmental treaties such as the Convention on International Trade in Endangered Species.[141] Nonbinding instruments may also recommend that states adopt certain trade measures to pursue the environmental objectives. For example, the Code of Conduct for Re-

---

verse effects of sea-level rise, 44/172 of 19 December 1989 on desertification) and to the Ministerial Declaration of the Second World Climate Conference adopted on 7 November 1990, on the United Nations Conference on Environment and Development, and resolutions 43/53 of 6 December 1988, 44/207 of 22 December 1989, 45/212 of 21 December 1990 and 46/169 of 19 December 1991 on protection of global climate for present and future generations of mankind.

[139] The preamble of UNCLOS references the UNGA Resolution 2749 (XXV) of 17 December 1970 on common heritage of mankind.

[140] The Preamble of the Fish Stocks Agreement references Agenda 21, chapter 17, programme area C.

[141] Convention on International Trade in Endangered Species of Wild Fauna and Flora (3 March 1973), 993 UNTS 243.

sponsible Fisheries (CCRF) and the related International Plan of Action to prevent, deter and eliminate illegal, unreported and unregulated fishing (IPOA-IUU) generally prescribe the use of trade measures for resource protection.[142] The International Plan of Action on IUU fishing for instance recommends that such "trade-related measures to reduce or eliminate trade in fish and fish products derived from IUU fishing could include the adoption of multilateral catch documentation and certification requirements, as well as other appropriate multilaterally-agreed measures such as import and export controls or prohibitions."[143]

Trade restrictions of this sort may raise a conflict with numerous provisions of the GATT, in particular Article III GATT prohibiting internal measures such as the prohibition of fish sales; Article XI: 1 GATT prohibiting quantative restriction; Article XIII GATT prescribing the non-discriminatory administration of quantative restrictions; and the most favoured nation principle stipulated in Article I GATT. As indicated, the measures can however be justified if they meet the requirements of the exceptions stipulated in Article XX GATT. In order to meet the two-tiered test of Article XX GATT, the concrete measure must implement policies that fall under one of the exception clauses of this Article and the manner of application of the measure must meet the general requirements of the chapeau of Article XX GATT.[144]

Regarding the first condition, most of the environmental measures prescribed for instance in the above mentioned International Plan of Action could either be considered measures "necessary to protect human, animal or plant life and health" under Article XX b) GATT,[145] or meas-

---

[142] Article 11.2 FAO Code of Conduct for Responsible Fisheries and IPOA-IUU, paras 66-76.

[143] IPOA-IUU, para. 69.

[144] United States — Import Prohibition of Certain Shrimp and Shrimp Products, Appellate Body Report of 12 October 1998, WT/DS58/AB/R, para. 152. In United States – Standards for Reformulated and Conventional Gasoline, the Appellate Body stressed that the chapeau analysis is not upon the specific content of the measure, but rather upon the manner of its application, compare United States – Standards for Reformulated and Conventional Gasoline, Appellate Body Report of 29 April 1996, WT/DS2/AB/R, para. 22.

[145] The provision does not include measures aimed to protect the environment per se, but most environmental measures will nevertheless fall under this provision, see P.-T. Stoll/L. Strack, "Article XX lit. b GATT 1994" in: R. Wolfrum/P.-T. Stoll/A. Seibert-Fohr (eds.), WTO – Technical Barriers and SPS Measures, 2007, 96-120 (para. 27). In United States — Restrictions on Imports of Tuna (Mexico), BISD 39S/155, paras 5.24-5.29 and United States – Re-

ures "relating to the conservation of exhaustible natural resources ..." of Article XX lit. g) GATT. The latter exception is not restricted to non-living resources, but also includes, as a result of a dynamic treaty interpretation, living resources that can be exhausted such as marine living resources, and arguably also clean air, the atmosphere or the ozone layer.[146] Even if Article XX g) GATT is interpreted not to cover any measure with extraterritorial effect, it does include protective measures as long as there is a sufficient nexus between the resource and the territory of the state in question, as is the case for migratory species and arguably whenever states protect ecosystems that expand beyond their own territory.[147]

Turning to the chapeau of Article XX GATT, it must be established that the application of the measure is not applied in a manner that it constitutes a "disguised restriction" on trade or an "arbitrary or unjustifiable discrimination between countries where the same conditions prevail."[148]

---

strictions on Imports of Tuna (EEC), WT/DS29/R, para. 5.30, the Panels accepted that protection of dolphin life was a policy that falls under the exemption of Article XX lit. B GATT.

146 A dynamic interpretation which must include modern understandings of living resources as exhaustible environmental resources was accepted in United States — Import Prohibition of Certain Shrimp and Shrimp Products, Appellate Body Report of 12 October 1998, WT/DS58/AB/R, para. 130.3; further, clean air was accepted as falling under the exception in United States — Import Prohibition of Certain Shrimp and Shrimp Products, Appellate Body Report of 12 October 1998, WT/DS58/AB/R, para. 13; compare on this N. Matz-Lück/ R. Wolfrum, "Article XX lit. g GATT" in: R. Wolfrum/P.-T. Stoll/A. Seibert-Fohr (eds.), WTO – Technical Barriers and SPS Measures, 2007, 141-157 (paras 19-21).

147 In contrast to a more limited approach of the Panel in United States — Restrictions on Imports of Tuna (Mexico), BISD 39S/155, the Panel in United States – Restrictions on Imports of Tuna (EEC), WT/DS29/R permitted the extraterritorial application of US policies to US nationals and US vessels, and the Appellate Body in United States — Import Prohibition of Certain Shrimp and Shrimp Products, Appellate Body Report of 12 October 1998, WT/DS58/ AB/R, para. 133 held that such a territorial nexus also existed in the case of migratory species, as long as the protected species occurred in the territory or in waters under the jurisdiction of the state in question. For a convincing general argument that any connection arising from ecosystems considerations suffices see N. Matz-Lück/R. Wolfrum, "Article XX lit. g GATT" in: R. Wolfrum/P.-T. Stoll/A. Seibert-Fohr (eds.), WTO – Technical Barriers and SPS Measures, 2007, 141-157 (para. 25).

148 Article XX GATT, chapeau.

This step in the analysis is intended to prevent member states from abusing or misusing the exceptions of Article XX GATT or Article XIV GATS.[149] Both the Panel and the Appellate Body in *United States – Import Prohibition of Certain Shrimp and Shrimp Products*[150] considered the unilateral application of the measures of the United States to constitute an "unjustifiable discrimination", and therefore found the manner of application of the import restrictions a contravention of the requirements of the chapeau.[151]

For its interpretation, the Panel turned to the object and purpose of the norm in accordance with Article 31 (1) of the Vienna Convention on the Law of Treaties. In this respect, it takes the object and purpose of the WTO Agreement as a point of orientation, since the GATT 1994 and hence Article XX GATT constitutes an integral part of that Agreement.[152] Pointing *inter alia* to the preamble of the WTO Agreement, the Panel stressed that the multilateral nature of the trading system would be jeopardised if states unilaterally pursued environmental policies. If one Member was allowed to do so, then other Members could also apply differing or conflicting requirements to the same product, and in consequence market access could become subject to an increasing number of conflicting policy requirements for the same product.[153]

Moreover, this finding according to the first Panel in *US-Shrimp* does not conflict with the objective of the WTO Agreement to pursue sustainable development which is also mentioned in its preamble.[154] This is where nonbinding instruments played a role in the interpretation. Both the Panel and the Appellate Body in *US-Shrimp* referred to nonbinding instruments alongside international treaty norms to support their argument that sustainable development and environmental protection

---

[149] United States – Standards for Reformulated and Conventional Gasoline, Appellate Body Report of 29 April 1996, WT/DS2/AB/R, para. 22.

[150] United States – Import prohibition of certain shrimp and shrimp products, WT/DS58/R of 15 May 1998 [hereinafter US-Shrimp, Panel Report]; United States – Import Prohibition of Certain Shrimp and Shrimp Products, Appellate Body Report of 12 October 1998, WT/DS58/AB/R, 38 ILM 118 (1999) [hereinafter US-Shrimp, Appellate Body Report].

[151] US-Shrimp, Appellate Body Report, paras 168 et seq.

[152] US-Shrimp, Panel Report, 7.41.

[153] US-Shrimp, Panel Report, 7.45.

[154] US-Shrimp, Panel Report, 7.52.

must be pursued through multilateral approaches. In addition to multilateral environmental treaties,[155] both dispute settlement bodies frequently referred to the Rio Declaration on Environment and Development[156] as well as the Agenda 21. These instruments both stress that unilateral measures for environmental protection should be avoided and based on international consensus as far as possible.[157]

Accordingly, the Panel and the Appellate Body establish on the basis of these principles that the application of a measure is more likely to be considered an "unjustified discrimination" when the state in question does not attempt to act multilaterally. Such multilateral action apparently must not necessarily comprise the actual conclusion of an international agreement, even though it seems clear that if trade restrictions are adopted pursuant to a multilateral environmental treaty, a Panel or the Appellate Body is likely to accept them.[158] The Appellate Body instead cautiously stressed the need for international negotiations on the matter, and that the failure to engage countries "in serious across-the-board negotiations with the objective of concluding bilateral or multilateral agreements"[159] would constitute discrimination of the Members left out of such negotiations. The Panel had held that negotiation of multilateral agreements or actions under multilaterally defined criteria are possible ways to "avoid threatening the multilateral trading system."[160] The attempt to negotiate and cooperate multilaterally can thus be seen as one

[155] E.g. the Convention on Biological Diversity (5 June 1992), 1760 UNTS 79; 31 ILM 818 (1992); the Convention on the Conservation of Migratory Species of Wild Animals (23 June 1979), 1651 UNTS 333; 19 ILM 15 (1980).

[156] Rio Declaration on Environment and Development, Doc. A/CONF. 151/5/Rev.1 of 13 June 1992, 31 ILM (1992) 874 [hereinafter Rio Declaration].

[157] Rio Declaration, principle 12; Agenda 21, para. 2.22(i).

[158] See e.g. P. Sands, Principles of International Environmental Law, 2003, 945-946; R. Wolfrum, "Means of Ensuring Compliance with and Enforcement of International Environmental Law", Recueil des cours de l'Académie de Droit International de La Haye 272 (1998), 13-154 (74); J. Scott, "International Trade and Environmental Governance: Relating Rules (and Standards) in the EU and the WTO", European Journal of International Law 15 (2004), 307-354 (311).

[159] US-Shrimp, Appellate Body Report, para. 166.

[160] US-Shrimp, Panel Report, para. 7.52.

of the preconditions for having recourse to the Article XX GATT exception.[161]

It is not a huge step from here to argue that when measures are adopted on the basis of nonbinding instruments, the condition of multilateral negotiations is fulfilled. The negotiations and adoption of international norms in multilateral fora of UN organisations which are open to all countries do not leave out any WTO member.[162] Where nonbinding instruments propose measures against non-participants in fisheries management schemes, for example, it could not be argued that such measures were not negotiated multilaterally, even if they are not (yet) enshrined in a binding treaty. And it does not matter whether a particular state has in fact consented to the measure in question. The interpretation of the chapeau adopted by the Panel and the Appellate Body, as mentioned, does not require the actual conclusion of a multilateral treaty; what matters is the good faith effort to reach agreement. Nor does it require the consent of states to a particular agreement. The Appellate Body in *US-Shrimp* refers explicitly to UNCLOS and the Convention on Biological Diversity even though the US had not consented to either of those treaties.[163]

The proposition that nonbinding instruments may be relied upon to demonstrate the multilateral nature of the measures is also supported by the fact that both the Panel and the Appellate Body in *US-Shrimp* cite a number of nonbinding instruments when testing whether the measures of the US were justifiable. Whether criteria are defined in a binding or a nonbinding instrument does not appear to make a difference for the Panel or the Appellate Body. In arguing that trade measures adopted pursuant to multilateral agreements would meet the test of the

---

[161] J. Pauwelyn, "WTO compassion or superiority complex?: What to make of the WTO waiver on 'conflict diamonds'", Michigan journal of international law 24 (2002-2003), 1177-1207 (1191); J. Scott, "International Trade and Environmental Governance: Relating Rules (and Standards) in the EU and the WTO", European Journal of International Law 15 (2004), 307-354 (337).

[162] This is also argued with respect to the nonbinding Kimberley Certification Scheme by J. Pauwelyn, "WTO compassion or superiority complex?: What to make of the WTO waiver on 'conflict diamonds'", Michigan journal of international law 24 (2002-2003), 1177-1207 (1191).

[163] J. Scott, "International Trade and Environmental Governance: Relating Rules (and Standards) in the EU and the WTO", European Journal of International Law 15 (2004), 307-354 (311).

Chapeau of Article XX GATT,[164] the Panel Report at one point also includes Agenda 21 as one example for "multilateral agreements".[165] More importantly, it also explicitly specifies the legally nonbinding international standards accepted under the WTO Agreements on Technical Barriers to Trade and on Sanitary and Phytosanitary Measures as one example for a desirable international coordination.[166] The Panel even makes reference to the standards and rules of the FAO Code of Conduct for Responsible Fisheries as further evidence for its conclusion that current fisheries standards do not prescribe unilateral measures as those applied by the United States. In the words of the Panel, "[T]his Code, even though it is not binding, is evidence of the methods currently favoured for the promotion and development of conservation methods ...".[167] If that code had however prescribed the kind of measure taken by the US, one can conclude *e contrario,* the unilateral nature of the measures would have been called into question. In other words, if states follow the prescription of international instruments, whether nonbinding or binding, this increases their chances that their measures will not conflict with WTO law and pass the test of the chapeau as established by *US-Shrimp*. Neither the Panel nor the Appellate Body in *US-Shrimp* suggested that the multilateral instruments must necessarily be legally binding.[168] To the contrary, in citing nonbinding instruments for their interpretation, and in referring to nonbinding international norms to support their findings, the Panel and the Appellate Body seem to suggest that nonbinding instruments may be also be used as indicators for multilateral standards as well.

The application of nonbinding instruments which reflect an international consensus can be justified, because following the recommendations of these instruments also prevents the multilateral trading system from being jeopardised by an unpredictable and possibly conflicting proliferation of national laws and standards. A measure that is clearly based upon such an international instrument would also deflect one of the main concerns voiced by the Appellate Body in *US-Shrimp*, namely

---

164 US-Shrimp, Report of the Appellate Body, para. 171.

165 US-Shrimp, Panel Report, para. 7.59.

166 US-Shrimp, Panel Report, para. 7.1.

167 US-Shrimp, Panel Report, para. 7.59, in particular footnote 277.

168 This is also stressed by E. Hey, "International Institutions" in: D. Bodansky/J. Brunnée/E. Hey (eds.), Oxford Handbook of International Environmental Law, 2007, 749-769 (758).

that domestic measures would be unpredictable for exporting states.[169] The argument for the use of nonbinding instruments as indicators for the multilateral approval of measures is furthermore supported by the approach taken in other WTO Agreements. Under the SPS and TBT Agreements, compliance with (nonbinding) international standards leads to a presumption of legality for certain measures.

Finally, the suggested reading would not disregard the principle that states can only be bound if they have consented to such an obligation, nor would it conflict with *pacta tertiis nec nocent nec prosunt*. The nonbinding requirements would not be incorporated into the WTO law, and trade restrictions are not automatically lawful if adopted pursuant to such an instrument.[170] Nonbinding instruments entailing recommendations for trade measures can serve as indicators for interpretation. By increasing the acceptability of trade measures in dispute settlement, the suggested reading provides an incentive for corporation through binding *and* nonbinding instruments in cases of transboundary or global environmental problems.

c) Supplementing treaty law through references in treaty law

References in treaty law to nonbinding instruments constitute a further important mechanism through which nonbinding instruments supplement treaty law. This mechanism is of particular importance for the context of this study, since nonbinding norms in this manner often gain immediate legal effect.

Sometimes, this is achieved by explicit references to one particular instrument. The CCRF is referred to explicitly in some regional fisheries treaties. Thus, the Agreement on the International Dolphin Conservation Program initiated by the Inter-American Tropical Tuna Commission provides in its general principles that "the Parties shall ... take measures to ensure the conservation of ecosystems as well as conservation and management measures ... and apply the precautionary approach, consistent with the relevant provisions of the FAO Code of

---

[169] US-Shrimp, Report of the Appellate Body, WT/DS58/R, 832, 849.

[170] J. Scott, "International Trade and Environmental Governance: Relating Rules (and Standards) in the EU and the WTO", European Journal of International Law 15 (2004), 307-354 (343).

Conduct for Responsible Fisheries and the United Nations Agreement on Straddling Fish Stocks and Highly Migratory Fish Stocks."[171]

Most often, however, the references are more general in nature. Treaties often refer to standards and guidelines of binding but also of nonbinding nature to supplement and concretise treaty provisions. For example, the Joint Convention on the Safety of Spent Fuel and Radioactive Waste Management provides that national law must give "due regard to internationally endorsed criteria and standards",[172] and requires Parties to have "due regard to internationally endorsed standards on radiation protection".[173] The references in this Convention arguably allow sufficiently flexible concretisations that reflect changing best practices while at the same time avoiding lengthy treaty making or amendment processes.[174]

The following section will briefly discuss some important references to nonbinding instruments in two highly significant international treaties, namely those of the World Trade Organization and the United Nations Convention on the Law of the Sea. In connection with the latter, the Fish Stocks Agreement will be analysed as an implementation agreement to UNCLOS that combines a linkage to nonbinding norms with a particularly strong enforcement mechanism that is rare in international environmental law. Furthermore, the references in the Agreements of the WTO, UNCLOS and the Fish Stocks Agreement to nonbinding instruments of the International Maritime Organization (IMO), the Food and Agriculture Organization (FAO), the International Standardisation Organisation or the Codex Alimentarius Commission are particularly significant due to the availability – at least to some extent – of compul-

---

[171] Agreement on the International Dolphin Conservation Program, Article IV sec. 1. The Agreement is available at http://www.iattc.org/PDFFiles2/AIDCP- (amended-Oct-2007).pdf.

[172] Joint Convention on the Safety of Spent Fuel Management and on the Safety of Radioactive Waste Management (24 September 1997), IAEA INFCIRC/546, Article 4 (iv) and Article 11 (iv) and (vii).

[173] Joint Convention on the Safety of Spent Fuel Management and on the Safety of Radioactive Waste Management (24 September 1997), IAEA INFCIRC/546, Article 24.

[174] B. Kellman, "Protection of Nuclear Materials" in: D. Shelton (ed.), Commitment and compliance: the role of non-binding norms in the international legal system, 2000, 486-505 (487).

sory dispute settlement under these treaties.[175] Acknowledgment in dispute settlement is one of the main avenues through which nonbinding instruments are hardened and gain effect.[176]

(1) World Trade Organization

While the GATT 1947 was originally mostly concerned with reciprocal tariff reductions, the World Trade Organization (WTO) today ialso functions as an important "linkage machine"[177] insofar as its rules and the respective policies of states often link trade with other issues such as intellectual property protection, government procurement, aspects of investment law or environmental protection.[178] Today the WTO and WTO law is in fact increasingly engaged in policy coordination.[179] One of the mechanisms how this is achieved is through references to externally adopted international standards and the presumption of legality that it attaches to adherence to these standards. WTO law bestows authority upon these standards, and thereby indirectly promotes the development and the use of such international norms, whether legally binding or not.

(i) References in the SPS and TBT Agreements

One of the most significant linkages of nonbinding norms with WTO law is the recognition of the international standards of the International

---

[175] Details on availability of dispute settlement will be provided in the following for each discussed reference.

[176] J.E. Alvarez, International Organizations as Law-makers, 2005, 504.

[177] J.E. Alvarez, "The WTO as linkage machine", The American journal of international law 96 (2002), 146-158 (147).

[178] The literature on this issue is vast. See generally J.P. Trachtman, "Institutional linkage: transcending "trade and ..."", The American journal of international law 96 (2002), 77-93; J. Pauwelyn, Conflict of norms in public international law: how WTO law relates to other rules of international law, 2003; more specifically M. Böckenförde, Grüne Gentechnik und Welthandel: das Biosafety-Protokoll und seine Auswirkungen auf das Regime der WTO, 2004; H.P. Hestermeyer, Human rights and the WTO: the case of patents and access to medicines 2007.

[179] J. Scott, "International Trade and Environmental Governance: Relating Rules (and Standards) in the EU and the WTO", European Journal of International Law 15 (2004), 307-354 (310).

Organization for Standardization and the Codex Alimentarius Commission[180] by both the WTO Agreement on Technical Barriers to Trade[181] (TBT Agreement) and the Agreement on the Application of Sanitary and Phytosanitary Measures[182] (SPS Agreement).[183] The SPS Agreement covers domestic food, plant, and livestock regulation;[184] the TBT Agreement is concerned with "technical regulations" such as labelling and product specification.[185] The rules of the SPS and TBT Agreements encourage WTO Members to base their domestic regulation on internationally harmonised standards so as to reduce the non-tariff trade barriers that may arise if each country adopted standards individually. Article 3.1 SPS accordingly provides that members "shall base" their SPS measures on international standards, and Article 2.4. TBT stipulates that they shall use such standards "as a basis for" their technical regulation.[186] In addition to this general obligation, the SPS and TBT Agreements create an incentive which favours those states that adopt conforming measures. The incentive comes in the form of the rebuttable presumption that measures which conform to the international standards are in compliance with the SPS Agreement or the GATT[187] or – in the case of the TBT Agreement – do not create unnecessary obsta-

---

180 On these organisations see already in Part 1, at A.IV., further above.

181 WTO Agreement on Technical Barriers to Trade (15 April 1994), Marrakesh Agreement Establishing the World Trade Organization, Annex 1A, 1868 UNTS 120 (1994).

182 WTO Agreement on the Application of Sanitary and Phytosanitary Measures (15 April 1994), Marrakesh Agreement Establishing the World Trade Organization, Annex 1A, 1867 UNTS 493 (1994).

183 The SPS Agreement in Recital 6 speaks of international standards, guidelines and recommendations; the TBT Agreement in Recital 4 speaks of international standards only, but standards are defined in Annex 1.2. as "rules, guidelines or characteristics for products or related processes and protection methods, with which compliance is not mandatory". The term "international standards" will be used for both the TBT and the SPS Agreements.

184 SPS Agreement, Article 1 and Annex A.1.

185 TBT Agreement, Articles 2 and Annex 1.1.Annex 1, para. 1.

186 Article 3.1 SPS Agreement, Article 2.4 TBT Agreement.

187 Article 3.2 SPS Agreement provides that "[S]anitary or phytosanitary measures which conform to international standards, guidelines or recommendations shall be deemed to be necessary to protect human, animal or plant life or health, and presumed to be consistent with the relevant provisions of this Agreement and of GATT 1994."

cles to international trade.[188] Similar provisions can be found in the North American Free Trade Agreement.[189]

According to the cautious interpretation of the Appellate Body, the above mentioned norms do not however bestow obligatory force upon the above mentioned international standards. With respect to the obligation to base domestic SPS measures on international standards, the Appellate Body stressed in *European Communities – Measures Concerning Meat and Meat Products (Hormones)* that the term "based on" in the Article 3.3. SPS Agreement is not equivalent to "conform to" as used under Article 3.2. SPS, since otherwise these standards would gain "obligatory force and effect" in the absence of an indication of "any intent on the part of the Members to do so".[190] Rather, as explained by the Appellate Body in *European Communities – Trade Description of Sardines*, states must ensure "a very strong and close relationship between" domestic measures and international standards.[191] Similarly, the jurisprudence of the Appellate Body with respect to the presumption in favour of conforming measures only provides these standards with a modest amount of authority.[192] A rebuttable presumption creates an incentive insofar as it shifts the burden of proof, so that it is for the complainant to demonstrate a failure of a particular state to comply with its obligations. However, this incentive is comparatively weak. According to the Appellate Body, the provisions of the SPS and TBT Agreements cannot be read in terms of rules and exceptions to rules.[193] Consequently, even without the presumption, it is never up to the Member accused of non-compliance first to demonstrate that its measures are in

---

188 Article 2.5 TBT Agreement. This presumption would then become relevant for the assessment of compliance with Article 2.2. TBT and Article XX GATT.

189 NAFTA, Articles 905 and 915.

190 European Communities – Measures Concerning Meat and Meat Products (Hormones), Appellate Body Report of 16 January 1998, WT/DS26/AB/R, WT/DS48/AB/R, paras 163-165.

191 European Communities – Trade Description of Sardines, Appellate Body Report of 26 September 2002, WT/DS231/AB/R, para. 245.

192 J. Scott, "International Trade and Environmental Governance: Relating Rules (and Standards) in the EU and the WTO", European Journal of International Law 15 (2004), 307-354 (330).

193 European Communities – Measures Concerning Meat and Meat Products (Hormones), Appellate Body Report of 16 January 1998, WT/DS26/AB/R, WT/DS48/AB/R, para. 169.

compliance, but rather the complaining Member must always first to establish a prima facie case of inconsistency. Thus, the shift of burden of proof as applied so far does not provide a major advantage.[194] It would be different if a Member without that shift had to face the initial burden of justifying its measures if these were not based upon international standards. At least so far, this has however not been the approach of the Appellate Body.

So what kind of nonbinding instruments are vested with authority by these Agreements? The SPS Agreement explicitly refers to the food safety standards issued by the Codex Alimentarius Commission, the International Office for Epizootics and the secretariat of the International Plant Protection Convention along with "appropriate standards" of "other relevant international organisations open for membership to all Members".[195] Among the mentioned institutions, the Codex Alimentarius Commission has proven to be most active in standard setting and has played the most relevant role under the SPS Agreement.[196] The TBT Agreement does not specifically mention which standards shall be "relevant international standards" for the purpose of Article 2.4 TBT; the Annex only refers to standards of "recognised bodies".[197] Recognised to date are the Codex Alimentarius Commission,[198] the International Organization for Standardization (ISO), the International Electrotechnical Commission and the World Health Organization and the

---

[194] J. Scott, "International Trade and Environmental Governance: Relating Rules (and Standards) in the EU and the WTO", European Journal of International Law 15 (2004), 307-354 (325-330).

[195] SPS Agreement, Annex A, para. 3 (a) (defining international standards for food safety as those created by Codex Alimentarius). The two other bodies named in the SPS Agreement are the secretariat of the International Plant Protection Convention, which issues standards for plant health, and the International Office of Epizootics, which issues standards for animal health, compare para. 3(b)–(c) of the Annex A to the SPS Agreement.

[196] D.G. Victor, "The Sanitary and Phytosanitary Agreement of the World Trade Organization: an assessment after five years", New York University journal of international law & politics 32 (2000), 865-937 (894-895).

[197] TBT Agreement, Annex, para. 1.2., for the recognition of the standards of the Codex Alimentarius Commission see e.g. Appellate Body Report, European Communities – Trade Description of Sardines, Appellate Body Report of 26 September 2002, WT/DS231/AB/R, paras 221 and 315(e).

[198] See e.g. European Communities – Trade Description of Sardines, WT/DS231/AB/R, Appellate Body Report of 26 September 2002, paras 221 and 315(e).

Food and Agriculture Organization, with the ISO being the most important one.[199]

Most of these standards are of rather technical nature. One should however not overlook the policy implications even of technical norms. Technical standards such as those developed by the Codex Alimentarius Commission and the ISO have social and environmental effects. Take for example the technical norms developed by the Codex Alimentarius Commission on maximum pesticide concentrations in foods.[200] Limitations to pesticide concentrations in food products have a direct impact on agricultural practices and the environment.

Furthermore, the ISO increasingly develops management standards with direct social policy implications. One area is corporate social responsibility. In particular the environmental management standards of the ISO 14000 series are of interest here. As described earlier in this study, these prescribe mostly procedural standards concerning the integration of environmental considerations into the planning, management, operation and the controlling processes of an enterprise or other organisation.[201] The recently developed ISO 14064 and ISO 14065 standards establish norms for the quantification, monitoring, reporting and validation of greenhouse gas emissions, an exercise that has direct implications for the assessment of emission levels and, consequently, for climate policy and regulation.

These standards do not simply define characteristics of products. Environmental management standards such as ISO 14001 define norms on decision making which affect production processes and methods, for example by establishing procedures for environmental impact assessment. They must also be considered relevant standards under the TBT Agreement, because this Agreement defines "standard" as a "document approved by a recognised body that provides, for common and repeated use, rules, guidelines or characteristics for products or related processes and production methods, with which compliance is not mandatory. It may also include or deal exclusively with terminology, symbols, packaging, marking or labelling requirements as they apply to a

---

199 M.J. Trebilcock/R. Howse, The regulation of international trade, 2005, 226.

200 For a short description of the Codex Alimentarius Commission, compare Part 1, at A.VI.1., further above.

201 Compare for a short introduction of ISO already Part 1, at A.VI.2. , further above.

product, process or production method."[202] In affecting and defining processes and production methods, the adoption of these environmental management standards therefore also creates the rebuttable presumption that measures in conformity with these standards do not constitute unnecessary obstacles to trade under the TBT Agreement.

(ii) New rules on fisheries subsidies in the framework of discussions for reform of the WTO Agreement on Subsidies and Countervailing Measures

The ongoing Doha Round negotiations on new rules on subsidies and countervailing measures appear to be producing yet another linkage between nonbinding instruments and the WTO which is highly significant for marine resource protection. Nonbinding fisheries instruments of the FAO such as the Code of Conduct for Responsible Fisheries that was discussed above thereby would become minimum standards furnished with legal authority through references similar to those known from TBT and SPS Agreements in the envisaged new rules of the Agreement on Subsidies and Countervailing Measures (WTO Subsidies Agreement).[203]

The negative effects of the extensive fisheries subsidies worldwide[204] on the sustainability of fisheries are difficult to assess and can hardly be generalised. Under perfectly managed fisheries regimes, subsidies cannot lead to overfishing. And furthermore, not all subsidies have negative effects. However, there is a consensus among experts worldwide that for a great number of fishery-related subsidies and under most existing fisheries regimes, fisheries subsidies tend to lead to increased fishing efforts.[205] And consequently, most subsidies have negative effects on the sustainability of the resource.

---

202 TBT Agreement, Annex 1, at 2.

203 Chen speaks of de facto bindingness, see C-J. Chen, Fisheries subsidies under international law, 2010, 165.

204 The World Bank estimated in 1998 that fisheries subsidies worldwide amounted to 18-25 billion dollar annually, compare Matteo Milazzo, Subsidies in World Fisheries: A Re-examination, World Bank Technical Paper 406 (1998).

205 WTO Committee on Trade and Environment, Report of the Expert Consultation on Economic Incentives and Responsible Fisheries of 18 June 2001, WTO Doc. WT/CTE/W/189, para. 45.

The global trade regime so far covers fisheries subsidies only to a limited extent and leaves a number of loopholes.[206] Although WTO Members are obliged to notify the WTO on their subsidies under Article 25 of the WTO Subsidies Agreement, the requirement to report remains unclear and of limited scope with respect to fisheries subsidies.[207] The definition of subsidies as a "financial contribution" in Article 1.1 (a) (1) of the WTO Subsidies Agreement does not for example clearly include instances in which governments acquire access to foreign exclusive economic zones without charging the fisheries industry. And only those reported subsidies which are related to export performance or which favour domestic over imported goods are actually prohibited, while others are allowed under the condition that they do not have adverse effects on the interests of another party.[208]

The working draft of the Chairman of the Negotiating Committee on Rules (Chairman Draft)[209] which was issued in May 2008 after roughly seven years of negotiations suggests new rules on subsidies which codify a number of recommendations of the FAO International Plan of Action for the Management of Fishing Capacity and the Code of Conduct for Responsible Fisheries (CCRF) and include explicit references to nonbinding fisheries instruments of the FAO such as the FAO CCRF.

The Consolidated Draft of the Chair of the Committee proposes a complex system of rules.[210] Generally speaking, it comprises specific prohibitions on industrial fisheries subsidies, and exempts those subsidies designed to improve fisheries management. For determining which subsidies are thus allowed, the draft points to international standards, norms and best practices suggested by international fisheries treaties, but also by nonbinding instruments adopted by the FAO, including in particular the FAO Code of Conduct for Responsible Fisheries.

---

206 On the current regime and the need for reform in detail C-J. Chen, Fisheries subsidies under international law, 2010, Chapters 1 and 2.

207 O.S. Stokke/C. Coffey, "Institutional Interplay and Responsible Fisheries: Combating Subsidies, Developing Precaution" in: S. Oberthür/T. Gehring (eds.), Institutional interaction in global environmental governance: synergy and conflict among international and EU policies, 2006, 127-155 (133).

208 WTO Agreement on Subsidies and Countervailing Measures, 15 April 1994, 1867 UNTS 14, Articles 3, 5 and 6.

209 WTO Doc. TN/RL/W/232 (28 May 2008), Annex C, available at www.wto.org/english/tratop_e/rulesneg_e/rules_may08_e.doc.

210 For a detailed in-depth analysis see C-J. Chen, Fisheries subsidies under international law, 2010, Chapter 3 at 113-165.

In concrete terms, the proposed new Article 3.1.c) of the WTO Subsidies Agreement on the one hand prohibits fisheries subsidies that contribute to overcapacity and overfishing as specified in a proposed Annex VIII to the WTO Subsidies Agreement. The prohibited subsidies include, inter alia, (1) subsidies conferred on the acquisition, construction, repair, renewal, renovation, modernisation of fishing vessels; (2) subsidies conferred on operating costs of fishing including fuel; (3) subsidies in respect of port infrastructure; (4) income support for fishermen; (5) price support for fisheries products; and (6) subsidies the benefits of which are conferred on any vessel engaged in illegal, unreported or unregulated fishing.[211] As provided in a footnote, "the terms "illegal fishing", "unreported fishing" and "unregulated fishing" shall have the same meaning as in paragraph 3 of the International Plan of Action to Prevent, Deter and Eliminate Illegal, Unreported and Unregulated Fishing of the United Nations Food and Agricultural Organisation."[212] In other words, this (nonbinding) FAO International Plan of Action would serves as to define prohibited subsidies.

The Annex further foresees a limited number of exceptions, including (1) subsidies for improving fishing vessel and crew safety; (2) subsidies for the adoption of environmentally friendly technologies and equipments; and (3) subsidies for early retirement of fishermen as a result of government policies to reduce fishing capacity or effort.[213] Most importantly in the present context, the subsidies can only be exempted on the condition that any member maintaining such subsidies operates a fisheries management system within its jurisdiction based on internationally recognised best practices for fisheries management.[214] By internationally recognised best practices, the proposed rules mean those established through relevant provisions of international treaty and nonbinding instruments. Explicitly mentioned are the Fish Stocks Agreement, the FAO Compliance Agreement, the FAO Code of Conduct for Responsible Fisheries, as well as technical guidelines and plans of action for the implementation of these instruments, or other related or successor instruments.[215]

---

[211] Chairman Draft, Annex VIII, Article I.

[212] Chairman Draft, footnote 81.

[213] Chairman Draft, Annex VIII, Article II.

[214] Chairman Draft, Annex VIII, Article V.

[215] The proposed provision in the Chairman draft reads: "Any Member granting or maintaining any subsidy as referred to in Article II or Article

Moreover, the proposed disciplines also take into account the transboundary effects of fishing and its subsidisation. According to the Draft, no Member shall through the use of subsidies cause depletion of or harm to straddling and highly migratory fish stocks whose range extends into the exclusive economic zone of another Member, or to stocks in which another Member has identifiable fishing interests. In determining whether or not such a situation exists, it is of particular importance whether the subsidising Member is implementing internationally-recognised best practices for fisheries management and conservation. Again, the Draft mentions international fisheries instruments including the nonbinding FAO Code of Conduct for Responsible Fisheries, as well the FAO Technical Guidelines and International Plans of Action, or other related or successor instruments.[216]

To sum up, compliance with international fisheries instruments, including nonbinding ones, would play a prominent role in determining

---

III.2(b) shall operate a fisheries management system regulating marine wild capture fishing within its jurisdiction, designed to prevent overfishing. Such management system shall be based on internationally-recognized best practices for fisheries management and conservation as reflected in the relevant provisions of international instruments aimed at ensuring the sustainable use and conservation of marine species, such as, inter alia, the Fish Stocks Agreement, the Code of Conduct, the Compliance Agreement, technical guidelines and plans of action (including criteria and precautionary reference points) for the implementation of these instruments, or other related or successor instruments."

[216] Article IV.1 of Annex VIII to the WTO Subsidies Agreement as proposed in the Chairman Draft reads: "No Member shall cause, through the use of any subsidy referred to in paragraphs 1 and 2 of Article 1, depletion of or harm to, or creation of overcapacity in respect of, (a) straddling or highly migratory fish stocks whose range extends into the EEZ of another Member; or (b) stocks in which another Member has identifiable fishing interests, including through user-specific quota allocations to individuals and groups under limited access privileges and other exclusive quota programmes. The existence of such situations shall be determined taking into account available pertinent information, including from other relevant international organizations. Such information shall include the status of the subsidizing Member's implementation of internationally-recognized best practices for fisheries management and conservation as reflected in the relevant provisions of international instruments aimed at the sustainable use and conservation of marine species, such as, inter alia, the Fish Stocks Agreement, the Code of Conduct, the Compliance Agreement, and technical guidelines and plans of action (including criteria and precautionary reference points) for the implementation of these instruments, or other related or successor instruments."

whether the subsidies of the fishing industry of any Member are exempted from the prohibitions. If the rules are adopted in accordance with the Draft, the WTO would grant authority to external institutions, including the FAO, to establish the relevant minimum standards that inform the prohibitions under the WTO. This mechanism not only coordinates WTO law with international treaty law and nonbinding instruments on resource protection, but also would greatly enhance the impact of these norms.

More generally, this would be a huge step forward in the development of WTO rules and in the interrelation of the WTO with other regimes and institutions.[217] The new rules provide a new means of coordinating environmental and trade interests, again by granting authority to external institutions as under the SPS and TBT Agreements. Furthermore, the new Annex on the probation of fisheries subsidies would establish for the first time WTO rules that not only aimed at addressing trade distortions, but also directly addressed issues of resource protection. The mechanism could be particularly effective since it directly addresses an environmental concern (as well as an economic one) through a clear prohibition. In contrast to Article XX GATT, environmental protection would not only play a role as a justification for regulation, but fisheries subsidies would generally be assumed to be harmful for environmental reasons. Members would then have to justify their use of such subsidies, and for this matter have to demonstrate that they have effective fisheries management in place and comply with international norms on sustainable development. In the end, WTO would either contribute to resource protection directly through the prohibition of the subsidies, or through an incentive for sustainable resource management where Members wish to continue subsidization due to domestic pressure.

The new rules are still being debated.[218] Among other contentious issues, difficulties also remain with respect to the references to instruments of nonbinding nature. One question is how the typically nonbinding nature and relatively general and flexible wording of those instruments could be reconciled with what would amount to binding

217 Similar C-J. Chen, Fisheries subsidies under international law, 2010, 165.

218 Thus, the Chair in December 2008 did not propose a new draft including rules on fisheries subsidies rules, but rather disseminated a text outlining issues that needed further discussion; see WTO Negotiating Group on Rules, 'New Draft Consolidated Chair Texts of the AD and SCM Agreements', TN/RL/W/236 of 19 December 2008, available at http://www.wto.org/english/tratop_e/rulesneg_e/rulesneg_e.htm.

conditionalities.[219] However, it appears to be the case that Members generally accept that subsidies should be exempted from the prohibitions when the Member is implementing a fisheries management scheme that is in line with international best practices. And they do accept that the permissible management practices should be defined by international fisheries instruments that include nonbinding ones. Irrespective of the details, the rules on fisheries subsidies if eventually adopted are very likely to include the above mentioned or similar exceptions with references to nonbinding international instruments. Compliance with the best practices proposed in the nonbinding instruments of the FAO will then become a decisive factor when determining the WTO compatibility of domestic subsidisation in the fisheries sector, and thus of general fisheries policies of Member states. Moreover, compliance with (nonbinding) FAO instruments could then be subject to the WTO's binding dispute resolution procedures.

The linkage of regulation on subsidies to fisheries management systems based on international best practices is likely to pose challenges for both the WTO and issuing organisations such as the FAO or regional fisheries management organisations. WTO dispute settlement panels would possibly have to decide on the adequacy of the fisheries management system of a member country in light of international instruments adopted by FAO and other institutions. Scrutiny of norm development processes at these institutions by WTO Panels but also through increased public attention may be a further consequence of these developments. The question of legitimacy of norm production at these institutions will and must receive increased attention, whether the norms are binding or not.[220]

(2) United Nations Convention on the Law of the Sea

It is one of the defining characteristics of the United Nations Convention on the Law of the Sea (UNCLOS) that it not only establishes substantive rules, but also serves as an "umbrella agreement" setting out general rules which are concretised through norms of binding and nonbinding instruments developed by other institutions. A typical legal technique used in the Convention for this purpose is to refer to interna-

[219] This challenge is mentioned by the Chairman in its "Fisheries Subsidies Roadmap", compare WTO Doc. TN/RL/W/236, 91 of 19 December 2008.

[220] On legitimacy see Part 3 of this study, further below.

tional rules and standards. A great number of these references are included in the UNCLOS provisions on fisheries and environmental protection. The provisions on fisheries oblige states to "take into account" "generally recommended international minimum standards" among other factors when determining the maximum sustainable yield on the high seas and in the EEZ.[221] The norms of Part XII on environmental protection repeatedly refer either to "internationally agreed rules, standards and recommended practices and procedures"[222], to "generally accepted international rules and standards"[223] or – with regard to enforcement provisions – to "applicable international rules and standards"[224] This technique of using an umbrella convention coupled with dynamic references allows for progressive adjustment to changing regulatory necessities and circumstances.[225]

(i) Applicability of nonbinding instruments

The above mentioned rules of UNCLOS on marine pollution can be understood as references to standards embodied in international treaty law, as for example contained in MARPOL or SOLAS in the case of norms on marine pollution.[226]

---

[221] UNCLOS, Article 61 (3), Article 119 (1) (a).

[222] UNCLOS, Articles 207 para. 1, 212 para. 1.

[223] UNCLOS, Articles 208 para. 3, 210 para. 6, 211 para. 2 (flag states) and para. 5 (coastal states).

[224] UNCLOS, Article 213 (enforcement of Article 207), 214 (enforcement of Article 208), 216 para. 1(enforcement of Article 210), 217 (enforcement of Article 211 by flag states), 218 para. 1 (enforcement of Article 211 by port states), 220 paras 1-3 (enforcement of Article 211 by coastal states), 222 (enforcement of Article 212). UNCLOS, Article 226 para. 1 a) also obliges states to rely on "generally accepted international rules and standards" or Article 226 para. 1 b) UNCLOS to rely on "applicable ... international rules and standards" for the investigation and release of vessels. The latter norms are not dealt specifically discussed here, as they do not oblige a state to adopt any of these norms, but build upon them for certain procedural legal consequences.

[225] C. Tomuschat, "Obligations Arising for States Without or Against Their Will", Recueil des Cours de l'Académie de Droit International 241 (1994), 195-374 (352).

[226] R. Wolfrum, "Die Entwicklung des Seerechts zum Recht der marinen Umwelt" in: P. Ehlers/W. Erbguth (eds.), Aktuelle Entwicklungen im Seerecht, 2000, 69-81 (78).

But could they also be understood as references to nonbinding instruments – similar to the references in the WTO Agreements? An affirmative answer to this question is not self-evident. After all, states would then be legally obliged to take into account or give effect to rules and standards that they did not specifically consent to or that they did not adopt with the will to be legally bound by them.

Considering the ordinary meaning of the terms in accordance with Article 31 (1) of the Vienna Convention on the Law of Treaties,[227] one could assume from the usage of the term "recommended" as opposed to "agreed" or the distinction between "standards" and "rules" that the standards referred to must not necessarily be binding. As indicated by the negotiating history of UNCLOS, the various terms in fact proved to be "one of the most difficult sections to harmonise".[228] The multitude of expressions used in the final text could be taken to indicate the intent of the parties to refer to a multitude of different international measures,[229] in particular if one considers that the intention of states parties was to avoid different wording in cases where the intended meaning was the same.[230] However, this argument is not fully conclusive, since the distinction between "recommended" and "agreed" could also simply indicate a different level of acceptance in practice.

Given that the above mentioned expressions do not have any definite ordinary meaning, one must seek to interpret the rules in light of the object and purpose of UNCLOS.[231] A narrow view considers the purpose of these rules of reference to simply refer to other instruments which are legally binding upon the respective state. This understanding therefore limits the content of the referenced rules and standards to those of customary law or to binding international legal instruments to

---

[227] Vienna Convention on the Law of Treaties (23 May 1969), 1155 UNTS 331, 8 ILM. 679 (1969), Article 31 (1).

[228] UN Doc. A/CONF.62/L.40 No. XX (1979), reprinted in UNCLOS III, XII Official Records, 101.

[229] B. Vukas, "Generally Accepted International Rules and Standards" in: A. H.A. Soons (ed.), Implementation of The Law of the Sea Convention Through International Institutions, 23 Law of the Sea Institute Proceedings, 1990, 405-421 (408).

[230] A/CONF.62/L.40 of 22 August 1979, in UNCLOS III, XII Official Records, 95.

[231] Vienna Convention on the Law of Treaties (23 May 1969), 1155 UNTS 331, 8 ILM 679 (1969), Article 31 (1).

which the respective state is a party.[232] This interpretation does however have difficulties explaining why these particularly worded references were included in the first place. The numerous references seem almost superfluous if they had been meant only to refer to those rules that are in any case binding upon a particular state.[233] It is difficult to see the added value of the inclusion of such references if it was conditioned upon the acceptance or ratification of the instrument containing the standards by concerned states, or on the customary law status of the rule, because once customary law, the respective rule would bind states even without the reference.

More importantly than this argument, the historical evolution of the duty to respect generally accepted international standards in UNCLOS indicates a different purpose of the rules in question than being a mere reference to other international legal obligations. This duty to respect standards is closely linked to one of the central goals of the traditional law of the sea, namely to seek universal adherence to what may be called "maritime rules of the road". These rules reflect practices of seafaring states which had not yet been accepted and implemented by all states, for example by inclusion in a treaty or convention. This technique was later expanded at UNCLOS III to include further regulatory activities and objectives, such as the protection of the environment.[234] Essentially, the original idea was to establish a predominance or primacy of international law, practices and regulations over national ones in order to achieve uniformity, but also, in particular with regard to en-

---

[232] A. Blanco-Bazán, "IMO Interface with the Law of the Sea Convention" in: M.H. Nordquist/J.N. Moore (eds.), Current Maritime Issues and the International Maritime Organization, 1999, 268-287 (278); T. Treves, "A handbook on the new law of the sea" in: R.-J. Dupuy/D. Vignes (eds.), A Handbook on the New Law of the Sea, 1991, 835-976 (874-877); W. van Reenen, "Rules of reference in the new Convention on the Law of the Sea, in particular in connection with the pollution of the sea by oil from tankers", Netherlands yearbook of international law 12 (1981), 3-44 (11-12).

[233] In this sense also B.H. Oxman, "The Duty to Respect Generally Accepted International Standards", New York University Journal of International Law and Politics 24 (1991-1992), 109-159 (146-147); see also C. Tietje, Internationalisiertes Verwaltungshandeln, 2001, 347.

[234] See for a detailed study of these developments B.H. Oxman, "The Duty to Respect Generally Accepted International Standards", New York University Journal of International Law and Politics 24 (1991-1992), 109-159 (121).

vironmental matters and resource protection, to ensure obedience to certain international minimum standards.[235]

By being referenced in UNCLOS, the rules and standards are thus made obligatory for states regardless of whether all states are parties to the respective instrument that contains these rules and standards; in other words regardless of whether they are binding upon a particular state.[236] Otherwise, states would be free to ignore these rules and norms if they did not consent to them. This interpretation effectively separates being obliged to follow or take into account these standards from formal consent to the instrument which contains these rules and standards. If the obligation therefore comes into existence regardless of whether a particular international minimum standard is binding on a particular state, it is only logical further to conclude that the legal nature of the instrument cannot be the decisive factor for its application. Consequently, even rules and standards contained in instruments that have not yet achieved binding status generally could qualify as rules and standards referred to by these references.[237] Instruments such as the FAO Code of

---

[235] Final Report of the Committee on Coastal State Jurisdiction relating to Marine Pollution, International Law Association (2000), at 37, available at www.ila-hq.org; N.C. Carstensen, Das Verhältnis des Seerechtsübereinkommens der Vereinten Nationen von 1982 zu fischereirechtlichen Übereinkommen und deren Streitbeilegungsvorschriften, 2005, 100; B.H. Oxman, "The Duty to Respect Generally Accepted International Standards", New York University Journal of International Law and Politics 24 (1991-1992), 109-159 (131).

[236] In this sense also D. Vignes, "La valeur juridique de certaines règles, normes ou pratiques mentionnées au TNCO comme 'généralement acceptées'", Annuaire Français de Droit International 25 (1979), 712-718 (716); R. Wolfrum, "Die Entwicklung des Seerechts zum Recht der marinen Umwelt" in: P. Ehlers/W. Erbguth (eds.), Aktuelle Entwicklungen im Seerecht, 2000, 69-81 (78); L.B. Sohn, "'Generally accepted' International Rules", Washington Law Review 61 (1986), 1073-1080 (1075); similarly also R. Wolfrum, "IMO Interface with the Law of the Sea Convention" in: M.H. Nordquist/J.N. Moore (eds.), Current Maritime Issues and the International Maritime Organization, 1999, 223-234 (231-232); R.R. Churchill/V. Lowe, The law of the sea, 1999, 107-108; International Law Association, Final Report of the Committee on Coastal State Jurisdiction relating to Marine Pollution (2000), 37; different A. Blanco-Bazán, "IMO Interface with the Law of the Sea Convention" in: M.H. Nordquist/J.N. Moore (eds.), Current Maritime Issues and the International Maritime Organization, 1999, 268-287 (278).

[237] B.H. Oxman, "The Duty to Respect Generally Accepted International Standards", New York University Journal of International Law and Politics 24 (1991-1992), 109-159 (110 and 141); R. Wolfrum, "The Role of the International

Conduct for Responsible Fisheries and the International Plans of Action adopted under its framework can thus for instance contribute to the interpretation and concretisation of the broad fisheries provisions of UNCLOS,[238] and environmental recommendations of the IMO to those of Part XII of UNCLOS.[239] In the fisheries context, states are however merely obliged to take such instruments into account, i.e. there is no direct obligation to implement the respective instruments.

This interpretation does not entirely disregard consent as the basis of obligation. By ratifying UNCLOS, states have formally consented to this legal mechanism.[240] With regard to each individual rule and standard, the mechanism however sidesteps to some extent traditional in-

---

Tribunal for the Law of the Sea" in: M.H. Nordquist/J.N. Moore (eds.), Current Fisheries Issues and the Food and Agriculture Organization of the United Nations, 2000, 369-385 (372); R. Wolfrum/V. Röben/F.L. Morrison, "Preservation of the Marine Environment" in: F.L. Morrison/R. Wolfrum (eds.), International, Regional and National Environmental Law, 2000, 225-283 (233); U. Beyerlin/T. Marauhn, Rechtsetzung und Rechtsdurchsetzung im Umweltvölkerrecht nach der Rio-Konferenz 1992, 1997, 55; International Law Association, Final Report of the Committee on Coastal State Jurisdiction relating to Marine Pollution (2000), at 38; C. Tietje, Internationalisiertes Verwaltungshandeln, 2001, 374; N.C. Carstensen, Das Verhältnis des Seerechtsübereinkommens der Vereinten Nationen von 1982 zu fischereirechtlichen Übereinkommen und deren Streitbeilegungsvorschriften, 2005, 100; A.E. Boyle, "Some Reflections on the Relationship of Treaties and Soft Law", International and Comparative Law Quarterly 48 (1999), 901-913 (906).

238 R.R. Churchill, "The Jurisprudence of the International Tribunal for the Law of the Sea Relating to Fisheries: Is There Much in the Net?", The International Journal of Marine and Coastal Law 22 (2007), 383-424 (424); R. Wolfrum/V. Röben/F.L. Morrison, "Preservation of the Marine Environment" in: F.L. Morrison/R. Wolfrum (eds.), International, Regional and National Environmental Law, 2000, 225-283 (233).

239 The applicability of nonbinding recommendations of the IMO under Part XII UNCLOS is widely accepted, compare e.g. A.E. Boyle, "Some Reflections on the Relationship of Treaties and Soft Law", International and Comparative Law Quarterly 48 (1999), 901-913 (906).

240 L.B. Sohn, "'Generally accepted' International Rules", Washington Law Review 61 (1986), 1073-1080 (1075); D. Vignes, "La valeur juridique de certaines règles, normes ou pratiques mentionnées au TNCO comme "généralement acceptées"", Annuaire Français de Droit International 25 (1979), 712-718 (718); N.C. Carstensen, Das Verhältnis des Seerechtsübereinkommens der Vereinten Nationen von 1982 zu fischereirechtlichen Übereinkommen und deren Streitbeilegungsvorschriften, 2005, 101.

ternational lawmaking with its time-consuming ratification and amendment procedures.[241] It therefore carries the advantage of flexibilising UNCLOS. Most importantly, a narrower interpretation is hardly reconcilable with the historical development of these references and the objective of UNCLOS, namely to give the widest regard possible to internationally accepted or recommended minimum rules and standards, and thereby to achieve some international harmonisation.[242] In light of this, only to accept those norms with customary law status cannot be an accurate interpretation,[243] because otherwise there would not be a need for a duty to conform to or take into account such rules.[244]

But one certainly needs to limits the scope of the references in order to avoid unclear and overburdening obligations of states. The corrective factor which delimits the scope of the references is however not specific formal consent but state practice.[245] The textual reference to such practice lies in the terms "generally recommended" and "generally accepted". Since states are bound without their formal consent, and in order not to prevent states from cooperating in multilateral fora for fear that they are immediately legally bound by any nonbinding instrument, these terms should not be interpreted too liberally.[246]

When seeking to determine the threshold, one cannot overlook the distinction between "generally accepted" and "generally recommended".

241 International Law Association, Final Report of the Committee on Coastal State Jurisdiction relating to Marine Pollution (2000), at 38.

242 A.E. Boyle, "Marine Pollution under the Law of the Sea Convention", American Journal of International Law 79 (1985), 347-372 (356).

243 This is however argued by W. van Reenen, "Rules of reference in the new Convention on the Law of the Sea, in particular in connection with the pollution of the sea by oil from tankers", Netherlands yearbook of international law 12 (1981), 3-44.

244 B.H. Oxman, "The Duty to Respect Generally Accepted International Standards", New York University Journal of International Law and Politics 24 (1991-1992), 109-159 (146-147).

245 B.H. Oxman, "The Duty to Respect Generally Accepted International Standards", New York University Journal of International Law and Politics 24 (1991-1992), 109-159 (110).

246 A too liberal interpretation could possibly give rise to the attempt by states to try to restrain the emergence or content of new standards, compare B.H. Oxman, "The Duty to Respect Generally Accepted International Standards", New York University Journal of International Law and Politics 24 (1991-1992), 109-159 (134).

It is suggested that for standards and rules to be "generally recommended", the act of adoption of the particular norm is sufficient, while for "generally accepted/agreed" rules and standards, there must be some form of complying practice in addition to wide support for the instrument in the relevant organisation. In the case of "recommended" standards and rules where the act of adoption suffices, the corrective element is that they must be "generally" recommended, i.e. that the instrument in question must have been adopted by a large majority of states. In applying these considerations to the CCRF adopted by the FAO, it can easily be assumed that the standards and rules of the CCRF meet this requirement, since the CCRF has been adopted by 170 states, including all major fishing nations. For a standard to be "generally accepted", the mere act of adoption cannot suffice. Otherwise there would not be a difference between the two types of obligation. Therefore, it is suggested that the corrective element for "generally accepted" standards and rules is not only their wide support in the act of adoption, but additionally widespread compliance.

But how much acceptance is necessary? For a standard to be considered generally accepted, there must be some kind of formal accepting act of adoption (ratification or otherwise) by a large majority of states, combined with implementation by those and other states.[247] But again, what is a "large majority" of states. The necessary amount of acceptance is difficult to determine. The International Law Commission (ILC), in first articulating the duty to respect internationally accepted international standards, could not settle on a description of the degree of acceptance needed.[248] Criteria such as tonnage or the number of maritime states were not accepted by those that saw a risk in vesting the principal maritime powers with quasi-legislative authority. On the other hand, a mere numerical approach was dismissed in light of arguments pointing out that the uneven distribution of expertise and interest forbids authority of the numerical majority. As a step towards a solution, both

---

[247] This has been proposed by B.H. Oxman, "The Duty to Respect Generally Accepted International Standards", New York University Journal of International Law and Politics 24 (1991-1992), 109-159 (141). Oxman additionally mentions compliance by private actors, but I would suggest that acceptance by states should be solely decisive, with compliance by private actors not being the precondition for validity but the outcome of the implementation of the norms in question.

[248] B.H. Oxman, "The Duty to Respect Generally Accepted International Standards", New York University Journal of International Law and Politics 24 (1991-1992), 109-159 (155).

quantitative and functional elements should be considered in combination.[249] This combination gives due regard to the purpose of the rules of reference to strive for international uniformity while paying deference to the interests of both coastal and flag states.

(ii) Application of FAO and IMO instruments

The references to nonbinding instruments under UNCLOS afford particular significance to fisheries instruments of the FAO and nonbinding recommendations and codes of the IMO dealing with marine pollution.[250] The linkages increase the likelihood of compliance with the nonbinding instruments.[251] But as will be seen, the rules potentially referencing fisheries instruments as those of the FAO differ from those referencing IMO instruments.

As mentioned, under the fisheries rules of UNCLOS states are obliged to "take into account" "generally recommended international minimum standards, whether subregional, regional or global"[252] when determining the amount of fish – the maximum sustainable yield – that provides the point of orientation for their management measures.[253] In other words, the management measures taken to avoid overexploitation in the EEZ are qualified by international minimum standards. A similarly mechanism can be found in the rules on fishing on the high seas.[254]

---

249 B.H. Oxman, "The Duty to Respect Generally Accepted International Standards", New York University Journal of International Law and Politics 24 (1991-1992), 109-159 (157); International Law Association, Final Report of the Committee on Coastal State Jurisdiction relating to Marine Pollution (2000), at 38; N.C. Carstensen, Das Verhältnis des Seerechtsübereinkommens der Vereinten Nationen von 1982 zu fischereirechtlichen Übereinkommen und deren Streitbeilegungsvorschriften, 2005, 104.

250 A.E. Boyle, "Some Reflections on the Relationship of Treaties and Soft Law", International and Comparative Law Quarterly 48 (1999), 901-913 (906); E. Brown Weiss, "Conclusions: Understanding Compliance with Soft Law" in: D. Shelton (ed.), Commitment and Compliance: The Role of Non-binding Norms in the International Legal System, 2000, 535-556 (537).

251 E. Brown Weiss, "Conclusions: Understanding Compliance with Soft Law" in: D. Shelton (ed.), Commitment and Compliance: The Role of Non-binding Norms in the International Legal System, 2000, 535-556 (537).

252 UNCLOS, Article 61 (3).

253 UNCLOS, Article 61 (2) and (3).

254 UNCLOS, Articles 117 and 119 para. 1 (a).

In light of the considerations presented above, these references can be understood as references to the FAO Code of Conduct for Responsible Fisheries (CCRF). The FAO CCRF was not only unanimously adopted, but also comprises a number of fisheries management rules that substantively fit the obligation.[255] But the linkage under these rules in UNCLOS is rather weak. First of all, the references are weakly worded. None of these references directly obliges states to apply international standards. International standards only constitute one of the factors to be "taken into account" for conservation measures. This leaves much space for states to manoeuvre. Furthermore, there are no specific rules for fisheries in coastal waters, even though most fishing is conducted in these waters. Even through the general duty to protect the marine environment[256] which applies to all waters also establishes a linkage between the fishery and environmental rules of UNCLOS,[257] more specific references to international standards regarding fisheries management are absent. To the extent that the FAO CCRF is taken into account, however, it imports precautionary and ecosystem considerations into the law of the sea, thereby remedying a lacuna in UNCLOS.

In contrast, the references in Part XII of UNCLOS are worded more strongly. For example, the central provisions on dumping, pollution from seabed activities and pollution by vessels require that laws and measures with regard to these sources of pollution should be no less effective or should conform to "generally accepted international rules and standards".[258] States are thus directly bound to observe these standards. In addition, they also apply to the territorial sea and are reinforced by the enforcement obligations in section 6 of Part XII. However, in areas

---

[255] Article 8 CCRF and the related FAO Technical Guidelines for Responsible Fisheries No. 1 (1996). For details on the content of these instruments see Part 1, at B.I.3.c), further above.

[256] UNCLOS, Article 193.

[257] International Tribunal for the Law of the Sea, Southern Bluefin Tuna Cases (New Zealand v. Japan; Australia v. Japan), Requests for Provisional Measures, Order of 27 August 1999, 38 ILM 1624 (1999), para. 70 (the "conservation of living resources of the sea is an element in the protection and preservation of the marine environment"); in this sense also R. Wolfrum, "The Role of the International Tribunal for the Law of the Sea" in: M.H. Nordquist/ J.N. Moore (eds.), Current Fisheries Issues and the Food and Agriculture Organization of the United Nations, 2000, 369-385 (380).

[258] UNCLOS, Article 210 (6), 208 (3), 211 (2) (flag states) and (5) (coastal states).

where coastal states have sole jurisdiction as is the case with respect to atmospheric or land-based pollution, the role of international standards resembles the provisions on fisheries.[259]

The relatively strong references in Part XII of UNCLOS to international standards are mostly to standards developed by the International Maritime Organization (IMO). They do not strengthen the applicability of nonbinding standards of the FAO. There are two reasons for this. One is the substantive content of the provisions of Part XII UNCLOS. Even if the general provisions on environmental protection[260] are in general applicable also to the formulation of fisheries policy and law despite the absence of a direct reference to fishing,[261] only the obligations to take measures against pollution by vessels and against atmospheric pollution substantively overlap with some of the provisions of the CCRF.[262] However, for the provisions dealing with the impact of vessels on the marine environment, the FAO does not have any regulatory competence. The provisions, by requiring that states act "through the competent organisation"[263], refer to the singular of "organisation". They thereby contrast with other references which oblige states to act "through competent international organisations".[264] The use of the singular of "organisation" indicates that the reference applies exclusively to standards of the IMO.[265] Only when used in the plural, may the organisations referred to include other organisations such as the FAO,

---

[259] The obligation of states to take measures against land-based or atmospheric pollution is only qualified by the condition that they must "take into account" "internationally agreed rules, standards and recommended practices and procedures", see Articles 207 (1), 212 (1) UNCLOS.

[260] UNCLOS, Articles 193 and 194 (4).

[261] R. Wolfrum, "The protection of the marine environment after the Rio Conference: progress or stalemate?" in: U. Beyerlin/M. Bothe/R. Hofmann/E.-U. Petersmann (eds.), Recht zwischen Umbruch und Bewahrung, 1995, 1003-1017 (1009).

[262] These are Articles 8.7 and 8.8 of the FAO Code of Conduct for Responsible Fisheries relating to the protection of the aquatic environment and the protection of the atmosphere.

[263] UNCLOS, Articles 211 (1), 217 (1), 218 (1).

[264] UNCLOS Articles 207 (4), 208 (5), 210 (4), 212 (3).

[265] R. Wolfrum, "IMO Interface with the Law of the Sea Convention" in: M.H. Nordquist/J.N. Moore (eds.), Current Maritime Issues and the International Maritime Organization, 1999, 223-234 (223); R.R. Churchill/V. Lowe, The law of the sea, 1999, 346-347.

UNEP or the Intergovernmental Oceanographic Commission of UNESCO.[266] As a result, the environmental protection provisions of UNCLOS in its Part XII only to a very limited extent, namely with regard to the marginal provisions in the CCRF on atmospheric pollution, could be seen as a reference to existing FAO instruments.

The references in UNCLOS thus support the standard-setting activities of the IMO to a much greater extent than those in the FAO relating to fisheries. The strength of the references and the privileged role of the IMO in the environmental provisions may be part of the reasons for the different effectiveness of the recommendations of the IMO and those of the FAO. While IMO recommendations are generally complied with, the CCRF of the FAO has not been implemented and followed to the same extent.[267]

(3) United Nations Fish Stocks Agreement

The Fish Stocks Agreement supplements and concretises obligations of cooperation in the high seas with respect to straddling and highly migratory fish stocks. The Fish Stocks Agreement is modelled upon the rules of UNCLOS when outlining some of the elements to be regarded at the adoption of management and conservation measures.[268]

Regarding highly migratory fish stocks and those straddling the high seas and the EEZ, the rules of the Fish Stocks Agreement include a similar reference as those just described for UNCLOS. In its Article 5 (b), the Fish Stocks Agreement requires states to take into account "generally recommended international minimum standards" when determining the maximum sustainable yield that serves as the target figure for management and conservation measures.[269] The parallel wording is

---

266 These are the only international organisations explicitly mentioned in UNCLOS, Annex VIII, Article 2 (2) (in connection with the lists of experts to be established by these international organizations).

267 For the implementation problems of the fisheries instruments, compare D. Barnhizer, "Waking from Sustainability's "Impossible Dream": The Decisionmaking Realities of Business and Government", Georgetown International Environmental Law Review 18 (2006), 595-690.

268 Article 5 (b) Fish Stocks Agreement requires states – just as Article 119 (1) a) UNCLOS – to take into account "generally recommended international minimum standards" when determining the maximum sustainable yield that serves as the target figure for management and conservation measures.

269 Article 5 (b) Fish Stocks Agreement.

owed to the fact that the Fish Stocks Agreement seeks to implement UNCLOS, namely its Article 64 UNCLOS.

While the reference mentioned is also comparatively weak, Article 10 (c) Fish Stocks Agreement even goes further than that. This provision obliges states to "adopt and apply any generally recommended international minimum standards for the responsible conduct of fishing operations" in order to fulfil their obligation to cooperate through subregional or regional fisheries management organisations. This is a direct obligation to adopt and apply such standards, and not only to take them into account.

Again, this reference to minimum standards for the responsible conduct of fishing operations can be understood as a reference to the FAO Code of Conduct for Responsible Fisheries.[270] The use of the term "responsible" can be seen as a reference to a concept of "responsible fisheries" established by the CCRF. Given that both the Fish Stocks Agreement and the CCRF have been developed in parallel and under close cooperation between the UN Conference on Highly Migratory Fish Stocks and the FAO,[271] it can be safely assumed that the parallel use of the word is not coincidental. Even more importantly however, at least at that time, one of the only international standards that existed in this field of application was that of the FAO Code of Conduct for Responsible Fisheries. If applied in regional fisheries organisations, the standards of the CCRF then take part in the extraordinary enforcement mechanisms of the Fish Stocks Agreement which basically exclude non-cooperating states from access to the resources.[272] The Fish Stocks Agreement thus gives much stronger support to the standards of the FAO than UNCLOS does.

d) Reinforcement through international dispute settlement

Generally speaking, it is a major disadvantage of nonbinding instruments that a breach does not give rise to the possibility of dispute settlement. However, where references in treaty law provide for a link to nonbinding instruments, compliance with nonbinding instruments may

---

[270] In this sense also V. Röben, "Proliferation of Actors" in: R. Wolfrum/ V. Röben (eds.), Developments of International Law in Treaty Making, 2005, 511-542 (528).

[271] See for the history of the elaboration process the case study at Part 1, B.I, further above.

[272] Article 8 (4) and 20 (7) Fish Stocks Agreement.

become the subject of enforcement through international dispute settlement where it exists. When international courts such as the ICJ or the International Tribunal for the Law of the Sea (ITLOS) or the panels and the Appellate Body of the WTO take nonbinding norms into account when interpreting and applying treaty provisions, these nonbinding norms are afforded increased impact in a particular case. But perhaps more importantly, already the mere possibility that compliance with nonbinding instruments improves the prospects of winning legal disputes constitutes an incentive for compliance with such norms.

(1) WTO

The comparatively strong dispute settlement mechanism of the WTO plays a prominent role in this regard. Disputes under WTO law are subject to the dispute settlement procedures established by the Dispute Settlement Understanding, albeit subject to special rules and procedures under some of the agreements listed in Appendix 2 of the WTO Dispute Settlement Understanding.[273] All of the agreements and rules of the WTO discussed above as possible points of entry for nonbinding norms can therefore be subject to binding dispute settlement.[274] Regarding the newly proposed rules on fisheries subsidies, the Chairman Draft also foresees dispute settlement through Article VIII of the newly proposed Annex VIII to the WTO Subsidies Agreement.[275] If a party does not appeal to the Appellate Body, the adoption of the panel reports is normally secured, because Article 16 DSU provides that a consensus of Parties must exist against the adoption of a panel report,[276] and this con-

---

[273] Understanding on rules and procedures governing the settlement of disputes, contained in Annex 2 of the WTO Agreement, Article 1.1 provides that the DSU applies to all disputes brought under the provisions of the agreements listed in Appendix 1 (including GATT and GATS); Article 1.2. DSU provides that the rules and procedures of the DSU apply to the agreements listed in Appendix 2 (including TBT, SPS and WTO Subsidies Agreements), subject to any special or additional rules and procedures contained in these agreements.

[274] Compare e.g. Articles XXII, XXIII GATT 1947; Article 11 SPS Agreement, Article 14 TBT Agreement.

[275] The proposed Article VIII provides for dispute settlement for disputes over prohibition of subsidies by referring to Article 4 WTO Subsidies Agreement, and additionally establishes some specific rules on disputes over fisheries subsidies which are applicable via Article 1.2 DSU.

[276] Article 16.4 DSU.

sensus would have to include the winning party to the dispute. An implementation and surveillance procedure furthermore ensures, through ongoing compliance reporting and the possibility for further dispute settlement, that losing defendants comply with the decisions.[277]

Within the limits indicated above, WTO law gives authority to nonbinding instruments or can be interpreted to that extent. Where the Appellate Body is permissive in its interpretation of Article XX GATT with respect to trade restrictions adopted pursuant to nonbinding instruments, this creates be a powerful incentive for multilateral cooperation. However, with their competence to interpret WTO law, the Panel or the Appellate Body also have discretion in determining to which extent they give authority to these instruments which were adopted outside the WTO. For instance, in the case of Article XX GATT, it remains a matter of interpretation whether and to what extent nonbinding international norms serve as indicators that a state does not act unilaterally. And as exemplified by the reluctance of panels and the Appellate Body to accept, for instance, international nonbinding standards under the SPS and TBT Agreements,[278] panels and in particular the Appellate Body have in fact made use of their discretion in determining to which extent international standards are in fact provided with authority through the references.

The Appellate Body could therefore play the role of a "gatekeeper"[279] that determines the details of the impact from external organisations and their standards.[280] With such a role, it could help mitigating one of the dangers that lies in increased authority of nonbinding instruments or rules of MEAs to which WTO members may not have consented to. While some departure from strict consent and sovereignty may be necessary to achieve flexibility and effectiveness of a legal regime, traditional rules on sovereignty and consent also protect against capture of

---

[277] DSU, Article 21, in particular Article 21.5.

[278] J. Scott, "International Trade and Environmental Governance: Relating Rules (and Standards) in the EU and the WTO", European Journal of International Law 15 (2004), 307-354 and the analysis earlier in this Part 2, at A.I.2, further above.

[279] J. Scott, "International Trade and Environmental Governance: Relating Rules (and Standards) in the EU and the WTO", European Journal of International Law 15 (2004), 307-354 (333 and 344).

[280] J. Scott, "International Trade and Environmental Governance: Relating Rules (and Standards) in the EU and the WTO", European Journal of International Law 15 (2004), 307-354 (310).

international organisations by specific interests, and they serve to protect the equality of states and therefore the weaker participants in the international system.[281] The dilemma between flexibility and effectiveness on the one hand and danger of power abuse on the other could be confronted at least to some extent by an Appellate Body if it controls the claim to authority of external norms under the WTO.[282] Thus, the Appellate Body could establish entrance conditions for standards and rules of external instruments which serve to protect against interest capture or inequalities, The Appellate Body, being a standing institution, could do so for example by developing a set of benchmarks through its case law against which it could evaluate other organisations and their instruments. These benchmarks could possibly take into account and evaluate whether the respective organisations work transparently, the extent of support by states for a particular norm in question, whether this support comprised both developing and developed states as well as the attitude and participation of non-state actors. Giving authority to nonbinding instruments under certain conditions of transparency and participation is likely to have repercussions on the institutions that adopt the norms in question.

Of course, these and similar suggestions that give the Appellate Body of the WTO the power to assess and influence the work of other (environmental) regimes and institutions could be subjected to the criticism that this unduly affords the WTO the competence to judge upon the matters and activities of other regimes. However, this must not necessarily be so, at least not more than is already the case. As rightly observed by Joanne Scott, the Appellate Body would not in fact evaluate the international instruments and institutions as such, but only evaluate the extend to which their authority can be extended into the WTO. It only assesses whether external norms and institutions should be provided with additional authority beyond their original scope and power.[283] The Appellate Body would not for example interfere with the

[281] B. Kingsbury, "Sovereignty and Inequality", European Journal of International Law 9 (1998), 599-625 (625).

[282] This is discussed, albeit for MEAs only, by J. Scott, "International Trade and Environmental Governance: Relating Rules (and Standards) in the EU and the WTO", European Journal of International Law 15 (2004), 307-354 (343-347).

[283] J. Scott, "International Trade and Environmental Governance: Relating Rules (and Standards) in the EU and the WTO", European Journal of International Law 15 (2004), 307-354 (349).

voluntary nature of these instruments; it would still be up to states to decide whether or not to apply these norms.

But if states and institutions count on the increased authority of these standards and on benefitting from compliance with them under WTO rules, then the Appellate Body could rightly evaluate on the basis of the norms of the WTO system whether and to which extent they should be accorded this power. This does not of course prohibit the factual influence of recognition by the Appellate Body – and thus of the WTO system – on other institutions. It is this *de facto* influence which must prompt the Appellate Body to use this power in a way that increases the responsiveness of the WTO system to external values and norms without improperly imposing itself on other institutions. It can do so through the prescription of procedural requirements instead of imposing substantive values, as done in *US-Shrimp*. And it should use the possibility to integrate other institutions into its decision making. It could do so through the varied mechanisms to obtain information and consult with experts from the relevant institutions.[284]

In sum, however, many of these questions regarding the adequate role of the WTO remain open for discussion and must be resolved through adequate rule-making. The problems are illustrated for example by the ongoing controversial discussions under the new fisheries subsidies rules on the question whether the FAO or the WTO should determine whether states have adequate management systems in place.[285]

(2) United Nations Convention on the Law of the Sea

Nonbinding instruments could also become relevant during dispute settlement procedures under the law of the sea whenever they fit under the references in the provisions of UNCLOS.[286] States parties to UNCLOS

---

[284] For example, the newly proposed Chairman Draft on rules on fisheries subsidies foresees a consultative role for the FAO on technical or scientific questions relating to fisheries in Annex VIII, Article VIII.4.

[285] This problem is still open for debate, as indicated by the "Fisheries Subsidies Roadmap" issued by the Chairman in December 2008, WTO Doc. TN/RL/W/236, 91.

[286] R. Wolfrum, "The Role of the International Tribunal for the Law of the Sea" in: M.H. Nordquist/J.N. Moore (eds.), Current Fisheries Issues and the Food and Agriculture Organization of the United Nations, 2000, 369-385 (372). For the question whether and when nonbinding instruments can be subsumed under the references, see the analysis above in this Part 2, at A.I.2c).

are free to choose from one or more of the following dispute resolution procedures: the International Court of Justice (ICJ), the International Tribunal for the Law of the Sea (ITLOS), arbitration or special arbitration.[287] If a state has not declared any preference or if the parties to the dispute have not accepted the same procedure – which is still the case for most State Parties – arbitration under Annex VII generally remains the default mechanism unless the Parties agree otherwise.[288] This means that arbitration is compulsory but that Parties can agree on a different way of settling their differences, either in advance or *ad hoc*.[289] The arbitral tribunal or court which has jurisdiction may also apply binding provisional measures.[290] ITLOS can prescribe such measures not only when it is the forum of choice but also pending the constitution of an arbitral tribunal, provided that "prima facie the tribunal which is to be constituted would have jurisdiction and that the urgency of the situation so requires."[291] This power to prescribe provisional measures can be immensely important for marine environmental and resource protection, as illustrated by the *Southern Bluefin Tuna Case* where provisional measures served the protection of resources and consequently the marine environment.[292]

When assessing the extent to which the nonbinding instruments of the IMO or the FAO can be reinforced through dispute settlement under UNCLOS, one must take into account some limits to the scope of jurisdiction. Most importantly, the jurisdiction of dispute settlement bodies is limited with respect to obligations of coastal states, as provided in Article 297 UNCLOS. While compulsory dispute settlement generally applies for those rules and standards of protection of the marine envi-

---

287 UNCLOS, Article 287 (1).

288 UNCLOS, Article 287 (3) and (5). Exceptions may apply pursuant to reservations or optional exceptions under Article 298 UNCLOS.

289 A. Boyle, "Dispute Settlement and the Law of the Sea Convention: Problems of Fragmentation and Jurisdiction", International and comparative law quarterly 46 (1997), 37-54 (40).

290 UNCLOS, Article 290 (1).

291 UNCLOS, Article 290 (5).

292 ITLOS, Southern Bluefin Tuna Cases (New Zealand v. Japan; Australia v. Japan), Provisional Measures,Order of 27 August 1999, 38 ILM 1624 (1999); for an assessment see R.R. Churchill, "International Tribunal for the Law of the Sea the Southern Bluefin Tuna cases (New Zealand v. Japan; Australia v. Japan): order for provisional measures of 27 August 1999", The international and comparative law quarterly 49 (2000), 979-990.

ronment contained in Part XII of UNCLOS,[293] and thus includes instruments in particular of the IMO, limitations therefore apply in respect to fisheries. As Article 297 paragraph 3 (a) UNCLOS clarifies, a coastal state is not obliged to accept dispute settlement in respect to its sovereign rights over living resources in the EEZ. This generally limits scrutiny of the fisheries obligations of coastal states in the EEZ by ITLOS or an arbitration Panel, as specified in Article 61 UNCLOS. It consequently also limits the impact of the respective international standards referred to in the UNCLOS, i.e. in this context those of the FAO. However, Article 297 paragraph 3 (a) UNCLOS need not necessarily be understood as a complete exclusion of jurisdiction. First of all, the language of Article 297 (3) (a) UNCLOS, in particular that coastal states "are not obliged to accept", implies that coastal states could waive the exception.[294] Furthermore, the restriction may also be read only to limit judicial scrutiny with regards to the discretionary powers and sovereign rights of the coastal states. This interpretation would leave open the possibility that the court or tribunal can assess the correct use of discretionary power by the coastal state, including whether it has taken international standards properly into account.[295] The dispute settlement body could then consider whether a coastal state has for example taken the recommendations of the FAO Code of Conduct for Responsible Fisheries into account. There remains of course also the difficult issue of standing. A violation of the interests of another state resulting from

---

[293] Article 297 para. 1 (c) of UNCLOS provides that dispute settlement applies "when it is alleged that a coastal State has acted in contravention of specified international rules and standards for the protection and preservation of the marine environment which are applicable to the coastal State and which have been established by this Convention or through a competent international organization or diplomatic conference in accordance with this Convention." The specified international rules and standards referred to are the rules contained in Part XII of the Convention, compare A.O. Adede, The System for Settlement of Disputes under the United Nations Convention on the Law of the Sea, 1987, 258.

[294] R.R. Churchill, "International Tribunal for the Law of the Sea the Southern Bluefin Tuna cases (New Zealand v. Japan; Australia v. Japan): order for provisional measures of 27 August 1999", The international and comparative law quarterly 49 (2000), 979-990 (988).

[295] For a discussion of this question see R. Wolfrum, "The Role of the International Tribunal for the Law of the Sea" in: M.H. Nordquist/J.N. Moore (eds.), Current Fisheries Issues and the Food and Agriculture Organization of the United Nations, 2000, 369-385 (374).

the neglect of the requirements of Article 61 UNCLOS, which would be the precondition for a dispute to arise, could only be imagined for neighbouring states or a distant fishing state having historic fishing rights in the EEZ.[296] For other states, much depends on whether the exception of Article 297 paragraph 3 (a) UNCLOS can be narrowly interpreted so as only to exclude disputes relating to stocks which live exclusively within the EEZ.[297] Such a narrow reading would increase the number of states which could initiate compulsory dispute settlement and consequently the judicial scrutiny of coastal states' measures. In the interest of equal access to justice and in the interest of the effective management of fish stocks this interpretation seems at least desirable.[298]

The Fish Stocks Agreement also incorporates the dispute settlement system of Part XV of UNCLOS through its Article 30.[299] However, the Fish Stocks Agreement extends the dispute settlement provisions of UNCLOS to all disputes between parties to the Fish Stocks Agreement concerning the interpretation or application of regional, subregional or global fisheries agreements relating to straddling fish stocks or highly migratory fish stocks to which the states in question are parties.[300] This effectively establishes compulsory dispute settlement between parties to the Fish Stocks Agreement even for fisheries management rules contained in treaties that do not include such a possibility.[301]

---

[296] R. Wolfrum, "The Role of the International Tribunal for the Law of the Sea" in: M.H. Nordquist/J.N. Moore (eds.), Current Fisheries Issues and the Food and Agriculture Organization of the United Nations, 2000, 369-385 (374).

[297] E.D. Brown, The International Law of the Sea, 1994, 228.

[298] A.E. Boyle, "Problems of Compulsory Jurisdiction and the Settlement of Disputes Relating to Straddling Fish Stocks" in: O.S. Stokke (ed.), Governing High Seas Fisheries: the interplay of global and regional regimes, 2001, 91-120 (100 and 113).

[299] Through Article 30 paragraph 1 of the Fish Stocks Agreement, Parties to the Fish Stocks Agreement made use of the possibility foreseen in Article 288 (2) UNCLOS to extend the jurisdiction of dispute settlement bodies under UNCLOS to questions of interpretation and application of another international agreement related to the purposes of the Convention, irrespective of whether a state is a Party to UNCLOS.

[300] Article 30, paragraph 2 Fish Stocks Agreement.

[301] T. Treves, "The Settlement of Disputes According to the Straddling Stocks Agreement of 1995" in: D. Freestone (ed.), International law and sustainable development: past achievements and future challenges, 1999, 253-269 (257).

3. Summary

Nonbinding instruments play a significant role in the development of international law. First of all, to the extent that principles contained in nonbinding instruments are implemented in national legal systems and courts, these instruments may also play a role in generating general principles of law. Second, the adoption of a nonbinding instrument and its implementation by states can facilitate the emergence of customary law norms. The likeliness that this happens depends on a number of characteristics of the instrument in question. Central for this role is whether or not it is broadly accepted by states and supported by follow-up mechanisms which institutionalise continuous discourse. The continuous production of nonbinding instruments by international institutions is part of a process in which customary law norms emerge from the adoption of binding and nonbinding norms in international fora and the responses of states to those norms. As explained, one must however not overlook that some characteristics of a number of nonbinding instruments constrain these processes. Among these are in particular explicit disclaimers and language pointing to the voluntary character of the particular instrument as well as norms addressed directly to private actors.

The most relevant contribution of nonbinding instruments to international law development is their role as precursors of treaty law. As explained, nonbinding instruments carry a number of advantages if employed as a first step towards international rules. These advantages include, *inter alia*, the easier integration of reluctant states and non-state actors into substantive discourse, the swift adoption and application of norms without the need for ratification and the possibility of avoiding domestic legal or political obstacles. In particular, as illustrated by the experience under the voluntary PIC system, nonbinding instruments allow flexible learning processes in situations of uncertainty about the scientific bases of a problem and/or the appropriate solutions for the problem.[302] This at least partly explains their particular relevance in environmental matters, as environmental issues are often characterised by uncertainty about problems and their solutions, and hence often require flexible learning processes.

As much as nonbinding instruments can be useful as generators of change and learning processes in new issue areas, states often appear to

---

[302] Similarly A.C. Kiss/D. Shelton, International environmental law, 2004, 89.

prefer adopting binding rules eventually, provided that this is politically feasible. At least for complex systems such as prior informed consent, states have eventually negotiated legally binding conventions even if all they did was basically to make the existing model binding. The legal certainty and predictability induced by binding instruments may represent a value in itself that limits the capacity of nonbinding instruments.

The impact on the development of binding law should however not only be measured by the extent to which particular norms have become binding or not. Nonbinding instruments such as the Rio Declaration, for instance, will likely never become binding but establish general normative frameworks of discourse which indirectly influence law and policy making at the international and domestic level.[303] They guide and constrain legal arguments, since states that have adopted a particular nonbinding instrument face political constraints to make an argument that clearly disregards the norms of this instrument.[304] Formal bindingness only plays a secondary role in this respect, even if it is a distinct factor that enhances the potential of norms to influence the behaviour of states.[305]

Legally nonbinding instruments also play a supplementary role in relation to treaty law. First, depending on their characteristics, nonbinding instruments may guide the interpretation and concretise indeterminate treaty law norms. In particular if nonbinding instruments contain precise norms, they can deliver much needed concretisations of those terms and provisions in treaties that due to reluctance of states to agree on

---

303 In this sense, for the specific context of the impact of the Stockholm Declaration, J. Brunnée, "The Stockholm Declaration and the Structure and Processes of International Environmental Law" in: M.H. Nordquist/J.N. Moore/S. Mahmoudi (eds.), The Stockholm Declaration and Law of the Marine Environment, 2003, 67-84 (80); see also D. Bodansky, "Customary (and not so customary) international environmental law", Indiana journal of global legal studies 3 (1995), 105-131 (119).

304 J. Brunnée, "The Stockholm Declaration and the Structure and Processes of International Environmental Law" in: M.H. Nordquist/J.N. Moore/S. Mahmoudi (eds.), The Stockholm Declaration and Law of the Marine Environment, 2003, 67-84 (80).

305 J. Brunnée, "The Stockholm Declaration and the Structure and Processes of International Environmental Law" in: M.H. Nordquist/J.N. Moore/S. Mahmoudi (eds.), The Stockholm Declaration and Law of the Marine Environment, 2003, 67-84 (79); D. Bodansky, "Customary (and not so customary) international environmental law", Indiana journal of global legal studies 3 (1995), 105-131 (119).

binding norms often remain vague and imprecise. Nonbinding instruments of resource protection may for instance also play a role in the interpretation of Article XX GATT, i.e. even beyond environmental treaties. Second, references in treaty law bestow direct legal effect on norms of nonbinding instruments. In particular the general references to international standards in important treaty law equipped with dispute settlement mechanisms such as the WTO law and UNCLOS are of significance in this respect. The environmentally relevant norms of international institutions such as the FAO, the IMO as well as the ISO and the Codex Alimentarius Commission may gain legal effect and authority in this way.

As seen, references to nonbinding instruments can contribute to more precision but also to universality of norms of behaviour.[306] Moreover, through their supplementary function, nonbinding instruments also strengthen the adaptability and flexibility of international law.[307] Flexibility and adaptability are crucial in a field defined by high levels of uncertainty and the need for constant adjustments to new scientific knowledge and learning. Referencing nonbinding instruments not only hardens the commitments, but carries the potential to adapt treaty law at least to some extent to changing circumstances.[308] Nonbinding instruments of a rather technical nature such as those of the IMO, ISO or the Codex Alimentarius set out continuously evolving minimum requirements and best practices for vessels, products or production processes. But also instruments of a more policy oriented nature introducing new progressive concepts and practices help to adapt widely accepted treaty law to advancement in knowledge and policy. A particularly interesting example is the proposed references to adequate fisheries man-

---

[306] B.H. Oxman, "The Duty to Respect Generally Accepted International Standards", New York University Journal of International Law and Politics 24 (1991-1992), 109-159 (110).

[307] C. Tietje, "The Changing Structure of International Treaties as an Aspect of an Emerging Global Governance Architecture", German Yearbook of International Law 42 (1999), 27-55 (53); H.E. Ott, Umweltregime im Völkerrecht: eine Untersuchung zu neuen Formen internationaler institutionalisierter Kooperation am Beispiel der Verträge zum Schutz der Ozonschicht und zur Kontrolle grenzüberschreitender Abfallverbringungen, 1998, 207 et seq.

[308] B.H. Oxman, "The Duty to Respect Generally Accepted International Standards", New York University Journal of International Law and Politics 24 (1991-1992), 109-159 (110); C. Tietje, "The Changing Structure of International Treaties as an Aspect of an Emerging Global Governance Architecture", German Yearbook of International Law 42 (1999), 27-55 (46).

agement systems under the proposed new rules of the WTO Agreement on Subsidies and Countervailing Measures. Nonbinding instruments adopted by the FAO would then play a decisive role alongside binding treaty law in assessing domestic fisheries policies of member states to the WTO. The opening up of the WTO and UNCLOS regimes to the nonbinding fisheries instruments of the FAO carries with it the potential to incorporate modern ecosystem and precautionary considerations which – notwithstanding some exceptions[309] – can hardly be found in these treaties so far.[310]

Furthermore, the supplementary role of nonbinding instruments can help to relax the strict consent requirements which are often burdensome for effective environmental cooperation. References within treaties to nonbinding instruments which can be adopted by majority and without the consent of a particular state may indirectly bind even those states which have not specifically consented. This deviates from the state-centred approach to international law and fortifies the influence of international bodies and expert bodies that elaborate and adopt these instruments.[311]

However, one cannot ignore that making treaties more flexible through references presents challenges to the principle of consent, with possible repercussions for the legitimacy of the law. References to nonbinding instruments adopted by international organisations also increase the relevance of the respective international organisations and thus increase the need for their proper legitimation.[312]

The interplay of nonbinding instruments with international environmental law rules may also be highly relevant for compliance by states with these norms. But again, this role as described in this section alone hardly captures the unique qualities of nonbinding instruments. Often, nonbinding instruments are influential upon state and private actors

---

309 Notable exceptions can be found e.g. in UNCLOS, Article 194 (5) and 196 (1).

310 For the law of the sea, this is for example stated by R. Wolfrum, "The Role of the International Tribunal for the Law of the Sea" in: M.H. Nordquist/J.N. Moore (eds.), Current Fisheries Issues and the Food and Agriculture Organization of the United Nations, 2000, 369-385 (372).

311 C. Tietje, "The Changing Structure of International Treaties as an Aspect of an Emerging Global Governance Architecture", German Yearbook of International Law 42 (1999), 27-55 (40-41).

312 These issues will be discussed in detail in Part 3, further below.

even when they do not become binding through references in treaty law or in the transitional period before they become binding; a period that may actually last decades.[313]

II. Inter-institutional cooperation through nonbinding instruments

International institutions increasingly rely on nonbinding instruments for inter-institutional cooperation and cross-sectoral norm setting. Nonbinding instruments hereby serve two main functions: they help establishing cross-cutting standards that recognise instruments of other institutions, and they are instrumental in influencing norm development and implementation activities of other institutions and thereby in integrating the views and norms of various institutions.

1. Establishment of cross-cutting standards

One of the most striking features of the nonbinding instruments analysed in this study is the way in which they incorporate and recognise instruments adopted by other institutions. In this manner they contribute to the establishment of cross-cutting standards. The techniques used to do this range from explicit and direct references to binding treaty law to the explicit or implicit incorporation of a number of different nonbinding or binding instruments and standards developed within other institutions.

Consider for instance the frequent references in the FAO Pesticide Code to other relevant binding and nonbinding instruments. The code stresses the need for conformity with WTO Agreements related to technical barriers to trade[314] and includes frequent references to Conventions as well as legally binding and nonbinding standards and guidelines of the World Health Organization, the OECD, the International Labour Organization, the International Civil Aviation Organization, the International Maritime Organization and the International Air Transport Association.[315] In its Annex 1, it also lists a number of envi-

313 J.E. Alvarez, International Organizations as Law-makers, 2005, 258.

314 FAO Pesticide Code (2002), Article 6.1.11.

315 FAO Pesticide Code (2002), Articles 7 and 8.

ronmental treaty and non-treaty instruments relevant for pesticide regulation.[316]

The FAO's Code of Conduct for Responsible Fisheries (CCRF) similarly aims to establish cross-cutting standards which give due consideration to instruments of other institutions. It incorporates the goals, objectives and concrete norms of various treaty and non-treaty instruments adopted outside the FAO.[317] The CCRF not only affords deference to treaty law by emphasising that it is to be interpreted and applied in conformity with relevant rules of international law, including UNCLOS and the relevant rules of the Fish Stocks Agreement,[318] the WTO Agreements including in particular the SPS and TBT Agreements,[319] and in light of other important international (nonbinding) instruments such as the 1992 Rio Declaration and Agenda 21.[320] It also contains a number of norms which directly incorporate rules of international treaty law such UNCLOS and – as "an integral part of the code" – the Compliance Agreement of the FAO.[321] States, but most notably also vessel owners and crews, are called upon to implement the provisions of the International Convention for the Prevention of Pollution from Ships of 1973, as modified by the Protocol of 1978[322] (MARPOL

[316] FAO Pesticide Code, (2002), Annex 1 (references include e.g. the Montreal Protocol on Substances that Deplete the Ozone Layer, the Stockholm Convention on Persistent Organic Pollutants Convention concerning Safety in the Use of Chemicals at Work and the Rio Declaration on Environment and Development).

[317] For the relationship and mutual reinforcement of the various fisheries instruments see D.J. Doulman, "Code of Conduct for Responsible Fisheries: Development and Implementation Considerations" in: M.H. Nordquist/J.N. Moore (eds.), Current Fisheries Issues and the Food and Agriculture Organization of the United Nations, 2000, 307-330 (313); R. Rayfuse, "The interrelationship between the global instruments of international fisheries law" in: E. Hey (ed.), Developments in international fisheries law, 1999, 107-158.

[318] CCRF, Articles 3.1 and 3.2 a) and b).

[319] CCRF, Article 11.2.1. References stressing the compatibility with WTO law or the duty to cooperate with the WTO are frequent in chapter 11 of the code, see CCRF, Articles 11.2.2, 11.2.4; 11.2.6, 11.2.14, 11.3.8.

[320] CCRF, Article 3.2 c).

[321] This is most prominently stated in Article 1.1 of the CCRF.

[322] International Convention for the Prevention of Pollution from Ships (2 November 1973), 12 ILM 1319 (1973), 1340 UNTS 184, and Protocol of 1978 Relating to the International Convention for the Prevention of Pollution from Ships (17 February 1978), 1340 UNTS 61, 17 ILM 546 (1978).

73/78).[323] The CCRF thus expands, albeit in a nonbinding way, the scope of the obligations of the treaties to which it refers beyond their membership and geographical scope. The same applies to instruments – often nonbinding – issued by other international organisations or expert bodies. Thus, states are called upon to ensure that states follow the instruments issued by the International Maritime Organization relating to the organisation of marine traffic, the protection of the marine environment, the prevention of damage to or loss of fishing gear and the removal of redundant offshore structures.[324] The norms on post-harvest practices recommend compliance with the quality standards of the FAO/WHO Codex Alimentarius Commission.[325] References are also frequent in those nonbinding instruments adopted under the framework of the CCRF to support the implementation of the CCRF. Technical Guidelines frequently include references to very specific nonbinding instruments of other international organisations such as ICES or the IMO.[326]

The same is true to some extent for the OECD Guidelines. As described above, the substantive norms of the OECD Guidelines integrate environmental, human rights, social labour-related and investment issues. They draw upon or refer to a number of treaty instruments and nonbinding instruments of other institutions. These linkages are sometimes less explicit than in previously mentioned instruments but the commentary of the OECD secretariat reveals that most norms are the result of influence from leading instruments developed in other fields.[327]

---

[323] CCRF, Article 8.7 (vessel owners and crews are called upon to obey the norms on equipment and cleaning procedures under MARPOL 73/78).

[324] CCRF, Articles 8.4.1. and 8.10.1.

[325] CCRF, Article 11.1.3.

[326] For example, the FAO "Technical Guidelines for Responsible Fisheries No 2: Precautionary Approach to Capture Fisheries and Species Introductions" of 1996 contain references to relevant Guidelines of ICES and the Ballast Water Control Forms of the International Maritime Organization. All FAO Technical Guidelines on the CCRF are available at http://www.fao.org/fishery/publications/technical-guidelines/en.

[327] The main "inspirations" for the environmental chapter V of the OECD Guidelines which are mentioned in the Commentary to the OECD Guidelines issued by the OECD secretariat are the Rio Declaration, Agenda 21 and the Aarhus Convention. The commentary is included in Part III of the booklet on the OECD Guidelines issued by the OECD, available from http://www.oecd.org/dataoecd/56/36/1922428.pdf.

The OECD Guidelines thus represent a cross-cutting collection of the most widely accepted principles and norms.

Generally speaking, references to other instruments can often be easier negotiated into a legally nonbinding instrument, because decision makers are usually less concerned about domestic ratification and short-term compliance costs. Even if all of the other instruments referred to in a particular nonbinding instruments were not directly implemented, nonbinding instruments help to establish and regularly revise ambitious standards for a specific issue by integrating all relevant concerns across institutional boundaries, irrespective of the legal status of the source instruments. If one assumes, as will be further outlined below, that a nonbinding instrument plays a role in shaping discourse and in initiating processes of norm formation and implementation activities, nonbinding instruments therefore have the potential to promulgate and globally promote cross-cutting norms and standards.

If a particular instrument is furthermore referenced in other instruments of other institutions, the cross-cutting norms of that particular instrument over time gain acceptance and forge international common understandings.[328] For instance, repeated references further the authority and support the goals of the CCRF so that it is becoming the main point of reference regarding fishery management issues. Thus, implementation of the CCRF is explicitly called for in the Johannesburg Plan of Implementation which was endorsed by the United Nations General Assembly.[329] The Johannesburg Plan also explicitly underscores the importance of the IPOAs and the FAO Technical Guidelines, even though they were "only" developed by the FAO secretariat without being backed by the FAO Conference. Numerous General Assembly Resolutions and other political declarations such as the Reykjavik Declaration of the FAO Conference repeatedly confirm the goals of the CCRF and the IPOAs.[330]

---

[328] P.-M. Dupuy, "Soft law and the international law of the environment", Michigan Journal of International Law 12 (1990-1991), 420-435 (424).

[329] World Summit on Sustainable Development, Johannesburg Plan of Implementation, para. 31 (c), endorsed by the United Nations General Assembly in UNGA Resolution A/RES/57/253 (20 December 2002).

[330] See for example UN Doc. A/60/31 (17 November 2005), paras 48 and 82, available at http://daccessdds.un.org/; para. 1 of the 'Reykjavik Declaration on responsible fisheries in the marine ecosystem', contained in Appendix 1 of the report of the Reykjavik Conference on responsible fisheries in the marine ecosystem of 2-13 November 2001, FAO Doc. C 2001/INF/25; the Rome Declara-

The contribution of nonbinding instruments to the gradual emergence of cross-cutting standards builds on instrument-based linkages of different institutions. Since different institutions speak to and for different interests, for example as industry finds better recognition in organisations such as the OECD while environmental NGOs have better linkages to UNEP or agricultural ministries within the FAO, these instrument-based linkages also soften clear-cut sectoral boundaries. The balancing of interests required to pursue the objectives of sustainable development in this manner takes place at least to some extent at the international level.

To be sure, it has to be stressed however that this integrationist potential of many nonbinding instruments does not solve or overcome fragmentation problems. Nor does it render superfluous the domestic balancing of interests through legal procedures such as environmental impact assessments. In fact, the recognition of various values, norms and objectives in nonbinding instruments does not solve any concrete norm or institution conflicts. The difficult decision-making that needs to reconcile various interests still needs to be done at the level of implementation. The political decisions therefore do not become less difficult. The contribution of nonbinding instruments in this respect merely consists in broadening the discourse through integration of various perspectives. And finally, while integrating a large amount of various norms, possibly of different sectors facilitates acceptance of the instrument in question, the actual normative prescription for a concrete situation may in fact become blurred, watered down or weakened.

2. The role of nonbinding instruments in inter-institutional cooperation

Once a particular nonbinding instrument adopted by one institution (the "source institution") becomes accepted in a certain field, it is often adopted as a point of reference by instruments of other organisations and institutions as a basis for their own implementation and policy making activities. Where this happens, the norms and expertise attached to the instrument find recognition in the other institutions. In this way, commitments undertaken by states in the context of the source institution affect norm setting and compliance-inducing activities in another

---

tion on the Implementation of the Code of Conduct for Responsible Fisheries, adopted at the FAO Ministerial Meeting on Fisheries in Rome (10-11 March 1999), in particular at d), e), g), i), l), available at www.fao.org//DOCREP/005/X2220e/X2220e00.htm.

institution.[331] These inter-institutional linkages often have positive repercussions for the effectiveness of the instrument adopted as well as for the activities of the source institutions.

Some examples presented in this section will clarify and substantiate this point on a less abstract level. These examples also demonstrate how such horizontal inter-institutional cooperation not only happens between international institutions with a potentially global reach, but more broadly also between international and regional organisations.

a) Influence on norm development in other institutions (FAO-WTO)

Nonbinding instruments adopted in the source institution influence norm making activities in other institutions. They may become relevant by supporting certain arguments and delegitimising others in the policy making and norm development processes in another institution.[332] The actors which may drive this development vary. Being directly involved in negotiations, states can make use of nonbinding instruments in support of their particular positions. But also other actors such as the representatives or experts of the source institution or environmental NGOs frequently use their observer status to influence negotiations or policy making, for example by supporting references to a nonbinding instrument.

One example for these processes is the role played by FAO fisheries instruments in the negotiations taking place within the Doha Round of the WTO on fisheries subsidies. As will be seen in this section, states in support of reforms of the subsidies rules, the FAO secretariat and NGOs such as the World Wildlife Fund in concert with UNEP have used nonbinding fisheries instruments of the FAO in support of their positions in the negotiations.

Nonbinding instruments adopted by the FAO have identified overcapacity as a major factor for overfishing. They therefore advocate the reduction of fisheries subsidies. The general objective is formulated in the

---

[331] O.S. Stokke/C. Coffey, "Institutional Interplay and Responsible Fisheries: Combating Subsidies, Developing Precaution" in: S. Oberthür/T. Gehring (eds.), Institutional interaction in global environmental governance: synergy and conflict among international and EU policies, 2006, 127-155 (131).

[332] J.B. Skjaerseth/O.S. Stokke/J. Wettestad, "Soft Law, Hard Law, and Effective Implementation of International Environmental Norms", Global Environmental Politics 6 (2006), 104-120 (117).

Code of Conduct for Responsible Fisheries which calls upon states to avoid excess fishing capacity and to make sure that economic conditions under which fishing industries operate promote responsible fisheries.[333] More specifically, the International Plan of Action for the Management of Fishing Capacity ("IPOA-Capacity") adopted under the framework of the CCRF recommends national plans for the management of fishing capacity which include assessments of "all factors, including subsidies, contributing to overcapacity",[334] and that "[s]tates should reduce and progressively eliminate all factors, including subsidies ... which contribute ... to the build-up of excessive fishing capacity thereby undermining the sustainability of marine living resources."[335]

The FAO instruments give support to the position of a number of states united in the so-called "Friends of Fish" group[336] which seek a reform of WTO rules on subsidies. Although the causalities are of course difficult to empirically prove, it is indicative of the significance of this fact that members of this group of states have consistently emphasised the existence of the FAO instruments in this area.[337] Further, the FAO secretariat used its position as an observer in the WTO Committee on Trade and Environment to present the IPOA-Capacity to the Committee in June 1999.[338] In addition to the Fish Stocks Agreement, the CCRF and the IPOA-Capacity of the FAO therefore seem to have been influential in setting the agenda by providing a legitimate argument in favour of reforms. The global fisheries norms adopted by the FAO and additional scientific work of the FAO made it particularly difficult for opponents to resist arguments for reform. [339] The linkage between subsidies and overcapacity made it to the forefront of the

[333] FAO CCRF, Article 7.2.2.

[334] International Plan of Action on Fishing Capacity, Article 25.

[335] International Plan of Action on Fishing Capacity, Article 26.

[336] This group includes such important fishing nations as Australia, Chile, Ecuador, Iceland, New Zealand, Peru, the Philippines and the United States.

[337] See e.g. WTO Doc. WT/CTE/W/121 (28 June 1999), para. 14 and Annex II; WTO Doc. WT/GC/W/303 (6 August 1999); WTO Doc. WT/CTE/W/154 (4 July 2000), para. 3.

[338] WTO Doc. WT/CTE/W/126 (12 October 1999), available at www.Docs online.wto.org.

[339] O.S. Stokke/C. Coffey, "Institutional Interplay and Responsible Fisheries: Combating Subsidies, Developing Precaution" in: S. Oberthür/T. Gehring (eds.), Institutional interaction in global environmental governance: synergy and conflict among international and EU policies, 2006, 127-155 (138-139),

agenda on rule reforms. Accordingly, the 2001 Doha Ministerial Declaration stated that in the context of the new round of trade negotiations, "participants shall also aim to clarify and improve WTO disciplines on fisheries subsidies, taking into account the importance of this sector to developing countries."[340] The negotiations eventually led to states broadly agreeing that rules must be developed which "strengthen disciplines on subsidies in the fisheries sector, including through the prohibition of certain forms of fisheries subsidies that contribute to overcapacity and over-fishing."[341]

This is not to be taken as argument that the FAO instruments were the only or decisive factor in this process. Perhaps even more decisive in terms of agenda setting was the argument made in particular by UNEP that fishing subsidies hurt the development process in developing countries which have to compete with subsidised fishing fleets from developed countries.[342] UNEP was also actively cooperating with the World Wildlife Fund by organising a number of conferences and expert meetings on the issue. These meetings were attended by over 120 experts from governments, international organisations, including the FAO and the WTO, as well as regional fisheries organisations and NGOs. However, the reports of these meetings indicate that the nonbinding instruments of the FAO have played a significant role as they buttressed arguments in favour of sustainability reforms.[343]

The negotiations on new rules on fisheries subsidies show that the nonbinding instruments of the FAO contributed to the emergence of more differentiated WTO rules on the issue. As outlined in a previous sec-

---

340 WTO, Ministerial Declaration, adopted in Doha on 14 November 2001, WT/MIN(01)/DEC/1, para. 28. The declaration is available at http://www.wto.org/english/thewto_e/minist_e/min01_e/mindecl_e.pdf.

341 WTO, Ministerial Declaration, adopted at 6th Ministerial Meeting in Hong Kong on 18 December 2005, WT/MIN(05)/DEC, Annex D, para. 8. The declaration is available at http://www.wto.org/english/theWTO_e/minist_e/min05_e/final_text_e.pdf.

342 O.S. Stokke/C. Coffey, "Institutional Interplay and Responsible Fisheries: Combating Subsidies, Developing Precaution" in: S. Oberthür/T. Gehring (eds.), Institutional interaction in global environmental governance: synergy and conflict among international and EU policies, 2006, 127-155 (137).

343 UNEP/WWF, Chair's summary on Symposium of 1-2 March 2007 on "Disciplining fisheries subsidies: incorporating sustainability at the WTO & Beyond", available at http://www.unep.ch/etb/events/pdf/Fish_Chairs_ Summary_final.pdf.

tion,[344] the FAO instruments are likely to form the threshold for acceptable subsidies under a new agreement. These rules are – due to the strength of the WTO regime – likely to strengthen implementation of the mentioned norms of the FAO instruments.[345] The draft of the Chairman of the Negotiating Committee proposed a number of new rules which – simply put – condition the acceptability of subsidies on compliance with best practices of fisheries management as defined by the Fish Stocks Agreement, the Compliance Agreement but also the CCRF, IPOAs and Technical Guidelines of the FAO.[346]

b) Influence on norm implementation activities of other institutions

The cross-institutional recognition of nonbinding standards through references to nonbinding instruments in international treaty law has already been discussed in this Part in section A.I.2. The references in UNCLOS to (binding and nonbinding) instruments of the FAO and the IMO as well as the references in the TBT and SPS and possibly the future WTO Subsidies Agreement of the WTO clearly open the respective treaty regime for standard setting activities of other institutions. In this way they recognise the expertise and particular motives of other institutions which typically represent other societal actors and interests. The amplification of the influence of these instruments through treaty law not only contributes to cross-institutional cooperation and interest diversification, but it also heightens the need to secure the legitimacy of the source institutions.[347]

Apart from references within treaties, there are also other ways in which international institutions rely on nonbinding instruments developed in other institutions in their standard setting and policy implementation activities. In terms of the impact of the international norms, these linkages become particularly relevant when the other organisation relying on these instruments disposes of means to enhance and ensure

---

344 See in this Part at A.I.3.a),further above.

345 J.B. Skjaerseth/O.S. Stokke/J. Wettestad, "Soft Law, Hard Law, and Effective Implementation of International Environmental Norms", Global Environmental Politics 6 (2006), 104-120 (114).

346 See already the discussion of references in the WTO discussed in this Part at A.I.2, further above.

347 For the legitimacy question and the need for transparent and inclusive procedures, see Part 3 of this study, further below.

compliance that the source institution does not have. Three such examples will be discussed shortly, namely the recognition of nonbinding norms by financial institutions (World Bank), by the United Nations Security Council, and by regional organisations. Finally, where the European Union takes up international nonbinding norms, this obviously greatly contributes to the effectiveness of these instruments vis-à-vis the member states of the EU.

(1) Example: World Bank

Particularly relevant in this context are linkages established between nonbinding instruments adopted by any international organisation with the instruments and activities of financial institutions. One outstanding example in this respect is the World Bank.

For instance, in response to the call of the World Summit on Sustainable Development's Johannesburg Plan of Implementation to address the critical state of world fisheries,[348] the World Bank established a Global Program on Sustainable Fisheries ("ProFish") which is financed by the World Bank Development Grant Facility. ProFish is a partnership between fishery sector donors, international financial institutions, developing countries, stakeholder organisations, and international agencies.[349] Borrowers can receive loans for implementing sustainable fisheries management practices. One of the main instruments guiding the activities of the program is the FAO Code of Conduct for Responsible Fisheries, along with related technical guidelines and Plans of Action which were referred to in the World Summit on Sustainable Development's Johannesburg Plan of Implementation.[350] One of the overall objectives of the program is to address fisheries governance on the national, regional and international level with a view to implement fisheries management as recommended in the FAO Code of Conduct for Responsible Fisheries.[351]

The Operational Procedures and Safeguards of the World Bank serve as a sort of enforcement mechanism for international standards developed

[348] WSSD, Johannesburg Plan of Implementation, Chapter IV, paras 29-34.

[349] World Bank website, http://go.worldbank.org/DM59TK1531 (permanent URL).

[350] WSSD, Johannesburg Plan of Implementation, para. 30 (c).

[351] World Bank website, www.worldbank.org/fish.

in other international institutions irrespective of their legal status.[352] As explained in Part 1,[353] the Operational Procedures and Safeguards are norms which determine whether a particular project will be financed. They also serve as benchmarks for the assessment and control of the World Bank's activities by the inspection panel.[354] In addition to environmental treaty law,[355] the Operational Policies refer to numerous nonbinding instruments. For example, the Operational Policies on pest management rely on so-called "operational principles" as described, *inter alia*, in the FAO Code of Conduct on the Distribution and Use of Pesticides in its revised version of 2003.[356] The Operational Procedures 4.09 on Pest Management furthermore require that financed pesticides must be manufactured, handled and disposed of in accordance with the FAO's Guidelines for Packaging and Storage of Pesticides, the Guidelines on Good Labelling Practice for Pesticides and the Guidelines for the Disposal of Waste Pesticide and Pesticide Containers on the Farm.[357] With the incorporation of the Operational Policies into loan and credit agreements between the Bank and borrowing, both the internal policies and the referenced nonbinding instruments become binding for the recipient state. On the other hand non-compliance with a particular policy or referenced instrument may lead to a negative deci-

[352] J.E. Alvarez, International Organizations as Law-makers, 2005, 237; D.A. Wirth, "Compliance with Non-Binding Norms of Trade and Finance" in: D. Shelton (ed.), Commitment and Compliance: The Role of Non-Binding Norms in the International Legal System, 2000, 330-344 (333-336).

[353] See Part 1, A.V.1., further above.

[354] For details, see L. Boisson de Chazournes, "Policy Guidance and Compliance: The World Bank Operational Standards" in: D. Shelton (ed.), Commitment and Compliance: The Role of Non-Binding Norms in the International Legal System, 2000, 281-302; C. Holstein, Der Umweltschutz in der Tätigkeit der Weltbankgruppe: Instrumente, rechtliches Mandat und Bedeutung für das internationale Umweltrecht, 2001, 59-89.

[355] Thus, no project is supported by the World Bank which appears to contravene international environmental obligations of the respective state, compare OP 4.01 Environmental Assessment, para. 1; OP 4.01 is available from http://go.worldbank.org/WL5H4IK9U0 (permanent URL).

[356] The World Bank Operational Manual, Operational Policies OP 4.00, Table A1, at Part C, para. 3, available at http://go.worldbank.org/XFBVTIUDK0 (permanent URL).

[357] Compare OP 4.09 on Pest Management (2004), Footnote 7, available from http://go.worldbank.org/QNORFLUFR0 (permanent URL).

sion of the World Bank on the financing of a project or withdrawal of funding.

(2) Example: the use of OECD Guidelines by the United Nations Security Council

Although the United Nations has so far not adopted a code of conduct for transnational enterprises,[358] it has relied on the OECD Guidelines for the assessment of the conduct of multinational corporations and initiated institutional cooperation between an expert body of the United Nations and the implementation bodies of the OECD Guidelines.

The incident in which this occurred concerned action of the United Nations Security Council on the illegal exploitation of natural resources in the Democratic Republic of Congo (DRC).[359] In this case, the Security Council directly addressed the conduct of multinational corporations by establishing an independent fact-finding body to investigate the conduct of multinational corporations in the DRC with respect to human rights and environmental standards.[360] Although primarily concerned with the linkages between illegal resource exploitation and the conflict in the DRC, the reports of the expert group also highlighted environmental issues such as illegal trade in endangered species taken from protected areas as well as illegal logging.[361] One of the reports includes in two Annexes (Annex I and III) lists of corporations which have – in the opinion and based on the information of the expert group – contravened the OECD Guidelines in their business activities in the DRC, either directly or through suppliers. For those corporations mentioned in Annex I, the expert group recommends that the home states issue financial and travel restrictions.[362] The 85 enterprises listed in Annex III were

[358] For the attempts to develop a comprehensive code of conduct at UN level, see already in Part 1, at B.III, further above.

[359] UN Security Council Res. 1457, U.N. Doc. S/RES/1457 (24 January 2003); Security Council Res. 1499, U.N. Doc. S/RES/1499 of 13 August 2003.

[360] Statement by the President of the Security Council, U.N. Doc. S/PRST/2000/20 of 2 June 2000.

[361] UN Security Council, Final report of the Panel of Experts on the Illegal Exploitation of Natural Resources and Other Forms of Wealth of the Democratic Republic of the Congo, U.N. Doc. S/2002/1146 of 16 October 2002, para. 33.

[362] UN Security Council, Final report of the Panel of Experts on the Illegal Exploitation of Natural Resources and Other Forms of Wealth of the Democ-

accused of not having observed the OECD Guidelines. The report distinguishes between those corporations based in signatory states of the Guidelines and those which are not, but which nevertheless contravened those norms. However, whether the home state of the enterprise officially adheres to the OECD Guidelines only plays a secondary role for the listing of the corporations and thus for its reputational damage. The report indicates that the "countries which are signatories to those Guidelines and other countries are morally obliged to ensure that their business enterprises adhere to and act on the Guidelines".[363] In other words, the OECD Guidelines are used as a general standard irrespective of their original scope and institutional boundaries.

The case not only highlights how the United Nations affirm the non-binding norms of an OECD instrument, but also provides an example of explicit institutional cooperation by the UN Security Council with the OECD Guidelines' implementation bodies. When endorsing the report of its expert group, the Security Council requested the expert group to cooperate with the implementation bodies of the OECD Guidelines by providing information to those bodies on the alleged contraventions of companies with the OECD Guidelines.[364] In other words, the Security Council not only attempts to assess implementation of the norms adopted at the OECD, but also aims to enforce these standards of OECD implementation bodies through listings and the authoritative request to those bodies to take action.

The response of the implementation bodies of the OECD remained however modest. Some companies reassessed their activities in the DRC.[365] A few of the National Contact Points initiated inquiries as a

ratic Republic of the Congo, U.N. Doc. S/2002/1146 of 16 October 2002, paras 174 et seq.

363 UN Security Council, Final report of the Panel of Experts on the Illegal Exploitation of Natural Resources and Other Forms of Wealth of the Democratic Republic of the Congo, U.N. Doc. S/2002/1146 of 16 October 2002, para. 177.

364 UN Security Council Res. 1457, U.N. Doc. S/RES/1457 of 24 January 2003, para. 14; reiterated in Security Council Res. 1499, U.N. Doc. S/RES/1499 (2003) of 13 August 2003, para. 4.

365 S.F. Vendzules, "The Struggle for Legitimacy in Environmental Standards Systems: The OECD Guidelines for Multinational Enterprises", 21 Colorado Journal of International Environmental Law & Policy 21 (2010), 451-489 (475).

follow-up to the report.[366] The expert group was invited to meetings of the OECD Committee on Multinational Enterprises (now Investment Committee) and some cooperation indeed took place. Irrespective of whether the cooperation in this case was actually successful, it is an example on how interinstitutional cooperation could potentially address cases where multinational corporations are involved in environmental exploitation and armed conflict.[367] In general terms, both organisations share their infrastructure and rely on each other's standards and expertise to pursue their goals to the extent that they overlap.

c) In particular: impact on regional organisations

This section takes a specific look at the potential impact on regional organisations. Again, such an impact is often clearly discernable despite the nonbinding nature of these instruments. When internationally agreed guidelines and codes of conduct are recognised by regional organisations, they first of all contribute to coordinating the activities of these organisations with the principles and standards contained in the instrument. Moreover, regional organisations are particularly significant for the effectiveness of international nonbinding norms. Nonbinding norms are reinforced when regional organisations take these norms into account or recognise them as guidance for their own activities. Because in contrast to most international organisations with a global reach, it is often easier to regulate issues and ensure compliance through regional organisations with fewer actors. The number of actors also has a positive effect on compliance, because reputational costs of noncompliance are higher when there are fewer actors, and compliance more directly improves the cost-benefit calculations.[368] Although it is certainly difficult to generalise, this finding from compliance research also applies to

---

[366] E. Morgera, "An Environmental Outlook on the OECD Guidelines for Multinational Enterprises: Comparative Advantages and Legitimacy in relation to Other International Initiatives", Georgetown International Environmental Law Review 18 (2006), 751-777 (773).

[367] E. Morgera, "An Environmental Outlook on the OECD Guidelines for Multinational Enterprises: Comparative Advantages and Legitimacy in relation to Other International Initiatives", Georgetown International Environmental Law Review 18 (2006), 751-777 (773).

[368] E. Brown Weiss, "Conclusions: Understanding Compliance with Soft Law" in: D. Shelton (ed.), Commitment and Compliance: The Role of Nonbinding Norms in the International Legal System, 2000, 535-556 (547).

nonbinding instruments of regional organisations. In addition, regional organisations often more easily than states respond to impulses from other multilateral organisations, and in contrast to global organisations more often have the competence to issue binding norms and to resort to enforcement mechanisms.[369]

(1) Example: regional fisheries management organisations

In the field of fisheries, cooperation within regional fisheries organisations is emerging as the key strategy to achieve sustainable fisheries. Among regional fisheries organisation, one must generally distinguish between regional fisheries bodies and regional fisheries management organisations. Regional fisheries bodies only have an advisory mandate, and act as facilitators of governmental cooperation. Regional fisheries management organisations, however, are often empowered to issue binding management measures, sometimes even by a two-thirds majority with opt-out possibility, and to restrict access to resources within their territory.[370] This does not mean that actual enforcement is secured. Implementation and enforcement vis-à-vis fishing vessels remain decentralised in the hands of states and national courts. Regional organisations can therefore not completely solve the difficulties of enforcement which exist in fisheries governance. But they provide essential elements in the international legal fisheries system by establishing the total allowable catch for their areas, allocating the resources through quota systems and enforcing through monitoring, control and surveillance measures. The Northwest Atlantic Fisheries Organization (NAFO)[371] is a particularly advanced example that can be taken as a model for future

---

[369] Due to its supranational nature, the European Union is of course the most effective in terms of enforcement among regional organizations. As it can legislate and directly enforce its norms, it will be dealt with in the section on direct legislative implementation of nonbinding instruments, in this Part 2, at B.II., further below.

[370] E.g. Agreement for the Establishment of the Indian Ocean Tuna Commission, Article IX para. 1. The Agreement is available at ftp://ftp.fao.org/FI/OCUMENT/iotc/Basic/IOTCA_E.pdf.

[371] The NAFO was founded through the Convention on Future Multilateral Cooperation in the Northwest Atlantic Fisheries (24 October 1978); UNTS 369; 34 ILM 1452 (1995).

development.[372] Authorised to establish measures binding upon member states by majority with a contracting out possibility,[373] the NAFO Fisheries Commission establishes total allowable catch and quota systems, which can be enforced through a number of enforcement measures such as inspections at sea and a licensing and electronic vessel monitoring system. In addition, it has also established a listings procedure to counter illegal, unreported and unregulated fishing through port state enforcement. NAFO establishes a list of vessels from State Parties and non-State Parties which are presumed to be engaged in illegal, unreported or unregulated fishing on the basis of information from State Parties. The Parties to the Convention are obliged to deny access to their ports, so that these fishing vessels cannot land their fish or reflag and change crews.[374]

To give one example, regional fisheries organisations, in particular regional fisheries management organisations, respond in manifold ways to the international nonbinding instruments of the FAO such as the CCRF. Some of these organisations have adapted their founding documents and measures to conform to the objectives of the Code of Conduct for Responsible Fisheries (CCRF). The Statute of the South West Indian Ocean Fisheries Commission – a regional fisheries body recently established by the FAO Council[375] – for example provides that "[T]he Commission has due regard for and promotes the application of the provisions of the FAO Code of Conduct on Responsible Fisheries, including the precautionary approach and the ecosystem approach to

---

372 The mechanisms of North Atlantic Fisheries Organisation are analysed with a specific focus on their legitimacy by R. Wolfrum, "Legitimacy of International Law and the Exercises of Administrative Functions: The Example of the International Seabed Authority, the International Maritime Organization (IMO) and International Fisheries Organizations" in: A. von Bogdandy/ R. Wolfrum/J. von Bernstorff/P. Dann/M. Goldmann (eds.), The Exercise of Public Authority by International Institutions: Advancing International Institutional Law 2010, 917-940 (938-939).

373 NAFO Convention, Article XI, para. 7, XII and XIV.

374 NAFO Conservation and Enforcement Measures 2009, NAFO FC Doc. 09/1 Serial No. N5614, Article 57, available at http://www.nafo.int/ about/frames/activities.html.

375 The South West Indian Ocean Fisheries Commission (SWIOFC) was established in 2004 by Resolution 1/127 of the FAO Council under Article VI (1) of the FAO Constitution, see for the Agreement www.fao.org/fishery/rfb/ SWIOFC.

fisheries management."[376] This is hardly surprising for those organisations which are established under the FAO Constitution with the approval of the Conference or the Council.[377] These organisations remain linked to the FAO governing bodies in a number of ways. This facilitates the influence of FAO policy on them. Proposals for amendments of their founding documents can generally be made by the Director-General.[378] Furthermore, they remain under the general oversight of the FAO Council, which has the power to disallow amendments.[379]

Non-FAO regional fishery bodies also take steps to implement the CCRF. The International Baltic Sea Fishery Commission has introduced the so-called "Baltic Agenda 21" which outlines objectives for sustainable fisheries on the basis of principles which are taken from, *inter alia*, international instruments like the CCRF. Further implementation efforts are undertaken by the North Atlantic Fishery Organisation with respect to the application of the precautionary approach, and by the Permanent South Pacific Commission, which is actively encouraging its members to include the CCRF in their national legislation. The Latin American Organisation for Development of Fisheries has established a regional programme for the implementation of international fishery instruments, including the CCRF, in cooperation with the Inter-American Development Bank.[380] The case of the newly created South West Indian Ocean Fisheries Commission illustrates that newly created regional fisheries organisations include express references to the CCRF instead of enumerating guiding principles.

But the CCRF also directly influences management strategies and thus management measures of regional fisheries management organisations. The code is often referred to as a justification for concrete resolutions and decisions, a fact that underlines the acceptance of the code as a gen-

[376] Statute of the South West Indian Ocean Fisheries Commission, para. 5, available at http://www.intfish.net/orgs/fisheries/swiofc.htm.

[377] FAO Constitution, Article 14 (2) b). The FAO regional fisheries organisations comprise, in addition to the South West Indian Ocean Commission (SWIOC) the Fishery Committee for the Eastern Central Atlantic (CECAF), the Western Central Atlantic Fishery Commission (WECAFC), the Indian Ocean Fishery Commission (IOFC), the Asia-Pacific Fishery Commission (APFIC), the General Fisheries Commission for the Mediterranean (GFCM), and the Indian Ocean Tuna Commission (IOTC).

[378] E.g. Indian Ocean Tuna Commission Agreement, Article XX, para. 2.

[379] E.g. Indian Ocean Tuna Commission Agreement, Article XX para. 3.

[380] This information is available from http://www.oldepesca.com/convenio.

eral best practice standard. Thus, the Indian Ocean Tuna Commission – a regional fisheries management organisation set up by the FAO – repeatedly bases its resolutions and decisions explicitly on the principles and recommendations of the CCRF. Its resolution 99/01 on overcapacity, juvenile overfishing and flags of convenience vessels mentions the CCRF as a principal source which "provides that States should take measures to prevent or eliminate excessive fishing capacity", and cites the International Plan of Action on Fishing Capacity as the source "calling for immediate action to reduce fishing capacity in major international fisheries".[381] Similarly, the recommendations and resolutions of the International Commission for the Conservation of Atlantic Tunas (ICCAT) since 1995 often refer to the CCRF and related instruments as justification for their measures.[382]

Less frequently, parts of the CCRF have also been transformed into binding measures by regional fisheries bodies.[383] For example, the implementation of the precautionary principle through the adoption of the CCRF's policy measures and methodologies is a common practice among regional fisheries bodies.[384] A particularly salient example is the implementation of the FAO's International Plan of Action on Illegal, Unregulated and Unreported Fishing (IPOA-IUU) – adopted under the framework of the CCRF – by the Commission for the Conservation of Southern Bluefin Tuna. The Commission has adopted a binding resolution on this issue which essentially establishes a system of authorised fishing based on a public record of authorised vessels.[385] Finally,

---

[381] Indian Ocean Tuna Commission (1999), Resolution 99/01.

[382] ICCAT Recommendation 04/10 (2004) refers to the FAO International Plan of Action on Sharks in its opening paragraph by "recalling that the United Nations Food and Agriculture Organization (FAO) International Plan of Action for Sharks calls on States, within the framework of their respective competencies and consistent with international law, to cooperate through regional fisheries organizations with a view to ensuring the sustainability of shark stocks as well as to adopt a National Plan of Action for the conservation and management of sharks", available from www.iccat.int.

[383] FAO, The Challenge of Renewal, Independent External Evaluation of the Food and Agriculture Organization, Working Draft (2007), para. 630, available at: http://www.fao.org/unfao/bodies/IEE-Working-Draft-Report/K0489E.pdf.

[384] FAO, COFI/2005/2, para. 39.

[385] Commission on the Conservation of Southern Bluefin Tuna, Resolution on 'Illegal, Unregulated and Unreported Fishing (IUU) and Establishment of a CCSBT Record of Vessel over 24 meters' of 7-10 October 2003, adopted at the Eleventh Annual Meeting of 19-22 October 2004, available at www.ccsbt.org.

some regional organisations have made efforts to implement the mechanisms suggested by the IPOA-IUU by adopting regulatory measures that range from the prohibition of fishing and inspections,[386] to trade measures as a last resort.[387]

The CCRF thus serves as a general guideline and standard for the instruments and measures of both regional fisheries bodies and regional fisheries management organisations. In sum, however, even though the CCRF is widely recognised and influential, actual translation into enforceable measures still remains limited.[388] Regional organisations often only decide by consensus which often marginalises scientific findings. Further, management organisations often do not dispose of the legal competences and capabilities needed to monitor and enforce their management measures, and thus largely depend on the discretion and capability of states to do so.

(2) Example: European Union

The example of the European Union (EU) illustrates how nonbinding instruments can gain force and authority through a strong regional organisation. The EU on the one hand has the political and economic leverage to become influential on other non-EU states' policies. On the other hand, the EU due to its supranational nature also has a strong direct impact on policy and law-making of its member states. To the extent that decision makers at the EU level recognise the objectives and measures of nonbinding instruments, these mechanisms have enormous potential to raise the impact and authority of nonbinding international instruments.

A clear case of linkage economic incentives and nonbinding international instruments for sustainable development is the recently adopted Economic Partnership Agreement between the CARIFORUM States and the European Union.[389] The conclusion of these Economic Partner-

---

[386] E.g. the Commission for the Conservation of Antarctic Marine Living Resources, compare FAO COFI/2005/2, para. 45.

[387] E.g. International Commission for the Conservation of Atlantic Tunas.

[388] G. Hosch, "Analysis of the implementation and impact of the FAO Code of Conduct for Responsible Fisheries since 1995", FAO Fisheries and Aquaculture Circular No. 1038 (FAO 2009), 53, available from www.fao.org/fi.

[389] Economic Partnership Agreement between the CARIFORUM States, of the one part, and the European Community and its member states, of the other

ship Agreements was decided in a framework agreement signed in Cotonou by the European Community and African, Caribbean and Pacific group of states as a successor to the Lomé Agreements. The aim of these agreements is to foster the gradual integration of the African, Caribbean and Pacific group of states into the global economy and to assist development in these countries. The Agreement also obliges Parties to follow sustainable management practices. In the part on sustainable development, the Agreement stresses that increasing the competitiveness of production and trade in agricultural and fishery products must be in line with sustainable management practices. For example, one of the goals is to maximise benefits from fisheries and agriculture in accordance with preserving the ecosystem, with reference to the precautionary principle and the management practices as defined in the FAO Code of Conduct for Responsible Fisheries.[390]

Furthermore, the European Union, with its law-making competence, is often instrumental in making nonbinding international instruments binding for its member states. Consider for instance the case of the (voluntary) implementation of the PIC procedure. The European Union passed implementing legislation long before the PIC Convention entered into force in response to the voluntary instruments containing the PIC procedure. The Regulation 2455/92 was passed already during the start-up phase of PIC.[391] This regulation, which has direct effect in member states, makes the international PIC procedure mandatory for all exporters.[392] Technically, this is achieved by directly incorporating the PIC list into Annex II of the Regulation.[393] With the first formal

---

part; Official Journal of the European Union (15 October 2008), L 289, 15/10/2008, p. 3.

[390] Article 37.3 of the Economic Partnership Agreement between CARIFORUM and the EU reads: "[T]he Parties recognise that the fisheries and marine ecosystems of the CARIFORUM States are complex, biologically diverse and fragile and that exploitation should take into account these factors through effective conservation and management of fisheries resources and related ecosystems based on sound scientific advice and on the precautionary principle as defined by the FAO Code of Conduct on Responsible Fisheries."

[391] See EEC Regulation 2455/92 (1992) concerning the export and import of certain dangerous chemicals, (Official Journal L 251, 29/08/1992, p. 13).

[392] See EEC Regulation 2455/92, Article 5 and Annex II.

[393] The mechanism did however not include a dynamic reference, but changes at the international level had to be accepted by formal amendments to the EC regulation, compare Article 11 (2) of EEC Regulation 2455/92.

amendment, the first PIC list as adopted by the FAO/UNEP Joint Expert Group was inserted into the Annex which at the time of adoption of the regulation had been left blank.[394] The voluntary procedure as well as the PIC lists adopted by the Joint Expert Group thereby became directly enforceable law in the European Community. As required by the nonbinding international instruments, the EC Regulation imposed direct obligations on importers to act in conformity with the international PIC system.[395]

With respect to fisheries, the issue is less clear. Even if the EU is still far from securing sustainable resource management in the fisheries sector,[396] some influence of nonbinding fisheries instruments can nevertheless be observed. Thus, Regulation 345/92 implemented directly the prohibition of large scale pelagic drift net fishing contained in UNGA Resolution 44/225[397] by establishing a moratorium for fishing with drift-nets longer than 2.5 kilometres.[398] In confirming the prohibition of the use of large driftnets through an EU Regulation, the European Court of Justice *inter alia* argued that it was validly based on, among other things, the UNGA Resolution.[399] Moreover, Regulation 2371/2002/EC outlining the conservation and sustainable exploitation of fisheries resources under the Common Fisheries Policy contains sev-

---

394 M.A. Mekouar, "Pesticides and Chemicals: The Requirement of Prior Informed Consent" in: D. Shelton (ed.), Commitment and Compliance: The Role of Non-Binding Norms in the International Legal System, 2000, 146-163 (158); M. Pallemaerts, "International Transfer of Restricted or Prohibited Substances, Regulation of Chemicals", Yearbook of International Environmental Law 3 (1992), 281-287 (285).

395 EEC Regulation 2455/92, Article 5, para. 4.

396 On fisheries policy of the European Union R. Wolfrum, "Die EG und das Meer: Versuch einer Neubewertung", Archiv des Völkerrechts 42 (2004), 67-79.

397 UN Doc. A/RES/44/225 (22 December 1989), para. 4(a).

398 See Council Regulation (EEC) No 345/92 of 27 January 1992, amending for the eleventh time Regulation (EEC) No 3094/86 which lays down certain technical measures for the conservation of fishery resources, Official Journal L 042, 18/02/1992 P. 0015 – 0023, amended by Council Regulation (EC) No 894/97 of 29 April 1997 laying down certain technical measures for the conservation of fishery resources, later repealed with the new common fisheries policy through EU Council Regulation 2371/2002/EC.

399 European Court of Justice, Mondiet, C-405/92 [1993] E.C.R. I-06133, in particular para. 35 and 42.

eral norms which correspond to Articles of the CCRF.[400] Also, some EU Council Regulations implement recommendations of UNGA Resolutions and the FAO Code of Conduct for Responsible Fisheries. For instance, the EU Council Regulation on the protection of vulnerable marine ecosystems in the high seas from the adverse impacts of bottom fishing gears adopted in July 2008 foresees a strict system of fishing permits, a vessel monitoring system and area closures to protect vulnerable marine ecosystems. The Council Regulation is explicitly justified as following the guidance of UNGA Resolution 61/105 and the Code of Conduct for Responsible Fisheries of the FAO.[401] Implementation of the EU Council Regulation concerning the establishment of a Community framework for the collection, management and use of data in the fisheries sector and support explicitly cites UNGA Resolutions and the FAO Code of Conduct for Responsible Fisheries. To base a legal act on nonbinding instruments was considered valid by the European Court of Justice in the *Mondiet* decision.[402]

3. Summary

By stressing institutional cooperation rather than norm conflict, this study tries to highlight, in line with an increasing number of scholars, that institutional interplay and cooperation emerge as defining elements in international environmental governance and law.[403] Already in 1993,

---

400 Article 2 Nr. 2 (i) of the respective EC Council Regulation basically mirrors Article 7.5.1 and 7.5.2 of the CCRF, see EC Council Regulation on the conservation and sustainable exploitation of fisheries resources under the Common Fisheries Policy, 2371/2002/EC, 20 December 2002, Official Journal, 2002 L 358, pp. 0059-0080, Article 2 Nr. 2 (i).

401 EC Council Regulation No 734/2008 of 15 July 2008 on the protection of vulnerable marine ecosystems in the high seas from the adverse impacts of bottom fishing gears, Official Journal of the European Union, L 201, 30/7/2008, p. 8–13, para. 5.

402 European Court of Justice, Mondiet, Case C-405/92, [1993] E.C.R. I-06133.

403 M. Schroeder, Die Koordinierung der internationalen Bemühungen zum Schutz der Umwelt, 2005, 183-259; O.S. Stokke, "The Interplay of International Regimes: Putting Effectiveness Theory to Work", FNI Report 14/2001 (2001), 1 et seq.; S. Oberthür/T. Gehring (eds.), Institutional interaction in global environmental governance: synergy and conflict among international and EU policies, 2006.

when summing up the results of various case studies, Robert O. Keohane and others found it "somewhat surprising, but heartening, to discover that in our cases, cooperation among agencies is more salient than interinstitutional conflict".[404]

If successful, two effects may result from the institutional interplay through nonbinding instruments. First, the recognition and use of nonbinding instruments by other institutions indirectly increases the influence of these norms on states and private actors. This is because institutional linkages or "regime interplay"[405], referring to situations in which the operations of one institution are affected by another,[406] ultimately contribute to the effectiveness of environmental norms and institutions.[407] Secondly, the use of nonbinding instruments as coordinating devices contributes to mitigating the problem of "fragmentation"[408] and thus to further the objectives of sustainable development.[409] The prob-

404 R.O. Keohane/P.M. Haas/M.A. Levy, "The Effectiveness of International Environmental Institutions" in: P.M. Haas/R.O. Keohane/M.A. Levy (eds.), Institutions for the earth: sources of effective international environmental protection, 1993, 3-24 (15).

405 For this notion compare O.S. Stokke, "The Interplay of International Regimes: Putting Effectiveness Theory to Work", FNI Report 14/2001 (2001), 1 et seq. (2).

406 O.S. Stokke, "The Interplay of International Regimes: Putting Effectiveness Theory to Work", FNI Report 14/2001 (2001), 1 et seq. (2).

407 Very instructive in this respect is O.S. Stokke, "The Interplay of International Regimes: Putting Effectiveness Theory to Work", FNI Report 14/2001 (2001), 1 et seq. The relevance of the factor of the "international environment" – which includes international organisations, financial institutions and nongovernmental organisations – for compliance by states was already stressed by E. Brown Weiss (ed.), Engaging countries: strengthening compliance with international environmental accords, 1998, 528.

408 Compare e.g. A. Fischer-Lescano/G. Teubner, "Regime collisions: the vain search for legal unity in the fragmentation of global law", Michigan journal of international law 25 (2004), 999-1046; M. Koskenniemi, Fragmentation of international law: difficulties arising from the diversification and expansion of international law. Report of the Study Group of the International Law Commission, 2007; more optimistic is G. Abi-Saab, "Fragmentation or unification: some concluding remarks", New York University journal of international law & politics 31 (1999), 919-933.

409 The importance of such research has also been emphasized by J.J. Kirton/M.J. Trebilcock, "Introduction: Hard Choices and Soft Law in Sustainable Global Governance" in: J.J. Kirton/M.J. Trebilcock (eds.), Hard Choices, Soft

lems arising from fragmentation and the ensuing need for the coordination of various international treaties and institutions have repeatedly been stressed and analysed.[410] International law pursues not only the delimitation and coordination of spheres of influence of states, but increasingly also the protection of societal interests which formerly were exclusively the subject of domestic law and policy making.[411] With the internationalisation of norm development, the balancing of various interests, formerly undertaken at the domestic level, suffers. Thus, there exists a certain tension between the need to integrate environmental, economic and social values called for in the quest for sustainable development[412] on the one hand and the often separated and fragmented

---

Law: Voluntary Standards in Global Trade, Environment and Social Governance, 2004, 3-29 (6).

[410] For the field of environmental law N. Matz, Wege zur Koordinierung völkerrechtlicher Verträge: völkervertragsrechtliche und institutionelle Ansätze, 2005, 340-389; R. Wolfrum/N. Matz, Conflicts in international environmental law, 2003, 159-209; on WTO and other rules of international law J. Pauwelyn, Conflict of norms in public international law: how WTO law relates to other rules of international law, 2003; specifically on the issue of trade and environment M.W. Gehring (ed.), Sustainable development in world trade law, 2005; O. Perez, Ecological sensitivity and global legal pluralism: rethinking the trade and environment conflict, 2004; R. Senti, Die WTO im Spannungsfeld zwischen Handel, Gesundheit, Arbeit und Umwelt: geltende Ordnung und Reformvorschläge, 2006; M. Böckenförde, Grüne Gentechnik und Welthandel: das Biosafety-Protokoll und seine Auswirkungen auf das Regime der WTO, 2004.

[411] W. Friedmann, The Changing Structure of International Law, 1964, 62 et seq.; J.H.H. Weiler, "The geology of international law – governance, democracy and legitimacy", Zeitschrift für ausländisches öffentliches Recht und Völkerrecht 64 (2004), 547-562 (550).

[412] The need for integration as a core element of the concept of sustainable development is stressed by International Court of Justice, Gabčikovo-Nagymaros Project (Hungary/Slovakia), Judgment of 25 September 1997, ICJ Reports 1997, p. 7, para. 140; compare also on the P. Sands, Principles of International Environmental Law, 2003, 252; S. Fritz, Integrierter Umweltschutz im Völkerrecht, 2009, 105; D.B. Magraw/L.D. Hawke, "Sustainable Development" in: D. Bodansky/J. Brunnée/E. Hey (eds.), Oxford Handbook of International Environmental Law, 2007, 614-638 (624); U. Beyerlin, "Different Types of Norms in International Environmental Law: Policies, Principles and Rules" in: J. Brunnée (ed.), Oxford Handbook of International Environmental Law, 2007, 425-448 (443-445) ("below the threshold of normative quality"); D. Hunter/J. Salzman/D. Zaelke, International Environmental Law and Policy, 2002, 342 ("principle is part of modern international law"); Fitzmaurice after careful review of literature and decisions comes to the conclusion that there is "no uni-

landscape of international institutions and treaties on the other.[413] Coordination of activities on the basis of international (binding or nonbinding) norms and the integration of different sets of norms mitigate norm and institutional conflicts, and help better to integrate the various aspects of sustainable development at the international level.

More specifically, three issues have emerged as particularly noteworthy. First, nonbinding instruments are used to establish cross-cutting standards by referencing and integrating a great number of different instruments, binding and nonbinding, of different institutions to an extent which is rarely found in treaty law. On any particular issue, all relevant norms can be integrated without risking non-ratification by states. Generally, states are more inclined to adopt even ambitious norms as they are not legally bound by them. The scope *ratione personae* or *ratione materiae* of treaty instruments can be easily expanded (e.g. Compliance Agreement and UNCLOS through the CCRF). By establishing and promoting cross-cutting standards that easily integrate standards developed by other institutions, these instruments have the potential to somewhat soften the strict sectoral separations between different institutions (e.g. references in OECD Guidelines to freedom of investment and principles of environmental law). Diversifying the sources of standards in this manner gives due consideration to other interests and may thus enhance the integration of different (economic and environmental) perspectives and expertise in the sense promoted by the integrative concept of sustainable development. Of course, greater diversity of views does not necessarily mean that nonbinding instruments sufficiently integrate and balance various views and interests. Institutional bias remains, as each institution is more influenced by particular parts of governments with particular mindsets, for example because of particularly strong connections with industry (OECD), agriculture interests (FAO) or environmental NGOs (UNEP). But as nonbinding instruments also allow for states to choose which practices to implement and how, they usually leave much room for additional balancing of diverse interests at the national level.

---

form and widely accepted notion of sustainable development", see M. Fitzmaurice, Contemporary issues in international environmental law, 2009, 86.

413 J.J. Kirton/M.J. Trebilcock, "Introduction: Hard Choices and Soft Law in Sustainable Global Governance" in: J.J. Kirton/M.J. Trebilcock (eds.), Hard Choices, Soft Law: Voluntary Standards in Global Trade, Environment and Social Governance, 2004, 3-29 (6).

Second, nonbinding instruments – possibly already representing a cross-cutting standard – provide the basis for policy and norm diffusion from one institution to another. It has been suggested that such institutional cooperation can mitigate conflicts of norms.[414] The analysis in this section of the interplay of various institutions, treaty regimes and actors suggests that nonbinding instruments play a significant and perhaps underestimated role in these interactions. Nonbinding instruments contribute to processes of cross-institutional cooperation and processes of incorporation of external expertise and interests into a different institution or private or public organisation. This may happen through argumentative support for certain positions of actors participating in norm development (e.g. fisheries subsidies case) or by the recognition of external standards as a basis for norm development and policy making by other institutions (e.g. the World Bank, or the Security Council). The nonbinding nature of the instruments does not constrain these processes, nor is such a role unique to nonbinding norms. Treaty instruments such as relevant rules of UNCLOS, for example, are also used. Frequently enough, however, the nonbinding instruments are used simply for the fact that adequate treaty instruments have not been adopted in a particular area, or have not been widely accepted. Nonbinding instruments supported unanimously or with large majorities by states are well suited as both arguments and as a basis on which further activities can legitimately build.

Finally, the analysis of linkages and interplay is a prerequisite for understanding the effectiveness of a particular instrument. As mentioned, the interplay of institutions and actors defines to a considerable extent the effectiveness of instruments and institutions to steer the behaviour of addressees.[415] Various factors through which institutions and instruments gain impact are amplified through linkages of institutions or other actors.

Among these factors are incentive structures, costs and the diffusion of ideas.[416] Incentive structures and costs play a role for compliance of actors with a rational and utilitarian disposition. If linked to the World Bank's financing policies, for example, the potential benefits directly in-

---

414 N. Matz, Wege zur Koordinierung völkerrechtlicher Verträge: völkervertragsrechtliche und institutionelle Ansätze, 2005, 340 et seq.

415 This is also emphasized by O.S. Stokke, "The Interplay of International Regimes: Putting Effectiveness Theory to Work", FNI Report 14/2001 (2001), 1 et seq.

416 For details on this, see the discussion in this Part at B.I., further below.

fluence cost-benefit calculations and can therefore be expected as a strong compliance-inducing factor. Fear of reputational and actual costs attached to non-compliance with WTO rules, possibly enforced through dispute settlement, serves as a strong incentive to comply with a nonbinding instrument if it is hardened through WTO norms. The second factor amplified through inter-institutional linkages relates to the diffusion and emulation of information and ideas. One institution can support another institution's effectiveness by drawing public attention at the international and national level to the questions addressed in the recipient regimes, or it may provide solutions and best practices for the problems a particular institution strives to address.[417] This has been the case with the WTO negotiations on fisheries subsidies where the FAO instruments helped to raise the issue of overcapacity and subsidies at international and domestic levels, while at the same time providing a set of best practices for the differentiation of "good" from "bad" subsidies. Nonbinding instruments are ideal tools for the diffusion of information and best practices, because they can be easily adapted, and states are less wary of adopting a broad list of policies and measures and to engage in discourse when commitments are nonbinding. And finally, regional organisations with particular competences to issue binding rules and monitor or enforce compliance such as some regional fisheries management organisations and the European Union are of foremost significance for the impact of the norms on the respective member states.

B. The state level: compliance with and implementation of nonbinding instruments by states

I. Nonbinding instruments and compliance by states

When looking at the vast literature on compliance with international law,[418] one can deduce a number of factors that are of significance for

---

[417] O.S. Stokke, "The Interplay of International Regimes: Putting Effectiveness Theory to Work", FNI Report 14/2001 (2001), 1 et seq. (10-11).

[418] Some of the most important studies also used as a basis of this study include E. Brown Weiss (ed.), Engaging countries: strengthening compliance with international environmental accords, 1998; J. Brunnée/S.J. Toope, "Persuasion and enforcement: explaining compliance with international law", The Finnish yearbook of international law 13 (2002), 273-295; A. Chayes/A. Handler

the question why an actor may or may not comply with international law. When simplified to make them operational, four motivations for compliance can be distinguished:

One motivation traditionally thought to be decisive for compliance is that of coercion.[419] In a world of asymmetrical power distribution, the argument goes, states and institutions may use their superiour power to coerce other actors into compliance. Since, amongst other reasons, coercion involves high and usually growing costs in the long-term, few social orders primarily rely on coercion for control. At the international level where centralised enforcement is absent, and in particular in international environmental law where sanctions are not available nor conducive to the protection of common goods, coercion can hardly be the key to compliance.[420] This does not exclude that enforcement mechanisms other than coercion may play a role. But at the international level, the success of enforcement is rather linked to cost-benefit calculations rather than traditional coercion against will.

This suggests the existence of yet another motivation, namely that actors comply because it is in their self-interest to do so. From a rational choice perspective, actors are expected to comply when they benefit

---

Chayes, The New Sovereignty: Compliance with International Regulatory Agreements, 1995; T.M. Franck, The power of legitimacy among nations, 1990; A.T. Guzman, "A Compliance-Based Theory Of International Law", California Law Review 90 (2002), 1823-1887; B. Kingsbury, "The concept of compliance as a function of competing conceptions of international law", Michigan journal of international law 19 (1998), 345-372; K. Raustiala, "Compliance & effectiveness in international regulatory cooperation", Case Western reserve journal of international law 32 (2000), 387-440; R. Wolfrum, "Means of Ensuring Compliance with and Enforcement of International Environmental Law", Recueil des cours de l'Académie de Droit International de La Haye 272 (1998), 13-154; R.B. Mitchell, "Compliance Theory: compliance, effectiveness, and behaviour change in international environmental law" in: D. Bodansky/J. Brunnée/E. Hey (eds.), Oxford Handbook of International Environmental Law, 2007, 893-921; M. Burgstaller, "Amenities and Pitfalls of a Reputational Theory of Compliance with International Law", Nordic Journal of International Law 76 (2007), 39-71.

[419] The existence of a sanction was seen as a fundamental element of law by traditional lawyers such as John Austin, compare J. Austin, The province of jurisprudence determined: the uses of the study of jurisprudence, 1954, 13.

[420] M. Burgstaller, "Amenities and Pitfalls of a Reputational Theory of Compliance with International Law", Nordic Journal of International Law 76 (2007), 39-71 (47-48).

from compliance as a result of rational cost-benefit calculations.[421] In these calculations, the costs of possible sanctions may play a role, even though direct sanctions are rare in international law and in particular in international environmental law.[422] Where they exit, they usually take the form of economic disincentives. Economic incentives and disincentives are often seen as essential in particular in cases where norms ask for significant changes of behaviour, since only such mechanisms can offset the net benefits that often derive from a violation of those rules.[423] In the international system, another relevant factor to be considered in those cost-benefit calculations is reputation.[424] It is usually in the interest of states to have a good reputation, and thus reputational costs and gains influence their decision whether to comply. Given the interdependence of states, one may usually expect that states have an interest in being perceived as cooperative, and therefore a decision to violate international legal norms will hurt their reputation in the long-term. However, states may also wish to gain from short term benefits, deciding not to develop a strong reputation, or prefer other types of reputation.[425]

Third, actors may comply because they perceive compliance to be the right thing to do, i.e. due to an "internal" motivation other than mere calculations of costs and benefits.[426] While conscious of the risk of over-

---

[421] A.T. Guzmán, How international law works: a rational choice theory, 2008.

[422] The compliance mechanism under the Kyoto Protocol however establishes enforcement measures that contain sanctioning elements such as penalties for non-compliance with emission reductions, see on this R. Wolfrum/J. Friedrich, "The Framework Convention on Climate Change and the Kyoto Protocol" in: U. Beyerlin/P.-T. Stoll/R. Wolfrum (eds.), Ensuring Compliance with Multilateral Environmental Agreements: A Dialogue between Practitioners and Academia, 2006, 53-68 (60-63).

[423] G.W. Downs, "Enforcement and the evolution of cooperation", Michigan journal of international law 19 (1998), 319-344 (322).

[424] See in particular A.T. Guzmán, "Reputation and international law", Georgia journal of international and comparative law 34 (2006), 379-391; for a discussion and critique M. Burgstaller, "Amenities and Pitfalls of a Reputational Theory of Compliance with International Law", Nordic Journal of International Law 76 (2007), 39-71 (60 et seq.).

[425] A.T. Guzman, "A Compliance-Based Theory Of International Law", California Law Review 90 (2002), 1823-1887 (1845-1850).

[426] This does of course not imply that cost-benefit calculation and the following factors may not complement each other.

simplifying this complex issue, two main groups of factors may be distinguished that may trigger compliance for this reason. On the one hand, there may be certain qualities of the norm that induce compliance and on the other hand certain processes that persuade actors to comply or socialise them accordingly. Thus, actors may comply with binding norms out of a sense of legal obligation.[427] Even though sceptics of the normative force of international law deny that this "legality" factor has any relevance for the behaviour of states,[428] it will be argued that it matters for the reasons discussed below. A further characteristic of a norm that is related but conceptually distinct is its legitimacy. Actors may obey norms that they perceive as legitimate, irrespective of the availability of coercive power.[429] Legitimacy, according to Franck, exerts a pull towards compliance. It is based on two pillars, namely certain qualities of the rule, in the words of Franck "the property of a rule or rule-making institutions" on the one hand, and the accordance of the rule or institution with principles of "right process" on the other.[430]

Process is also stressed in the managerial model of Abram Chayes and Antonia Handler Chayes. In their model, which also incorporates rationalist thinking, states are generally willing to comply with legal rules. But often enough, external obstacles to compliance must be overcome through the establishment of a compliance management strategy which enhances transparency and information exchange, dispute settlement and capacity building.[431] Central to the response to these problems – often referred to as a managerial approach – is according to Chayes and Chayes institutionalised discourse and persuasion.[432] In contrast to

---

[427] This is stressed by traditional international lawyers and theorists, see e.g. H.L.A. Hart, The concept of law, 1961, 23-24.

[428] J.L. Goldsmith/E.A. Posner, The limits of international law, 2005.

[429] T.M. Franck, "Legitimacy in the international system", American Journal of International Law 82 (1988), 705 – 759 (705); T.M. Franck, The power of legitimacy among nations, 1990, 25 and 34; E. Brown Weiss, "Conclusions: Understanding Compliance with Soft Law" in: D. Shelton (ed.), Commitment and Compliance: The Role of Non-binding Norms in the International Legal System, 2000, 535-556 (543).

[430] T.M. Franck, The power of legitimacy among nations, 1990, 24.

[431] A. Chayes/A. Handler Chayes, The New Sovereignty: Compliance with International Regulatory Agreements, 1995, 3 et seq.

[432] According to Chayes and Chayes "[T]he fundamental instrument for maintaining compliance with treaties at an acceptable level is an iterative process of discourse among the parties, the treaty organization, and the wider pub-

Chayes and Chayes who ultimately build on the need for good standing in the international community as the decisive motivation for states, other scholars consider that institutionalised discourse and its framing in shared understanding may ultimately even affect the identities and interests of states. As a consequence, states would adhere to norms not out of obedience but because they have adopted their interests and identity to the shared understandings expressed in international norms.[433]

A fourth factor that increasingly finds supporters stresses the significance of socialisation in institutions. According to this approach, socialisation processes can well explain the soft power exerted by international institutions upon actors.[434] Membership in institutions and institutionalised follow-up processes lead to compliance because states undergo a process of socialisation or "acculturation" whereby they submit to social pressures of their social environment and its culture.[435]

Instead of relying on one particular factor and underlying theory of compliance, I will in the following analysis assume that each of these

---

lic", and "[P]ersuasion and argument are the principal engines of this process ...", compare A. Chayes/A. Handler Chayes, The New Sovereignty: Compliance with International Regulatory Agreements, 1995, 25-26.

433 Successful persuasion in the eyes of Jutta Brunnée and Stephen Toope derives from the "internal morality of law", which means in their account that the rules should be compatible with one another, should ask reasonable things of the addressees, should be transparent and relatively predictable, and officials should treat known rules as shaping their exercise of discretion, compare J. Brunnée/S.J. Toope, "International law and constructivism: elements of an interactional theory of international law", Columbia journal of transnational law 39 (2000), 19-74; these types of constructivist explanations are also stressed as an important explanation for long-term compliance by S. Oeter, "Towards a richer institutionalism for international law and policy", University of Illinois law review (2008), 61-70 (66 et seq.).

434 J.E. Alvarez, International Organizations as Law-makers, 2005, 626; R. Goodman/D. Jinks, "How to influence states: socialization and international human rights law", Duke law journal 54 (2004), 621-703; H.H. Koh, "Internalization through socialization", Duke law journal (2005), 975-982.

435 "Acculturation" is defined by Jinks and Goodman as the "general process of adopting the beliefs and behavioural patterns of the surrounding culture", compare R. Goodman/D. Jinks, "How to influence states: socialization and international human rights law", Duke law journal 54 (2004), 621-703 (638); for a positive assessment of this approach see also J.E. Alvarez, "Do states socialize?", Duke law journal 54 (2005), 961-974.

factors identified by scholarship highlights a different aspect that may generally be of significance.

Compliance in the environmental field is at least so far best achieved through a combination of enforcement and economic mechanisms arising from rationalist explanations with mechanisms allowing for persuasion and socialisation.[436] The point is then not to discuss various approaches to compliance, but to show which factor may also be at play for nonbinding instruments. And indeed, economic mechanisms providing for incentives and disincentives play an important role in particular in areas such as international environmental law where traditional enforcement and reciprocity are largely lacking.[437]

1. Nonbinding status as a compliance-enhancing factor

The binding or nonbinding status of an instrument by itself is one important (though not the only) factor that influences the ability of any instruments to induce compliance.[438] Even all other things being equal, states will more likely comply with binding than with nonbinding norms, for the reasons outlined below.

The centrality of the distinction between binding and nonbinding status is constantly confirmed by state practice. In fact, one of the recurring themes in international negotiations on a new instrument for environmental protection is the discussion on whether the instrument in ques-

---

[436] Many authors similarly stress the importance of combining in particular rationalist and constructivist approaches to compliance, compare e.g. J. Brunnée, "The Kyoto Protocol: Testing Ground for Compliance Theories?", Zeitschrift für ausländisches öffentliches Recht und Völkerrecht 63 (2003), 255-280 (279); R. Goodman/D. Jinks, "How to influence states: socialization and international human rights law", Duke law journal 54 (2004), 621-703 (627).

[437] The potential of economic mechanisms is analysed in R. Wolfrum (ed.), Enforcing environmental standards: economic mechanisms as viable means?, 1996.

[438] In this sense also D. Bodansky, The art and craft of international environmental law, 2010, 102; K.W. Abbott/D. Snidal, "Hard and Soft Law in International Governance", International Organization 54 (2000), 421-456 (426); K. Raustiala, "Form and Substance in International Agreements", American Journal of International Law 99 (2005), 581-614 (592); similarly also E. Brown Weiss, "Conclusions: Understanding Compliance with Soft Law" in: D. Shelton (ed.), Commitment and Compliance: The Role of Non-binding Norms in the International Legal System, 2000, 535-556.

tion should be a binding or nonbinding one. It matters for states whether they will be under a binding obligation or not. They therefore usually prefer a binding instrument over a nonbinding one when they wish to have strong rules that affect the behaviour of other states.[439] And states that do not clearly support a particular action often attempt to negotiate nonbinding instruments instead of a binding treaty. When the U.S. government was generally reluctant of taking measures to address climate change, it aimed to eschew binding measures under the Kyoto Protocol and instead promoted voluntary action. Instead of agreeing on a new binding treaty, states at the COP 15 to the UNFCCC in 2010 could only agree on the nonbinding "Copenhagen Accord".[440] The discussions on a global instrument on the protection of forests also illustrate this tendency.[441] During the negotiations developing countries with large forests consistently opposed binding rules. On the other hand, those states which are in favour of strong rules usually argue for binding norms. Thus, those developing countries keen on establishing a functioning access-and-benefit sharing system under the Convention on Biological Diversity argue for binding rules to concretise the general obligations of the Convention. Overall, there seems to be a perception among most states that binding international rules are stronger and that the binding nature of an instrument positively affects compliance.

There exist numerous possible explanations for why states perceive binding norms to be stronger, and why they generally attach more importance to a binding norm. One possible explanation may indeed be the fear of enforcement, for example through international or national courts. Indeed, the decision of the United States Court of Appeals for the District of Columbia Circuit in *Natural Resources Defense Council*

---

439 E. Brown Weiss, "Conclusions: Understanding Compliance with Soft Law" in: D. Shelton (ed.), Commitment and Compliance: The Role of Nonbinding Norms in the International Legal System, 2000, 535-556 (551).

440 Copenhagen Accord, adopted by the UNFCCC Conference of the Parties at its fifteenth session, Decision 2/CP.15 (2009), available at http://unfccc.int/resource/docs/2009/cop15/eng/11a01.pdf.

441 Consider for example the discussion of a non-legally binding instrument on forests, where developing states opposing a strong instrument also opposed its binding status.

v. *Environmental Protection Agency*[442] largely turned on whether a decision of the Conference of the Parties to the Montreal Protocol was considered to be law and therefore binding on the US Government under the Clean Air Act or not.[443] But at least in environmental law where traditional enforcement does not play a prominent role, the consistent reluctance of laggard states to consent to binding rules can hardly be explained in terms of enforcement alone. It is therefore suggested that other reasons are at play here.

First of all, only legally binding rules must be ratified and implemented into national law. Governments may have an interest in resisting binding norms if they are unsure of domestic political support.[444] Apart from that, one could also argue that binding norms may be perceived as stronger because of the legal obligation to implement them. But this explanation only shifts the question from the effect of a norm being legally binding to the question why states respond to the legal obligation of implementation.

Another explanation is that binding norms have, as a result of common understandings of states on the basic premises of international law, a certain inherent quality that creates a sense of obligation.[445] This sense of obligation may have different roots, but is often seen to derive from the fact that actors are socialised to adhere to the law.[446] When a norm is

---

[442] United States Court of Appeals for the District of Columbia Circuit, Natural Resource Defense Council v. Environmental Protection Agency, 464 F.3d 1 (D.C. Cir. 2006).

[443] *Ibid.* The Court found that the COP decisions are not law in the sense of the Clean Air Act, and therefore the EPA was not bound to act in accordance with the decision.

[444] The lack of support in the US Senate is of course one of the main reasons for the US government's position to which is often the attempt to eschew binding treaty law, as for instance in the case of ratification of the Kyoto Protocol or a subsequent treaty on climate change.

[445] D. Bodansky, The art and craft of international environmental law, 2010, 101-102; similar, albeit with a different complex understanding of the concept of obligation not based on form, but on Lon Fuller's criteria of criteria of legality, J. Brunnée/S.J. Toope, Legitimacy and legality in international law: an interactional account, 2010.

[446] In one possible explanation, such a sense of obligation derives from the fact that actors are conditioned or socialized to obey the law, and develop a habit of obedience, compare H.L.A. Hart, The concept of law, 1961, 23-24, 49-64; O.R. Young, Compliance and public authority, 1979, 23-24.

legally binding and thus mandatory, an actor's internal sense of obligation will raise the weight of the norm, even though he may not comply for other reasons.

Another argument regards the quality of the norm itself as the decisive element. Some authors argue that the perceived legitimacy of a norm elicits the normative belief that a norm ought to be obeyed.[447] Thomas M. Franck, one of the first scholars to argue that rules and institutions exert a "compliance pull" when they are legitimate and based on right process,[448] suggested four central conditions that determine the perceived legitimacy of a norms, two of which are directly related to the status of a norm. These are the *symbolic validation* of a norm and *adherence* to secondary rules of "right process".[449]

Symbolic validation for him refers to the validating communication cues which communicate the authenticity of a rule maker or a rule.[450] This can be nicely applied to the question of the relevance of a norm being legally binding. By agreeing and declaring that a norm is legally binding, states attach a particular cue or symbol to the norm which indicates that it is valid and has authority. It makes a difference whether one says that one is bound by a particular norm or not, and this distinction is attached to the rule in question. The symbolic act of formally signing a treaty, its domestic approval and its ratification in contrast to the formal mere one-time adoption by states should make a difference. As a consequence, the legally binding norm carries a distinct authority that is superiour to those norms which have not received the same validation.

Moreover, and related, nonbinding instruments have a lower degree of legitimacy due to the fact that they do not in the same way as binding

---

[447] I. Hurd, "Legitimacy and Authority in International Politics", International Organization 53 (1999), 379-408 (381).

[448] Franck defines legitimacy as "a property of a rule or rulemaking institution which exerts a pull towards compliance ... because those addressed believe that the rule or institution has come into being and operates in accordance with generally accepted principles of right process", see T.M. Franck, The power of legitimacy among nations, 1990, 24.

[449] T.M. Franck, The power of legitimacy among nations, 1990, 91 et seq. and 183 et seq. The other two conditions are determinacy, defined by Franck as the ability of the text of the rule to transmit a clear message, and coherence of rules, which for Franck refers to the characteristic that rules emanate from principles of general application, compare *ibid.*, 52 and 150 et seq.

[450] T.M. Franck, The power of legitimacy among nations, 1990, 91.

norms stem from a particular pre-defined process of law-making. For Franck and others, adherence to underlying rules of law-making and its "right process" is the precondition for their legitimacy and their compliance pull.[451] Adherence to underlying rules of rule making is in turn to a great extent rooted in the rules of *pacta sunt servanda* and in the position of states as individual members of the community of states. *Pacta sunt servanda*, as expressed in Article 26 of the Vienna Convention on the Law of Treaties,[452] is a fundamental rule of international law and part of its minimum core content.[453] It describes the common acceptance by the participants in the legal system that binding promises must be kept. And it provides one underlying reason why the characteristic of being legally binding improves the authority of a norm. Improved authority also stems from the fact that the generally accepted process of making international law entails the requirement that the norm in question will be put to domestic ratification or acceptance procedures. Participants in the system are thus reassured that implementing activity will follow.

From the perspective of the other participants in the legal system, the higher authority and legitimacy of a norm can be equated with an increased credibility of binding commitments.[454] By agreeing to be bound by a norm, states demonstrate to other participants the seriousness of their intention to keep the promise.[455] If states believe that binding commitments are more credible, they are also more inclined to com-

---

[451] T.M. Franck, The power of legitimacy among nations, 1990, 183 et seq; K. Raustiala, "Form and Substance in International Agreements", American Journal of International Law 99 (2005), 581-614 (592)

[452] Vienna Convention on the Law of Treaties, (23 May 1969), 1155 UNTS 331, 8 I.L.M. 679 (1969).

[453] G. Dahm/J. Delbrück/R. Wolfrum, Völkerrecht, Band I/1, Die Grundlagen. Die Völkerrechtssubjekte, 1989, 41.

[454] K. Raustiala, "Form and Substance in International Agreements", American Journal of International Law 99 (2005), 581-614 (592); similarly D. Bodansky, "Customary (and not so customary) international environmental law", Indiana journal of global legal studies 3 (1995), 105-131 (118); M. Bothe, "Legal and Non-Legal Norms – A Meaningful Distinction in International Relations?", Netherlands yearbook of international law 11 (1980), 65-95 (78).

[455] C. Lipson, "Why are some International Agreements Informal", International Organization 45 (1991), 495-538 (508); K. Raustiala, "Form and Substance in International Agreements", American Journal of International Law 99 (2005), 581-614 (592).

ply.[456] This higher authority and credibility can also be translated into higher reputational costs. Breaking binding rules despite higher expectations of compliance by others necessarily entails higher reputational costs than does not adhering to nonbinding ones.

This reputational factor matters for the cost/benefit calculations stressed by some that analyse compliance in terms of rational choice.[457] From a rational choice perspective, a higher degree of compliance with binding norms can be expected if one assumes that an actor strives for a positive reputation as a reliable and rule-complying partner.

This does not mean that nonbinding instruments are devoid of any authority, credibility or reputational impact. Their formal procedures of adoption and the legitimacy of the forum adopting such instruments also bestow legitimacy and authority upon them. And even nonbinding instruments raise expectations of compliance. Just as treaty norms, they establish a binary code, not of legal/illegal but of right/wrong, by which future behaviour can be judged. To dishonour the expectations raised by the adoption of a code of conduct also affects the standing of a state among the members of the particular organisation or more generally in the international community. Even though violations carry less severe reputational costs, some nevertheless occur.[458] It hurts a state's reputation of being cooperative and trustworthy if it adopts a nonbinding instrument, but does not pay attention to it. And a noncompliant state may not be able to rely on the cooperation of other states on the basis of nonbinding instruments, nor could a state extract concessions from other states for adopting such instruments.[459]

In conclusion it has to be stressed that the binding status of a norm can be expected to matter for compliance. As stressed before, however, the nonbinding or binding status is only one factor among many others that induce compliance. In particular situations nonbinding instruments

---

[456] K.W. Abbott/D. Snidal, "Hard and Soft Law in International Governance", International Organization 54 (2000), 421-456 (429-430); K. Raustiala, "Form and Substance in International Agreements", American Journal of International Law 99 (2005), 581-614 (592).

[457] For a comprehensive rational choice analysis that takes into account the factor of reputation compare A.T. Guzmán, How international law works: a rational choice theory, 2008.

[458] A.T. Guzman, "A Compliance-Based Theory Of International Law", California Law Review 90 (2002), 1823-1887 (1880-1881).

[459] A.T. Guzman, "A Compliance-Based Theory Of International Law", California Law Review 90 (2002), 1823-1887 (1880-1881).

may be better suited for achieving cooperation. A nonbinding instrument can for example be the preferable option where the need for ratification and domestic implementation risks limiting the number of participants or where it risks eroding the norms and their potential to become customary law through low ratification numbers.[460] And while a binding instrument may remain without effect if it is not ratified, nonbinding instruments which do not require ratification provide flexibility for implementation, i.e. they give states the possibility of deciding to what extent and with regard to which aspects they wish to comply.[461] The legitimacy problems that derive from this detachment from national ratification procedures will be dealt with below in Part 3.

2. *Norm characteristics and compliance*

A number of factors that are generally seen as enhancing the behavioural impact of norms have to do with their content and substance. Thomas Franck has argued that the coherence and determinacy of rules are two elements of their legitimacy, which in turn enhances compliance. Coherence refers to the emanation of rules from principles of general application; determinacy refers to the property that they transmit a clear message.[462] To the extent that this holds true, there is no apparent reason why this does not also apply to norms of a nonbinding nature. Similarly, when addressees feel that the norms addressed to them, binding or not, are equitable and fair, they will more likely follow those norms.[463]

Moreover, it is sometimes assumed, notably by states wishing to negotiate strong instruments, that the more precisely worded is a norm, the

---

[460] For example, the Convention on the Law of the Non-Navigational Uses of International Watercourses (21 May 1997), 36 ILM 700 (1997), is with 17 ratifications by 2010 even 13 years after its adoption not even close to the 35 ratifications required for entry into force.

[461] E. Brown Weiss, "Conclusions: Understanding Compliance with Soft Law" in: D. Shelton (ed.), Commitment and Compliance: The Role of Nonbinding Norms in the International Legal System, 2000, 535-556 (552).

[462] T.M. Franck, Fairness in international law and institutions, 1995, 31 et seq.

[463] E. Brown Weiss, "Conclusions: Understanding Compliance with Soft Law" in: D. Shelton (ed.), Commitment and Compliance: The Role of Nonbinding Norms in the International Legal System, 2000, 535-556 (547).

more influential it is.[464] According to this logic, it would make an immediate difference whether nonbinding instruments only provide for general objectives and principles or precisely define targets, measures and best practices. As mentioned, where the political climate is not conducive to precise binding norms, then it would according to this logic be better to negotiate nonbinding instruments if this – as is indeed often the case – allows for more precise norms.

Research shows however that it is not that simple.[465] Precision alone is not a factor that can balance out or override other factors that speak against compliance. In fact, ambiguous and general norms may be even more or at least similarly effective as precise ones. Precision can, however, help to further compliance indirectly. The more precisely a norm defines what should be done, the easier it is to determine whether or not the addressee has indeed implemented the norm in question or not. Consequently, non-state actors or secretariats could more easily determine if a state complies with the norm in question or not. It may therefore at times be better to negotiate a nonbinding instrument with a strong reporting and monitoring mechanism or NGO participation than a binding one of vague and general substance.

3. Norm development processes and compliance

The characteristics of typical norm development processes of nonbinding instruments often comprise two elements which enhance compliance with the norms if compared to treaty law. First, as seen above, international institutions generally strive to achieve unanimity or at least consensus when adopting a nonbinding instrument, even if the respective rules of procedure also allow for majority voting. This tendency in practice has positive repercussions for compliance with the instrument in question. In particular in areas such as the protection of the envi-

---

[464] This is identified as one of the myths of compliance by E. Brown Weiss, "Understanding compliance with international environmental agreements: the Baker's dozen myths", University of Richmond law review 32 (1999), 1555-1589 (1572).

[465] E. Brown Weiss, "Understanding compliance with international environmental agreements: the Baker's dozen myths", University of Richmond law review 32 (1999), 1555-1589 (1573); H.K. Jacobson/E. Brown Weiss, "A Framework for Analysis" in: H.K. Jacobson/E. Brown Weiss (eds.), Engaging Countries: Strengthening Compliance with International Environmental Accords, 1998, 1-18.

ronment, compliance and effectiveness depend on securing universal acceptance in norm making processes and the maintenance of this acceptance through continuous discourse over time.[466] Broad acceptance by states also enhance the legitimacy of a norm and thus its potential impact.[467]

Second, it has been observed in this study that the turn to nonbinding instruments generally facilitates the integration of non-state actors not only in follow-up mechanisms, but notably also to the processes of norm development and revision.[468] As demonstrated in the case of the voluntary PIC procedure, nonbinding initiatives in fact are often prime vehicles for non-governmental organisations to increase their impact on norm development within international organisations. Moreover, revision of such instruments and the elaboration of supporting documents which often take place at lower political levels open numerous possibilities for direct NGO input.[469] The participation of non-state actors can be beneficial for the effectiveness of a particular instrument in various respects. In terms of compliance, including non-state actors which have the ability to participate in both international and domestic public debate in the norm development process has the potential to improve processes of domestic implementation, because these actors are likely to pressure and persuade decision-makers at the domestic level to implement those norms which they have helped to create at the international

[466] R. Wolfrum, "Vorbereitende Willensbildung und Entscheidungsprozeß beim Abschluß multilateraler völkerrechtlicher Verträge" in: J. Ipsen/ E. Schmidt-Jortzig (eds.), Recht-Staat-Gemeinwohl: Festschrift für Dietrich Rauschning, 2001, 407-418 (407-408 and 418).

[467] E. Brown Weiss, "Conclusions: Understanding Compliance with Soft Law" in: D. Shelton (ed.), Commitment and Compliance: The Role of Nonbinding Norms in the International Legal System, 2000, 535-556 (537 and 543).

[468] This is also confirmed by other research on nonbinding instruments, compare e.g. J.J. Kirton/M.J. Trebilcock, "Introduction: Hard Choices and Soft Law in Sustainable Global Governance" in: J.J. Kirton/M.J. Trebilcock (eds.), Hard Choices, Soft Law: Voluntary Standards in Global Trade, Environment and Social Governance, 2004, 3-29 (5).

[469] Perhaps the best illustration of this was the review of the OECD Guidelines in 2000 which allowed for informal but nevertheless direct participation by trade unions, industry associations and NGOs. Similarly remarkable was the direct input of NGOs and industry in decision-making by the FAO/UNEP expert group on PIC.

level.[470] In other words, these actors *inter alia* help to internalise these norms by linking international rules and internal policies.[471] Furthermore, to include the addressees of a particular instrument in the process of norm development, as done for instance in the latest reviews of the OECD Guidelines in 2000 and 2010, additionally enhances the acceptance of rules and possibly compliance.[472]

4. International means of enhancing compliance by states

a) Economic incentives and disincentives

As mentioned, from the perspective of a self-interested actor it is important that the benefits of compliance outweigh the costs, in particular economic ones. Compliance is less likely the costlier it is.[473] Given the relative weakness of traditional enforcement in international environmental law, economic mechanisms are widely used in international environmental law as a tool to ensure and enhance compliance by states.[474]

---

[470] The significance of the involvement of so-called transnational society for compliance has been highlighted early by W. Friedmann, The Changing Structure of International Law, 1964. (PAGE) For the role of these actors in persuasive processes compare J. Brunnée/S.J. Toope, "International law and constructivism: elements of an interactional theory of international law", Columbia journal of transnational law 39 (2000), 19-74 (70).

[471] J.E. Alvarez, "International Organizations: Then and Now", American Journal of International Law 100 (2006), 324-347 (336); Harold Hongju Koh claims that internalisation of international norms in what he calls "transnational legal process" can provide the necessary link between externally existing rules and internal voluntary obedience, compare H.H. Koh, "Why do nations obey international law?", The Yale law journal 106 (1997), 2599-2659.

[472] P. Muchlinski, "Human Rights, social responsibility and the regulation of international business: The development of international standards by intergovernmental organisations", Non-State Actors and International Law 3 (2003), 123-152.

[473] E. Brown Weiss, "Conclusions: Understanding Compliance with Soft Law" in: D. Shelton (ed.), Commitment and Compliance: The Role of Non-binding Norms in the International Legal System, 2000, 535-556 (538).

[474] For an overview of measures of select agreements P.K. Wouters, "Trade Measures in Multilateral Environmental Agreements" in: M. Bothe/P.H. Sand (eds.), La politique de l'environnement: De la réglementation aux instruments économique, 2003, 159-189 (162 et seq.); compare also in detail M. Bothe, "The Evaluation of Enforcement Mechanisms in International Environmental Law: An Overview" in: R. Wolfrum (ed.), Enforcing environmental standards: eco-

Trade restrictions relating to certain hazardous substances or endangered species for example allow states to use border controls to implement environmental objectives and prevent international trade benefits from worsening environmental problems. Considering that compliance is to a significant extent the result of cost-benefit calculations of states, economic mechanisms are also employed as incentives or disincentives. This builds on the theory that a rational self-interested actor may only comply if the benefits of compliance outweigh the costs of non-compliance.[475] Incentives and disincentives can thus play a role in tipping the balance in favour of compliance.[476] International treaties therefore often aim to outweigh the costs of trade restrictions by creating some kind of economic benefit for compliant states. Examples include the restriction of access to markets, technology, funding or resources for compliant members and prohibition of trade with non-participants[477] or the linkage of the economic costs of environmental

---

nomic mechanisms as viable means?, 1996, 13-38 (35 et seq.); P.H. Sand, "Sticks, Carrots, and Games" in: M. Bothe/P.H. Sand (eds.), La politique de l'environnement: De la réglementation aux instruments économique, 2003, 3-36 (16-28); R. Wolfrum, "Means of Ensuring Compliance with and Enforcement of International Environmental Law", Recueil des cours de l'Académie de Droit International de La Haye 272 (1998), 13-154 (58 et seq. and 110 et seq.).

475 Costs and benefit calculations as determinants of compliance instead of enforcement were already stressed by L. Henkin, How nations behave: law and foreign policy, 1979, 45 et seq.; see also G.W. Downs, "Enforcement and the evolution of cooperation", Michigan journal of international law 19 (1998), 319-344.

476 G.W. Downs, "Enforcement and the evolution of cooperation", Michigan journal of international law 19 (1998), 319-344; O.S. Stokke, "The Interplay of International Regimes: Putting Effectiveness Theory to Work", FNI Report 14/2001 (2001), 1 et seq (9); the underlying rational choice theory has been translated into a sceptical view of international law by J.L. Goldsmith/E.A. Posner, The limits of international law, 2005. Goldsmith and Posner however underestimate the power of international rules which can be explained even from a rational choice perspective, compare for this A.T. Guzmán, How international law works: a rational choice theory, 2008; generally on enforcement through economic mechanisms in international environmental law R. Wolfrum (ed.), Enforcing environmental standards: economic mechanisms as viable means?, 1996.

477 E.g. Article V of the Agreement on the Conservation of Polar Bears (15 November 1973), 13 ILM 13 (1974); Article III, IV and V of the Convention on International Trade in Endangered Species of Wild Fauna and Flora (3 March 1973), 993 UNTS 243; Article 3, para. 2 (c) of the Convention for the

commitments with benefits which are generated from the use of the protected resource.[478]

(1) Restriction of market access to compliant states

These intrusive mechanisms are most often based on binding international law. Nonbinding norms are hardly ever used to establish an exclusive privilege or incentive for adhering states which could be subsequently withdrawn. In most cases, states seem to favour the legal security of binding international law for the basis of economic mechanisms. However, economic tools such as trade restrictions may also be used for enhancing compliance with nonbinding instruments as well. This is illustrated by the Kimberley Process Certification Scheme for Rough Diamonds (KPCS). The KPCS seeks to suppress the trade in so-called conflict or blood diamonds. It was adopted by means of a nonbinding ministerial declaration, the Interlaken Declaration,[479] which was endorsed in General Assembly and Security Council resolutions.[480] The language used in the documents ("recommend", "declare") clearly indicates that the KPCS is not an international treaty, but a nonbinding instrument. According to the KPCS, participants should ensure that only rough diamonds that are accompanied by a certificate are imported and exported;[481] and participants should neither import rough diamonds from non-participants nor export rough diamonds to non-participants.[482] In other words, the KPCS establishes a system of trade

---

Prohibition of Fishing with Long Driftnets in the South Pacific (24 November 1989), 1899 UNTS 3, 29 ILM 1454 (1990); Article 4 (1) of the Montreal Protocol on Substances that Deplete the Ozone Layer (16 September 1987), 26 ILM 1550 (1987).

478 As for example in the case of access and benefit sharing under the Convention on Biological Diversity, compare R. Wolfrum, "Means of Ensuring Compliance with and Enforcement of International Environmental Law", Recueil des cours de l'Académie de Droit International de La Haye 272 (1998), 13-154 (111).

479 Interlaken Declaration of 5 November 2002 on the Kimberley Process Certification Scheme for Rough Diamonds, available from http://www.kimberleyprocess.com/documents.

480 See e.g. UN Doc. A/Res/57/302 (2003) and UN Doc. S/Res/1459 (2003).

481 Kimberley Process Scheme for Rough Diamonds, Section III, (a) and (b).

481 Kimberley Process Scheme for Rough Diamonds, Section IV (a).

482 Kimberley Process Scheme for Rough Diamonds, Section III, (c).

restrictions and related restriction of access to diamond trade for participants through a nonbinding instrument.

However, the linkage of nonbinding instruments to treaty law may result in indirect economic incentives where the treaty in question contains one. One example is that of the Fish Stocks Agreement. The Fish Stocks Agreement mainly concretises the duty to cooperate contained in the UNCLOS by obliging states to cooperate either directly or through fisheries management organisations.[483] The incentive mechanism of the Fish Stocks Agreement directly addresses the "free rider" problem by excluding states which do not adopt the prescribed management measures from access to resources. Only members of a management organisation or those states which are not members but agree to apply its management measures enjoy access to the fishery resources.[484] While the mechanism certainly has great potential,[485] its effective implementation hinges upon wide acceptance of the Fish Stocks Agreement because non-Parties can hardly be excluded from the resources in practice. As seen above, the Fish Stocks Agreement contains references to "generally recommended international minimum standards for the responsible conduct of fishing operations"[486] which can be understood as reference to the FAO Code of Conduct for Responsible Fisheries.[487] The reference therefore effectively "hardens" the norms of

---

483 Fish Stocks Agreement, Article 8; for details, see D. König, "The Protection of Marine Living Resources – The 1995 Agreement on Straddling Fish Stocks and Highly Migratory Fish Stocks " in: T. Zhenghua/R. Wolfrum (eds.), Implementing International Environmental Law in Germany and China, 2001, 75-84; G. Vigneron, "Compliance and International Environmental Agreements: A Case Study of the 1995 United Nations Straddling Fish Stocks Agreement", Georgetown International Environmental Law Review 10 (1998), 581-624.

484 Fish Stocks Agreement, Article 8 (4).

485 G. Vigneron, "Compliance and International Environmental Agreements: A Case Study of the 1995 United Nations Straddling Fish Stocks Agreement", Georgetown International Environmental Law Review 10 (1998), 581-624 (613).

486 Fish Stocks Agreement, Article 10 c).

487 The Fish Stocks Agreement was elaborated in parallel to the CCRF, often by the same delegates, so that references to "responsible fishing" should be understood as referring to the concept of "responsible fishing" which is introduced and concretised in the FAO CCRF. This is confirmed by one of the participants in the development process who states that the reference to "responsible fishing" in the UN Fish Stock Agreement's preamble is meant to be a refer-

the nonbinding CCRF. As a consequence, the Fish Stocks Agreement enforcement mechanisms serve indirectly as an enforcement mechanism for nonbinding standards, at least to the extent that states comply with their obligations under the Fish Stocks Agreement.

(2) Loan policies of financial institutions

Effective incentives can also be established through the loan policies of international financial institutions such as the World Bank, the International Monetary Fund or the Global Environment Facility. The World Bank makes compliance with its Operational Procedures a condition for access to financial resources.[488] As decisions on the eligibility of projects of borrowers are guided by these nonbinding standards, and as borrowers depend on the support for their development, there exists a strong incentive to comply with these standards. [489] If one stresses the possibility that received funding such as World Bank loans can be cancelled and revoked in the case of non-compliance, the mechanism also constitutes a negative incentive.[490] Where other nonbinding instruments are referenced in the Operational Procedures, they are included in this incentive system. One example has been dealt with above, namely the linkage of the World Bank Operational Procedures with nonbinding instruments of the FAO.[491]

Similarly, the main environmental financial institution – the Global Environment Facility (GEF) – frequently refers to nonbinding policy in-

---

ence to the FAO CCRF, see W. Edeson, "Soft and Hard Law Aspects of Fisheries Issues: Some Recent Global and Regional Approaches" in: M.H. Nordquist/J.N. Moore/S. Mahmoudi (eds.), The Stockholm Declaration and Law of the Marine Environment, 2003, 165-182 (170).

[488] See on the Operational Procedures already Part 1, at A.V., above.

[489] In detail on this mechanism L. Boisson de Chazournes, "Policy Guidance and Compliance: The World Bank Operational Standards" in: D. Shelton (ed.), Commitment and Compliance: The Role of Non-Binding Norms in the International Legal System, 2000, 281-302; D.A. Wirth, "Compliance with Non-Binding Norms of Trade and Finance" in: D. Shelton (ed.), Commitment and Compliance: The Role of Non-Binding Norms in the International Legal System, 2000, 330-344 (335).

[490] N. Matz, "Financial and Other Incentives for Complying with MEA Obligations" in: U. Beyerlin/P.-T. Stoll/R. Wolfrum (eds.), Ensuring Compliance with Multilateral Environmental Agreements 2006, 301-318 (313).

[491] See Part 2, at A.II.2.b)(1), above.

struments in its planning framework for the design of projects called the Operational Program. For example, the Operational Program for the GEF focal area of the protection of international waters stresses that the GEF activities will be consistent with Agenda 21, and also mentions the FAO Code of Conduct for Responsible Fisheries as one of the instruments which should be considered.[492] The CCRF is also more specifically mentioned as an "institutional tool" that should guide the so-called Large Marine Ecosystem Projects which are part of the water-based Operational Program.[493]

(3) Export credit guarantees

The granting of export credit guarantees on the condition of compliance with nonbinding international environmental norms is a way of providing an economic incentive for better compliance of private investors and ultimately of states with those standards. The assessment procedures that precede the decision whether a particular project is eligible for export credits are in most OECD states governed by the recommendations of the OECD Recommendation on Common Approaches on the Environment and Officially Supported Export Credits adopted by the OECD Council on 12 June 2007.[494] In Germany, for instance, the environmental assessment procedure which identifies possible environmental risks is directly governed by the OECD's Recommendation.[495] The OECD Recommendation also recommends that Members should benchmark projects in the environmental review against the Operational Procedures of the World Bank as a minimum requirement.[496] The environmental assessment for the granting of export cred-

---

[492] Operational Strategy of the Global Environment Facility, Chapter 4, International Waters, available at http://www.gefweb.org/public/opstrat/complete.htm.

[493] *Ibid.*

[494] OECD Council, Revised Council Recommendation on Common Approaches on the Environment and Officially Supported Export Credits (12 June 2007), OECD Doc. TAD/ECG(2007)9.

[495] Information of the German Federal Ministry for Economics and Technology at http://www.agaportal.de/en/aga/nachhaltigkeit/umwelt.html.

[496] The Revised OECD Council Recommendation on Common Approaches on the Environment and Officially Supported Export Credits (12 June 2007), OECD Doc. TAD/ECG(2007)9, in para. 12 provides that "[W]hen undertaking a review ... for all projects, Members should benchmark projects against host

its is thus governed to a great extent by nonbinding standards developed by the World Bank.[497]

c) Compliance management

Instead of sanctions or other confrontational means, international environmental law often relies on non-confrontational so-called compliance mechanisms.[498] In the environmental field, the significance of these mechanisms is increased by the irreversibility of environmental damages, the non-availability of reciprocal enforcement and the inadequacy of the law of state responsibility for addressing environmental problems.[499]

The proposition that non-confrontational compliance mechanisms can provide a compliance-enhancing strategy which at least mitigates some of the mentioned insufficiencies is rooted in the theoretical assumption that most states have a "propensity to comply"[500] with international law even in the absence of enforcement. According to this view, non-compliance is rarely a question of lack of will, but rather the result of

---

country standards and either against the relevant aspects of all ten World Bank Safeguard Policies or, where appropriate for private sector limited or non-recourse project finance cases, against the relevant aspects of all eight International Finance Corporation Performance Standards, or where such institutions are supporting the project, against the relevant aspects of the standards of the Regional Development Banks, or against any relevant internationally recognised standards, such as European Community standards, that are more stringent than those standards referenced above."

[497] See for details on these standards the Part 1 of this study, at A.V. , further above.

[498] For a comprehensive overview of compliance mechanisms of various treaty regimes U. Beyerlin/P.-T. Stoll/R. Wolfrum (eds.), Ensuring compliance with multilateral environmental agreements: a dialogue between practitioners and academia, 2006; for a more theoretical discussion of compliance mechanisms J. Klabbers, "Compliance Procedures" in: D. Bodansky/J. Brunnée/ E. Hey (eds.), The Oxford Handbook of International Environmental Law, 2007, 995-1009.

[499] D. Shelton, "Law, Non-Law and the Problem of 'Soft Law'" in: D. Shelton (ed.), Commitment and Compliance: The Role of Non-Binding Norms in the International Legal System, 2000, 1-18 (16).

[500] A. Chayes/A. Handler Chayes, The New Sovereignty: Compliance with International Regulatory Agreements, 1995, 3.

the ambiguity of norms, the lack of capacity of a state to comply as well as unforeseen social, political or economic changes.[501] Rather than confrontational enforcement or sanctions, these root causes could be best addressed by a managerial strategy.[502] Central elements of such a strategy include informal dispute settlement, capacity building and transparency, the latter established mainly through data collection, monitoring and reporting complemented by NGO supervision and monitoring.[503] Measures to ensure transparency are in turn significant in bringing to bear the reputational factor.[504] When the conduct of states is made transparent for all participants, states must then justify non-compliant behaviour in formal and informal fora. Ideally, this "justificatory discourse"[505] takes place in an institutionalised form where each state has regularly to report on and defend its efforts. In cases of non-compliance, the reasons for non-compliance are then evaluated and addressed in a largely cooperative atmosphere, for example through norm concretisation, the adjustment of targets for specific states or by capacity building. Through this cooperative discursive process, states are ideally persuaded rather than forced to comply.[506]

Compliance management tools such as reporting, compliance assistance and continuous non-confrontational discourse are not only found in Multilateral Environmental Treaties. They are also employed as one of the main means of directly enhancing compliance with nonbinding instruments. As illustrated in the following, the nonbinding legal status of

---

501 A. Chayes/A. Handler Chayes, The New Sovereignty: Compliance with International Regulatory Agreements, 1995, 9-17.

502 A. Chayes/A. Handler Chayes, The New Sovereignty: Compliance with International Regulatory Agreements, 1995, 25; for a critique see Guzman, "A Compliance-based theory of International Law"; 1830-1833.

503 A. Chayes/A. Handler Chayes, The New Sovereignty: Compliance with International Regulatory Agreements, 1995, 25 and 207.

504 E. Brown Weiss, "Conclusions: Understanding Compliance with Soft Law" in: D. Shelton (ed.), Commitment and Compliance: The Role of Nonbinding Norms in the International Legal System, 2000, 535-556 (541).

505 A. Chayes/A. Handler Chayes, The New Sovereignty: Compliance with International Regulatory Agreements, 1995, 118-123.

506 A. Chayes/A. Handler Chayes, The New Sovereignty: Compliance with International Regulatory Agreements, 1995, 25.

an instrument does not preclude the establishment of such mechanisms.[507]

(1) Reporting

Reporting mechanisms have become a central feature of most environmental law treaties.[508] But as will be seen, reporting on implementation is also increasingly employed by international organisations as a follow-up to the adoption of nonbinding instruments. This is well known for example from reporting mechanisms on recommendations of the International Labour Organization. But also in international environmental matters, nonbinding recommendations and instruments often comprise a reporting commitment. All FAO codes of conduct provide for such a mechanism.[509] The Non-legally Binding Instrument on all Types of Forests adopted in 2007 asks states to report on implementation as part of national progress reports to the United Nations Forum on Forests,[510] and the so-called Overarching Policy Strategy of the recently established (nonbinding) Strategic Approach to International Chemical Management requests that states and other stakeholders report on progress in implementation of the approach to the International Conference on Chemicals Management.[511] Where the founding treaty of an or-

---

[507] J.B. Skjaerseth/O.S. Stokke/J. Wettestad, "Soft Law, Hard Law, and Effective Implementation of International Environmental Norms", Global Environmental Politics 6 (2006), 104-120 (117).

[508] P. Sands, Principles of International Environmental Law, 2003, 832; J. Wettestad, "Monitoring and Verification" in: D. Bodansky/J. Brunnée/E. Hey (eds.), The Oxford Handbook of International Environmental Law, 2007, 974-993 (977-987); R. Wolfrum, "Means of Ensuring Compliance with and Enforcement of International Environmental Law", Recueil des cours de l'Académie de Droit International de La Haye 272 (1998), 13-154 (37-43)

[509] See e.g. FAO International Code of Conduct for Plant Germplasm Collecting and Transfer, Article 15.1.; FAO International Code of Conduct on the Distribution and Use of Pesticides, Article 12.7.

[510] The "Non-legally binding instrument on all types of forests", adopted through General Assembly Resolution A/RES/62/98 of 17 December 2007, Annex, para. 9.

[511] Strategic Approach to International Chemical Management, Overarching Policy Strategy, at para. 24; available at http://www.saicm.org/documents/saicm %20texts/SAICM_publication_ENG.pdf.

ganisation provides for a binding reporting obligation,[512] member states could even be legally obliged to report on the implementation of norms by which they are not bound.[513] In these cases, the nonbinding instruments cease to be simple recommendations but represent "qualified recommendations".[514]

The FAO, for instance, establishes reporting mechanisms for all of its nonbinding instruments. Thus, even where international organisations do not have the authority to adopt binding norms, states are – through their membership of the organisation – drawn into processes of reporting and discussion of the implementation of norms that they are not obliged to implement. Similarly to most treaty instruments, reporting in the case of the FAO takes the form of a self-assessment of states through questionnaires.[515] In these questionnaires, states are asked to report on progress made with the implementation of the instrument.[516] The FAO secretariat collects the reports of states and then summarises the results of the reports. In the case of the FAO CCRF, the results provide the input for the progress report on implementation presented by the secretariat to COFI biannually and subsequently published.[517] However, these summary reports do not mention individual states but rather describe the collective performance. Thus, in line with the volun-

---

[512] FAO Constitution, Article 11 para. 3; ILO Constitution, Article 19 para. 6 (e).

[513] For the FAO this is also confirmed by G.G.R. Blom, "Institutional and legal aspects of the Food and Agriculture Organization of the United Nations (FAO)", Revue de Droit International 74 (1996), 227-332 (264); see generally S.A. Metaxas, Entreprises Transnationales et Codes de Conduite: Cadre Juridique et Questions d'Effectivité, 1988, 116.

[514] In relation to recommendations of international organizations, this is e.g. argued by G. Dahm/J. Delbrück/R. Wolfrum, Völkerrecht, Band I/1, Die Grundlagen. Die Völkerrechtssubjekte, 1989, 71.

[515] FAO, Analysis of Government Responses to the Second Questionnaire on the State of Implementation of the International Code of Conduct on the Distribution and Use of Pesticides (1993), available at www.fao.org; FAO Council Report of its Hundred and Twelfth Session, 1997, CL 112/REP, para. 29; Report of the twenty-second session of the Committee on Fisheries, 1997, FAO Fisheries Report No. 562 FIPL/R562 (En), para. 29.

[516] FAO Conference Resolution 10/85, para. 3; FAO Pesticide Code, Articles 12.6 (version of 1985) and 12.7 (version of 2003).

[517] Report of the Twenty-Second Session of the Committee on Fisheries, Rome 17-20 March 1997, FAO Fisheries Report No. 562 FIPL/R562, para. 29.

tary nature of the commitments, states do not have to fear being publicly accused of non-compliance.

As indicated by the example of the FAO CCRF, states indeed participate in these processes. In the case of the FAO CCRF, the five progress reports issued between 1999 and 2009 show that a significant number of countries responded to the questionnaires sent to them by the secretariat, but that the numbers varied considerably from one year to another.[518] After a high turnout in 2000 with 113 states reporting, only 33 per cent of states handed back the questionnaire in 2008. This indicates that voluntary reporting cannot be expected to remain stable over years. The latest rate of 33 per cent puts the usefulness of the entire exercise into doubt. Reporting on the FAO Pesticide Code was substantial but equally not comprehensive. Altogether 99 out of 177 countries that were members of the FAO at the time responded to the questionnaires.[519]

Generally speaking, however, these mechanisms carry significant potential for enhancing compliance. First of all, an important function of reporting and monitoring mechanisms is to increase transparency in a regime.[520] Transparency in turn is one of the main factors for enhancing compliance with norms.[521] It not only facilitates the coordination of states pursuing a common objective, but also provides reassurance that

---

[518] While the number of reporting countries first increased to reach the significant level of 103 reporting countries in 2000 and remained equally elevated in 2002 (105), the number dropped considerably in 2004 (49), remounted in 2006 to 70 but dropped again in 2009 to only 33 percent of FAO members reporting. See the consecutive progress reports on the implementation of the code of conduct for responsible fisheries and related plans of action at the biannual meetings of the Committee on Fisheries; all reports are available at http://www.fao.org/fi/body/cofi/cofi.asp.

[519] FAO, Analysis of Government Responses to the Second Questionnaire on the State of Implementation of the International Code of Conduct on the Distribution and Use of Pesticides (1993), in the introduction.

[520] Generally on monitoring and verification in international environmental law J. Wettestad, "Monitoring and Verification" in: D. Bodansky/J. Brunnée/E. Hey (eds.), The Oxford Handbook of International Environmental Law, 2007, 974-993.

[521] M. Bothe, "The Evaluation of Enforcement Mechanisms in International Environmental Law: An Overview" in: R. Wolfrum (ed.), Enforcing environmental standards: economic mechanisms as viable means?, 1996, 13-38 (19); A. Chayes/A. Handler Chayes, The New Sovereignty: Compliance with International Regulatory Agreements, 1995, 22.

other parties are complying. This deters non-compliance; the latter particularly in constellations of "contingent cooperation" where actors only comply on the condition that others also comply.[522] As mentioned, transparency is also the precondition for the functioning of reputation as a factor for compliance.[523] The mentioned reporting mechanisms generally improve transparency. However, when reports are published without references to individual compliance as is most frequently the case with respect to nonbinding instruments, transparency is limited to the performance of the collective of states. Individual performance then remains largely obscure. Although this impedes that individual states have to defend themselves in what Chayes and Chayes have called "justificatory discourse, even these general assessment reports can help to build trust that most other participants are moving in the same direction. Governments receive at least some reassurance on the practice of other states, and may therefore be more inclined to make efforts to comply.

Solely defining the function of a reporting and monitoring system in terms of transparency would however neglect other important aspects. Irrespective of the binding or nonbinding nature of the instrument, the repeated reporting exercise also enhances compliance by keeping the issue continuously on the national and international agenda and subjecting progress to scrutiny at both levels.[524] At the national level, these processes – if undertaken in a transparent manner – can strengthen the position of domestic actors and NGOs in favour of implementation and compliance. They provide these actors with opportunities to challenge governmental positions or otherwise exert pressure.[525] According

---

522 A. Chayes/A. Handler Chayes, The New Sovereignty: Compliance with International Regulatory Agreements, 1995, 22 and 135; K. Raustiala, "Compliance & effectiveness in international regulatory cooperation", Case Western reserve journal of international law 32 (2000), 387-440 (416).

523 E. Brown Weiss, "Conclusions: Understanding Compliance with Soft Law" in: D. Shelton (ed.), Commitment and Compliance: The Role of Nonbinding Norms in the International Legal System, 2000, 535-556 (541).

524 Compare e.g. A.A. Fatouros, "On the implementation of international codes of conduct: an analysis of future experience", American University Law Review 30 (1980 – 1981), 941-972 (963).

525 David G. Victor, Kal Raustiala & Eugene B. Skolnikoff, "Systems of Implementation Review", in D.G. Victor/K. Raustiala/E.B. Skolnikoff (eds.), The Implementation and Effectiveness of International Environmental Commitments, 1998, 51; K. Raustiala, "Compliance & effectiveness in international

to a FAO analysis in the context of the reporting on the FAO Pesticide Code, a number of governments experienced the reporting exercise as an opportunity to review their programmes and institute improvements on the use and control of harmful products.[526] As far as the international level is concerned, reporting to the FAO certainly helps to establish continuous dialogue and debate which is structured by the norms and objectives of the reported on instruments. If linked to a follow-up procedure in which the individual state must defend its conduct, reporting additionally institutionalises a form of justificatory discourse in a public forum. Such justificatory discourse is a key element for the persuasive processes of non-contentious compliance management and the influence of norms.[527] If reports are not individualised, however, justificatory discourse between a single reporting state and a political body of the institution which has been so effective in human rights and environmental treaty law[528] largely fails.

The reporting exercise on nonbinding instruments however also contributes to learning processes. It enables continuous feedback processes from the national to the international level through which learning and flexible adaptation to new issues becomes possible.[529] Sharing information with other states gives governments the opportunity to learn from each others' experience in dealing with specific implementation problems. Deficits in domestic capabilities, the need for improved international assistance and areas where further normative efforts are required more adequately to address specific problems can be identified. A clear example of such learning processes is the voluntary PIC system. The FAO/UNEP Joint Expert Group provided a forum where the information on implementation and conditions of use was processed and evalu-

---

regulatory cooperation", Case Western reserve journal of international law 32 (2000), 387-440 (416).

526 FAO, Analysis of Government Responses to the Second Questionnaire on the State of Implementation of the International Code of Conduct on the Distribution and Use of Pesticides (1993).

527 A. Chayes/A. Handler Chayes, The New Sovereignty: Compliance with International Regulatory Agreements, 1995, 26.

528 M. Bothe, "The Evaluation of Enforcement Mechanisms in International Environmental Law: An Overview" in: R. Wolfrum (ed.), Enforcing environmental standards: economic mechanisms as viable means?, 1996, 13-38 (23).

529 K. Raustiala/D.G. Victor, "Conclusions" in: D.G. Victor/K. Raustiala/ E.B. Skolnikoff (eds.), The Implementation and Effectiveness of International Environmental Commitments, 1998, 659-708 (687).

ated. Experiences from capacity building projects and the results of reporting were discussed in the meetings of the group. The FAO/UNEP Joint Expert Group was able to learn from the experiences and – facilitated by the flexibility of the nonbinding instruments – could adjust the procedure or expand the PIC list accordingly.[530]

(2) Monitoring and verification

The nonbinding instruments analysed in this study do not provide for formalised processes of independent monitoring of implementation or mechanisms of inspection. At least in the context of nonbinding instruments, states are generally not willing to develop strict transparency measures through verification and external monitoring. With respect to monitoring, nonbinding instruments thus differ from most modern environmental treaties which increasingly contain complex monitoring systems based on expert bodies that review evidence from reporting.[531] Nonbinding instruments rarely foresee monitoring of individual compliance or intrusive inspections in practice. It can be assumed that such mechanisms are seen to be incompatible with their voluntary nature. To the extent that state-sponsored public monitoring is lacking, highly specialised environmental non-governmental organisations (NGOs) in practice play an even more important role as watchdogs for compliance with these instruments.[532]

However, peer review and monitoring could also be perceived for nonbinding instruments. Nonbinding instruments are particularly conducive to this exercise if they contain concrete verifiable objectives. This is illustrated by the UNGA Resolution calling for a moratorium on drift net fishing on the high seas by 31 December 1992. The UNGA resolu-

---

530 D.G. Victor, "'Learning by Doing' in the Nonbinding International Regime to Manage Trade in Hazardous Chemicals and Pesticides" in: D.G. Victor/K. Raustiala/E.B. Skolnikoff (eds.), The Implementation and Effectiveness of International Environmental Commitments: Theory and Practice, 1998, 221-281 (260).

531 Compare for example the highly complex monitoring system of the Kyoto Protocol linked to expert group assessments. On the compliance procedure of the Kyoto Protocol R. Wolfrum/J. Friedrich, "The Framework Convention on Climate Change and the Kyoto Protocol" in: U. Beyerlin/P.-T. Stoll/R. Wolfrum (eds.), Ensuring Compliance with Multilateral Environmental Agreements: A Dialogue between Practitioners and Academia, 2006, 53-68.

532 See for details the analysis of the role of NGOs in this Part, at B.I.5b).

tion called upon members, international and non-governmental organisations as well as scientific institutions to submit to the Secretary-General information concerning activities or conduct inconsistent with the moratorium.[533] The continuous monitoring of implementation efforts and breaches through reports of the Secretary-General to the UNGA between 1992 and 1998 which were based upon information submitted by international organisations such as the FAO, NGOs, regional fisheries organisations and a number of active states, including the United States of America, contributed to the overall success of the moratorium by exposing state practice to public scrutiny.[534]

(3) Compliance Assistance

Compliance with international environmental law is not only a question of political will, but also one of capacity.[535] In Agenda 21, states recognised that "[T]he ability of a country to follow a sustainable path is determined to a large extent by the capacity of its people and its institutions ..."[536] This is not any different for nonbinding instruments.[537] Since it is the ultimate objective of all international instruments in the environmental sector to affect the behaviour of corporations and individuals, compliance with the environmental objectives of both binding and nonbinding instruments requires not only implementing legislation, but also administrative and enforcement efforts. In particular developing states frequently lack the financial resources, technical and scientific expertise and the administrative capacity to live up to these requirements. For example, without training and education of local peasants, and improvement of regulatory capacities at administrative levels,

---

[533] UN Doc. A/RES/46/215 (22 December 1991), para. 6.

[534] D.R. Rothwell, "The General Assembly Ban on Driftnet Fishing" in: D. Shelton (ed.), Commitment and Compliance: The Role of Non-Binding Norms in the International System, 2000, 121-146 (131-135, 145).

[535] M. Bothe, "The Evaluation of Enforcement Mechanisms in International Environmental Law: An Overview" in: R. Wolfrum (ed.), Enforcing environmental standards: economic mechanisms as viable means?, 1996, 13-38 (17); A. Chayes/A. Handler Chayes, The New Sovereignty: Compliance with International Regulatory Agreements, 1995, 13.

[536] Agenda 21, UN Doc. A/Conf. 151/26, Chapter 37, para. 37.1.

[537] E. Brown Weiss, "Conclusions: Understanding Compliance with Soft Law" in: D. Shelton (ed.), Commitment and Compliance: The Role of Nonbinding Norms in the International Legal System, 2000, 535-556 (552).

the goals pursued by the PIC procedure are hardly attainable; a fact that highlights the importance of financial and technical support used for education and capacity building.[538]

Given that capacity to comply is an important factor, mechanisms of financial and technical assistance or capacity building[539] generally play a large role in international environmental law.[540] They have become central features of non-confrontational compliance management approaches.[541] Mechanisms and institutions providing for financial assistance[542] and technology transfer[543] are common in most multilateral environmental agreements and institutions.[544] International organisations

---

538 R.L. Paarlberg, "Managing Pesticide Use in Developing Countries" in: P.M. Haas/R.O. Keohane/M.A. Levy (eds.), Institutions for the Earth: Sources of Effective International Environmental Protection, 1993, 309-350 (347); J. Ross, "Legally Binding Prior Informed Consent", Colorado Journal of International Environmental Law and Policy 10 (1999), 499-529 (517).

539 On the meaning of capacity-building in the context of international environmental law and compliance L. Gündling, "Compliance Assistance in International Environmental Law: Capacity-Building Through Financial and Technology Transfer", Zeitschrift für ausländisches öffentliches Recht und Völkerrecht 56 (1996), 796-809 (800).

540 L. Boisson de Chazournes, "Technical and Financial Assistance and Compliance: the Interplay" in: U. Beyerlin/P.-T. Stoll/R. Wolfrum (eds.), Ensuring Compliance with Multilateral Environmental Agreements, 2006, 273-300 (277 et seq.); M. Bothe, "The Evaluation of Enforcement Mechanisms in International Environmental Law: An Overview" in: R. Wolfrum (ed.), Enforcing environmental standards: economic mechanisms as viable means?, 1996, 13-38 (17-18); P.H. Sand, "Institution-Building to Assist Compliance with International Environmental Law: Perspectives", Zeitschrift für ausländisches öffentliches Recht und Völkerrecht 56 (1996), 774-795 (780-786).

541 A. Chayes/A. Handler Chayes, The New Sovereignty: Compliance with International Regulatory Agreements, 1995.

542 E.g. UN Framework Convention on Climate Change, Article 3.1.; Kyoto Protocol, Article 10; Convention on Biological Diversity, Articles 18.2 and 19.2; Convention to Combat Desertification, Article 5.a and 6.b; Stockholm Convention on Persistent Organic Pollutants, Article 12.

543 E.g. Montreal Protocol, Article 10 (a); Convention on Biological Diversity, Article 18; United Nations Framework Convention on Climate Change, Articles 4.1 (c), 4., 4.5, 4.7-4.10; 9.2 (c), 11 and 11.1; Kyoto Protocol, Article 10 and 11; UNCLOS, Articles 62, 143-144 and 266-277 and Annex III, Basic Conditions of Prospecting, Exploration and Exploitation, Article 5.

544 Compare L. Boisson de Chazournes, "Technical and Financial Assistance" in: D. Bodansky/J. Brunnée/E. Hey (eds.), Oxford Handbook of Inter-

such as the FAO or the IMO and international financial institutions such as the World Bank have over the years turned into major channels for financial and technical assistance to developing countries.[545] Usually specific funds are created or the multilateral environmental agreement cooperates through the Conference of the Parties with a financial institution such as the Global Environment Facility or the World Bank.[546] In addition to enabling states to comply, these mechanisms also provide incentives for states to become parties to environmental agreements or to return to compliance.[547]

These mechanisms do not depend on the binding or nonbinding status of the underlying commitment, but can also be used in relation to nonbinding norms. And indeed, capacity building and other measures have become one of the main tools for international institutions to enhance compliance with nonbinding instruments. Nonbinding instruments also frequently serve as a basis for the formulation and design of capacity building projects and for mechanisms of financial and technical assistance.[548] Technical and financial assistance undertaken by the FAO and

---

national Environmental Law, 2007, 947-973; R. Wolfrum, "Means of Ensuring Compliance with and Enforcement of International Environmental Law", Recueil des cours de l'Académie de Droit International de La Haye 272 (1998), 13-154 (117-137).

545 L. Boisson de Chazournes, "Technical and Financial Assistance and Compliance: the Interplay" in: U. Beyerlin/P.-T. Stoll/R. Wolfrum (eds.), Ensuring Compliance with Multilateral Environmental Agreements, 2006, 273-300 (274); P.H. Sand, "Institution-Building to Assist Compliance with International Environmental Law: Perspectives", Zeitschrift für ausländisches öffentliches Recht und Völkerrecht 56 (1996), 774-795 (780-781).

546 On these financial institutions N. Matz, "Environmental financing: function and coherence of financial mechanisms in international environmental agreements", Max Planck Yearbook of United Nations Law 6 (2002), 473-534.

547 For the distinction between assistance as a means to induce and assistance as a means to restore compliance see L. Boisson de Chazournes, "Technical and Financial Assistance and Compliance: the Interplay" in: U. Beyerlin/P.-T. Stoll/R. Wolfrum (eds.), Ensuring Compliance with Multilateral Environmental Agreements, 2006, 273-300 (277); L. Boisson de Chazournes, "Technical and Financial Assistance" in: D. Bodansky/J. Brunnée/E. Hey (eds.), Oxford Handbook of International Environmental Law, 2007, 947-973 (972).

548 W. Edeson, "The International Plan of Action on Illegal, Unreported and Unregulated Fishing: The Legal Context of a Non-Legally Binding Instrument", The International Journal of Marine and Coastal Law 16 (2001), 603 et seq. (623).

other organisations as well as NGOs and industry were also some of the main reasons for compliance with the voluntary PIC procedure.[549] Further, the FAO's pesticide management program includes capacity building measures directed at supporting the implementation of the FAO Code of Conduct on the Use and Distribution of Pesticides in general. Through the FAO's Technical Cooperation Programme, the FAO funds small scale projects to address specific problems in agriculture, forestry and fisheries, including the implementation of various FAO codes of conduct.[550] Similarly, the Fisheries Department of the FAO secretariat, on the basis of a broad and unspecific mandate from the FAO Conference,[551] provides the institutional platform, executive know-how and funding to help local communities and developing states to implement the provisions of the CCRF. Through the so-called Programme of Global Partnerships for Responsible Fisheries or 'FishCode', the FAO provides technical assistance, undertakes training and human-capacity development which help fishermen and states to adopt responsible management and conservation practices and to design improved legal and institutional arrangements.[552] These activities are financed through the FishCode Trust Fund which draws on donations of individual countries as well as regular programme resources of the FAO. Since the adoption of the FAO CCRF, the FAO's legal advisory service concentrates on assisting governments in the formulation and

---

549 M.A. Mekouar, "Pesticides and Chemicals: The Requirement of Prior Informed Consent" in: D. Shelton (ed.), Commitment and Compliance: The Role of Non-Binding Norms in the International Legal System, 2000, 146-163 (155).

550 An example is the project on the "Implementation of the International Code of Conduct on the Distribution and Control of Pesticides – Sahel region", see for details http://www.fao.org/ag/agp/agpp/pesticid/manage/sahel.htm; for details on the Technical Cooperation Programme of the FAO see the website at http://www.fao.org/tc/tcp/index_en.as.

551 The mandate to the FAO secretariat is simply to "give advice to developing countries and establish the Interregional Assistance Programme" without any further terms of reference, see FAO Conference Resolution 4/95, para. 4.

552 For an assessment of the achievements of the programme compare S.C. Venema, "The FISHCODE Project: Achievements" in: M.H. Nordquist/J.N. Moore (eds.), Current Fisheries Issues and the Food and Agriculture Organization of the United Nations, 2000, 341-359; for further details on the FishCode Programme see FAO Committee on Fisheries, 'Progress in the Implementation of the 1995 Code of Conduct for Responsible Fisheries, Related International Plans of Action and Strategy', COFI 2007/2, paras 57-60; information on the FishCode is available at www.fao.org/figis.

revision of fisheries legislation[553] or in the drafting or reform of regional treaty instruments that are in conformity with the requirements of CCRF.[554] This can lead to important legal developments in the respective issue area. For example, the recent focus of the FAO secretariat on inland fisheries has led to important inland fisheries agreements such as those for Lake Tanganyika and Lake Victoria.[555] The FAO also makes use of its network of regional offices or FAO regional fishery bodies to organise regional workshops and expert consultations which address specific regional problems of implementation.[556] Another initiative, the Sustainable Fisheries Livelihoods Programme undertaken by the FAO in cooperation with individual donor countries, focuses on providing assistance to local fisheries communities.[557]

Other examples directly related to the case studies include the capacity building programmes of the United Nations Institute for Training and Research which were designed to implement the voluntary PIC system in developing countries.[558] An example is the World Bank, which as co-

---

553 For example, the FAO secretariat has assisted in the revision of pertinent legislation of Angola, Namibia, Malaysia, The Maldives, Vietnam, Barbados, Saint Lucia, Antigua and Barbuda.

554 For more information see http://www.fao.org/Legal/advserv/fish-e.htm.

555 Convention on the Sustainable Management of Lake Tanganyika, 12 June 2003; Convention for the Establishment of the Lake Victoria Fisheries Organization (30 June 1994); all agreements available at http://faolex.fao.org/ faolex/.

556 The progress report to COFI of 2003 lists 22 workshops, expert consultations and conferences related to different issues of the CCRF that had been organized and taken place within just 2 years, see COFI/2003/3Rev.1, Committee on Fisheries, Twenty-fifth session of 24-28 February 2003, "Progress in the Implementation of the Code of Conduct for Responsible Fisheries and Related International Plans of Action", para. 16; for one of the workshops see e.g. FAO Regional Office for Asia and the Pacific, "Proceedings Asia Regional Workshop on Implementation, Monitoring and Observance International Code of Conduct on the Distribution and Use of Pesticides", 2005, available at http://www.fao.org/docrep/008/af340e/af340e0i.htm.

557 Established in 1999, the programme represents a partnership between the Department for International Development of the United Kingdom and Northern Ireland which provides the funds, the FAO who serves as the executive agency, and 25 West-African states. For more information, see the SFLP-website at http://www.sflp.org/.

558 The participation of developing countries in regional and national workshops was extensive. According to David Victor, ninety-two developing countries have participated in regional and sub-regional workshops on PIC as con-

operates with the FAO in the ProFish Programme and links its Operational Policies to binding or nonbinding standards of other institutions.[559] Capacity building and thus compliance assistance is also undertaken by NGOs as will be seen further below in this section.[560]

(4) Norm concretizations and interpretations

A further tool employed by international institutions to enhance compliance with nonbinding instruments is to provide informal guidance for implementation. International institutions issue nonbinding instruments to guide the implementation of general nonbinding norms.

One recurring form is the concretisation and interpretation of norms through political bodies, but this is also frequently done by the secretariats of international organisation. The OECD secretariat for instance provides "commentaries" on the OECD Guidelines which are not officially part of the Guidelines but are nevertheless attached to the official booklet containing the OECD Guidelines and relevant procedures.[561] Furthermore, as mentioned earlier, the FAO Fisheries Department – sometimes but not always in cooperation with governments, other organisations and NGOs – regularly produces so-called Technical Guidelines and supplementary documents.[562] These instruments contain general explanations on how certain provisions of the CCRF should be understood, and usually include specific suggestions, best practices and

---

ducted by UNITAR/UNEP/FAO, compare D.G. Victor, "'Learning by Doing' in the Nonbinding International Regime to Manage Trade in Hazardous Chemicals and Pesticides" in: D.G. Victor/K. Raustiala/E.B. Skolnikoff (eds.), The Implementation and Effectiveness of International Environmental Commitments: Theory and Practice, 1998, 221-281 (254-255). UNITAR continues in this tradition and still plays an important role in capacity building efforts in the field of chemical and pesticide management. Its Programme in Chemical and Waste Management is – among other objectives – directed at the implementation of the PIC Convention in cooperation with UNEP and the FAO. See for details http://www.unitar.org/cwg/cwmoverview.html.

[559] The details have already been discussed above in this Part 2, at A.II.2.b)(1).

[560] See in this Part 2, at B.I.5.b), further below.

[561] The booklet on the OECD Guidelines issued by the OECD secretariat is available from http://www.oecd.org/dataoecd/56/36/1922428.pdf.

[562] All FAO Technical Guidelines and accompanying Supplements are available at: www.fao.org/fi.

other recommendations. They also comprise annexes that include guidance on specific technical subjects.[563] Although not adopted by any political body of the FAO, the Technical Guidelines are widely accepted. This is most prominently underscored by the fact that the Technical Guidelines are mentioned in the Johannesburg Plan of Implementation.[564] Recently, the FAO secretariat has even started to develop supplements to Technical Guidelines called "companion documents" which outline how actors can comply with the Guidelines, and thus reach an even higher degree of specificity.[565]

These interpretational exercises and concretisations, even if nonbinding, serve to overcome the ambiguity and vagueness of norms. The analysis has shown that international nonbinding instruments, in particular if adopted at the highest political level, often contain a number of indeterminate and vague norms. Concretisations and the codification of best practices through additional guidelines fulfil two functions which potentially enhance compliance. First, just as for treaty law, ambiguity may at times also be a reason for an otherwise willing actor not properly to implement a particular norm.[566] States willing to implement international norms receive guidance on how properly to implement a particular norm, possibly by means of debates on the proper implementation of a nonbinding instrument arising in the wake of the adoption of the concretising instrument. Second, compliance with a vague norm in this way becomes verifiable, which is a precondition for pressuring states to comply through compliance monitoring.

---

563 For example, the FAO Technical Guidelines for Responsible Fisheries on Fishing Operations contain an Annex III which outlines a "Proposed System for the Marking of Fishing Gear".

564 World Summit on Sustainable Development, Johannesburg Plan of Implementation, para. 31 (c).

565 E.g. FAO document on "Compliance to FAO Technical Guidelines for Responsible Fisheries: Health management for responsible movement of live aquatic animals" as announced in FAO Technical Guidelines for Responsible Fisheries No. 5 Aquaculture Development, Suppl. 2.

566 A. Chayes/A. Handler Chayes, The New Sovereignty: Compliance with International Regulatory Agreements, 1995, 24.

5. Actors

a) The central role of international institutions

International institutions are the key to compliance for most nonbinding instruments that aim to change the status quo. In the absence of international enforcement, institutionalisation that secures long-term interaction and discourse is central for the potential impact of nonbinding instruments. Voluntary reporting mechanisms, capacity building and financial assistance, informal guidance by various sets of best practices as well as constant collection and dissemination of information can be important compliance-inducing factors.

The central role of institutions for compliance can be explained by reference to fundamental conceptions of why states behave in a certain way. First of all, states can only pursue their goals in the modern interdependent world through their membership and participation in international cooperative treaty regimes or other institutions.[567] The condition for continuous beneficial participation particularly in the long run is to remain a member in good standing in the international system as a whole, and the wish for status is the reason for bringing behaviour into conformity.[568] It is even suggested by some commentators, in particular Chayes and Chayes, that the concept of sovereignty today is changing from a principle defined in terms of autonomy to one that describes the ability to participate effectively in international institutions.[569] Similarly, the legitimacy-based perspective of Franck emphasises that the ultimate motivation for compliance is the desire to be a member of the international community, and to benefit from this status.[570] These concepts confirm the assessment made by Wolfgang Friedmann as early as 1964 that "[i]n the international law of co-operation, the sanction of exclusion or non-participation in joint activities replaces the punitive sanc-

567 A. Chayes/A. Handler Chayes, The New Sovereignty: Compliance with International Regulatory Agreements, 1995, 27.

568 A. Chayes/A. Handler Chayes, The New Sovereignty: Compliance with International Regulatory Agreements, 1995, 27; T.M. Franck, The power of legitimacy among nations, 1990, 38.

569 J.E. Alvarez, "International Organizations: Then and Now", American Journal of International Law 100 (2006), 324-347 (335); A. Chayes/A. Handler Chayes, The New Sovereignty: Compliance with International Regulatory Agreements, 1995, 27.

570 T.M. Franck, The power of legitimacy among nations, 1990, 38.

tion. ..."[571] As a consequence, states are under pressure to play by the rules of the respective organisation, because leaving the organisation is hardly feasible.[572] This also applies to nonbinding instruments, although to a lesser extent than binding ones. A non-compliant state risks its reputation and good standing as a cooperative member, and therefore the influence it can exert to achieve its goals, when it flatly disregards or openly denies its commitment to a code of conduct or a particular set of guidelines. By contrast, as pointed out by Brown-Weiss, compliance with nonbinding instruments entails a particular chance for states to strengthen their reputation as a cooperative member, because it is a demonstration that the state complies even if it is not legally obliged to do so.[573]

Reputation-based compliance pressure may however only explain one part of how institutions influence the behaviour of actors. As mentioned above, state actors are not only complying out of rational interest, but may according also be persuaded to act in accordance with norms through interactional discursive processes that take place on the basis and around legal norms in what Jutta Brunnée and Stephen Toope call "communities of practice".[574] By creating and furnishing such communities through continuous interaction on the basis of norms, for instance at COPs or through implementation review processes, institutions shape and foster compliance by the participants with the norms in question.[575] However, it makes a difference whether such discourse is of

---

[571] W. Friedmann, The Changing Structure of International Law, 1964, preface.

[572] G. Ulfstein, "Comment on J. Brunnée's "Reweaving the Fabric of International Law? Patterns of Consent in Environmental Framework Agreements"" in: V. Röben (ed.), Developments of International Law in Treaty Making, 2005, 145-153 (150-151).

[573] E. Brown Weiss, "Conclusions: Understanding Compliance with Soft Law" in: D. Shelton (ed.), Commitment and Compliance: The Role of Nonbinding Norms in the International Legal System, 2000, 535-556 (542).

[574] J. Brunnée/S.J. Toope, Legitimacy and legality in international law: an interactional account, 2010, at 100.

[575] *Ibid.*, at 100; compare also J. Brunnée/S.J. Toope, "Persuasion and enforcement: explaining compliance with international law", The Finnish yearbook of international law 13 (2002), 273-295; J. Brunnée/S.J. Toope, "International law and constructivism: elements of an interactional theory of international law", Columbia journal of transnational law 39 (2000), 19-74; J. Brunnée, "Of Sense and Sensibility: Reflections on International Regimes as Tools for Environmental Protection", International and Comparative Law Quarterly 53

legal nature or not, i.e. whether participants discusses the legality of a certain action or not. With this caveat, arguments about the significance of continuous and repeated norm-based discourse can be transposed to nonbinding instruments. It can then give one explanation why interaction will foster compliance, but that it still makes a difference whether actors develop a sense of legal obligation or not.[576] In fact, they may very well explain why in the absence of enforcement, and in the absence of clear reputational risks, such norms nevertheless often are accepted and implemented. Institutionalisation enables continuous and repeated interaction, for instance in the Committee of Fisheries or its subcommittees. If this institutionalisation includes a follow-up mechanism, for example in the form of periodic reporting, the issues will remain on the agenda of domestic and international decision-making bodies. This generates continuous discourse on the issues in question. By pushing for the establishment of nonbinding progressive norms and in providing the fora, international institutions enable and shape the discourse between actors, and thereby ultimately affect perceptions about what is right and proper conduct within an issue-area.[577]

In fact, nonbinding instruments may even be particularly useful to trigger such processes.[578] Entry barriers for participants are low, even in more advanced stages of the process. And the nonbinding nature relieves participants in negotiations from considering the implications of a possible binding commitment and its acceptability by legislators. They can interact and find best practices in a non-contentious atmosphere without having to implement right away. The risk of alienating single states or groups of states is low, and this may be important considering

---

(2004), 351-367; see also for a similar direction already F. Kratochwil, Rules, norms and decisions: on the conditions of practical and legal reasoning in international relations and domestic affairs, 1989.

576 J. Brunnée/S.J. Toope, Legitimacy and legality in international law: an interactional account, 2010, 98 et seq.

577 O.S. Stokke, "The Interplay of International Regimes: Putting Effectiveness Theory to Work", FNI Report 14/2001 (2001), 1 et seq. (9).

578 H.S. Dashwood, "Corporate Social Responsibility and the Evolution of International Norms" in: J.J. Kirton/M.J. Trebilcock (eds.), Hard Choices, Soft Law: Voluntary Standards in Global Trade, Environment and Social Governance, 2004, 189-201 (198).

that the participation of all actors is of utmost significance for effectively addressing global environmental problems.[579]

Socialisation as a compliance inducing factor may further explain the influence of institutionalisation on compliance with nonbinding instruments. Recent scholarship on norm diffusion and compliance stresses the significance of socialisation processes in such institutions.[580] International norms – whether nonbinding or binding – are internalised and constrain states not only as a matter of rational calculations, but also because states strive to participate in the international ethos that the respective norms represent.[581] In other words, international organisations and regimes have their own soft power which is rooted in the values that they represent.[582] In particular for state actors, codes of conduct represent the shared values of the members of the organisation, and this may enhance their acceptance and their implementation by the members, or by those wishing to become members.

Finally, institutionalisation at the international level strengthens transnational networks which have an interest in cooperating on the basis of nonbinding instruments below the threshold of treaty law.[583] Coalitions with foreign governmental agencies, NGOs, industry or even the secretariats of the respective institutions can be formed at the international level even with a view to countering differing interests of other governmental agencies of the same state.[584] The interests of the environ-

---

579 This latter point is also claimed to be a general advantage of nonbinding instruments by A.E. Boyle, "Some Reflections on the Relationship of Treaties and Soft Law", International and Comparative Law Quarterly 48 (1999), 901-913 (912).

580 R. Goodman/D. Jinks, "How to influence states: socialization and international human rights law", Duke law journal 54 (2004), 621-703; J.E. Alvarez, International Organizations as Law-makers, 2005, 624; H.H. Koh, "Internalization through socialization", Duke law journal (2005), 975-982; J.E. Alvarez, "Do states socialize?", Duke law journal 54 (2005), 961-974.

581 José Alvarez explicitly refers to both soft and hard law norms as contributing to socialization processes, compare J.E. Alvarez, "International Organizations: Then and Now", American Journal of International Law 100 (2006), 324-347 (338).

582 J.E. Alvarez, International Organizations as Law-makers, 2005, 626.

583 For a comprehensive description of network cooperation and their domestic impact, see A.-M. Slaughter, A New World Order, 2004.

584 R.O. Keohane/J.S. Nye, "Transgovernmental Relations and International Organizations", World Politics 27 (1974), 39-62 (50-54).

mental ministry for example often differ from that of the ministry of economy. Nonbinding instruments can serve as tools for these government agents in support of international environmental objectives to agree on norms within an international institution. The possibility of adopting nonbinding norms in lower-level bodies without implicating the foreign ministry and without triggering questions of domestic implementation at the time of adoption can facilitate such processes.[585] Moreover, nonbinding instruments help governmental officials to agree on international rules with their like-minded counterparts in other states, and to implement those norms on the domestic level. The nonbinding guiding instrument serves as a set of best practices jointly elaborated and adopted by an authoritative forum from which actors can easily copy if they wish.[586] The instruments are sufficiently flexible to provide guidance in uncertain and complex issue areas as they can be easily adapted to changing circumstances, scientific progress or regional and national particularities.[587]

b) The role of non-governmental organisations

Non-governmental organisations ("NGOs"), in particular environmental NGOs but also industry NGOs, play a significant role in enhancing compliance with nonbinding instruments.[588]

With respect to treaty law, it is widely accepted that the work of non-governmental organisations often improves implementation and compliance.[589] Generally speaking, environmental NGOs greatly contribute to monitoring the performance of addressees. They assemble and publish independent information on the performance of states and private actors and transmit this information to the national and international

585 Similar considerations in the context of nonbinding standard setting are made by J.E. Alvarez, International Organizations as Law-makers, 2005, 247.

586 A.-M. Slaughter, A New World Order, 2004, 179-180.

587 A.-M. Slaughter, A New World Order, 2004, 178 and 181.

588 E. Brown Weiss, "Conclusions: Understanding Compliance with Soft Law" in: D. Shelton (ed.), Commitment and Compliance: The Role of Nonbinding Norms in the International Legal System, 2000, 535-556 (553).

589 S. Riedinger, Die Rolle nichtstaatlicher Organisationen bei der Entwicklung und Durchsetzung internationalen Umweltrechts, 2001, 246-258; J. Ebbesson, "Public Participation" in: D. Bodansky/J. Brunnée/E. Hey (eds.), Oxford Handbook of International Environmental Law 2007, 681-703 (688).

fora concerned with the assessment of state performance.[590] Their particular contributions often serve as a means of verifying the environmental data provided by states and thereby introduce an element of independent control on state reporting.[591] NGOs also set up capacity building activities, provide technical expertise and financial assistance to states, and thereby play an increasingly important role in the effective implementation.[592]

None of these activities of NGOs depends on the legal status of an instrument. Nonbinding instruments and related processes due to their informality often provide better opportunities than treaty law for NGOs to become influential in international norm development and norm implementation processes.[593] Their inclusion in these processes increases the chance that they are actually implemented.[594]

In fact, the comparative weakness of enforcement and compliance mechanisms in the context of nonbinding instruments renders the work of NGOs even more significant for the effectiveness of these instruments. As seen above, formal reporting systems for nonbinding instruments do not comprise individualised compliance control nor do they generally provide for independent monitoring and verification. NGOs are the only actors which can fill this gap. Their role is enhanced when international institutions deliberately integrate them into their monitoring and reporting activities. As already mentioned, NGOs played an important role in the implementation of the UNGA moratorium on driftnet fishing by submitting information on breaches of the morato-

---

590 M. Bothe, "The Evaluation of Enforcement Mechanisms in International Environmental Law: An Overview" in: R. Wolfrum (ed.), Enforcing environmental standards: economic mechanisms as viable means?, 1996, 13-38 (25); A. Chayes/A. Handler Chayes, The New Sovereignty: Compliance with International Regulatory Agreements, 1995, 171-172; K. Raustiala, "The "participatory revolution" in international environmental law", The Harvard environmental law review 21 (1997), 537-586 (562).

591 S. Riedinger, Die Rolle nichtstaatlicher Organisationen bei der Entwicklung und Durchsetzung internationalen Umweltrechts, 2001, 259-274.

592 Financial assistance by NGOs comprises for instance debt-for-nature swaps and the establishment of so-called National Environmental Funds, compare S. Riedinger, Die Rolle nichtstaatlicher Organisationen bei der Entwicklung und Durchsetzung internationalen Umweltrechts, 2001, 350-354.

593 A.C. Kiss/D. Shelton, International environmental law, 2004, 89.

594 V. Röben, "Proliferation of Actors" in: R. Wolfrum/V. Röben (eds.), Developments of International Law in Treaty Making, 2005, 511-542 (525).

rium to the UN Secretary-General as requested by the UNGA resolution.[595] The FAO Code of Conduct for Responsible Fisheries for example stresses that all states, whether members or non-members of FAO as well as governmental and non-governmental organisations, should cooperate with the FAO in its monitoring work.[596] The questionnaires are being sent not only to states and regional fisheries organisations, but also to NGOs. NGOs are usually providing independent information on implementation and general compliance.[597] The most recent version of the FAO Code of Conduct on the Use and Distribution of Pesticides extended the voluntary reporting procedure by "inviting" the pesticide industry, NGOs and "other interested parties" to monitor implementation activities and report accordingly to the Director-General.[598] The significant knowledge of NGOs on implementation of the Pesticide Code stems from experiences and contacts at the local level and in remote areas. NGOs can therefore highlight problems and violations that may not be mentioned in official reports. At the time of the voluntary PIC system, NGOs, most prominently the Pesticide Action Network (PAN), have monitored the implementation of the system. NGOs also contributed valuable expertise to the decision-making process of the FAO/UNEP Joint Expert Committee on the PIC procedure which they attended as observers. Their assessment of the "conditions of use" of certain pesticides in developing countries had considerable impact on the decision-making of the Joint Expert Group, which extended the PIC list according to the information received by NGOs.

In more general terms, assuming that they are sufficiently precise, nonbinding instruments are generally well-suited to serve as a point of reference and initiation for the environmental efforts of NGOs. The adoption of nonbinding instruments allows NGOs to base their claims on an objective standard which was accepted by the relevant actors and therefore provides them with formidable support for their activities. Moreover, nonbinding instruments can be used by these actors to initiate public discourse as well as naming and shaming campaigns by exposing the gaps between the international commitments of states or private ac-

---

[595] UN Doc. A/RES/46/215 (22 December 1991), para. 6. Compare the analysis of D.R. Rothwell, "The General Assembly Ban on Driftnet Fishing" in: D. Shelton (ed.), Commitment and Compliance: The Role of Non-Binding Norms in the International System, 2000, 121-146 (134).

[596] CCRF, Article 4.2.

[597] Consider e.g. FAO Doc. COFI/2009/2, paras 56-61.

[598] FAO Pesticide Code (revised version, 2003), Articles 12.8 and 12.9.

tors and their actual conduct.[599] Concerned about reputation in their constituencies, decision-makers can be forced to bow to public pressure and comply regardless whether the commitment is technically binding or not. A loss of reputation not only in the eyes of the public, but also in the eyes of other states can derive from the "breach" of soft law instruments,[600] even though the reputational consequences may be less severe than in the case of a treaty.[601]

As financial resources are often limited in particular for nonbinding initiatives, NGOs also play an important role with regard to compliance assistance. For example, NGOs undertake a range of activities to promote and raise awareness of the FAO Code of Conduct for Responsible Fisheries. They organise conferences and workshops as well as training programmes for fish workers, translate the CCRF into local languages or directly work as consultants in consultative bodies established by FAO Members.[602] The progress reports of the Committee on Fisheries also indicate that NGOs were involved in assisting states in their implementation activities. These included sharing information on best practices and the establishment of codes of practice, as well as the monitoring of illegal and unregulated fishing at sea.[603]

---

[599] In this sense also K.W. Abbott/D. Snidal, "Hard and Soft Law in International Governance", International Organization 54 (2000), 421-456 (452); H. Neuhold, "The Inadequacy of Law-Making by International Treaties: "Soft Law" as an Alternative" in: V. Röben (ed.), Developments of International Law in Treaty Making, 2005, 39-52 (51).

[600] Neuhold mentions publicity and mobilisation of shame as mechanisms which can increase the pressure to comply even with soft law instruments, see H. Neuhold, "The Inadequacy of Law-Making by International Treaties: "Soft Law" as an Alternative" in: V. Röben (ed.), Developments of International Law in Treaty Making, 2005, 39-52 (51).

[601] A.T. Guzman, "A Compliance-Based Theory Of International Law", California Law Review 90 (2002), 1823-1887 (1879-1880).

[602] FAO Committee on Fisheries (2009), 'Progress in the Implementation of the 1995 Code of Conduct for Responsible Fisheries, Related International Plans of Action and Strategy', COFI 2009/2, para. 56; FAO, Committee on Fisheries (2007), 'Progress in the implementation of the 1995 Code of Conduct for Responsible Fisheries, related International Plans of Action and Strategy', COFI/2007/2, para. 53.

[603] FAO Committee on Fisheries, Twenty-seventh session 2007, Progress in the Implementation of the 1995 Code of Conduct for Responsible Fisheries, Related International Plans of Action and Strategy, COFI 2007/2, para. 56.

Taken together, NGO activity in the case of nonbinding instruments frequently is decisive for their success. The consequence of this is that the success of a nonbinding instrument is to a certain extent dependent on the interests and financial resources of non-state actors. In areas which enjoy the attention of the media and the public, compliance control by NGOs might be a strong mechanism that outperforms more formalised monitoring through expert bodies or other mechanisms. But it is certainly of limited effect in other areas and cannot guarantee stability in the long-term and when public attention, and thus NGO attention, shifts away to other issues.

6. *Summary*

Despite the lack of legal obligation and direct enforcement mechanisms,[604] a number of factors that are linked to the specific characteristics of nonbinding instruments may enhance compliance by states.

First of all, mechanisms providing economic incentives for compliance are also available for some nonbinding instruments, and are not the reserve of international treaty law. Where linkages of nonbinding instruments to standards of financial institutions such as the World Bank or the Global Environment Facility exist, there is a great incentive to comply with nonbinding instruments irrespective of their legal nature.

Compliance management mechanisms are increasingly established by international institutions even for nonbinding instruments. These correspond to the suggestions of scholars that perceive compliance management as the key to compliance. Reporting mechanisms enhance transparency and institutionalise repeated discourse at the international and national level. Compliance assistance by international institutions is widely used to address the capacity problems of developing countries and helps them bring their environmental law and policy into compliance with the recommended norms. Norm concretisations and guiding documents are produced with a view to addressing ambiguities and helping the reform of legal systems and governance structures at the national level.

As will be seen in the analysis of the limits of nonbinding instruments at the end of this chapter, international institutions however do not employ the whole range of tools of a sound compliance management strat-

---

604 On the implications of these limitations, see further below in this Part, at B.III.2.

egy. The analysed systems commonly lack a procedure of justificatory discourse due to the absence of individualised assessment of compliance before an international body. Further weaknesses exist with respect to monitoring and verification. When states agree on a nonbinding instrument instead of binding rules, they usually do not support a strong compliance mechanism.

The shortcomings of nonbinding instruments are to some extent offset by regional organisations or non-state actors. As seen above, regional fisheries management organisations function to a certain degree as implementing agencies of nonbinding instruments. Their (binding or nonbinding) standards and enforcement mechanisms are sometimes influenced by the international nonbinding instruments issued by a global institution such as the FAO. NGOs can also fill some of the gaps in compliance management, and thus they gain particular significance in nonbinding initiatives. Generally speaking, access of NGOs to norm elaboration and implementation activities is greater in the context of nonbinding instruments. Some compliance mechanisms specifically call upon regional organisations, NGOs and industry to join the common effort to implement a particular standard. NGOs respond with independent reporting, monitoring and naming and shaming campaigns.

Finally, nonbinding instruments cannot only be assessed on the basis of the criteria applied to treaty law. The specific potential of nonbinding instruments develops through the increasing institutionalisation in international organisations or other international institutions such as Conferences of the Parties and sub-committees. Continuous discourse and long-term repeated interaction of actors initiate processes of learning, persuasion and socialisation. Because the interdependence of states increases the significance of reputation and the need to remain a respected member in these institutions, nonbinding instruments may shape state behaviour in the long-term.

II. Implementation in national legal systems[605]

It has often been postulated that legally nonbinding commitments directly give rise to change of behaviour by states through internal legislative changes.[606] However, rarely have these been subjected to systematic analysis.[607] Such analysis is important to understand the potential of nonbinding instruments. To identify those avenues through which international nonbinding norms enter the domestic legal orders is furthermore a prerequisite for a meaningful assessment of the legitimacy of these activities.[608]

1. Legislative implementation

Legislative implementation is the key method for the implementation of most of the international nonbinding instruments discussed. Generally speaking, however, nonbinding instruments cannot guarantee that domestic implementation will take place, since nonbinding instruments do not trigger domestic legislative implementation efforts automatically. Even though the adoption of nonbinding instruments can thus not guarantee that domestic processes of implementation will be put into motion, nonbinding instruments *may* and often do lead to legislative implementation efforts.

---

605 Parts of the following section on the implementation of nonbinding instruments into the national legal system have been previously published, see J. Friedrich, J. Lohse, "Revisiting the Junctures of International and Domestic Administration in Times of New Forms of Governance: Modes of Implementing Standards for Sustainable Development and their Legitimacy Challenges", 2 European Journal of Legal Studies No. 1 (2008), 49-86.

606 Schachter refers to this as the internal effect of nonbinding commitments; see O. Schachter, "The Twilight Existence of Nonbinding International Agreements", American Journal of International Law 71 (1977), 296-304 (303); see also C. Tietje, Internationalisiertes Verwaltungshandeln, 2001, in particular 621-639; D. Shelton (ed.), Commitment and Compliance: The Role of Non-binding Norms in the International Legal System, 2000.

607 Tietje addresses these issues but does not focus on them, see C. Tietje, Internationalisiertes Verwaltungshandeln, 2001, in particular 621-639; the overview of nonbinding instruments provided in J.E. Alvarez, International Organizations as Law-makers, 2005, focuses on the international level, but does not give a detailed account of the possible linkages of international and domestic for those instruments.

608 Compare the legitimacy assessment in Part 3, below.

a) The need for legislative implementation

Nonbinding norms cannot bind the executive or courts. They have to be implemented into national law in order to have legal status. Legislation is the only way the legislature can oblige the national administration to respect and implement the nonbinding international standards, and to ensure that such norms are enforced through the judiciary. In contrast to international treaties, nonbinding codes of conduct do not immediately create enforceable rights and obligations in the national legal order (monist system) nor are states obliged to transpose them into domestic law as they would with international treaties in a dualist system. Thus the national executive is not legally bound by them.[609]

In the German legal order, nonbinding instruments cannot be considered acts stemming from an international organisation with supranational legislative competencies within the meaning of Article 24 German Basic Law. Nor do they generally constitute general rules of international law directly incorporated into German national law through Article 25, although exceptions may exist where they indeed reflect such general rules. Finally, nonbinding instruments do not – as do international treaties – enter the German legal order through a legislative act of the parliament by means of Article 59 II of the German Basic Law. Attempts to apply either Article 25 or Article 59 by way of analogy may possibly be construed theoretically,[610] but at least generally speaking do not find expression in the practice of German governmental and judicial institutions.[611]

However, legislative implementation is usually a requirement for the application of nonbinding instruments by national executives, at least insofar as executives – as in the German legal system – may generally only act on the basis of competencies being properly delegated from the legislature to the executive. Indirect forms of implementation are possible even in the absence of specific legislation, but they require suitable

---

609 For the German legal order see C. Tietje, Internationalisiertes Verwaltungshandeln, 2001, 608-613 and 622-623.

610 C. Engel, Völkerrecht als Tatbestandsmerkmal deutscher Normen, 1989, 224 et seq.

611 See e.g. Bundesverfassungsgericht (Federal Constitutional Court of Germany), BVerfGE 68, 1 et seq.; for details compare the discussion by U. Dieckert, Die Bedeutung unverbindlicher Entschließungen internationaler Organisationen für das innerstaatliche Recht der Bundesrepublik Deutschland, 1993, 67 et seq.

discretion being delegated to the executive either by the constitution or through legislative act. If one wanted to guarantee their application, legislation would need to be adopted in order to transform nonbinding instruments into binding national law. Without a specific legislative act, action in accordance with the respective code of conduct would not be guaranteed but would be dependent on the inclinations of government officials and the breadth of their discretionary power. Transposing nonbinding international standards into binding national law thus carries the advantage of preventing legal uncertainty and guarantees a level playing field for private actors.

The incorporation of nonbinding norms into national law is indispensable when codes contain prescriptions of results and objectives which can be reached by different means. For example, the FAO Code of Conduct for Responsible Fisheries in Article 7.6.9. reads "States should take appropriate measures to minimise waste, discards, catch by lost and abandoned gear ...", leaving the way how to do that to the domestic legislator.[612] A similar need for legislative incorporation exists when states – perhaps for reasons of capacity – prefer to 'pick and choose' from the practices of the code when enacting national legislation.[613]

Much less leeway for the domestic is left when an international nonbinding instrument stipulates a certain conduct of the addressees.[614] Take for example the UNEP Liability Guidelines adopted in 2010 which clearly stipulate the liability of an operator for activities dangerous to the environment, and which prescribe in detail the possible ex-

---

612 On obligations of result in international law and the consequential permissiveness in implementation see R. Wolfrum, "Obligation of Result Versus Obligation of Conduct: Some Thoughts About the Implementation of International Obligations" in: M.H. Arsanjani/J.K. Cogan/R.D. Sloane/S. Wiessner (eds.), Looking to the Future: Essays on International Law in Honor of W. Michael Reisman, 2010, Chapter 20.

613 Compare K.M. Meessen, "Internationale Verhaltenskodizes und Sittenwidrigkeitsklauseln", Neue Juristische Wochenschrift 34 (1981), 1131-1132 (1131).

614 See on these types of norms already Part 1, at C.II and compare the considerations by R. Wolfrum, "Obligation of Result Versus Obligation of Conduct: Some Thoughts About the Implementation of International Obligations", in: M.H. Arsanjani/J.K. Cogan/R.D. Sloane/S. Wiessner (eds.), Looking to the Future: Essays on International Law in Honor of W. Michael Reisman, 2010, Chapter 20.

ceptions to such liability.[615] Compliance can in these cases be assessed rather easily, so that states wishing to demonstrate compliance must legislate and implement legislation exactly as prescribed. In other words, the trend towards prescriptions of concrete conduct and actions in nonbinding instruments raises the need for legislative implementation and reduces to the leeway of how to do that.

b) Specific legislative implementation

The degree to which states implement nonbinding instruments is often remarkable. The nonbinding nature of many of these instruments apparently does not keep states from issuing conforming legislation.[616] The codes of conduct analysed above are no exception. For example, the PIC procedure was implemented by most adopting states within a relatively short period of time.[617] Not only did states – as already mentioned – formally participate in the PIC procedure by designating national authorities, but they also widely adopted implementing regulations.[618] Similarly, the reporting exercise of the FAO has revealed that

---

615 UNEP Guidelines for the development of national legislation on liability, response action and compensation for damage caused by activities dangerous to the environment, UNEP/GCSS.XI/L.5 (2010), in Guideline 6 under the heading "exoneration from liability" stipulates: "Without prejudice to additional exonerations provided for in domestic law, the operator should not be liable, or in the case of (c) below not liable to the degree not apportioned to him or her, if the operator proves that the damage was caused: (a) By an act of God/force majeure (caused by natural phenomena of an exceptional, inevitable and uncontrollable nature); (b) By armed conflict, hostilities, civil war, insurrections or terrorist attacks; (c) Wholly or in part by an act or omission by a third party, notwithstanding safety measures appropriate to the type of activity concerned; (d) As a result of compliance with compulsory measures imposed by a public authority."

616 P. Kunig, "The Relevance of Resolutions and Declarations of International Organizations for Municipal Law" in: G.I. Tunkin/R. Wolfrum (eds.), International Law and Municipal Law, 1988, 59-78 (65); C. Tietje, Internationalisiertes Verwaltungshandeln, 2001, 622.

617 K. Kummer, "Prior Informed Consent for Chemicals in International Trade: The 1998 Rotterdam Convention", Review of European Community and International Environmental Law 8 (1999), 323-330 (324).

618 To name just two out of numerous examples, Madagascar discontinued its use of DDT for agricultural practices, and Colombia prohibited the import, production and use of Dieldrin in 1993, compare for details B. Dinham, "The

95 per cent of the reporting FAO Members reported to have legislation and policies in place which are partially or totally in conformity with the CCRF, and 9 out of 10 states reported to be either in conformity or working towards conformity in both policy and legal domains.[619] For instance, 70 per cent of the FAO member states are using the "vessel monitoring systems" recommended by the CCRF.[620] Empirical studies confirm the impact of the CCRF on domestic legal frameworks in many important fisheries regions.[621] However, as mentioned above, legislative changes do not always translate into actual environmental improvements, and major implementation problems exist in particular in developing countries despite legislative changes.[622] The main constraints

---

Success of a Voluntary Code in Reducing Pesticide Hazards in Developing Countries", Green Globe Yearbook of International Co-operation on Environment and Development 3 (1996), 29-36 (33); K. Kummer, "Prior Informed Consent for Chemicals in International Trade: The 1998 Rotterdam Convention", Review of European Community and International Environmental Law 8 (1999), 323-330 (324); M.A. Mekouar, "Pesticides and Chemicals: The Requirement of Prior Informed Consent" in: D. Shelton (ed.), Commitment and Compliance: The Role of Non-Binding Norms in the International Legal System, 2000, 146-163 (157).

619 FAO, Committee on Fisheries, COFI/2007/2, Progress in the implementation of the 1995 Code of Conduct for Responsible Fisheries, related International Plans of Action and Strategy, para. 6.

620 FAO Newsroom, Global Code for sustainable fishing turns 10 – FAO calls for renewed efforts to improve fisheries management on 10th anniversary of code's adoption, 31 October 2005, available at www.fao.org/newsroom/en/news/2005/1000112/index.html.

621 G. Hosch/G. Ferraro/P. Failler, "The 1995 FAO Code of Conduct for Responsible Fisheries: Adopting, implementing or scoring results?", Marine Policy 35 (2011), 189-200; T. J. Pitcher/D. Kalikoski/G. Pramod (eds.), Evaluations of Compliance with the FAO (UN) Code of Conduct for Responsible Fisheries, 2006, available at http://www.fisheries.ubc.ca/archive/publications/reports/14-2.pdf.

622 On the problems of implementing the FAO CCRF, see T.J. Pitcher/D. Kalikoski/G. Pramod/K. Short, "Not honouring the Code", Nature 457 (2009), 658-659; S.M.N. Alam/C. Kwei Lin/A. Yakupitiyage/H. Demaine/M.J. Phillips, "Compliance of Bangladesh shrimp culture with FAO code of conduct for responsible fisheries a development challenge", Ocean and Coastal Management 48 (2005), 177-188 (186); D. Barnhizer, "Waking from Sustainability's "Impossible Dream": The Decisionmaking Realities of Business and Government", Georgetown International Environmental Law Review 18 (2006), 595-690 (677). The largest implementation problems persist with regards to the im-

on more rapid progress are insufficient resources and institutional capacity as well as lack of awareness in developing countries.[623] Nevertheless, these statistics suggest that the code of conduct has been embraced widely by governments at least at the level of policy making and legislation.[624] A recent independent expert evaluation on the effectiveness of the work of the FAO goes even further. According to this detailed study, the CCRF and the implementing instruments have had "a very considerable impact" on worldwide fisheries management by both developing and developed states.[625]

The influence of nonbinding instruments on legislative activities especially in developing countries is more specifically highlighted by numerous examples of legislative efforts. For example, in the area of pesticide regulation, the government of the Republic of Korea adopted a full pesticides registration scheme pursuant to Article 3 and Article 6.1.2. of the FAO Pesticides Code.[626] A particularly far reaching example of the influence of the CCRF on national legislation is provided by the norms of the new Tanzanian Fisheries Act of 2003. It contains frequent references to "responsible fishing" as a recurring objective of various obligations. These references can only be understood as incorporating the objectives of the CCRF. Article 1 paragraph 2 of the Act defines the term "responsible fisheries" as "the principles and standards applicable to

---

plementation of the ecosystem and precautionary approach as well as the overexploitation of stocks, compare FAO, Committee on Fisheries, COFI/2009/2, Progress in the implementation of the 1995 Code of Conduct for Responsible Fisheries, related International Plans of Action and Strategy, paras 33–36.

623 FAO, The State of World Fisheries and Aquaculture 2006, Part II, available at www.fao.org/fi.

624 COFI/2005/2, paras 33 to 36; for a positive assessment in this sense see also D.J. Doulman, "Code of Conduct for Responsible Fisheries: Development and Implementation Considerations" in: M.H. Nordquist/J.N. Moore (eds.), Current Fisheries Issues and the Food and Agriculture Organization of the United Nations, 2000, 307-330 (327).

625 FAO, The Challenge of Renewal, Independent External Evaluation of the Food and Agriculture Organization, Working Draft (2007), para. 425, available at http://www.fao.org/unfao/bodies/IEE-Working-Draft-Report/ K0489E.pdf.

626 S.-M. Hong, "Report on the Republic of Korea" in: FAO, Regional Office for Asia and the Pacific, Proceedings of the Asia Regional Workshop on the Implementation, Monitoring and Observance, International Code of Conduct on the Distribution and Use of Pesticides, Bangkok, Thailand, 26-28 July 2005, RAP Publication 2005/29, available at www.fao.org/docrep/008/af340e/af340e0i.htm.

conservation, management and development of all fisheries and it covers the capture, processing and trade of fish and fishery products, fishing operations, aqua culture, fisheries research and integration of fisheries into coastal area management".[627] This definition follows precisely the description of the content and scope of the CCRF which is given in the CCRF itself.[628] The reference to responsible fishing is for example used in two places in the enumeration of intrusive administrative measures that can be taken by the relevant government Minister. The conditions that the Minister has to impose on industry must ensure that due regard is given to traditional practices that are consistent with "responsible fisheries".[629] Further, Article 17 (n) gives the Minister the power to replace fishing gear which does not meet the performance standard of "responsible fishing".[630] Finally, Article 9 paragraph 1 (k) (i) of the Act gives the Director of the Ministry the broad power to "strengthen regional and international collaboration" by – *inter alia* – "supporting responsible fishery practices within the country".[631] This enables the Director to orientate the provision of public services and other state activity according to the objectives of the CCRF. Most importantly perhaps, the Director – in exercising his responsibility to adopt fisheries management measures to maintain or restore stocks at levels capable of producing the maximum sustainable yield – must also take into account the "promotion of responsible fisheries" as one of the relevant factors.[632] Apart from implementing the nonbinding code, this example from domestic law also shows that states indeed take into account the CCRF as being part of "generally recommended international minimum stan-

---

[627] Tanzania Fisheries Act (2003), Article 1 para. 2, available at http://faolex.fao.org/faolex.

[628] According to Article 1.3. CCRF, the Code provides "principles and standards applicable to the conservation, management and development of all fisheries. It also covers the capture, processing and trade of fish and fishery products, fishing operations, aquaculture, fisheries research and the integration of fisheries into coastal area management."

[629] Tanzania Fisheries Act (2003), Article 17 (p), available at http://faolex.fao.org/faolex.

[630] Tanzania Fisheries Act (2003), Article 17 (n).

[631] Tanzania Fisheries Act (2003), Article 9 para. 1 (k) (i).

[632] Tanzania Fisheries Act (2003), Article 9 para. 2 (b).

dards"[633] when implementing relevant UNCLOS provisions for fisheries management.[634]

States frequently do not only establish new substantive rules in accordance with the requirements of the codes, but also new institutions and agencies competent to enforce and implement them. State compliance with the international requirements to set up or designate national authorities for international procedures is remarkably high both in the case of the OECD Guidelines and the PIC procedure. Thus, all OECD members have established National Contact Points (NCPs) as the implementing agencies for the international implementation procedures of the OECD Guidelines.[635] While the establishment of NCPs was a result of binding OECD decisions, the tendency of states to comply even with nonbinding institutional requirements is exemplified by the high number of states that had designated national authorities for the PIC procedure shortly after the adoption of the voluntary PIC system. As already mentioned, over 150 countries were participating in the PIC procedure eight years after it was first included in the nonbinding instruments: a number higher even than the 131 states that had ratified the PIC Convention by April 2010.[636]

Developing countries in particular often lack the institutional framework for proper administration of environmental issues. The number of developing countries without an approved legislative authority to regulate the distribution and use of pesticides has significantly decreased between 1986 and 1993 after the Pesticides Code had been elaborated.[637] Institutional short-comings have also been identified as a fundamental problem for the implementation of the FAO CCRF.[638] In addressing

---

[633] E.g. UNCLOS, Articles 61 para. 3 and 119 para. 1 (a).

[634] The issue of references in UNCLOS has been discussed in detail in this Part 2 at A.I.2c)(2), further above.

[635] For a list of contact points that had been established by the end of 2009, see http://www.oecd.org/dataoecd/17/44/1900962.pdf.

[636] N.S. Zahedi, "Implementing the Rotterdam Convention: the challenges of transforming aspirational goals into effective controls on hazardous pesticide exports to developing countries", Georgetown international environmental law review 11 (1999), 707-739 (709).

[637] FAO, Analysis of Government Responses to the Second Questionnaire on the State of Implementation of the International Code of Conduct on the Distribution and Use of Pesticides (1993), Article 3.

[638] FAO, National governance of fisheries, available at http://www.fao.org/fi/website/FIRetrieveAction.do? dom=topic&fid=12261.

this problem, the already mentioned Indian Comprehensive Marine Fishing Policy repeatedly points to the requirement for building agencies competent to enforce and implement legislation in conformity with the FAO CCRF.[639] The Indian government accordingly established a Coastal Aquaculture Authority in order to promote environment-friendly and responsible aquaculture.[640]

Depending on the subject area, nonbinding recommendations stemming from global organisations often represent a common denominator which is qualitatively lower than the existing rules in developed states.[641] In these cases, legislation of developed states serves as a model for the international norms. Developed states then use nonbinding instruments to attempt international harmonisation on the basis of their standards. For example, the FAO Pesticide Code was substantively influenced *inter alia* by German experiences and advanced legislation in the field.[642] In the fisheries sector, the Canadian change of attitude towards more sustainable exploitation of marine resources occurred before the CCRF was adopted, and in fact Canada has been one of the main drivers in its negotiation.[643] The "Implementation Plan for the

[639] Indian Comprehensive Marine Fishing Policy (2004), Articles 5.1, 5.4 and 9.0. The policy was issued by the Indian Ministry of Agriculture, Department of Animal Husbandry & Dairying and is available at: http://dahd.nic.in/fishpolicy.htm; compare also the FAO, 'Making Global Governance Work for Small-Scale Fisheries', FAO-Document No. 07, Rome 2006, available at www.fao.org.

[640] Indian Costal Aquaculture Authority Act (2005), The Gazette of India (23 June 2005), Part II, Section 1; on the motivation underlying the establishment of this authority see the Annual Report 2004-2005 issued by the Government of India, Department of Animal Husbandry & Dairying, Ministry of Agriculture, para. 5.4.3.2., available at http://www.dahd.nic.in/rep/ann2005.htm.

[641] This is confirmed on the basis of interviews by U. Dieckert, Die Bedeutung unverbindlicher Entschließungen internationaler Organisationen für das innerstaatliche Recht der Bundesrepublik Deutschland, 1993, 81.

[642] This observation is made – based on information by the responsible ministry in Germany – by U. Dieckert, Die Bedeutung unverbindlicher Entschließungen internationaler Organisationen für das innerstaatliche Recht der Bundesrepublik Deutschland, 1993, 88; the fact that the FAO Pesticide Code was issued before the new Plant Protection Act is, according to Dieckert, not a sign to the contrary.

[643] Fisheries and Oceans Canada, Responsible Fisheries Summary, Code of Conduct for Responsible Fishing Operations (2003), available through Fisheries and Oceans Canada.

Code of Conduct for Responsible Fisheries" issued by the U.S. National Marine Fisheries Service in 1997 stresses that the ideas and goals of the CCRF had already been part of U.S. fisheries legislation before the negotiation of the CCRF.[644] Nevertheless, the existence of the CCRF prompted the U.S. to question the state of its fisheries legislation. The U.S. Magnuson-Stevens Fishery Conservation and Management Act of 1976[645] had to be amended by the Sustainable Fisheries Act, 1996[646] in order to capture most of the principles of responsible fisheries in Articles 6 through 8 of the CCRF.

c) Different techniques: programmed legislation or dynamic references

Two legislative techniques for the implementation of nonbinding instruments can be distinguished. The first is the establishment of a legal framework through legislative acts based on the substantive content of the international instruments. This has been referred to as "parallel legislation"[647], a term which indicates that the legislative response frequently takes international norms as a model and forms domestic law in accordance with them.[648] As the complete adoption of an international instrument's content is rather rare, the term "programmed legislation" seems more adequate. It reflects that international nonbinding instruments often at least predetermine the outcome of legislative processes both in terms of wording or/and substance, but do not simply copy the

---

644 National Marine Fisheries Service of the United States Department of Commerce, Implementation Plan for the Code of Conduct for Responsible Fisheries (1997) available at http://www.nmfs.noaa.gov/plan.html.

645 United States Magnuson-Stevens Fishery Conservation and Management Act, 1976, 16 U.S.C. 1801-1882 (13 April 1976).

646 United States Sustainable Fisheries Act, 1996, Public Law 104–297 (11 October 1996).

647 P. Kunig, "The Relevance of Resolutions and Declarations of International Organizations for Municipal Law" in: G.I. Tunkin/R. Wolfrum (eds.), International Law and Municipal Law, 1988, 59-78 (65); C. Tietje, Internationalisiertes Verwaltungshandeln, 2001, 622.

648 P. Kunig, "The Relevance of Resolutions and Declarations of International Organizations for Municipal Law" in: G.I. Tunkin/R. Wolfrum (eds.), International Law and Municipal Law, 1988, 59-78 (65).

international instruments.[649] The second legislative technique employed is the incorporation of the norms via references. In contrast to programmed legislation, references in domestic legislation in these cases directly point to an external instrument and processes for the content of the substantive rules. References where the legislative act refers to the international instrument in its most recent form instead of a particular version are in German legal doctrine and hereinafter referred to as "dynamic references", with a view to underscore that the substance of the referenced norm may "dynamically" change. In particular such dynamic references may pose particular legitimacy challenges, because in these cases the decision over the content of the norm in question is in fact left to the international level, and therefore both the role of the domestic legislature as well as that of the opposition to contest such a decision is generally reduced.[650]

One problem of programmed legislation besides the issue of legitimation is the loss of flexibility. Once the transposed codes are transposed into municipal law, modification is difficult and cannot keep pace with international changes·.[651] The alternative option of using dynamic references ensures that the flexibility of the instrument is safeguarded. Dynamic references can be seen as better reflecting the legal ideal of the principle of international cooperation,[652] because the process of norm elaboration and their further development is left to the international level and national unilateral measures are prevented.[653] One such dynamic reference to an instrument discussed in this study can be found in the German Plant Protection Act.[654] By stipulating that the FAO Pesticide Code is to be taken into consideration when exporting of pes-

---

649 U. Dieckert, Die Bedeutung unverbindlicher Entschließungen internationaler Organisationen für das innerstaatliche Recht der Bundesrepublik Deutschland, 1993, 80.

650 For a discussion of the legitimacy issues involved in legislative implementation, see the chapter on legitimacy in Part 3 further below.

651 C. Engel, Völkerrecht als Tatbestandsmerkmal deutscher Normen, 1989, 247-249.

652 C. Tietje, Internationalisiertes Verwaltungshandeln, 2001, 605.

653 E. Rehbinder, "Das neue Pflanzenschutzgesetz", Natur und Recht 9 (1987), 68-71 (71) (stating that one of the incentives for the reference to the FAO Pesticides Code and the choice for the PIC system was to prevent an isolated German approach with a view to protect German industry).

654 German Plant Protection Act (Pflanzenschutzgesetz), BGBl. 1986, I-1505, revised enactment of 1 July 1998, BGBl. 1998, I-971, 1527, 3512.

ticides, the entire code of conduct in its most recent version is made directly applicable in national law.[655] Instead of prohibiting exports, Germany thus adhered to the voluntary international norms that did not foresee such a prohibition.[656] Furthermore, the provision remains rather soft. It only stipulates that the code should be "taken into account". By contrast dynamic references in the German Nature Protection Law to international environmental treaties such as CITES sanction violations as criminal offenses.[657] While infringements of most of the provisions of the German Plant Protection Act give rise to sanctions such as fines, the paragraph containing the dynamic reference to the FAO Pesticide Code is explicitly excluded.[658] This could be interpreted as reflecting the voluntary character of the international instrument.[659] Another example – this time from a developing country – can be found in the reformed fisheries law of Tanzania.[660] It directly refers to the CCRF when outlining the responsibilities of local authorities.[661] Even though the reference

655 Paragraph 23 (1) sentence 2 of the German Plant Protection Act (Pflanzenschutzgesetz) stipulates that the FAO Pesticide Code should be taken into account in addition to numerous other conditions when pesticides are exported (in the original wording: "... im übrigen ... bei der Ausfuhr internationale Vereinbarungen, insbesondere der Verhaltenskodex für das Inverkehrbringen und die Anwendung von Pflanzenschutz- und Schädlingsbekämpfungsmitteln der FAO, berücksichtigt werden ...".)

656 E. Rehbinder, "Das neue Pflanzenschutzgesetz", Natur und Recht 9 (1987), 68-71 (71).

657 Compare e.g. German Nature Protection Act, BGBl. 2002 I-1193, § 66 paras 2 and 3 (Bundesnaturschutzgesetz) which penalises trade of species protected under the Convention on International Trade in Endangered Species of Wild Fauna and Flora with imprisonment of up to 5 years.

658 Compare § 40 of the German Plant Protection Act, BGBl. 1998, I-971.

659 U. Dieckert, Die Bedeutung unverbindlicher Entschließungen internationaler Organisationen für das innerstaatliche Recht der Bundesrepublik Deutschland, 1993, 100.

660 Tanzania Fisheries Act (2003) (No. 22 of 2003), available at http://faolex.fao.org/faolex.

661 Article 8 para. 3 of the Tanzania Fisheries Act of 2003 states: "Where the Director is of the opinion on the basis of information given to him by any officer exercising functions under this Act or otherwise that a local authority with the responsibility to exercise functions in accordance with the provisions of this Act, Code of Conduct for responsible fisheries, aqua culture development, fishing operations, fisheries management, fish utilization and marketing has mis-

remains somewhat ambiguous, it explicitly and dynamically references the CCRF either as part of the responsibilities of local authorities or as the foundation and delimitation of these responsibilities of the local authority.

2. *Implementation in the absence of specific implementing legislation*

Legislative action is not the only way in which nonbinding norms can enter the domestic legal sphere. In fact, one of the distinct characteristics of nonbinding instruments remains the possibility of domestic implementation without ratification and domestic legislation, achieved through executive action. International nonbinding instruments may directly serve as guiding norms for executive action without any instrument-based legislation. The nonbinding instruments may serve as the basis for governmental or administrative policy making, or they may be directly applied by the executive in the absence of adequate domestic regulation.[662]

But in the absence of a specific legislative act, nonbinding instruments can at least under German law only guide administrative activities to the extent that the there exists a legally authorised room of the executive administration to act.[663] In the German legal system, the doctrine of the Gesetzesvorbehalt, which demands that the administration must be able to base any significant exercise of authority on a legal basis, safeguards the linkage of administrative action to formal legislative proc-

managed the functions of this Act, the Director may ...", Tanzania Fisheries Act of 2003 is available at http://faolex.fao.org/faolex.

662 B.R. Palikhe, "Report on Nepal" in: FAO, Proceedings of the Asia Regional Workshop on the Implementation, Monitoring and Observance, International Code of Conduct on the Distribution and Use of Pesticides, Bangkok, Thailand, 26-28 July 2005, RAP Publication 2005/29, available at www.fao.org/docrep/008/af340e/af340e0i.htm.

663 P. Kunig, "Deutsches Verwaltungshandeln und Empfehlungen internationaler Organisationen" in: K. Hailbronner/G. Ress/T. Stein (eds.), Staat und Völkerrechtsordnung – Festschrift für Karl Doehring, 1989, 529-551 (539); U. Dieckert, Die Bedeutung unverbindlicher Entschließungen internationaler Organisationen für das innerstaatliche Recht der Bundesrepublik Deutschland, 1993, 196 et seq.

esses.[664] As a consequence, the administration may autonomously implement nonbinding instruments where a general legislative act leaves discretionary leeway or through administrative acts where the norms in question would not infringe basic freedoms. Insofar as a domestic legal order provides for such leeway, opting for a nonbinding instrument at the international level regularly entails a potential shift of power to the executive.[665] The executive, often represented at the international level by those officials also concerned with implementation, is thus empowered both to adopt and implement international instruments without the involvement of the legislature. Given the close relationship between administrative discretion and the application of nonbinding instruments through administrative decision making, the potential impact of nonbinding instruments and the related shift to the executive may depend on the legal system of a particular state. The German legal system – at least generally speaking – is marked by reliance of the administration on material requirements which are set by the legislature. Other administrative systems such as those of the United Kingdom, France or the United States of America rely to a much greater extent on prerogative power of the executive. In these cases, administrative activities are typically only restricted by a minimum of material requirements.[666] Hence, the space for the direct application of nonbinding instruments via the exercise of administrative discretion is even larger in these legal systems.[667]

---

[664] Compare Bundesverfassungsgericht, BVerfGE 40, 237, 248 et seq., BVerfGE 84, 212, 226 ("Wesentlichkeitstheorie"); compare H. Maurer, Allgemeines Verwaltungsrecht, 2009, § 6 mn. 3.

[665] On deparliamentarisation and internationalisation see e.g. W. Kahl, "Parlamentarische Steuerung der internationalen Verwaltungsvorgänge" in: H.-H. Trute/T. Groß/H.C. Röhl/C. Möllers (eds.), Allgemeines Verwaltungsrecht – Zur Tragfähigkeit eines Konzepts, 2008, 71-106 (79). The issue will be discussed in detail in Part 3, further below.

[666] R. Breuer, "Zunehmende Vielgestaltigkeit der Instrumente im deutschen und europäischen Umweltrecht – Probleme der Stimmigkeit und des Zusammenwirkens", Neue Zeitschrift für Verwaltungsrecht 16 (1997), 833-845 (837) (pointing out that most states like France and England usually rely on wider margins of discretion for their administration than Germany does).

[667] Compare e.g. J.M. Beermann, Administrative law, 2006, 3-4 ("Congress often instructs an agency in very broad terms and leaves important matters to agency discretion, including – the requirements for obtaining the various benefits, licenses, and permits administered or required by federal law in numerous areas").

In search of administrative leeway which allows international nonbinding instruments to be applied directly, one must at least from a German legal perspective distinguish between various administrative activities. These are: the elaboration of general policies typically undertaken by the government; the adoption of internal administrative guidelines; administrative decision-making in the form of either interventionist administration ("Eingriffsverwaltung") or distributive administration ("Leistungsverwaltung"), and finally non-regulatory informal administration ("Informelles Verwaltungshandeln"). The extent of leeway and discretion given to executive officials and administrators differs from one of these categories to another, and with it the potential impact of international nonbinding instruments.

a) Policy making

Nonbinding instruments influence states already at the level of policy making.[668] In practice, this is one of the main avenues through which these instruments gain influence within states. The government uses them when developing and adopting their plans of action and general policies directed at shaping legislative and administrative processes. Nonbinding instruments are thus used as a source of authority for new strategies. Take for example the way in which the fisheries policies elaborated by the U.S. Department of Commerce were guided by the FAO Code of Conduct for Responsible Fisheries (CCRF) and the related International Plans of Action. After the CCRF had been adopted at the international level, the US National Marine Fisheries Services of the U.S. Department of Commerce devised its "Implementation Plan for the Code of Conduct for Responsible Fisheries"[669] modelled after the CCRF as well as several National Plans of Action (NPOA)[670] which all correspond both in substance and title to the four International

---

[668] In this manner also the analysis of FAO experts, see e.g. D.J. Doulman, "Code of Conduct for Responsible Fisheries: Development and Implementation Considerations" in: M.H. Nordquist/J.N. Moore (eds.), Current Fisheries Issues and the Food and Agriculture Organization of the United Nations, 2000, 307-330.

[669] This plan is available at http://www.nmfs.noaa.gov/.

[670] These are the US NPOA-Capacity, the US NPOA-Longline, the US NPOA Sharks and the US National Plan of Action on Illegal, Unregulated and Unreported Fishing, all available at www.nmfs.noaa.gov.

Plans of Action adopted by the FAO under the framework of the FAO Code of Conduct for Responsible Fisheries.[671]

Examples from developing countries that indicate implementation efforts with respect to FAO instruments at the level of policy making are manifold. For example, the Mexican Government implements the International Plans of Action of the FAO by means of National Plans of Action.[672]

Similarly, the Indian Comprehensive Marine Fishing Policy which mirrors every component of the FAO Code of Conduct for Responsible Fisheries represents a change in paradigms in Indian fisheries policy.[673] The Indian policy also clearly illustrates how such policies may contribute to the effective implementation of the international instruments. It clearly "calls upon" local governments of the coastal states of India to translate policy into action by adopting legislation and implementation mechanisms in accordance with international standards. It does so without distinguishing between legally binding and nonbinding instruments. The FAO CCRF is considered to be an integral part of these international standards. With respect to legislation, the Indian Comprehensive Marine Fishing Policy demands that national fishery laws should be harmonised with international standards,[674] and foresees a number of legal acts to underscore a change in policy through the introduction of new institutions, instruments and agencies competent to enforce and implement legislation in accordance with the CCRF.[675]

---

[671] For the various policies implemented by the United States as a response to meet the requirements of the FAO CCRF consider also P.D. Dalton, "Implementing the International Code of Conduct for Responsible Fisheries" in: M.H. Nordquist/J.N. Moore (eds.), Current Fisheries Issues and the Food and Agriculture Organization of the United Nations, 2000, 331-340.

[672] E.g. the Mexican Plan of Action to implement the IPOA-Sharks, CONAPESCA-INP (2004), 'Plan de Acción Nacional para el Manejo y Conservación de Tiburones, Rayas y Especies Afines en México', available at ftp://ftp.fao.org/FI/DOCUMENT/IPOAS/national/mexico/PANMCT_VERSIONFINAL.pdf.

[673] Indian Comprehensive Marine Fishing Policy (2004), in the Foreword; the policy is available at http://dahd.nic.in/fishpolicy.htm.

[674] Indian Comprehensive Marine Fishing Policy, Article 9.3.

[675] Indian Comprehensive Marine Fishing Policy, Articles 5.1, 5.4 and 9.0. See also the specific FAO publication FAO, 'Making Global Governance Work for Small-Scale Fisheries, New Directions in Fisheries' (No. 09, 2007), available at http://www.fao.org/fi/.

b) Administrative legal rules, directives and internal guidelines

Nonbinding international instruments may be implemented by executive officials through further sets of domestically adopted codified norms such as administrative directives as well as internal administrative guidelines.

In most states, the executive branch is entitled to enact legal rules in order to concretise parliamentary acts, specify technical requirements or regulate a limited area. Based on a general authorisation those decrees, regulations or ordinances sometimes mirror international nonbinding instruments even if the primary legislation does not explicitly refer to them. In Cambodia, for instance, the implementation of a pesticides management scheme responding to requirements of the FAO Pesticides Code was mainly effectuated by administrative decrees and declarations[676] which lay down – together with the Prakas No. 245[677] – detailed procedures for all aspects of pesticide control such as registration, control and export in line with the FAO Pesticide Code.[678] It not only in-

[676] Cambodian Ministry of Agriculture, Forestry and Fisheries of Cambodia, Sub-decree No. 69 on the standard and management of agricultural material of 21 October 2002, Annex 1 and Prakas (Ministerial Declaration) No. 245 on the implementation of Sub-decree No. 69 of 21 October 2002 (2002).

[677] Cambodian Ministry of Agriculture, Forestry and Fisheries, Prakas (Ministerial Declaration) No. 245 on the implementation of Sub-decree No. 69, 21 October 2002, Ministry of Agriculture, Forestry and Fisheries 2002; the information is given by L. Socheata, "Report on Cambodia" in: FAO, 'Proceedings of the Asia Regional Workshop on the Implementation, Monitoring and Observance, International Code of Conduct on the Distribution and Use of Pesticides, Bangkok, Thailand, 26-28 July 2005', RAP Publication 2005/29 (2005), available at http://www.fao.org/docrep/008/af340e/af340e07.htm#bm07.

[678] The same is true for Vietnam: the Ordinance on Plant Protection and Quarantine, Công Báo No. 37, 8 October 2001, pp. 3-10 (available through FAOLEX at: http://faolex.fao.org/docs/pdf/vie35158.pdf) has been revised in order to include pesticide regulations in accordance with the FAO Pesticide Code, and Decision No 121/1999 on permitted pesticides (25 August 1999, Official Gazette No. 44 (30-11-1999), 13-18) has been enacted due to the FAO Pesticides Code. "Report on Viet Nam" in: FAO, 'Proceedings of the Asia Regional Workshop on the Implementation, Monitoring and Observance, International Code of Conduct on the Distribution and Use of Pesticides' (2005), RAP Publication 2005/29 (Bangkok 2005) ("based on the guidelines of the FAO Code of Conduct the registration scheme was revised in line with international scheme"), the report is available at www.fao.org/docrep/008/af340e/af340e0i.htm.

troduces the practices suggested in the FAO Code of Conduct, but its wording in many instances closely follows that of the international model.[679]

In particular internal administrative directives or guidelines represent an important mechanism for the implementation of international non-binding instruments. Although internal administrative guidelines are only binding upon administrators, they have indirect legal effects for citizens and corporations.[680] Depending on their specific function, they can determine the interpretation and application of legal norms, direct the exercise of discretion by the administration, and in exceptional cases may provide substantive guidance for decision-making. The German Federal Administrative Court (Bundesverwaltungsgericht) also accepts that the administration may rely on concretising administrative guidelines ("normkonkretisierende Verwaltungsvorschrift") in areas of environmental law and other highly technical issues which are only to a limited extent subject to judicial scrutiny.[681] Such internal guidelines are a tool for administrators with which they can push the uniform implementation of international norms and standards even if legislative incorporation has not occurred. Higher level officials who have participated in the elaboration of codes of conduct on the international level can therefore ensure that the ideas of the code of conduct are disseminated on the national level via lower administrative agencies.

---

[679] In the Cambodian Sub-decree No. 69, Article 20, the Cambodian uses nearly the exact wording of Article 8.1.2. of the Pesticides Code; compare L. Socheata, "Report on Cambodia" in: FAO, 'Proceedings of the Asia Regional Workshop on the Implementation, Monitoring and Observance, International Code of Conduct on the Distribution and Use of Pesticides, Bangkok, Thailand, 26-28 July 2005', RAP Publication 2005/29 (2005), available at http://www.fao.org/docrep/008/af340e/af340e07.htm#bm07.

[680] In German legal scholarship and practice, this external factual effect also gives rise to a legal right of the affected citizen to rely on the internal administrative guideline, and thus an external legal effect, by relying on the equal rights clause. The administration is thus seen to bind itself through its institutional practice. See e.g. Bundesverwaltungsgericht in BVerwGE 100, 335, 339 (guidelines for foreigners), in BVerwGE 104, 220, 223 (on subsidies guidelines); for an overview of the heavily discussed topic only H. Maurer, Allgemeines Verwaltungsrecht, 2009, § 24 mn. 20 et seq.

[681] BVerwGE 72, 300, 320 et seq. 114, 342, 344 et seq.; for an overview of the discussion H. Maurer, Allgemeines Verwaltungsrecht, 2009, § 24 mn. 25-26a (with further references).

In particular technical rules and standards are regularly implemented in this manner. The (nonbinding) recommendations of the Commission of Lake Constance, for example, have been implemented and comprehensively applied by the riparian states and the German Länder of Bavaria and Baden-Württemberg.[682] Also, recommendations of the Helsinki Commission have been implemented by general administrative directives by the German government.[683]

One interesting example that illustrates the implementation of the FAO Code of Conduct for Responsible Fisheries through an administrative directive stems from Brazil. There the CCRF of the FAO is implemented via internal administrative directives. Mandated through legislative act,[684] the Brasilian Special Secretary for Aquaculture and Fisheries (Secretaria Especial de Aqüicultura e Pesca) has the power to formulate policies and directives for the development of fisheries and aquaculture production. The Special Secretary has recently enacted an internal administrative directive (Instrução Normativa) establishing the criteria for the development of local plans for marine protected areas (Planos Locais de Desenvolvimento de Maricultura). The directive includes a specific reference to the FAO CCRF. The international code of conduct thus determines the conditions and restrictions which these planned areas should meet.[685] Other internal directives specifically refer to all recommendations or to one Article of the FAO CCRF in their preambles.[686]

---

[682] A.K. Skala, Internationale technische Regeln und Standards zum Umweltschutz – ihre Entstehungsarten und rechtlichen Wirkungen, 1982, 122 and 240.

[683] Compare for these examples U. Dieckert, Die Bedeutung unverbindlicher Entschließungen internationaler Organisationen für das innerstaatliche Recht der Bundesrepublik Deutschland, 1993, 102 et seq.

[684] Article 23 of Law 10.683, available at http://www.panalto.gov.br/CCIVIL/leis/2003/L10.683.htm.

[685] Article 2 IX of the Instrução Normativa No 17 of 23 September 2005, available at http://www.ipaam.br/legislacao.

[686] Compare e.g. the Portaria no 145-N of 29 October 1998; Instrução Normativa N°-125 of 18 October 2006 (referring specifically to Article 8 of the CCRF); Instrução Normativa N° 14 of 18 February 2004 (referring to the principles of the CCRF), all available at http://www.ipaam.br/legislacao.

c) Administrative decision-making

(1) Interventionist administration (Eingriffsverwaltung)

The typical means of interventionist administrative action, i.e. action where the administration intrudes into the protected sphere of the citizen and constraints its freedom or property,[687] are prohibitions, licensing and permits coupled with monitoring and control of the relevant activity. In the context of interventionist administration, codes of conduct or other nonbinding norms can serve as standards for these decisions where the legal basis provides for discretion.

This could for example be the case wherever administrators are granted the power to optimise administrative decisions by attaching certain conditions to the decision. These kinds of decisions are quite numerous and important in environmental law. German environmental law, for instance, frequently empowers the administration to attach conditions and additional obligations to permits or licences.[688] The conditions attached to the permit could require the operator of the undertaking to fulfil the requirements of an international code of conduct, provided that they are sufficiently precise. Fishers could for example be required to use only environmentally safe fishing gear,[689] or distributors of pesticides to reduce risks to human health in accordance with the suggestions of these instruments.[690] Nonbinding instruments can be of use for administrative risk evaluation as well as for actions suitable to prevent the occurrence of a hazard. They might also serve as a yardstick when an authorising agency conducts an environmental impact assessment. An undertaking in accordance with the requirements of a specific nonbinding instrument could be deemed not to constitute a hazard for the environment. However, in order to be considered sufficient to pass the impact assessment, some kind of legislative approval of the code of

---

687 H. Maurer, Allgemeines Verwaltungsrecht, 2009, § 1 mn. 20.

688 Under German law, the Federal Administrative Procedure Act (Verwaltungsverfahrensgesetz), 23 January 2002, BGBl. 2002, p. I-102, § 36 as well as Federal Pollution Control Act, § 12; Federal Water Resources Act (Wasserhaushaltsgesetz), 19 August 2002, BGBl. 2002 , p. I-3245, § 4 all allow that conditions and obligations may be attached to a permit in order to make an undertaking licensable. It lies with the administrative discretion to decide whether a condition is indispensable and what the content of such a condition might be as long as it is not disproportionate.

689 Article 6.6. FAO CCRF.

690 Article 3.10. FAO Pesticide Code.

conduct as a standard would be needed at least in German administrative law.[691]

Second, nonbinding instruments can impact administrative decisions where these are taken on the basis of rules that contain wide and indeterminate legal terms.[692] Even if there is no explicit margin of discretion, codes of conduct can guide the interpretation and concretisation of indeterminate legal terms.[693] Whenever statutes make use of wide and indeterminate concepts like 'state of the art'[694], 'detrimental impacts on the environment'[695] or 'good practice'[696], the administration is called

---

691 In German law on environmental impact assessment, this is widely accepted with respect to the influence of non-legislative standards, compare W. Erbguth/A. Schink, Gesetz über die Umweltverträglichkeitsprüfung: Kommentar, 1996, 194-195.

692 P. Kunig, "The Relevance of Resolutions and Declarations of International Organizations for Municipal Law" in: G.I. Tunkin/R. Wolfrum (eds.), International Law and Municipal Law, 1988, 59-78 (68 ff.); P. Kunig, "Deutsches Verwaltungshandeln und Empfehlungen internationaler Organisationen" in: K. Hailbronner/G. Ress/T. Stein (eds.), Staat und Völkerrechtsordnung – Festschrift für Karl Doehring, 1989, 529-551 (539 ff.).

693 FAO, 'Analysis of Government Responses to the Second Questionnaire on the State of Implementation of the International Code of Conduct on the Distribution and Use of Pesticides' (1993), Articles 3-5: relying on the Code of Conduct for guidance. See also Pesticides Code, Article 1.2.: "may judge whether their proposed actions and the actions of others constitute acceptable practices."

694 German Federal Pollution Control Act (Bundesimmissionsschutzgesetz), BGBl. 2002, p. I-3830, § 5 para. 1 Nr. 2 ("Stand der Technik").

695 German Federal Pollution Control Act, § 5 para. 1 Nr. 1 ("schädliche Umwelteinwirkungen").

696 German Plant Protection Act, § 2a para. 1 phrase 1 ("gute fachliche Praxis"); § 6 para. 1: "Bei der Anwendung von Pflanzenschutzmitteln ist nach guter fachlicher Praxis [emphasis added] zu verfahren. Pflanzenschutzmittel dürfen nicht angewandt werden, soweit der Anwender damit rechnen muss, dass ihre Anwendung im Einzelfall schädliche Auswirkungen auf die Gesundheit von Mensch und Tier oder auf Grundwasser oder sonstige erhebliche schädliche Auswirkungen, insbesondere auf den Naturhaushalt, hat. Die zuständige Behörde kann Maßnahmen anordnen, die zur Erfüllung der in den Sätzen 1 und 2 genannten Anforderungen erforderlich sind." ("The use of pesticides should be guided by good practise [emphasis added]. Pesticides must not be used if it is foreseeable for the user that this might cause detriment for human or animal health or groundwater or other severe harm, especially to nature.

upon to interpret these concepts when applying the law. In order to guarantee consistent administrative practice and avoid arbitrariness, administrators need to rely on some kind of standard. This standard is mostly provided by a higher administrative body. International standards adopted in international organisations represent a science-based international consensus on best practices in a given field. They are therefore ideal standards for these kinds of evaluations. If not already incorporated into administrative guidelines, they could at least be used as indicators for what constitutes the "right" behaviour and best practices in a given field. If such a function of nonbinding norms is accepted in the case of technical guidelines elaborated by private expert associations,[697] it should even more so be acceptable in the case of international nonbinding instruments supported by the respective state government at the international level.[698]

This function of international nonbinding norms has also been stressed in a recent Swiss case regarding the use of the "Manual on development and use of FAO and WHO specifications for pesticides".[699] The Swiss *Rekurskommission für Chemikalien* found that an agency is entitled to refer to the nonbinding international manual of the FAO and WHO in order to guarantee a consistent interpretation of the law, because they

The competent authority may order measures to ensure the fulfilment of the above-mentioned requirements.").

[697] Thus, technical guidelines are accepted as having at least an indicatory role in the determination of pollution limits under the German Pollution Act, compare only H.D. Jarass, Bundes-Immissionsschutzgesetz: Kommentar, 2005, § 48 mn. 63.

[698] This has been the outcome of several judgments of German administrative courts; see e.g. Judgement of the Administrative Court Frankfurt of 19 June 1988, Neue Juristische Wochenschrift, 1988, 3032-3035, 3033, concerning the UN Charta of the Rights of the Child, which only entered into force as a binding convention in 1990. In this decision the court stated that soft-law can be taken into account as an additional means of interpretation. See also Administrative Court Berlin, Decision of 22 January 1996, Neue Zeitschrift für Verwaltungsrecht, 1996, Beilage, 51. For an analysis see C. Tietje, Internationalisiertes Verwaltungshandeln, 2001, 633-637.

[699] FAO, Manual on development and use of FAO and WHO specifications for pesticides, FAO/WHO Joint Meeting on Pesticide Specifications (JMPS), Rome 2002, which has been incorporated into the FAO Pesticides Code.

reflect an international consensus on the latest state of scientific and technical knowledge.[700]

As the above discussion indicates, the impact of international instruments directly corresponds to their specific quality and the composition of bodies that have adopted them. Sufficient professional quality and expertise may even make them more suitable than national ones to serve as concretising standards for "best practices" provisions, for example.[701] Moreover, specific and technical standards more easily serve to fill the frequently employed references to best practices in national environmental law, than general more ethical norms of nonbinding instruments which are better suited as points of orientation for general policies and plans of action.

(2) Distributive administration (Leistungsverwaltung)

Distributive administration refers to administrative decisions which grant benefits and assistance to citizens. In this area of decision-making international nonbinding norms can be of the utmost relevance. One reason for this is the large discretion given to the administration for such an administrative function. For example, in Germany, the judiciary has consistently opined that in contrast to interventionist administration, distributive administration must not necessarily be based on a specific legislative act. Although criticised by some German legal schol-

---

700 Eidgenössische Rekurskommission für Chemikalien, Judgment of 28 February 2006, CHEM 05.002, 16, available at http://www.vpb.admin.ch.

701 These requirements include that the international standards must not be less stringent than applicable national ones, as well as a certain correlation between the object and purpose of the national and the international norm, see for these and other preconditions for the application U. Dieckert, Die Bedeutung unverbindlicher Entschließungen internationaler Organisationen für das innerstaatliche Recht der Bundesrepublik Deutschland, 1993, 201-203.

ars,[702] courts consider the parliamentary control exercised through the allocation of the budget as sufficient to secure parliamentary control.[703]

Being thus less predetermined by legislative acts, distributive administrative decisions provide an opportunity for administrators to apply standards from international nonbinding instruments as criteria for granting benefits to private actors. In contrast to other nonbinding norms such as multilateral declarations, nonbinding instruments such as codes of conduct partly (FAO CCRF) or entirely (OECD Guidelines) address private actors directly, and are thus particularly well-suited to be directly applicable as criteria for subsidisation or other benefits for private activities. A State may for example establish a link between compliance with international nonbinding norms and the provision of State credits.

Among the various forms of distributive administration, the granting of export credits for external trade to the exporting industry provides one of the rare tools with which administrations may be able to influence private companies abroad. According to the OECD, 29 out of 39 States adhering to the OECD Guidelines for Multinational Enterprises used them in procedures for the granting of export credits and investment guarantees as of 2007.[704] Remarkable as they are, these linkages are however comparatively loose. The granting of export credits is hardly ever specifically conditioned upon compliance with the OECD Guide-

---

[702] P. Kunig, "The Relevance of Resolutions and Declarations of International Organizations for Municipal Law" in: G.I. Tunkin/R. Wolfrum (eds.), International Law and Municipal Law, 1988, 59-78 (72); H. Maurer, Allgemeines Verwaltungsrecht, 2009, § 6 mn. 19-23; U. Dieckert, Die Bedeutung unverbindlicher Entschließungen internationaler Organisationen für das innerstaatliche Recht der Bundesrepublik Deutschland, 1993, 118 et seq.

[703] According to the Federal Administrative Court of Germany, the decision over the budget by the parliament is a sufficient expression of the will of the parliament, and thus the administration could establish criteria for the granting of subsidies, for example, even without parliamentary acts as long as the fundamental rights of third persons are not infringed, compare Bundesverwaltungsgericht (German Federal Administrative Court), BVerwGE 6, 282, 287 et seq.; BVerwGE 90, 112, 126.

[704] OECD, Annual Meeting of National Contact Points 2007, Report of the Chair, 19-20 June 2007, para. 23 and Table 1, available at http://www.oecd.org/dataoecd/23/26/39319743.pdf.

lines. Export credit agencies in various countries by and large simply call the OECD Guidelines to the attention of exporters.[705]

A clear conditionality is established in most OECD countries between the granting of export credits and compliance with the standards of the OECD Recommendation on Common Approaches on Environment and Officially Supported Export Credits[706] – a nonbinding recommendation revised and adopted by the OECD Council in 2007.[707] Although not binding upon OECD members, the OECD Recommendation may thus directly be implemented by the executive of a particular country without any specific parliamentary decision.[708] In practice, states take concrete administrative decisions in order to ensure compliance with the mentioned standards. In a recent decision the export credit agencies of Germany, Austria and Switzerland temporarily suspended their support for the project of the controversial construction of a dam for a hydroelectric power plant in Ilisu, Turkey, due to non-compliance by the Turkish authorities with the mentioned standards.[709]

The implementation of the OECD Recommendation is however at least to some extent the consequence of its indirect recognition in WTO law. In the renewable energy and water projects sectors, compliance with the OECD Recommendation is a prerequisite for the eligibility of a

---

[705] In some cases, however, the linkages are more formalized. For instance, applicants to export credits or investment guarantees in the Netherlands only qualify for these benefits if they make an express statement that they are aware of the OECD Guidelines and that they will endeavour to comply with them to the best of their ability, compare OECD, Annual Meeting of National Contact Points 2007, Report of the Chair, 19-20 June 2007, Table 1, available at http://www.oecd.org/dataoecd/23/26/39319743.pdf.

[706] OECD, Revised Council Recommendation on Common Approaches on the Environment and Officially Supported Export Credits adopted by the OECD Council of 12 June 2007, TAD/ECG(2007)9, available at http://www.olis.oecd.org/olis/2007doc.nsf/linkto/tad-ecg(2007)9.

[707] For details, see the analysis in Part 1, at A.IV.h), further above.

[708] It is telling that the German export credit agency on its website speaks of "... the Common Approaches as internationally binding set of rules dealing with the subject of environment and officially supported export credits" [emphasis added] despite the nonbinding character of these OECD Recommendations, compare http://www.agaportal.de/en/aga/grundzuege/umweltaspekte.html.

[709] See Financial Times of 23 December 2008, "Berlin suspends support for dam", by Chris Bryant, available at http://www.ft.com.

project under the financial terms and conditions of the respective sector agreement of the OECD Arrangement on Guidelines for Officially Supported Export Credits.[710] Conformity with the OECD Arrangement in turn secures the acceptance of the export credit practice under the WTO Subsidies Agreement). Pursuant to the latter, an export credit practice shall not be considered an export subsidy otherwise prohibited by the Agreement if it is in conformity with the provisions "of an international undertaking on export credits to which at least twelve original members were parties as of 1 January 1979 or a successor undertaking which has been adopted by those original Members)".[711] As confirmed by the WTO Appellate Body in *Brazil – Aircraft*,[712] this is a reference to the OECD Arrangement on Guidelines for Officially Supported Export Credits in its most recent version,[713] which includes the sector agreements.[714]

Another example of implementation through distributive administration is the U.S. Fishing Capacity Reduction Program. It foresees funding for "a vessel and permit buyback program" in order to reduce excess fishing capacity as required by Article 7.6 of the FAO CCRF.[715] Administrative distributive decision-making can further be influenced

[710] OECD Arrangement on Officially Supported Export Credits, revised 2009, TAD/PG(2009)6 of 16 February 2009, Annex IV, para. 1 in conjunction with Appendix 1, available at http://www.oecd.org/topic/0,3373,en_2649_34171_1_1_1_1_37431,00.html.

[711] WTO Agreement on Subsidies and Countervailing Measures, Annex 1, para. (k) in its entirety reads "Provided, however, that if a Member is a party to an international undertaking on official export credits to which at least twelve original Members to this Agreement are parties as of 1 January 1979 (or a successor undertaking which has been adopted by those original Members), or if in practice a Member applies the interest rates provisions of the relevant undertaking, an export credit practice which is in conformity with those provisions shall not be considered an export subsidy prohibited by this Agreement."

[712] Brazil – Aircraft, WTO Appellate Body Report adopted 20 August 1999, WT/DS46/AB/R, para. 180.

[713] J. Pauwelyn, Conflict of norms in public international law: how WTO law relates to other rules of international law, 2003, 348.

[714] That the sector Agreements also qualify as successor undertakings was confirmed by the WTO Panel in Canada – Measures Affecting the Export of Civilian Aircraft – Recourse by Brazil to Article 21.5 of the DSU, Panel Report, WT/DS70/RW of 9 May 2000, footnote 63.

[715] United States Department of Commerce, Implementation Plan for the Code of Conduct for Responsible Fisheries, at D.3 (1).

by codes of conduct in the sense that these codes can lead to a cessation of governmental support. Thus, subsidies for the use of pesticides were phased out in India in order to implement the FAO Pesticide Code.[716]

d) Non-regulatory informal administration

Particularly in subject areas such as environmental protection with high degrees of complexity, uncertainty and dynamically changing circumstances, the ability of traditional administrations to respond effectively and adequately solely through command-and-control type regulation has been frequently called into question.[717] Non-regulatory or informal forms of law and administration are therefore advocated as supplementary strategies.[718] Typical forms of non-regulatory administration which are increasingly employed at least in the U.S. and Europe include the provision of information, raising awareness, warnings, consultations, and general recommendations, and administrators seek to influence private action and decisions by providing citizens with the information and knowledge available to the administration.[719]

---

[716] Compare M. Singh, "Report on India" in: FAO, Regional Office for Asia and the Pacific, RAP Publication 2005/29 (Bangkok 2005), Proceedings of the Asia regional workshop, "Implementation, Monitoring and Observance, International Code of Conduct on the Distribution and Use of Pesticides, Bangkok, Thailand, 26-28 July 2005; available at www.fao.org/docrep/008/af340e/af340e0i.htm.

[717] For United States law, compare R.B. Stewart, "A new generation of environmental regulation?", Capital University Law Review 29 (2001), 21-182; for German and European law, see R. Breuer, "Zunehmende Vielgestaltigkeit der Instrumente im deutschen und europäischen Umweltrecht – Probleme der Stimmigkeit und des Zusammenwirkens", Neue Zeitschrift für Verwaltungsrecht 16 (1997), 833-845 (835); J. Oster, "Das informell-kooperative Verwaltungshandeln im Umweltrecht: Begriffliche Abgrenzung, Erscheinungsformen und rechtliche Bewertung", Natur und Recht 30 (2008), 845-850; see further S.E. Gaines, "Reflexive Law as a Legal Paradigm for Sustainable Development", Buffalo Environmental Law Journal 10 (2002-2003), 1-24; E. Orts, "Reflexive Environmental Law ", Northwestern University Law Review 89 (1995), 1227; S.E. Gaines, "Reflexive Law as a Legal Paradigm for Sustainable Development", Buffalo Environmental Law Journal 10 (2002-2003), 1-24 (24)

[718] *Ibid.*

[719] J. Oster, "Das informell-kooperative Verwaltungshandeln im Umweltrecht: Begriffliche Abgrenzung, Erscheinungsformen und rechtliche Bewertung", Natur und Recht 30 (2008), 845-850.

National administrations in this manner often pursue a similar strategy to international organisations with their informal activities.[720] The nonbinding instruments of this study are part of the endeavour of international institutions to influence both states and individuals in a managerial manner by *inter alia* providing information, issuing recommendations and raising awareness. Informal administrative activities thus frequently represent a readily available way to reproduce the international approach on the national level.

(1) Adoption of national voluntary instruments

Following a trend of moving beyond mere command-and-control regulation, municipal legislatures often refrain from enacting legislation in accordance with international nonbinding instruments. They rather extrapolate the voluntary, flexible, bottom-up approach of international codes of conduct, which count on the participation and support of local private actors, by encouraging or even relying on codes of conduct.

One example is the Canadian Code of Conduct for Responsible Fishing Operations.[721] The Canadian government here relied on a self-regulatory code and supported its institutions. The reliance on private self-regulation is seen as a fundamental change in Canadian fisheries management.[722] The code was developed by local fishermen and follows the wish of Canadian fishermen and industry for a nonbinding non-regulatory approach.[723] It has been endorsed by 60 Canadian fisheries organisations representing 80 per cent of the total catch. Interestingly enough, the initiative did not simply copy the Code of Conduct on Responsible Fisheries of the FAO, but emerged from a quasi-parliamentary process based on principles of democratic participation of all affected members of the national fisheries organisations. After a consensus code was first agreed upon by 60 representatives from all

[720] This parallelism is also highlighted by J.E. Alvarez, International Organizations as Law-makers, 2005, 256.

[721] Canadian Code of Conduct for Responsible Fishing Operations, available at http://www.dfo-mpo.gc.ca/communic/fish_man/publication_e.htm.

[722] Fisheries and Oceans Canada, 'Backgrounder: United Nations Food and Agriculture Organization, Code of Conduct for Responsible Fisheries', available at: www.dfo-mpo.gc.ca/overfishing-surpeche/media/bk_fao_e.htm.

[723] Proceedings of the Standing Senate Committee on Fisheries, Issue 1 – Evidence, 20 March 2001, www.parl.gc.ca.

fishing sectors, it had to be adopted by all local fisheries organisations; mainly by direct voting of fishermen within their organisation.[724] Ratification and implementation were first overseen by the Canadian Responsible Fisheries Board and a secretariat, which were funded and supported by the government.[725] Since May 2003, the responsibility for the implementation of the Canadian code lies solely with the Responsible Fisheries Federation, a private board composed of representatives of different Canadian fisheries and fishery associations without any involvement by the state.

Similarly, at the EU level in the context of the reform of the Common Fisheries Policy, the EU Commission has developed a voluntary European Code of Sustainable and Responsible Fisheries Practices directed at its fishing sector. The code is based on the framework of the CCRF and encourages responsible fishery practices within the EU.[726]

(2) Promotional activities, information and warnings

Governments and their administrations frequently undertake informative and promotional activities with a view to inform citizens and industry about the requirements of codes of conduct and other nonbinding instruments. One typical administrative activity is the dissemination of translated versions of the international instruments, or awareness-raising measures such as posters, brochures and workshops.[727] In Germany, the responsible ministries have translated and disseminated among industry almost all international codes of conduct which contain norms directed at private persons.[728] In the case of the OECD Guide-

---

[724] *Ibid.*

[725] Department of Fisheries and Oceans Canada, 'Responsible Fisheries Summary: Code of Conduct for Responsible Fishing Operations', December 2003, available through Department of Fisheries and Oceans Canada.

[726] European Commission, European Code of Sustainable and Responsible Fisheries Practices (2003), available at http://govdocs.aquake.org/cgi/reprint/2004/1017/10170060.pdf.

[727] See M. Singh, "Report on India" in: FAO, Regional Office for Asia and the Pacific, Proceedings of the Asia regional workshop on the International Code of Conduct on the Distribution and Use of Pesticides, Bangkok, Thailand, 26-28 July 2005, RAP Publication 2005/29 (Bangkok 2005), available at http://www.fao.org/docrep/008/af340e/af340e00.htm.

[728] This includes the ILO Tripartite Declaration, the FAO Codes, the RBP-Code, the WHO Breastmilk Code, and the OECD Guidelines.

lines, the German Federal Ministry for Economics and Technology asked all German business associations to inform their members about the Guidelines.[729] Another interesting example is the joint declaration concerning guidelines to investors annexed to the Association Agreement of 2002 between Chile and the European Union which states that "[T]he Parties remind their multinational enterprises of their recommendation to observe the OECD Guidelines for Multinational Enterprises, wherever they operate."[730] The approach is also reflected in efforts taken by India and Cambodia for the implementation of the FAO Pesticide Code.[731]

(3) Labelling

The use of labelling as a means to inform consumers is a now commonplace in environmental policy. International codes of conduct or other international nonbinding instruments are increasingly used as standards and criteria for labelling programs by governments and administrations. The German Federal Ministry for Food, Agriculture and Consumer Protection, for example, aims to introduce an environmental label on the basis of the FAO Guidelines for the Eco-Labelling of Fish and Fishery Products from Marine Capture Fisheries,[732] which are to a large extent based on the CCRF.[733] Similarly, and perhaps more importantly,

---

[729] This information is provided by U. Dieckert, Die Bedeutung unverbindlicher Entschließungen internationaler Organisationen für das innerstaatliche Recht der Bundesrepublik Deutschland, 1993, 130.

[730] Agreement establishing an association between the European Community and its member states, of the one part, and the Republic of Chile, of the other part – Final act, EU Official Journal L 352 , 30/12/2002, p. 0003-1450. The Agreement entered into force on 1 March 2005. It is available at http://eur-lex.europa.eu.

[731] Compare the reports of M. Singh, "Report on India" and L. Socheata, "Report on Cambodia" in: FAO, Regional Office for Asia and the Pacific, RAP Publication 2005/29 (Bangkok 2005), Proceedings of the Asia regional workshop, "Implementation, Monitoring and Observance, International Code of Conduct on the Distribution and Use of Pesticides, Bangkok, Thailand, 26-28 July 2005; available at www.fao.org/docrep/008/af340e/af340e0i.htm.

[732] See the information at www.portal-fischerei.de.

[733] FAO Guidelines for the Eco-labelling of Fish and Fishery Products from Marine Capture Fisheries (2005). References to the CCRF are frequently made throughout the document, and it is stressed in para. 2.1. that eco-labelling schemes should generally be consistent with – inter alia – the CCRF.

the Commission of the European Union has initiated a discussion on a European Eco-label for Fisheries, and clarifies that the minimum requirements for such a certificate should be based on the standards of the FAO Code of Conduct for Responsible Fisheries.[734]

3. Judiciary

One of the main deficits of nonbinding international norms as compared to treaties is that legislative implementation is not secured or guaranteed. Where they have not been implemented through legislative acts, enforcement through national courts is not available. In the absence of specific implementing legislation, nonbinding instruments cannot provide the legal basis for a judicial decision.[735] As for administrative implementation, much therefore depends on proper legislative implementation of international nonbinding norms.

Even if they are not implemented through legislation, however, nonbinding instruments may have an indirect impact on domestic tribunals. As for administrative decision-making, courts may directly apply nonbinding international norms in the application and interpretation of blanket clauses of national law.[736]

Codes of conduct of corporate responsibility such as the OECD Guidelines for instance are well-suited for application in private law suits on the basis of private law statutes which establish a linkage to the international norms.[737] Thus, a widely accepted code of conduct could

---

[734] European Commission, "Communication from the Commission to the Council, the European Parliament and the European Economic and Social Committee: Launching a debate on a Community approach towards eco-labelling schemes for fisheries products", COM(2005)275 final of 29 June 2005.

[735] U. Dieckert, Die Bedeutung unverbindlicher Entschließungen internationaler Organisationen für das innerstaatliche Recht der Bundesrepublik Deutschland, 1993, 229.

[736] B.-O. Bryde, Internationale Verhaltensregeln für Private – Völkerrechtliche und verfassungsrechtliche Aspekte, 1981, 34; U. Dieckert, Die Bedeutung unverbindlicher Entschließungen internationaler Organisationen für das innerstaatliche Recht der Bundesrepublik Deutschland, 1993, 210 et seq; B. Simma/A. Heinemann, "Codes of Conduct" in: W. Korff et. al. (eds.), Handbuch der Wirtschaftsethik, Band 2: Ethik wirtschaftlicher Ordnungen, 1999, 403-418 (414).

[737] N. Horn, "International Rules for Multinational Enterprises: the ICC, OECD, and ILO initiatives", American University Law Review 30 (1980-

be used as evidence that a firm has not exercised reasonable care or due diligence.[738] Moreover, if private companies advertise their adherence to a certain standard, legal actions are conceivable in which consumers could claim that they bought a product on the assumption that a company complied with certain standards of conduct that were in fact not obeyed. To cite one example from the United States, in the case *Nike versus Kasky*,[739] the Californian citizen Marc Kasky for example sued Nike on behalf of the general public of the State of California for unfair and deceptive practices under California's Unfair Competition Law[740] and False Advertising Law[741] for allegedly false statements on the part of Nike concerning working conditions under which Nike products were manufactured. Although the trial was settled without any decision on the merits, the Californian Supreme Court implicitly agreed that such a claim was possible[742] when sending the case back to the lower court to settle whether any false representations had in fact been made. One can moreover conceive of court proceedings where a company may bring a claim against another corporation arguing that non-compliance with an international norm despite advertisement or certification to the contrary has contributed to a competitive advantage.[743] In Germany, the content of section 3 paragraph 1 of the German Unfair

1981), 923-940 (938); P. Sanders, "Implementing International Codes of Conduct for Multinational Enterprises", American Journal of Comparative Law 30 (1982), 241-254 (244).

738 P. Muchlinski, "Human Rights, social responsibility and the regulation of international business: The development of international standards by intergovernmental organisations", Non-State Actors and International Law 3 (2003), 123-152 (129).

739 Californian Supreme Court, Nike Inc. v. Kasky, 27 Cal. 4th 939, 946, 45 P. 3d 243, 247 (2002); the United States Supreme Court in Nike, Inc. v. Kasky, 539 U.S. 654 (2003) dismissed an appeal by Nike due to the fact that the judgment of the Californian Supreme Court was not final, that neither Party had standing to invoke federal jurisdiction and that the adjudication of novel constitutional questions would have been premature in this case.

740 United States Cal. Bus. & Prof. Code Ann. §17200 et seq. (West 1997).

741 United States Cal. Bus. & Prof. Code Ann.§17500 et seq. (West 1997).

742 In particular by arguing that the matter at hand concerned commercial speech and not public speech, and therefore was not protected by the First Amendment of the United States Constitution, compare 27 Cal. 4th 939, 946, 45 P. 3d 243, 247 (2002).

743 K.M. Meessen, "Internationale Verhaltenskodizes und Sittenwidrigkeitsklauseln", Neue Juristische Wochenschrift 34 (1981), 1131-1132.

Competition Act (Gesetz gegen den unlauteren Wettbewerb)[744] of the German Unfair Competition Act which prohibits "unfair acts of competition which are liable to have more than an insubstantial impact on competition to the detriment of competitors, consumers or other market participants"[745], German courts may arguably rely on an international *ordre public* as expressed in international nonbinding instruments.[746]

One of the main ways in which nonbinding instruments can influence national law through the judiciary is of course their application as tools for the interpretation and application of general clauses in national law.[747] Nonbinding norms can for example provide useful guidance for the interpretation of constitutional civil rights. This possibility has been explicitly confirmed for resolutions of the United Nations General Assembly by the German administrative court of Frankfurt in a decision concerning child adoption practices.[748] The court ruled that despite its nonbinding nature and even though it had not acquired customary law status, the Declaration of the Rights of the Child[749] could be used for the interpretation of national law, in this case in relation to general

744 German Unfair Competition Act, 3 July 2004 (Gesetz gegen den unlauteren Wettbewerb), BGBl. I 2004 32/1414.

745 § 3 (1) of the German Unfair Competition Act (Gesetz gegen den Unlauteren Wettbewerb) in German reads: "Unlautere geschäftliche Handlungen sind unzulässig, wenn sie geeignet sind, die Interessen von Mitbewerbern, Verbrauchern oder sonstigen Marktteilnehmern spürbar zu beeinträchtigen."

746 K.M. Meessen, "Internationale Verhaltenskodizes und Sittenwidrigkeitsklauseln", Neue Juristische Wochenschrift 34 (1981), 1131-1132 (1132); I. Seidl-Hohenveldern, "International Economic 'Soft Law'", Recueil des cours de l'Académie de Droit International de La Haye 163 (1979), 169-246 (200-202).

747 B.-O. Bryde, Internationale Verhaltensregeln für Private – Völkerrechtliche und verfassungsrechtliche Aspekte, 1981, 34; C.-T. Ebenroth, Code of Conduct – Ansätze zur vertraglichen Gestaltung internationaler Investitionen, 1987, 75; B. Simma/A. Heinemann, "Codes of Conduct" in: W. Korff et al. (eds.), Handbuch der Wirtschaftsethik, Band 2: Ethik wirtschaftlicher Ordnungen, 1999, 403-418 (414).

748 Judgement of the Administrative Court Frankfurt of 19 June 1988, Neue Juristische Wochenschrift, 1988, 3032-3035, 3032.

749 Declaration of the Rights of the Child, UNGA Resolution 1386 (XIV) of 20 November 1959. The Declaration was a precursor of the Convention on the Rights of the Child (20 November 1989), GA res. 44/25, annex, 44 UN GAOR Supp. (No. 49) at 167, U.N. Doc. A/44/49 (1989); 1577 UNTS 3; 28 ILM 1456 (1989).

clauses of police law in conjunction with the human dignity guarantee of the German Basic Law.[750]

In particular, blanket clauses which refer to public morals or public policy may potentially serve as points of entry for international norms, because they deliberately appeal to collective values which go beyond the mere legality of a private activity.[751] In determining the content of such clauses, courts look to formalised indicators of collective ethical standards.[752] In cases with an international dimension, collective values may be interpreted with reference to international instruments which reflect the consensus of the community of states on a given issue.[753]

In this fashion, international standards have already been recognised for the interpretation of *boni mores* clauses such as § 138 BGB of the German Civil Code, according to which a legal transaction that contravenes public morals is void.[754] In the Nigerian Cultural Property Case decided by the German Federal Supreme Court in Civil Matters in 1972, nonbinding instruments arguably were used to interpret this broad provision.[755] The case concerned the collection of an insurance policy in Germany for some African artefacts some of which had been lost on the way from Port Harcourt to Hamburg. The export was prohibited un-

---

750 Judgement of the Administrative Court Frankfurt of 19 June 1988, Neue Juristische Wochenschrift, 1988, 3032-3035, 3032.

751 Such as §§ 138, 826 of the German Civil Code (Bürgerliches Gesetzbuch) or § 1 of the German Law against restrictive business practices (Gesetz gegen unlauteren Wettbewerb).

752 K.M. Meessen, "Internationale Verhaltenskodizes und Sittenwidrigkeitsklauseln", Neue Juristische Wochenschrift 34 (1981), 1131-1132 (1131).

753 A. Bleckmann, "Sittenwidrigkeit wegen Verstoßes gegen den ordre public international", Zeitschrift für ausländisches öffentliches Recht und Völkerrecht 34 (1974), 112-132; U. Dieckert, Die Bedeutung unverbindlicher Entschließungen internationaler Organisationen für das innerstaatliche Recht der Bundesrepublik Deutschland, 1993, 219.

754 German Civil Code (Bürgerliches Gesetzbuch), § 138, a translation can be found at http://www.gesetze-im-internet.de/englisch_bgb/index.html.

755 Bundesgerichtshof, BGHZ 59, 82 (1972); for discussions of the case in the context of the effect of nonbinding norms see N. Horn, "Codes of conduct for MNEs and Transnational Lex Mercatoria: An international Process of Learning and Law Making" in: N. Horn (ed.), Legal Problems of Codes of Conduct for Multinational Enterprises, 1980, 45-81 (58); F. Kratochwil, Rules, norms and decisions: on the conditions of practical and legal reasoning in international relations and domestic affairs, 1989, 203 et seq.

der Nigerian law, but this prohibition could not be brought to bear due to the applicable international private law rule denying extraterritorial effect to foreign public law. The insurance carrier argued however that the transportation violated § 138 of the German Civil Code (Bürgerliches Gesetzbuch). The Court followed this argument. It found that the export contravened the public morality clause and was therefore void due to its incompatibility with "fundamental convictions" of the international community pertaining to the right of each country to protect its cultural heritage. In determining these convictions and interests of the international community, the Court explicitly cited a nonbinding UNESCO recommendation and the prohibitions of Article 2 of the Convention on the Means of Prohibiting and Preventing the Illicit Import, Export and Transfer of Ownership of Cultural Property[756] which had not been ratified in Germany at the time. Even though the Court in the actual analysis frequently referred to the text of the international convention and neglected the UNESCO recommendation,[757] the case shows that nonbinding instruments have at least the potential to impact domestic level court decisions, as otherwise the court would not have referred to the recommendation at all.[758]

The extent to which courts use these instruments can however hardly be generalised. One should be careful not to neglect the differences between different international instruments.[759] The objectives and content of the international instrument must suit the exigencies of the national law norm. And certainly, not all of the rules of international nonbinding instruments necessarily reflect "fundamental convictions" of the international community or common interests.[760] Criteria suited for distin-

---

[756] Convention on the Means of Prohibiting and Preventing the Illicit Import, Export and Transfer of Ownership of Cultural Property (14 November 1970), 823 UNTS 231.

[757] Klabbers sees the case therefore primarily as an illustration of the anticipatory application of a treaty. Compare J. Klabbers, The concept of treaty in international law, 1996, 162.

[758] F. Kratochwil, Rules, norms and decisions: on the conditions of practical and legal reasoning in international relations and domestic affairs, 1989, 204.

[759] In favour of a differentiation U. Dieckert, Die Bedeutung unverbindlicher Entschließungen internationaler Organisationen für das innerstaatliche Recht der Bundesrepublik Deutschland, 1993, 227.

[760] Hailbronner goes further and rejects the influence of norms of codes of conduct which are not general principles of law, see K. Hailbronner, "Völkerrechtliche und staatsrechtliche Überlegungen zu Verhaltenskodizes für transna-

guishing between various instruments could be the competence, expertise and representativeness of the international organisation, the quality of the instrument as well as the international and national acceptance of the norms.[761] The typology of different nonbinding international instruments as outlined in a preliminary style in the first chapter could provide useful additional distinguishing criteria. Thus, it may be useful in a concrete case to distinguish technical standards from those expressing value choices. Further one should distinguish concrete rules from vague political objectives, those addressed to states from those addressed to private actors and those adopted unanimously by organisations with global reach from those of regional organisations.[762]

4. Summary

The analysis of the implementation of nonbinding instruments in national legal systems first of all shows that nonbinding instruments are to a considerable extent implemented into national law or influence administrative decision-making despite or even because of their nonbinding character. Implementation of international norms therefore does not exclusively depend on whether an instrument is binding.[763] However, legislative implementation is not guaranteed, and although this requires further empirical research, treaty law implementation should still gain more attention of legislators. When the legislature does not act, the implementation and enforcement of nonbinding instruments depends on the discretion granted to administrators and the way that courts use nonbinding instruments to interpret general clauses in the law. Such in-

---

tionale Unternehmen" in: I. v. Münch (ed.), Staatsrecht – Völkerrecht – Europarecht: Festschrift für Hans-Jürgen Schlochauer, 1981, 329-362 (355).

[761] U. Dieckert, Die Bedeutung unverbindlicher Entschließungen internationaler Organisationen für das innerstaatliche Recht der Bundesrepublik Deutschland, 1993, 229 et seq.

[762] See for these distinctions already Part 1, at C.II.

[763] Similar K. Raustiala, "Form and Substance in International Agreements", American Journal of International Law 99 (2005), 581-614 (605); A. Peters/ I. Pagotto, "Soft Law as a New Mode of Governance: A Legal Perspective", New Modes of Governance Project 04/D11 (2006), (28); D.M. Johnston, "Book Review of Commitment and Compliance: The Role of Non-Binding Norms in the International Legal System", American Journal of International Law 95 (2001), 709-714 (who mentions studies such as the reviewed one as examples of persuasive demonstrations of this point).

direct forms of implementation thus exist, but generally speaking proper implementation should require legislative action in most cases. If they are not implemented via legislation, the norms often cannot directly be enforced through administrative acts, since administrations in many legal systems require legal authorisation in order to do so. And obviously, courts cannot directly enforce international nonbinding standards in the absence of any linkage to domestic law. The adoption of nonbinding instruments in consequence cannot give rise to legal security for states and private actors that the respective norms will indeed be implemented.

But nonbinding instruments may even in the absence of specific legislative acts become influential domestically. Numerous methods exist for administrators to implement international nonbinding norms directly. Within their respective competencies, governments may base their policies on international nonbinding instruments. Executive officials and administrators may issue internal guidelines or administrative rules in accordance with nonbinding international instruments, or may even take decisions that effectively implement the respective norms. Interventionist administrative decision making is most restricted by legislative acts, but administrators may nevertheless be guided by nonbinding instruments, for example when concretising indeterminate legal terms. More discretion for administrators to implement nonbinding standards often exists where the legal order allows for distributive administrative decisions without specific legislative acts. The export credit system is an example of distributive administration where the administration implements international nonbinding standards without any specific legislative act in Germany. And finally, informal non-regulatory administrative activities can implement and promote international nonbinding instruments by using them as points of orientation.

Apart from legal uncertainty on whether implementation will happen or not, one characteristic of nonbinding instruments is that they may – within the limits of the legal order – be implemented without specific legislative acts, and thus without parliamentary approval. The tendency to employ nonbinding instruments at the international level thus entails the possibility of a shift of decision-making power to the executive not only on the international, but also on the national level. This entails potential challenges for the legitimacy of nonbinding instruments and their implementation. The different legitimacy challenges arising form the various modes of implementation will be discussed in more detail in Part 3.

III. Limits of the regulatory capacity of nonbinding instruments regarding states

1. Limited utility to provide legal certainty and predictability

The nonbinding status of the instruments discussed in this study translates into a limited ability of the act of adoption of a nonbinding instrument to provide reassurance to other states that a state will actually abide by its norms. Legally binding treaty law norms are usually subject to some kind of national legislative approval before international ratification. Therefore, in the case of treaty law, a state can be relatively certain that another state will if necessary strive to bring its domestic law into compliance, at least if it does not want to risk a breach or being accused of non-compliance with severe reputational costs. In addition, a state has a legally binding commitment even after a change in government or of ministers, whereas a voluntary commitment is more easily abandoned with such a change. All of these factors, taken together, describe the particular ability of binding international law norms to provide predictability and legal security to other state actors.[764] This strength of legally binding norms is at the same time the weakness of nonbinding instruments and the downside of their flexibility which has been stressed in this study.

As a consequence, one can generally expect that the international regulation of any issue requiring high levels of reassurance and predictability over a long time is better addressed by means of treaty law. Moreover, the reassurance and credibility created by legally binding norms may be highly significant for regulating issues of cooperation where the costs of compliance and the benefits from defection are particularly high, whereas simple matters of coordination may often be sufficiently addressed by nonbinding ones.[765]

Practice in some ways supports these assumptions. For the international regulation of common goods such as fishing or emission trading where the incentive of free-riding is particularly high, states in favour of

---

[764] J.J. Kirton/M.J. Trebilcock, "Introduction: Hard Choices and Soft Law in Sustainable Global Governance" in: J.J. Kirton/M.J. Trebilcock (eds.), Hard Choices, Soft Law: Voluntary Standards in Global Trade, Environment and Social Governance, 2004, 3-29 (6).

[765] This distinction is made by K.W. Abbott/D. Snidal, "Hard and Soft Law in International Governance", International Organization 54 (2000), 421-456 (429).

the environmental objectives often insist on binding rules even where enforcement is not readily available. It appears that binding norms are often seen to be able to provide the reassurance and legal security to maintain cooperation over a long time in issue areas where changes of behaviour are required and reassurance needed. For example, the allocation of fisheries resources through quota systems among states cooperating in a regional fisheries organisation is usually done through binding law. Where this is first based on voluntary measures, which sometimes happens in the regional context, binding treaties are soon negotiated to give the measures a stronger footing. For instance, the three states of Japan, New Zealand and Australia established a voluntary agreement in 1985 providing for a total allowable catch of southern bluefin tuna and established quotas of allowable catches for each of the participants. In 1993 the three states however decided to formalise this agreement by establishing the Convention for the Conservation of Southern Bluefin Tuna.[766]

Similarly, states resort to binding instruments to establish systems of authorisation and control of fishing fleets, for example by closing port and conducting on-board inspections. The nonbinding International Plan of Action on Illegal, Unregulated and Unreported Fishing adopted under the framework of the FAO CCRF was apparently perceived as inadequate to address fishing by member states of the FAO effectively. In a meeting of COFI in 2007, they unanimously decided to develop a binding instrument instead.[767] With respect to the PIC procedure, there also seemed to be a desire among states to base the voluntary system on prior informed consent on legally binding ground even though it was largely functioning well.

2. *Implications of the lack of dispute settlement and enforcement*

Non-compliance with nonbinding instruments is not a violation of international law and therefore does not lead to state responsibility. Consequently, dispute settlement cannot be applied directly to nonbinding

[766] Convention for the Conservation of Southern Bluefin Tuna (10 May 1993) 1819 UNTS 360, available at http://www.ccsbt.org/docs/pdf/about_the_commission/convention.pdf.

[767] On the inadequacy of the nonbinding International Plan of Action as lacking the necessary "teeth" J.K. Ferrell, "Controlling flags of convenience: one measure to stop overfishing of collapsing fish stocks", Lewis & Clark Law School Environmental Law 35 (2005), 323-390 (379-380).

instruments, nor can states be held liable for damages through national or international state liability. Can this most obvious limitation of non-binding instruments be translated into a clear weakness in terms of compliance in comparison to environmental treaty law?

First of all, whether and to what extent actual enforcement is decisive for compliance has long been debated. In criticising the managerial approach of Chayes and Chayes, other scholars have warned against neglecting "enforcement" strategies.[768] The arguments of G.W. Downs and his colleagues are rooted in rationalist cost-benefit calculations. They point out that a number if not most environmental problems now and in the future will require multilateral strategies which can deter non-compliance by offsetting the net benefits which a violator of the rules could gain from non-compliance. The greater the benefits a state can gain from defection, the greater the necessity for deterrence in the form of a threat of punishment.[769] This means the more that states are required to change their behaviour and the greater the incentive to free ride, the greater the need for enforcement, incentives and disincentives as counterbalances.[770] This argument appears highly plausible and leads to the conclusion that one should not build on one mechanism and neglect economic interest structures. Thus, I perceive the discussion of managerial versus enforcement approaches as in fact highlighting the need to combine both managerial as well as the incentive mechanisms for environmental regulation in particular in those areas where strong short term economic interests encourage free riding by states. The compliance mechanism of the Kyoto Protocol which establishes both a facilitative and a so-called "enforcement branch" appears as a noteworthy development in this regard.[771] Where enforcement mechanisms ex-

---

[768] G.W. Downs, "Enforcement and the evolution of cooperation", Michigan journal of international law 19 (1998), 319-344 (321); G.W. Downs/K.W. Danish/P.N. Barsoom, "The transformational model of international regime design: triumph of hope or experience?", Columbia journal of transnational law 38 (2000), 465-514.

[769] G.W. Downs, "Enforcement and the evolution of cooperation", Michigan journal of international law 19 (1998), 319-344 (324).

[770] G.W. Downs/D.M. Rocke/P.N. Barsoom, "Is the good news about compliance good news about cooperation?", International Organization 50 (1996), 379-406 (388-392); G.W. Downs/K.W. Danish/P.N. Barsoom, "The transformational model of international regime design: triumph of hope or experience?", Columbia journal of transnational law 38 (2000), 465-514 (333).

[771] Procedures and mechanisms relating to compliance under the Kyoto Protocol, at V. The procedures are entailed in FCCC/KP/CMP/2005/8/Add.3, De-

ist, they are perhaps even more important for preventing non-compliance than for remedying breaches.[772]

But apart from that, are traditional reciprocal enforcement, state responsibility and the ascertainment of a norm in dispute settlement actually important factors for compliance with norms in international environmental law, and consequently are nonbinding instruments limited in their capacity? In assessing this question, one can differentiate between different regulatory problems: those where damage is done to the environment of another state and those where damage is done to the environment in areas or matters beyond national jurisdiction.

In a bilateral setting where the environment of another state is damaged, treaty suspensions not only hurt both states, but retaliation in kind is usually excluded by the fact that one state is naturally advantaged. One need only think about typical constellations of transboundary pollution which depend on natural conditions such as wind directions, the direction of water flow, etc. In these cases, state responsibility law and international dispute settlement have been significant in international legal practice. As indicated by important decisions in particular of the International Court of Justice[773] international dispute settlement has played a role for settling these problems, and dispute settlement here also had a considerable impact on the development of international environmental law.[774]

---

cision 27/CMP.1, Annex; for details see R. Wolfrum/J. Friedrich, "The Framework Convention on Climate Change and the Kyoto Protocol" in: U. Beyerlin/P.-T. Stoll/R. Wolfrum (eds.), Ensuring Compliance with Multilateral Environmental Agreements: A Dialogue between Practitioners and Academia, 2006, 53-68.

772 R. Wolfrum, "The protection of the environment through international courts and tribunals" in: A. Fischer-Lescano (ed.), Frieden in Freiheit: Festschrift für Michael Bothe zum 70. Geburtstag, 2008, 807-817.

773 E.g. ICJ, Gabčikovo-Nagymaros Project (Hungary/Slovakia), Judgment of 25 September 1997, ICJ Reports 1997, p. 7; ICJ, Pulp Mills on the River Uruguay (Argentina v. Uruguay), Judgment of 20 April 2010.

774 R. Wolfrum, "The protection of the environment through international courts and tribunals" in: A. Fischer-Lescano (ed.), Frieden in Freiheit: Festschrift für Michael Bothe zum 70. Geburtstag, 2008, 807-817; C.P.R.R. Romano, "International dispute settlement" in: D. Bodansky/J. Brunnée/E. Hey (eds.), The Oxford Handbook of International Environmental Law, 2007, 1036-1056.

Where damage to the environment beyond national jurisdiction is at stake, including emissions into the atmosphere, the issue is more complex. Not only does the object and purpose of global environmental protection render traditional enforcement such as treaty suspensions, countermeasures or retaliation in kind inadequate in these cases.[775] Any such act would basically mean for a state to cease to take environmental measures on its own, or pollute as much as the other. In consequence, these enforcement measures could mean to potentially hurt the environment and the common objective but only in a limited way the state responsible for a breach. More importantly, one can observe that reliance on state responsibility and its ascertainment in current dispute settlement systems is still largely inadequate and insufficient.[776] Again, however, this general finding requires clarification.

In some areas such as allocation of resources in the fisheries areas where individual rights are directly affected, state responsibility and adjudication by international courts may have considerable potential,[777] even if the jurisdiction of courts and monitoring needs further strengthening in this area.[778] The competence of the adjudicative bodies under UNCLOS, including the International Tribunal of the Law of the Sea, to issue provisional measures for environmental protection as applied in the

[775] R. Wolfrum, "Means of Ensuring Compliance with and Enforcement of International Environmental Law", Recueil des cours de l'Académie de Droit International de La Haye 272 (1998), 13-154 (56-57).

[776] See generally P. Birnie/A. Boyle, International law and the environment, 2002, 200; U. Beyerlin, Umweltvölkerrecht, 2000, mn. 462-464; G. Dahm/ J. Delbrück/R. Wolfrum, Völkerrecht, Band I/3, Die Formen völkerrechtlichen Handelns, die inhaltliche Ordnung der internationalen Gemeinschaft, 2002, 977 et seq; M. Koskenniemi, "Breach of Treaty or Non-Compliance? Reflections on the Enforcement of the Montreal Protocol", Yearbook of International Environmental Law 3 (1992), 123-162 (125-128); R. Wolfrum, "The protection of the environment through international courts and tribunals" in: A. Fischer-Lescano (ed.), Frieden in Freiheit: Festschrift für Michael Bothe zum 70. Geburtstag, 2008, 807-817 (817).

[777] R. Wolfrum, "Means of Ensuring Compliance with and Enforcement of International Environmental Law", Recueil des cours de l'Académie de Droit International de La Haye 272 (1998), 13-154 (79).

[778] In the case of fisheries and marine pollution, this pertains in particular to the International Tribunal of the Law of the Sea which has great potential in this respect, compare T. Stephens, International Courts and Environmental Protection, 2009. States however often prefer opting for arbitration and rarely include references to ITLOS in regional fisheries agreements.

Southern Bluefin Tuna Cases[779] can be seen as the tentative beginning of a system of control of fisheries resource allocation. In the end, international administrative systems governing the allocation of commonly used resources, as for example quota systems and port state controls must be accompanied by some kind of judicial control.

In other areas, however, the law on state responsibility and the procedural law of most dispute settlement bodies appear so far insufficiently developed. Although the reluctance of states to bring environmental claims to international courts seems to have lessened since the 1990s and environmental issues have played an increasingly important role in international jurisprudence,[780] environmental international jurisprudence remains so far limited to some cases.[781] States are by and large still hesitant to submit to compulsory jurisdiction or to refer environmental and resource issues to a judicial or arbitration body where such a possibility exists.[782] As will be indicated, however, the obstacles are not in-

---

[779] ITLOS, Bluefin Tuna Cases (New Zealand v. Japan; Australia v. Japan), Request for Provisional Measures, Order of 27 August 1999, available at http://www.itlos.org/case_documents/2001/document_en_116.pdf; see for an analysis R.R. Churchill, "International Tribunal for the Law of the Sea the Southern Bluefin Tuna cases (New Zealand v. Japan; Australia v. Japan): order for provisional measures of 27 August 1999", The international and comparative law quarterly 49 (2000), 979-990 (in particular 987); R. Wolfrum, "The protection of the environment through international courts and tribunals" in: A. Fischer-Lescano (ed.), Frieden in Freiheit: Festschrift für Michael Bothe zum 70. Geburtstag, 2008, 807-817 (816-817).

[780] On the significance of international litigation for the protection of the environment and the development of international environmental law P. Sands, "Litigating Environmental Disputes: Courts, Tribunals and the Progressive Development of International Environmental Law" in: T.M. Ndiaye/R. Wolfrum (eds.), Law of the Sea, Environmental Law and Settlement of Disputes, 2007, 313-325; R. Wolfrum, "The protection of the environment through international courts and tribunals" in: A. Fischer-Lescano (ed.), Frieden in Freiheit: Festschrift für Michael Bothe zum 70. Geburtstag, 2008, 807-817.

[781] A detailed analysis of international environmental jurisprudence is given by T. Stephens, International Courts and Environmental Protection, 2009.

[782] M. Bothe, "The Evaluation of Enforcement Mechanisms in International Environmental Law: An Overview" in: R. Wolfrum (ed.), Enforcing environmental standards: economic mechanisms as viable means?, 1996, 13-38 (29); R. Wolfrum, "Means of Ensuring Compliance with and Enforcement of International Environmental Law", Recueil des cours de l'Académie de Droit International de La Haye 272 (1998), 13-154 (97 et seq.); U. Beyerlin, Umweltvölkerrecht, 2000, mn. 464.

surmountable. Similar to the role of the International Tribunal for the Law of the Sea within the area of marine environmental protection and the law of the sea,[783] there is great potential and the necessity for dispute settlement even within environmental MEAs.[784]

First of all, the exigencies of effective environmental protection can hardly be met through a law on state responsibility that is ascertained after the fact, since environmental damage is often irrevocable and irreparable. However, procedures on provisional measures, for example as provided for under Article 290 UNCLOS and applied by the International Tribunal for the Law of the Sea in the *Southern Bluefin Tuna Case*, can provide some however imperfect remedy in this respect. Most remarkably, in arguing why the matter was one of urgency and therefore fell within its jurisdiction, the ITLOS stresses that "the conservation of the living resources of the sea is an element in the protection and preservation of the marine environment".[785] ITLOS thus in this case claimed jurisdiction not only to protect the rights of the parties but also to prevent serious harm to the marine environment.[786]

Second, the law on state responsibility so far presupposes a breach or violation, and therefore does not pay sufficient respect to the danger from legal but hazardous activities. That being said, there exists the possibility of developing international law into that direction. A significant step forward has been made by the International Law Commission in developing the ILC Draft Articles on Transboundary Harm from Hazardous Activities.

Third, it is difficult to legally define and factually identify environmental damages; they are typically incremental, and involve complex

---

[783] Compare D.R. Rothwell, "The Contribution of ITLOS to Oceans Governance through Marine Environmental Dispute Resolution" in: T.M. Ndiaye/R. Wolfrum (eds.), Law of the Sea, Environmental Law and Settlement of Disputes, 2007, 1007-1024.

[784] Necessity in particular insofar as authoritative dispute settlement is needed to further develop international environmental law, compare on this T. Stephens, International Courts and Environmental Protection, 2009, 364.

[785] ITLOS, Southern Bluefin Tuna Cases (New Zealand v. Japan; Australia v. Japan), Request for Provisional Measures, Order of 27 August 1999, para. 70.

[786] R.R. Churchill, "International Tribunal for the Law of the Sea the Southern Bluefin Tuna cases (New Zealand v. Japan; Australia v. Japan): order for provisional measures of 27 August 1999", The international and comparative law quarterly 49 (2000), 979-990 (980).

causalities which are difficult to prove.[787] International dispute settlement and courts will not gain a more significant role as long as there do not exist adequate procedures to compensate better for environmental damage and thereby also protect the environment per se.[788] In this respect, the recommendations of the United Nations Compensation Commission indicate a way forward, because they for the first time considered general ecological damage as part of compensation claims.[789]

Fourth, there is the issue of standing which prevents the effective assertion of state responsibility for violations of community interests and common goods.[790] Traditionally, "only the party to whom an international obligation is due can bring a claim in respect of its breach."[791] Invoking state responsibility for protecting community interests would presuppose the possibility of an *actio popularis* where an obligation of *erga omnes* character exists. These are obligations are "by their very nature the concern of all states" and "all states have a legal interest in their

---

[787] I. Brownlie, Principles of public international law, 2008, 276-277.

[788] Wolfrum therefore proposes to provide individuals or groups or non-governmental organizations with the possibility to trigger proceedings and to widen amicus curiae possibilities, compare R. Wolfrum, "The protection of the environment through international courts and tribunals", in: A. Fischer-Lescano (ed.), Frieden in Freiheit: Festschrift für Michael Bothe zum 70. Geburtstag, 2008, 807-817 (817).

[789] Compare UN Doc. S/RES/687 of 3 April 1991, para. 16; United Nations Compensation Commission, Report and recommendations made by the Panel of Commissioners concerning the fifth instalment of 'F4' claims, UN Doc. S/AC.26/2005/10 (2005), available at http://www.uncc.ch.; see also P.H. Sand, "Compensation for environmental damage from the 1991 Gulf War", Environmental Policy and Law 35 (2005), 244-249 (249).

[790] U. Beyerlin/T. Marauhn, Rechtsetzung und Rechtsdurchsetzung im Umweltvölkerrecht nach der Rio-Konferenz 1992, 1997, 83; P. Birnie/A. Boyle, International law and the environment, 2002, 199; M. Bothe, "The Evaluation of Enforcement Mechanisms in International Environmental Law: An Overview" in: R. Wolfrum (ed.), Enforcing environmental standards: economic mechanisms as viable means?, 1996, 13-38 (33); R. Wolfrum, "Means of Ensuring Compliance with and Enforcement of International Environmental Law", Recueil des cours de l'Académie de Droit International de La Haye 272 (1998), 13-154 (98).

[791] International Court of Justice, Reparations for Injuries suffered in the Service of the United Nations, Advisory Opinion of 11 April 1949, I.C.J Reports 1949, 174, at 181-182.

protection".[792] Article 48 of the ILC Draft Articles on the Responsibility of States for Internationally Wrongful Acts also provide the possibility that a state other than the injured state can invoke the responsibility of another state for obligations *erga omnes partes,* i.e. obligations due to a group of states which are established for the protection of the collective interest of the group [Article 48 (1) (a)], as well as for obligations *erga omnes* proper owed to the international community.[793] The Commentary on the ILC Articles in this respect specifically mentions environmental obligations. The ILC thus confirmed that a party to a multilateral environmental treaty could for example invoke a breach of *erga omnes* obligations in the case of non-compliance.

Practice from human rights treaties and environmental law shows however that states generally do not resort to these judicial procedures without being injured individually. In particular confrontational sanctions with a stigmatising effect are inadequate in a field of law the effectiveness of which depends on long-term cooperation by the greatest possible number of states and on preventive action.[794] In multilateral constellations addressing the protection of common goods, such confrontational measures – besides being inadequate for their high political costs[795] – are often counterproductive as they erode common objectives

---

[792] ICJ, Case concerning the Barcelona Traction, Light and Power Company, Limited (Belgium v. Spain) (Second Phase), Judgment of 5 February 1970, ICJ Reports 1970, 3, at 32; confirmed by ICJ, Legal Consequences of the Construction of a Wall in the Occupied Palestinian Territory, Advisory Opinion of 9 July 2004, ICJ Reports 2004, 136, paras 155 and 159.

[793] The ILC Draft Articles on the Responsibility of States for Internationally Wrongful Acts, 2001 [53 UN GAOR Supp. (No. 10) at 43, U.N. Doc. A/56/10 (2001)], in Article 48 provide that "[A]ny State other than an injured State is entitled to invoke the responsibility of another State ... if (a) The obligation breached is owed to a group of States including that State, and is established for the protection of a collective interest of the group; or (b) The obligation breached is owed to the international community as a whole. The Draft Articles have been commended to the attention of governments by the United Nations General Assembly to states in UN Doc. A/RES/56/83 (2002), para. 3.

[794] U. Beyerlin/T. Marauhn, Rechtsetzung und Rechtsdurchsetzung im Umweltvölkerrecht nach der Rio-Konferenz 1992, 1997, 74; J. Brunnée, "The Stockholm Declaration and the Structure and Processes of International Environmental Law" in: M.H. Nordquist/J.N. Moore/S. Mahmoudi (eds.), The Stockholm Declaration and Law of the Marine Environment, 2003, 67-84 (79).

[795] A. Chayes/A. Handler Chayes, The New Sovereignty: Compliance with International Regulatory Agreements, 1995, 29-108.

and harm all Parties equally.[796] Instead of resorting to dispute settlement, states establish prevention-oriented non-confrontational compliance mechanisms in Multilateral Environmental Agreements.[797] Even though these Agreements typically include optional clauses for arbitration or adjudication of the International Court of Justice,[798] there has not yet been a single case where it was actually used.

Despite the legal and practical difficulties and inadequacies that exist in enforcing binding international environmental law, it should not be concluded that enforcement and international judiciary are not necessary, nor should one settle for nonbinding solutions based on the argument that enforcement is not available in any case. Rather, decision makers should strive to establish adequate enforcement measures and dispute settlement mechanisms that complement other softer forms of compliance management. The compliance mechanisms of the Kyoto Protocol which include penalties as well as the mentioned developments in the law on state responsibility and dispute settlement systems, and the establishment of liability regimes[799] such as the Liability Annex to the Antarctic Treaty[800], however imperfect,[801] all show that enforce-

---

[796] D. Hunter/J. Salzman/D. Zaelke, International Environmental Law and Policy, 2002, 457; U. Beyerlin, "State Community Interests and Institution-Building in International Environmental Law", Zeitschrift für ausländisches öffentliches Recht und Völkerrecht 56 (1996), 602-627 (610).

[797] U. Beyerlin, Umweltvölkerrecht, 2000, mn. 466 and 467 et seq.

[798] Article 11 of the Convention for the Protection of the Ozone Layer, 1513 UNTS 323, 26 ILM 1529 (1987); Article 14 UN United Nations Framework Convention on Climate Change, 1771 UNTS 107 31, ILM 849 (1992); Article 28 of the United Nations Convention to Combat Desertification in Countries Experiencing Serious Drought and/or Desertification, Particularly in Africa, 1954 UNTS 3; 33 ILM 1328 (1994).

[799] On liability as a means of enforcement see R. Wolfrum/C. Langenfeld/P. Minnerop (eds.), Environmental Liability in International Law: Towards a Coherent Conception, 2005.

[800] The Annex on Liability Arising From Environmental Emergencies, adopted by the 28th Antarctic Treaty Consultative Meeting in Stockholm on June 14, 2005 as Annex VI to the Protocol on Environmental Protection to the Antarctic Treaty (4 October 1991), 30 ILM 1455 (1991), has yet to enter into force; for an assessment S. Vöneky, "The Liability Annex to the Protocol on Environmental Protection to the Antarctic Treaty" in: D. König/P.-T. Stoll/V. Röben/N. Matz-Lück (eds.), International Law Today: New Challenges and the Need for Reform?, 2008, 165-197.

ment possibilities can be improved and are often necessary requirements for effective regimes. Nonbinding norms are clearly limited in this respect, and therefore cannot be considered general alternatives to treaty law.

3. Lack of inspections and verifications

Nonbinding instruments cannot provide a legal basis for on-site verifications and inspections without the consent of a state. Intrusive control measures that touch upon the sovereignty of states such as on-site inspections and other forms of verification are however generally rare in environmental treaty law which is largely designed in non-confrontational terms.[802] Only very few multilateral environmental treaties provide for the possibility of an "inquiry" with the consent of the respective party,[803] but secondary rules provide for similar possibilities in some other cases.[804] Notable exceptions which allow for inspections without prior consent of the concerned party are the Antarctic Treaty

---

[801] The Liability Annex to the Protocol on Environmental Protection to the Antarctic Treaty is a great step forward, although other steps must follow, see S. Vöneky, "The Liability Annex to the Protocol on Environmental Protection to the Antarctic Treaty" in: D. König/P.-T. Stoll/V. Röben/N. Matz-Lück (eds.), International Law Today: New Challenges and the Need for Reform?, 2008, 165-197 (192-194).

[802] S. Oeter, "Inspection in international law – Monitoring compliance and the problem of implementation in international law", Netherlands yearbook of international law 28 (1997), 101-169 (164); R. Wolfrum, "Means of Ensuring Compliance with and Enforcement of International Environmental Law", Recueil des cours de l'Académie de Droit International de La Haye 272 (1998), 13-154 (44).

[803] Convention on International Trade in Endangered Species of Wild Fauna and Flora (3 March 1973), 993 UNTS 243, Article XIII sec. 2.

[804] See e.g. the possibility for information gathering by the Implementation Committee with the consent of the concerned party under the UNECE Convention on Long-Range Transboundary Air Pollution (13 November 1979), 1302 UNTS 217, 18 ILM 1442 (1979). The mandate to the Implementation Committee was given by the decision of the Executive Body of the Convention on the procedural rules for the Implementation Committee, EC/EB/AIR/75, Annex, para. 6 b). Another example is the possibility for inspections through the Compliance Committee of the Alpine Convention (7 November 1991), 31 ILM 767 (1992), as provided for through decision VII/4 para. 3.1.3. of the Alpine Conference.

system[805] and inspections of vessels under UNCLOS.[806] Similarly progressive are modern fisheries agreements, which strongly depend on effective control of quotas. These agreements increasingly foresee such measures, as most notably illustrated by Articles 21 to 23 of the Fish Stocks Agreement.[807] Because such controls are highly significant for the effectiveness of fisheries agreements,[808] nonbinding instruments are particularly deficient due to their inability to be the basis for such inspections – a deficiency which can only be remedied through linkages to treaty law.

4. Limited utility for long-term incentive structures

Nonbinding instruments are only to a limited extent useful as a sole basis for economic incentives and disincentives.[809] Of course, where financial assistance is linked to compliance with nonbinding standards, a powerful tool is created irrespective of whether the underlying standard

---

[805] Article VII of the Antarctic Treaty (1 December 1959), 402 UNTS 71, 19 ILM 860 (1980); Article XIV of the Protocol on Environmental Protection to the Antarctic Treaty (4 October 1991), 30 ILM 1455 (1991).

[806] See UNCLOS, Articles 73, 220 (2), (5) and 226.

[807] The Fish Stocks Agreement in Articles 21 to 23 provides for the rare possibility that at State Party may board and inspect ships flying the flag of another State Party with a view to enforce regional, sub-regional or global management and conservation measures even if the respective flag state is not a member of such an organization, compare e.g. R. Wolfrum, "Means of Ensuring Compliance with and Enforcement of International Environmental Law", Recueil des cours de l'Académie de Droit International de La Haye 272 (1998), 13-154 (46-47); C.J. Carr, "Recent developments in compliance and enforcement for international fisheries", Ecology law quarterly 24 (1997), 847-860 (852).

[808] D. König, "The Protection of Marine Living Resources – The 1995 Agreement on Straddling Fish Stocks and Highly Migratory Fish Stocks" in: T. Zhenghua/R. Wolfrum (eds.), Implementing International Environmental Law in Germany and China, 2001, 75-84 (82-83).

[809] An exception is the linkage of development assistance to nonbinding operational procedures as done by the World Bank and other multilateral development banks. Given their dependence on the financial support, the respective countries do not really have a choice whether or not to accept the operational procedures. These conditionalities could therefore rather be seen as enforcement mechanisms.

is binding or nonbinding.[810] This refers to conditionalities established through the Operational Procedures and loan policy of the World Bank, or to linkages of trade preference schemes to nonbinding instruments. But the foundation for the incentive hereby does not derive immediately from the commitments of the nonbinding instrument. In the case of the World Bank conditionalities, for instance, a binding loan agreement with the World Bank provides the recipient country with financial support for their projects if only they comply with the standards.

It will be argued in this section that nonbinding instruments alone, without such linkages, are only of limited value as a basis for economic incentives other than compliance assistance. The reason for this can be found in the basic concept of an incentive-based tool. The concept is to change the cost-benefit calculation of an actor by guaranteeing or promising some kind of benefit for its compliance with the underlying norm. In contrast to compliance assistance, the benefit from the incentive will only materialise in the form of a reward after an actor is in compliance.[811] In order to uphold this promise of balancing the environmental costs on the one hand and the benefits deriving from the incentive on the other over the necessary period of time, the underlying promise must be reliable and stable. In other words, the element of a promise and the temporal dimension raise the importance of the stability, certainty and credibility. Market-related incentives and disincentives therefore presuppose a certain long-term stability and certainty of the underlying commitments. As shown previously, stability, certainty and credibility are however advantages of legally binding instruments over nonbinding ones. Treaty instruments should therefore generally be better suited for these regulatory purposes.

Two examples from international environmental law can serve to illustrate this point. First, consider the incentive system of the Convention on Biological Diversity and the International Treaty on Plant Genetic Resources for Food and Agriculture. Ideally, both create a "network of

---

810 Similarly, albeit with the restriction to binding and nonbinding norms under multilateral environmental agreements L. Boisson de Chazournes, "Technical and Financial Assistance" in: D. Bodansky/J. Brunnée/E. Hey (eds.), Oxford Handbook of International Environmental Law, 2007, 947-973 (970).

811 N. Matz, "Financial and Other Incentives for Complying with MEA Obligations" in: U. Beyerlin/P.-T. Stoll/R. Wolfrum (eds.), Ensuring Compliance with Multilateral Environmental Agreements 2006, 301-318 (306-307).

trade-offs"[812] at the centre of which lays a complex access and benefit sharing system and mechanisms of financial and technology transfer.[813] States which have the resources are obliged to protect their resources and allow access to them, and in turn they receive not only the sovereign right over those resources but also financial support and the right to have access to biotechnology as well as to a share of the benefits.[814] These benefits work as an incentive for states rich of biodiversity to protect this diversity in conformity with the provisions.[815]

Clearly, the incentive and the underlying reciprocal relationships directly depend on credible and stable commitments of both developing and developed states. Only if biodiversity is conserved can access be granted and benefit sharing agreements be negotiated.[816] If host states could however not expect to gain from conservation in the future, the incentive would be lost. Moreover, as protection of biodiversity is prior to benefits from the use of the resources and generally relies on long-term prospects, the stability and credibility of commitments is of key importance to developing countries which need to be sure that their costly conservation efforts will be rewarded in the future by benefits from the results of research. Similarly, developed states only have an incentive to negotiate rules for benefit sharing and to provide financial assistance if they can be sure that the developing states will indeed use these means for protection and allow access. It may thus be unsurprising that both international instruments, the Convention on Biological

---

[812] R. Wolfrum/N. Matz, "The Interplay of the United Nations Convention on the Law of the Sea and the Convention on Biological Diversity", Max Planck United Nations Yearbook 4 (2000), 445-480 (460).

[813] Articles 15, 16 and 19 Convention on Biological Diversity; Articles 12 and 13 of the International Treaty on Plant Genetic Resources for Food and Agriculture.

[814] For the Convention on Biological Diversity R. Wolfrum, "The Convention on Biological Diversity: Using State Jurisdiction as a Means of Ensuring Compliance" in: R. Wolfrum (ed.), Enforcing Environmental Standards: Economic Mechanisms as Viable Means?, 1996, 373-393 (373 et seq.)

[815] R. Wolfrum, "The Convention on Biological Diversity: Using State Jurisdiction as a Means of Ensuring Compliance" in: R. Wolfrum (ed.), Enforcing Environmental Standards: Economic Mechanisms as Viable Means?, 1996, 373-393 (392).

[816] N. Matz, "Financial and Other Incentives for Complying with MEA Obligations" in: U. Beyerlin/P.-T. Stoll/R. Wolfrum (eds.), Ensuring Compliance with Multilateral Environmental Agreements 2006, 301-318 (310).

Diversity and the International Treaty on the Protection of Plant Genetic Resources, are legally binding instruments. In addition, in the negotiations on an international instrument for on Access to Genetic Resources and the Sharing of Benefits (ABS), provider countries pushed for a legally binding concretisation of the obligations under the Convention on Biological Diversity, i.e. a legally binding regime.[817] This stance is informed by their aim to increase the implementation of obligations in the Convention on Biological Diversity by user countries, and thus to achieve distributional justice in the international context.[818] The argument seems to be driven by the perception that a legally binding instrument is more suited to uphold the balance between preservation and benefits discussed above as the main incentive structure. The main argument by (mainly user) states in support of a nonbinding instrument, namely that the development of such an instrument is premature,[819] does not call this argument into question. It rather points to one of the advantages of a nonbinding instrument, namely flexibility. A nonbinding instrument may be necessary in terms of substance as long as uncertainty on underlying issues such as the economic value of genetic resources prevails.[820] Eventually, in line with the analysis suggested here, the 10th Conference of Parties to the Convention on Biological Diversity adopted the legally binding Nagoya Protocol on Access to Genetic Resources and the Fair and Equitable Sharing of Benefits Arising from their Utilization to the Convention on Biological Diversity[821] to establish legally binding rules on access and benefit sharing. The main rules of the Nagoya Protocol are mostly of procedural rather than substantive nature. It establishes prior informed consent proce-

---

817 Compare the report from the discussion at the COP 9 in Bonn in Earth Negotiation Bulletin, Vol. 9 No. 452, available at http://www.iisd.ca/download/pdf/enb09452e.pdf.

818 M. Dross/F. Wolff, "Do We Need a New Access and Benefit sharing Instrument?", Yearbook of International Environmental Law 15 (2004), 95-118 (112).

819 M. Dross/F. Wolff, "Do We Need a New Access and Benefit sharing Instrument?", Yearbook of International Environmental Law 15 (2004), 95-118 (114).

820 A. von Hahn, "Implementation and Further Development of the Biodiversity Convention", Zeitschrift für ausländisches öffentliches Recht und Völkerrecht 63 (2003), 295-312 (312).

821 The text of the Protocol is available at http://www.cbd.int/abs/doc/protocol/nagoya-protocol-en.pdf.

dures, and parties must establish the national legal framework and agreements that ensure legally certainty for access and benefit sharing and mutually agreed terms of access.[822] Thus, states hereby resort to legally binding rules to establish legal certainty regarding the overall structure and procedure of the mechanism, which consequently also secures the incentive structure mentioned above by legally linking access to benefit sharing. The nonbinding instruments such as the Bonn Guidelines contain substantive norms that may substantively inform the mutually agreed terms of access. Such substantive norms have to be flexible due to differences between local and regional situations and due to the need for adjusting such rules over time. The procedural rules of the Protocol and the concretisations of substantive rules in the nonbinding Bonn Guidelines are complementary.[823]

Although less controversial, the development of the International Treaty on Plant Genetic Resources is also telling in this respect.[824] FAO member states resorted to the extremely lengthy[825] process of developing a legally binding instrument[826] rather than simply introducing an ABS regime into the existing but nonbinding International Undertaking on Plant Genetic Resources for Food and Agriculture.

The second example which supports the main argument is the market-based tool of emission trading. The emissions trading system of the Kyoto Protocol, for example, establishes a market of emission reduction units whereby – simply speaking – emission rights are allocated and a market is established in which one may sell any unit one does not need. Given a certain price per unit, this creates an incentive to use less

---

[822] Nagoya Protocol on Access to Genetic Resources and the Fair and Equitable Sharing of Benefits Arising from their Utilization to the Convention on Biological Diversity, Art. 6.

[823] This complementarity is also stressed by U. Beyerlin/T. Marauhn, International Environmental Law, 2011, at 198.

[824] Generally on the development and implementation of the treaty E. Tsioumani, "International Treaty on Plant Genetic Resources for Food and Agriculture: Legal and Policy Questions from Adoption to Implementation", Yearbook of International Environmental Law 15 (2004), 119-144.

[825] The negotiations lasted 7 years, with an additional 3 years until the treaty entered into force.

[826] FAO Conference Report, Thirtieth Session, 12-23 November 1999, C/99/REP, para. 64, available at http://www.fao.org/docrep/X4041e/X4041e00.htm.

than the allocated amount and sell the surplus.[827] Now, emission trading markets can only properly function if states actually strive to comply with their commitments and regulate their corporations accordingly. Again, a participating state needs to be sure before regulating its industry that other states participating in emissions trading will not for example allocate more than the agreed amount, or suddenly reconsider their commitment, for example after the next elections. Only if meaningful reductions are undertaken by participating states could the market price attain a level at which an incentive for further reductions is created. And finally, the market also depends on the promise that states will buy emission reduction units on the global market if they are not in compliance at the end of a commitment period which may – as in the case of the Kyoto Protocol – lay ten years or more in the future. The general belief that states will honour their treaty commitments as well as the need for ratification by a certain number of participants before a treaty enters into force provides much more legal security than a non-binding instrument. Once a treaty is ratified and implemented via legislative action, the likeliness of a sudden drop-out reduces considerably. Nonbinding instruments on the other hand cannot provide for the stability and certainty needed for these mechanisms to function properly, because a state then reserves the right not to implement or not to buy at the end of the commitment period if in non-compliance.

5. Limitations due to non-individualized reporting and compliance review

Transparency and justificatory discourse are two of the main ingredients of a functioning compliance mechanism. There is no apparent reason why this should be different for nonbinding instruments. On the contrary, given their limited authority and credibility if compared to most binding rules, compliance management is the key to compliance for nonbinding initiatives.

The effectiveness of follow-up mechanisms for nonbinding instruments is this respect. In contrast to the reporting systems under most Multi-

---

[827] Emission trading as an economic incentive is also discussed by D.M. Driesen, "Is Emission Trading an Economic Incentive Program?", Washington and Lee Law Review 55 (1998), 289-350; N. Matz, "Financial and Other Incentives for Complying with MEA Obligations" in: U. Beyerlin/P.-T. Stoll/R. Wolfrum (eds.), Ensuring Compliance with Multilateral Environmental Agreements 2006, 301-318.

lateral Environmental Agreements, reporting systems established for nonbinding instruments are rarely designed as an evaluation of individual state performance.[828] One objective of the non-compliance procedures under treaties is to identify how individual states are performing, and how to address their non-compliance. Even comparatively soft compliance mechanisms, as for instance the one foreseen under the PIC Convention, comprise individualised non-compliance discussions and empower the Secretary-General to publish cases of non-compliance.[829] In stark contrast to this, the reporting mechanisms established for the support of nonbinding instruments often only provide for the collection of information about the collective performance of actors; the individual reports of states are not discussed in any institutional body.[830] Such a procedure therefore does not give rise to individualised justificatory discourse which is central for compliance under the managerial approach.[831] Where a state must not publicly present and defend a report, the state runs a lower risk of hurting its reputation and good standing within the community and among members of the particular institution.

---

[828] For example, under the compliance mechanism of the Kyoto Protocol states have to prepare national greenhouse inventories and regularly report on their performance. The reports are then reviewed by experts and – if problems with compliance are detected – discussed in the Compliance Committee, which may under certain conditions through its enforcement branch issue statements of non-compliance or decide on penalties, see COP/CMP Decision 27/CMP.1 on 'Procedures and mechanisms relating to compliance under the Kyoto Protocol'.

[829] See Draft text of the procedures and mechanisms on compliance with the Rotterdam Convention, U.N. Doc. UNEP/FAO/RC/COP.3/26 of 10 November 2006, available at www.pic.int, para. 19 (e).

[830] Compare e.g. FAO, 'Analysis of Government Responses to the Second Questionnaire on the State of Implementation of the International Code of conduct on the Distribution and Use of Pesticides' (2003), available at www.fao.org; for more details on the outcome of the reporting exercise, see further below; the difference is also highlighted by K. Raustiala, "Form and Substance in International Agreements", American Journal of International Law 99 (2005), 581-614 (605).

[831] On justificatory discourse A. Chayes/A. Handler Chayes, The New Sovereignty: Compliance with International Regulatory Agreements, 1995, 118-123; similarly as here K. Raustiala, "Form and Substance in International Agreements", American Journal of International Law 99 (2005), 581-614 (607).

6. Limited compliance assistance

Compliance assistance and financial support can be linked to nonbinding instruments. This is subject, however, to some limitations. Compliance assistance is dependent on adequate funding and resources. Arguably, as actors would be less likely to ignore binding funding obligations, financial mechanisms and compliance assistance are better secured through binding obligations, at least in the long term.[832] A binding obligation to contribute finances secures the flow of such finances, since often domestic budget decisions are more easily made or directly dependent on binding international obligations, whereas nonbinding ones may or may not be followed up. A shift to treaty making does not however automatically lead to a binding financial mechanism, as shown by the ongoing negotiations under the PIC Convention. The text of the Convention does not establish a financial mechanism, and only includes a softly worded obligation regarding technical assistance.[833] The Conference of the Parties of the PIC Convention has introduced a voluntary special trust fund, but Parties have not agreed upon a binding financial mechanism as of 2010.

C. Private level: direct impact on private actors

International legal doctrine does so far not accept the legal personality of private actors. Private actors can generally not be directly obligated by international treaty law. To nevertheless address the growing importance of private actors, international institutions increasingly seek to regulate private actors on the international level in other ways. Examples are the listing of terrorist suspects by the Security Council, the listing of ships undertaking illegal, unreported or unregulated fishing ac-

---

[832] Victor therefore concludes that one of the major advantages of the treaty option is a binding financial mechanism, see D.G. Victor, "'Learning by Doing' in the Nonbinding International Regime to Manage Trade in Hazardous Chemicals and Pesticides" in: D.G. Victor/K. Raustiala/E.B. Skolnikoff (eds.), The Implementation and Effectiveness of International Environmental Commitments: Theory and Practice, 1998, 221-281 (254).

[833] Article 16 of the PIC Convention obligates Parties "to cooperate in promoting technical assistance", and recommends that states with more advanced regulatory capacities "should provide technical assistance" to other Parties.

tivities by regional fisheries organisations,[834] and the determination of refugee status by the UN High Commissioner for Refugees. In other cases – some of which fall within the ambit of this study – they establish generally applicable norms aimed at regulating private actors, be they multinational corporations or fishermen. Apart from their nonbinding character, this feature of some of the instruments analysed in this study is perhaps the most salient difference distinguishing them from treaty law norms. When private actors are addressed directly by international institutions the role of states as the implementing agents is diminished, and the distinction between the international and the national sphere becomes blurred.

Now, one could dismiss this development simply by pointing to the voluntary nature of the norms which may amount to mere public relations as long as states are not directly involved. However, private actors such as multinational corporations typically cannot afford to ignore these instruments: their existence alone may lead to behavioural changes by some private actors due to reputational concerns. In addition, a number of mechanisms exist which have the potential to enhance the effectiveness of these instruments in regulating private actors.

I. Means of enhancing compliance by private actors

As mentioned above, research on compliance has identified various motivations for why states comply with norms.[835] These are *inter alia* coercion, compliance out of self-interest due to positive cost-benefit calculations, the sense of obligation arising from the legitimacy of norms and persuasive process, and finally socialisation. Some of these considerations can be generalised for any actor, including private business. Unlike states, however, private business corporations can in the long

---

[834] Listings of ships presumably engaged in IUU fishing is e.g. undertaken by the North Atlantic Fisheries Organisation, as discussed as a form of exercise of authority by R. Wolfrum, "Legitimacy of International Law and the Exercise of Administrative Functions: The Example of the International Seabed Authority, the International Maritime Organization (IMO) and International Fisheries Organizations" in: A. von Bogdandy/R. Wolfrum/J. von Bernstorff/P. Dann/M. Goldmann (eds.), The Exercise of Public Authority by International Institutions: Advancing International Institutional Law 2010, 917-940 (938-939).

[835] See already further above in this Part 2, at B.I.

term only survive if they are profitable. Consequently, economic cost-benefit calculations can be expected to be of overriding importance for these particular actors.

In their cost-benefit calculations, enforcement and sanctions of course play an important part. An increasingly important additional factor is reputation or the "image" of a company. Reputation in this sense refers to the perception of external audiences about the performance of a company.[836] In an increasingly complex and competitive business environment, reputation becomes a highly valuable asset and key to a successful business strategy at least for consumer-oriented companies. Adherence to environmental standards can signal environmental responsibility of a company to its consumers, business partners and investors. Given a change in consumption patterns in industrialised countries, adherence to environmental standards may in fact even become an indispensable tool to improve sales numbers and access new markets for those companies that directly produce for private consumption. Environmental leadership may thus appear as a beneficial marketing strategy. Consumer-dependent and service-oriented companies also need to avoid negative publicity in order to avoid reputational costs. Naming and shaming campaigns of environmental NGOs build on these fears. Due to the threat of negative publicity, NGOs can sometimes even convince companies and their competitors to co-operate with NGOs and implement nonbinding environmental norms. Multinational corporations or entire industry sectors also often extend this pressure to suppliers, for example through supplier agreements or the choice to only do business with certified suppliers.

Reputational concerns are however not the only reason why companies may comply with nonbinding norms as a result of cost-benefit calculations. Following voluntary environmental standards can improve efficiency and lower costs, for example when resources are used more efficiently or consumption is reduced. The introduction of and compliance with environmental standards may also reduce transaction costs through internal or industry-wide harmonisation. Recent studies also indicate a positive correlation between financial performance on the one hand and social and environmental performance on the other.[837] An-

[836] E. Brown Weiss, "Conclusions: Understanding Compliance with Soft Law" in: D. Shelton (ed.), Commitment and Compliance: The Role of Nonbinding Norms in the International Legal System, 2000, 535-556 (541).

[837] M. Orlitzky/F.L. Schmidt/S.L. Rynes, "Corporate Social and Financial Performance: A Meta-analysis", Organization Studies 24 (2003), 403-441.

other factor in their calculation of costs and benefits is often that compliance with non-binding standards allows them to avoid more stringent and binding governmental regulation.[838] And finally, businesses pioneering in environmentally friendly technology can secure a share of the expanding market in this field.

Companies may also change their behaviour because of a sense of obligation – which can also be termed persuasion – as well as socialisation. These factors are likely to be strengthened if companies participate in the making of the standard, and if the standard-setting is linked to institutionalised discourse between companies and environmental stakeholders as is attempted by modern multi-stakeholder initiatives. Ideally, companies would eventually believe a standard to be legitimate and comply out of that belief,[839] or become compliant due to socialisation deriving from peer pressure or similar social phenomena within industry associations. Given that companies have to be profitable in order to survive, these processes should not be expected to take place entirely independently from economic considerations.

1. *International complaint mechanisms*

International institutions not only increasingly adopt voluntary norms addressed to private actors, but also establish formalised procedures designed to directly implement them. Particularly remarkable in this respect are non-contentious transnational complaint procedures where private actors are able to bring complaints against business corporations. These mechanisms are provided for in the governance system of the United Nations Global Compact,[840] the "dispute procedure"[841] of

---

[838] D. Shelton, "Compliance with international human rights soft law", Studies of Transnational Legal Policy 29 (1997), 119-143 (129); H. Keller, "Codes of Conduct and their Implementation: the Question of Legitimacy" in: R. Wolfrum/V. Röben (eds.), Legitimacy in International Law, 2008, 219-298 (273).

[839] I. Hurd, "Legitimacy and Authority in International Politics", International Organization 53 (1999), 379-408 (387); H. Keller, "Codes of Conduct and their Implementation: the Question of Legitimacy" in: R. Wolfrum/V. Röben (eds.), Legitimacy in International Law, 2008, 219-298 (275).

[840] Complaints containing "credible allegations of systematic or egregious abuses of the Global Compact's overall aims and principles can be brought to the attention of the Global Compact Office which then can take various steps

the International Labour Organization in the specific context of the Tripartite Declaration[842] and the "specific instances" procedure of the OECD Guidelines. In particular the OECD procedure is increasingly used as one possibility by NGOs which are often excluded from bringing international claims.[843] For example, Greenpeace Germany has filed a complaint with the National Contact Point of Germany against the Vattenfall AB in which Vattenfall Europe is accused of violating the environmental guidelines of the OECD Guidelines by attempting to build a coal-fired plant in Moorburg close to Hamburg, and for its attempt to allegedly undermine German environmental law by filing a complaint with the International Centre for the Settlement of Investment Disputes.[844]

The OECD procedure will in the following be analysed more closely as an exemplary case for all these similarly structured procedures. At first sight, the general approach of the OECD procedures appears to be one of mediation. The National Contact Points (NCPs) – designated as recipients of complaints – indeed mainly act as mediators.[845] The language of the procedural rules diligently avoids any adversarial terminology such as 'dispute' or 'decision'.[846] Even if considered only to constitute

---

to resolve the issue. Compare for details the Global Compact website at http://www.unglobalcompact.org/AboutTheGC/integrity.html.

841 For details see http://www.ilo.org/public/english/employment/multi/tripartite/interpretation.htm; for a detailed discussion see J.M. Diller, "Social conduct in transnational enterprise operations: the role of the International Labour Organization" in: R. Blanpain (ed.), Multinational enterprises and the social challenges of the XXIst Century, 2000, 17-28.

842 International Labour Organization 'Tripartite Declaration of Principles concerning Multinational Enterprises and Social Policy' (adopted 16 November 1977) (1978) 17 ILM 422, available at http://www.ilo.org/public/english/employment/multi/tripartite/declaration.htm.

843 According to OECD Watch, 85 complaints have been brought since 2001, see www.oecdwatch.org.

844 On this complaint see http://oecdwatch.org.

845 For details see the analysis of the OECD Guidelines in Part 1, at B.III., further above.

846 The National Contact Points and the Investment Committee are not formally concerned with disputes, but with the resolution of "issues" or "specific instances". They do not issue decisions, but issue "statements", and possibly "recommendations", and the Investment Committees renders "clarifications of the Guidelines".

mediation, such a procedure can, however, be of importance.[847] Where it takes place, mediation not only serves to address compliance problems deriving from the vagueness of norms or lack of information. It also carries the potential to initiate a dialogue between industry and concerned public or private actors on the matter. Ideally, the mechanism contributes to resolving differences in viewpoint.[848] The mediation process thus supplements adversarial judicial proceedings or provides for a means of rationalised dialogue in areas where judicial dispute settlement is not available.

If the parties involved in a mediation procedure before a National Contact Point (NCP) cannot reach agreement, the NCP must issue a public statement and may if appropriate make recommendations on the implementation of the OECD Guidelines.[849] When mediation fails, the company must therefore still fear reputational costs resulting from a possibly negative public statement of the NCP. As this indicates, the term mediation does not capture the nature of such a procedure. The non-adversarial language cannot conceal that there will always be some kind of opposition of interest or of opinion between the "complainant" and the enterprise. It can even be argued that the notion of non-adversarial mediation implies that there is some kind of dispute.[850] Furthermore, the NCP may issue public statements and recommendations on the issues brought to its attention in the case of continuous disagreement between a complainant and an enterprise. If taken seriously by the respective NCP, this possibility implies more than a mere institutionalisation of dialogue. Given that a dispute is at the heart of any "issue", a public statement and/or a recommendation on the appropri-

---

847 K. Hailbronner, "Völkerrechtliche und staatsrechtliche Überlegungen zu Verhaltenskodizes für transnationale Unternehmen" in: I. v. Münch (ed.), Staatsrecht – Völkerrecht – Europarecht: Festschrift für Hans-Jürgen Schlochauer, 1981, 329-362 (333).

848 A. Böhmer, "The Revised 2000 OECD Guidelines for Multinational Enterprises – Challenges and Prospects after 4 Years of Implementation", Policy Papers on Transnational Economic Law 3 (2004), (6).

849 Procedural Guidance, para. I.C.4 (the Procedural Guidance is part of the booklet issued by the OECD on the OECD Guidelines, available at http://www.oecd.org/dataoecd/43/29/48004323.pdf.

850 P.M. Protopsaltis, "La mise en oeuvre des Principes directeurs de l'OCDE à l'intention des entreprises multinationales: Réflexions sur le nouveau mandat des Points de contact nationaux", International Law FORUM du droit international 7 (2005), 251-260 (255).

ate implementation of the Guidelines most often will imply a decision on the matter, i.e. whether one could observe a violation or not. Consequently, even if not always used in this manner, the procedure may serve a dispute resolution function.[851] The competence of the NCPs in this respect is indirectly acknowledged by the oversight exercised by the Investment Committee of the OECD.[852] The international procedural rules provide that the "[C]ommittee shall not reach conclusions on the conduct of individual enterprises."[853] A similar provision does not exist for the activities of the National Contact Points. But even when the Investment Committee clarifies the content of the Guidelines "in the event of doubt about the interpretation of the Guidelines in particular circumstances"[854], it is difficult to see how it could do so without in fact taking a decision on the issue. Any meaningful statement would almost automatically imply a position on whether the conduct of the enterprise in a particular case was in conformity with the Guidelines or not.[855] Nevertheless, this kind of procedure does not amount to one of a quasi-judicial nature as sometimes suggested.[856] To consider this type of procedure quasi-judicial would disregard the political, non-binding and

---

851 J.A. Zerk, Multinationals and Corporate Social Responsibility: Limitations and Opportunities in International Law, 2008, 252.

852 P.M. Protopsaltis, "La mise en oeuvre des Principes directeurs de l'OCDE à l'intention des entreprises multinationales: Réflexions sur le nouveau mandat des Points de contact nationaux", International Law FORUM du droit international 7 (2005), 251-260 (256 – 257).

853 Implementation Procedures, Part II, para. 4.

854 Implementation Procedures, Part II, para. 1.

855 Similarly P.M. Protopsaltis, "La mise en oeuvre des Principes directeurs de l'OCDE à l'intention des entreprises multinationales: Réflexions sur le nouveau mandat des Points de contact nationaux", International Law FORUM du droit international 7 (2005), 251-260 (259); I. Seidl-Hohenveldern, "International Economic 'Soft Law'", Recueil des Cours de l'Académie de Droit International 163 (1979), 169-246 (208).

856 A quasi-judicial character of the overall procedure is assumed by P.M. Protopsaltis, "La mise en oeuvre des Principes directeurs de l'OCDE à l'intention des entreprises multinationales: Réflexions sur le nouveau mandat des Points de contact nationaux", International Law FORUM du droit international 7 (2005), 251-260 (260); I. Seidl-Hohenveldern, "International Economic 'Soft Law'", Recueil des Cours de l'Académie de Droit International 163 (1979), 169-246 (208).

administrative nature of the process.[857] Contrary to judicial proceedings, the procedures are for example not conducted by independent persons but by political actors. The NCPs are either composed of civil servants of a particular country or of representatives from government and private institutions but never completely independent from governmental agencies. Often, the NCPs are located with the Ministry for Economics which has an interest in protecting the multinational corporations of that particular country. NCPs can dismiss complaints, and it is within their discretion not to issue a final statement. Finally, even the conclusion that a company did not sufficiently respect the OECD Guidelines in a public statement of the NCP cannot be compared to a judicial opinion which clearly determines the legality or illegality of the activities. The factual relevance of the procedure does not render the procedure quasi-judicial.[858] The outcome of the specific instances procedure remains formally nonbinding for the enterprise. The procedure should therefore more adequately be described as a "non-judicial review procedure"[859] or – by emphasising its potential relevance for compliance – as an (non-judicial) complaint mechanism.

Even though it cannot be an alternative to judicial proceedings and access to courts, the procedure certainly has merits with respect to mediation by providing a forum for discussion between business and interest groups.[860] Such discussions and the decisions of the NCPs additionally trigger an ongoing and evolving debate on what constitutes responsible business behaviour.[861] And it undoubtedly raises the accountability of

857 R. Blanpain, The OECD Guidelines for Multinational Enterprises and Labour Relations 1976-1979, 1979, 275.

858 The factual relevance is however considered as an indicator for the "quasi-judicial" character by P.M. Protopsaltis, "La mise en oeuvre des Principes directeurs de l'OCDE à l'intention des entreprises multinationales: Réflexions sur le nouveau mandat des Points de contact nationaux", International Law FORUM du droit international 7 (2005), 251-260 (257 and 260).

859 J.G. Ruggie, "Current Development: Business and Human Rights: The Evolving International Agenda", American Journal of International Law 101 (2007), 819-839 (834).

860 S.F. Vendzules, "The Struggle for Legitimacy in Environmental Standards Systems: The OECD Guidelines for Multinational Enterprises", 21 Colorado Journal of International Environmental Law & Policy 21 (2010), 451-489 (478).

861 S.F. Vendzules, "The Struggle for Legitimacy in Environmental Standards Systems: The OECD Guidelines for Multinational Enterprises", 21 Colorado Journal of International Environmental Law & Policy 21 (2010), 451-489 (478).

enterprises as it provides NGOs with a tool to raise issues of concern in an official complaint. The complaint procedure factually forces companies concerned about their reputation to defend themselves in a justificatory discourse even though they are not legally bound to do so. No possibility exists for a company to avoid the procedure except by simply refusing to participate. However there would almost certainly be a severe public shaming campaign of NGOs against companies that refuse to participate at all. Non-adherence to environmental standards or, for that matter, non-adherence to this procedure, can expose the environmental irresponsibility of a company to conscious consumers, ethically oriented investors or business partners which are concerned about their own reputation.[862] Enterprises can therefore hardly avoid the complaint procedures if they do not want to risk reputational costs. And indeed, no business enterprise has in practice flatly refused to cooperate once an NCP has accepted a particular complaint.[863] In most of the complaint procedures brought since 2001, business corporations have at least agreed to and participated in mediation procedures even though not legally obliged to do so.[864] In the practice of the OECD specific instances procedure, companies usually choose to defend themselves rather than to make the perfectly valid legal argument that they are not obliged by either the voluntary international norms or the procedure.

However, as remarkable and sophisticated this novel procedure may be, there are clear limits to what can it can achieve. In practice, a great number of cases have been pending for several years.[865] The effectiveness of the procedure hinges upon the capabilities of NGOs to monitor

---

862 Positive and negative reputational incentives as drivers for compliance are for instance stressed by J. Morrison/N. Roht-Arriaza, "Private and Quasi-Private Standard Setting" in: D. Bodansky/J. Brunnée/E. Hey (eds.), Oxford Handbook of International Environmental Law, 2007, 498-527 (516).

863 P. Sanders, "Implementing International Codes of Conduct for Multinational Enterprises", American Journal of Comparative Law 30 (1982), 241-254 (244).

864 Compare the case database with detailed information on the procedures at http://oecdwatch.org/cases.

865 No progress has been made in over four years in a case concerning the complaint of an NGO claiming that Alcoa Alumínios S.A. and Companhia Brasileira de Alumínio had used fraudulent environmental impact assessment to construct the Barra Grande hydroelectric plant in Brazil, see National Contact Point Brazil, at http://www.fazenda.gov.br/multinacionaispcn/.

companies and bring such claims. Furthermore, nothing ensures that National Contact Points in all participating countries view the issues with the same degree of seriousness. Their composition can vary widely, as can their independence from political or economic processes. Further, without any investigating power, it is difficult for NCPs to assess the merits of complaints, and the OECD Guidelines are held in such general terms as to provide a number of loopholes. There are a few cases where the respective NCP has indeed found a company to be in violation of the OECD Guidelines. One exceptional case concerned human rights violations. The National Contact Point of the United Kingdom in 2008 issued a statement in which it agreed with the allegation of the NGO *Global Witness* that *Afrimex Ltd.*, a British corporation, acted in violation of the OECD Guidelines by bribing a rebel group in the Democratic Republic of the Congo and purchasing minerals from mines that employed child and forced labour.[866] Since the state cannot enforce such a decision, it is not possible to ensure that it is actually complied with. Finally, although the decision of the National Contact Point can be reviewed by the Investment Committee, this does not amount to a real appeal procedure since the participants involved in the case cannot bring such a claim. And from the perspective of both companies and complainants, the procedures do not guarantee procedural rights despite potentially severe consequences for companies.

2. Membership systems and institutionalised discourse between private actors

The Global Compact Initiative of the United Nations illustrates how international nonbinding norms can be linked to a membership system.[867] Only those companies which agree to comply with the set of

---

[866] See for a discussion the ASIL insight at http://www.asil.org/insights 090123.cfm; S.F. Vendzules, "The Struggle for Legitimacy in Environmental Standards Systems: The OECD Guidelines for Multinational Enterprises", 21 Colorado Journal of International Environmental Law & Policy 21 (2010), 451-489 (476-477).

[867] For details, compare the website www.globalcompact.org. For an in-depth analysis of the governance structures of the Global Compact, see K. Nowrot, "The New Governance Structure of the Global Compact – Transforming a "Learning Network" into a Federalized and Parliamentarized Transnational Regulatory Regime", Beiträge zum Transnationalen Wirtschaftsrecht 47 (2005), 1-38.

norms – in this case the 10 environmental and social principles of the Global Compact – can become "participants", and only those who are members can use their membership and adherence as a marketing tool. The United Nations allow companies that are members of the Global Compact to use the UN logo. Membership in turn is conditioned upon compliance with the Principles of the Global Compact. Such a mechanism also allows the UN Global Compact Office[868] to establish so-called "integrity measures". These are in fact sanctions for non-compliance with the reporting obligations or with the standards in general. As a consequence of non-compliance, participants can be publicly listed if not in compliance. In the case of the Global Compact, such listings appear on the website. This can entail reputational costs for the company and may give rise to further naming and shaming by NGOs. Moreover, if companies do not improve their record, membership can be withdrawn with the consequence that companies lose the benefit of using the UN logo. This sanction is also available when participants refuse to participate in the resolution of allegations of abuse brought to the attention of the Global Compact Office.[869]

As well as this noteworthy mechanism, the United Nations Global Compact is also remarkable for its level of institutionalisation. Although entirely nonbinding, it has developed over the years from a mere learning network into a full-scale institution.[870] It now comprises six institutional entities, including the Global Compact Leaders Summit, Local Networks, the Annual Local Networks Forum, the Global Compact Board, the Global Compact Office, and the United Nations Inter-Agency Team. In contrast to traditional international organisations, private actors such as business associations, non-governmental organisations and trade unions are widely represented in the governance structure. Of the 20 members of the Global Compact Board – the main advisory body which can make recommendations to the Global Compact Office – 17 represent private actors: 11 for business, 2 for labour, 4 for NGOs. The other posts are reserved for the UN Secretary

---

868 The Global Compact Office is part of the Office of the UN Secretary-General.

869 See on the governance system of the United Nations Global Compact the Global Compact website at www.unglobalcompact.org.

870 See on this development in particular K. Nowrot, "The New Governance Structure of the Global Compact – Transforming a "Learning Network" into a Federalized and Parliamentarized Transnational Regulatory Regime", Beiträge zum Transnationalen Wirtschaftsrecht 47 (2005), 1-38.

General and heads of the Global Compact Office and the Global Compact non-profit entity. The Local Networks and Annual Networks Fora establish an institutionalised dialogue between participants in the Global Compact, NGOs as well as business and labour associations. If one assumes with compliance research that institutionalised dialogue and interaction between actors can trigger learning processes through socialisation and persuasion as well as increase the acceptance of the norms,[871] the institutionalisation of the Global Compact can directly affect the performance of private companies.

However, there are limits. Leaving aside for a moment the question of the need for implementation by states, the effectiveness of such a system always remains directly dependent on at least two factors which may vary considerably from one case to the next. First of all, a functioning monitoring system is required, and so far that depends on the work of NGOs. In the case of the Global Compact, monitoring of compliance is partly ensured through NGOs which can bring a complaint to the Global Compact Board similar to the one under the OECD Guidelines for Multinational Enterprises. Allegations of abuse by an enterprise can be brought by NGOs or other interested persons to the attention of the Global Compact Office which then attempts to resolve the issue.[872] Such a mechanism largely depends on the resources, the will and the access to information of non-governmental organisations which could bring such complaints. And secondly, as with any largely self-regulatory mechanism, it heavily depends on the participating company's sensitivity to reputational costs as well as public awareness of both the issues and the existence of the mechanism.

3. Listings

Nonbinding instruments directed at private actors provide a standard against which the behaviour of companies can be assessed. An illustrative example of how this can directly affect private corporations is the listing of corporations not in compliance with the OECD Guidelines

---

[871] Compare for the respective literature on state compliance J. Brunnée/S.J. Toope, "Persuasion and enforcement: explaining compliance with international law", The Finnish yearbook of international law 13 (2002), 273-295; R. Goodman/D. Jinks, "How to influence states: socialization and international human rights law", Duke law journal 54 (2004), 621-703.

[872] Compare the Global Compact website at http://www.unglobalcompact.org/AboutTheGC/integrity.html.

by the U.N. Security Council's Report on illegal exploitation of nature resources in the Democratic Republic of Congo (DRC).[873] The expert group installed by the UN Security Council to assess the situation in the DRC used the OECD Guidelines as a benchmark for assessing the conduct of corporations in this country.[874] Enterprises are thus not spared the reputational costs of being publicly listed even though the OECD Guidelines are voluntary. Even those corporations whose home state had not signed the Guidelines were listed, albeit with an indication that the home state had not signed. As could be expected, the report raised much media attention and forced companies to acknowledge their responsibility.[875] The reputational costs suffered through this incident again show that companies cannot ignore voluntary international instruments. It can be assumed that incidents such as this one generally increase compliance with nonbinding instruments.

4. Naming and shaming by NGOs

Consumer-dependent and service-oriented companies need to avoid negative publicity in order not to incur reputational costs. Naming and shaming campaigns of environmental NGOs build on these fears. Due to the threat of negative publicity, NGOs can successfully convince companies and their competitors to cooperate with them. As a result, NGOs may develop codes of conduct or other standards equipped with certification systems, and/or provide independent monitoring for companies. The latter may be the preferred choice since they can increase their reputation and avoid negative publicity.[876] Moreover, the fear of a spill over of reputational damage from the actions of an individual

---

[873] This case has also been discussed in more detail above, compare Part 2, A.II.2. b) (2).

[874] U.N. Security Council , Final report of the Panel of Experts on the Illegal Exploitation of Natural Resources and Other Forms of Wealth of the Democratic Republic of the Congo, U.N. Doc. S/2002/1146 of 16 October 2002, Annex III.

[875] E. Morgera, "An Environmental Outlook on the OECD Guidelines for Multinational Enterprises: Comparative Advantages and Legitimacy in relation to Other International Initiatives", Georgetown International Environmental Law Review 18 (2006), 751-777 (772).

[876] P.J. Spiro, "Non-governmental organizations and civil society" in: D. Bodansky/J. Brunnée/E. Hey (eds.), Oxford Handbook of International Environmental Law, 2007, 770-790 (784-789).

company to the entire industry also drives the development and enforcement of voluntary norms by industry and business associations.[877] Multinational corporations or entire industry sectors regularly pass this pressure on to suppliers, for example through supplier agreements or the choice only to do business with certified suppliers.

Internationally legitimated and therefore highly authoritative standards of behaviour for private actors are ideal tools on which NGOs can draw either for naming and shaming campaigns of NGOs against corporations or for increasing the legitimacy of their own standards. For instance, the existence of the FAO Pesticide code enabled NGOs to shame corporations and thereby pressure them to act responsibly.[878] An example of this was the campaign of the NGO "Pesticide Action Network Germany" (PAN Germany) directed at the Swiss chemical corporation Syngenta. Syngenta is a leading Swiss company in the production of agrochemicals and is also a member of the industry association CropLife International which actively promotes adherence to the FAO Pesticide Code. PAN Germany accused Syngenta of violating Article 11.2 of the FAO Pesticide Code by advertising in an inappropriate manner for their herbicide Paraquat in Thailand.[879] In their statements and argumentation, the NGO used statements of the industry association CropLife International in which the support of industry for the FAO Pesticide Code is praised as a sign for the corporate responsibility in sustainable development.[880] When the simple announcement of the accused contravention did not show any success, the NGO established an "Online-Court" on its website and invited the public to "judge" the behaviour of the company through voting "guilty" or "non-guilty".[881]

---

[877] H. Keller, "Codes of Conduct and their Implementation: the Question of Legitimacy" in: R. Wolfrum/V. Röben (eds.), Legitimacy in International Law, 2008, 219-298 (275).

[878] R.L. Paarlberg, "Managing Pesticide Use in Developing Countries" in: P.M. Haas/R.O. Keohane/M.A. Levy (eds.), Institutions for the Earth: Sources of Effective International Environmental Protection, 1993, 309-350 (349).

[879] Pestizid Action Network, Press release, "Pestizid-Werbung von Syngenta verstößt gegen internationalen Verhaltenskodex", 28 July 2004, available at www.pan-germany.org.

[880] *Ibid.*

[881] Compare http://www.paraquat.ch/home.cfm. For a critical report on the online court R. Suter, "Online-Justiz gegen Syngenta-Pflanzengift" of 26 September 2006, available at http://www.onlinereports.ch/2006/SyngentaParaquat.htm.

Being internationally renowned and having been openly embraced by industry, the FAO Pesticide Code has provided NGOs a fairly good basis for generating public concern and for threatening industry's valuable reputation. Industrial actors are forced to defend themselves publicly, and therefore a mechanism of accountability is established through the watchdog function of the NGO. However, as effective as naming and shaming may be, one should not forget that its effectiveness depends on the ability and resources of NGOs to monitor the behaviour of companies as well as on the sensitivity of the public to the particular issue. In contrast with institutionalised complaint mechanisms such as those of the OECD Guidelines, the public shaming campaign does not allow the accused company to defend its activities.

5. *Marketing and labelling*

The importance private companies attach to a good reputation in competitive consumer-oriented markets can be an important driving force for compliance with nonbinding international norms. Environmental leadership increasingly appears as a beneficial marketing strategy. Adherence to international environmental standards can signal the environmental responsibility of a company to consumers and investors concerned about environmental issues. Adherence to environmental standards may thus become a tool to improve sales numbers and secure access to new markets for those companies that directly produce for private consumption. Nonbinding instruments are particularly useful as marketing tools, because compliance indicates ethical business behaviour beyond legal requirements that apply to all competitors.[882] International nonbinding instruments in addition to private standards and ideal basis for and can be reinforced through eco-labelling by NGOs.[883]

Just to give some examples, the instruments considered in the case studies in this study are frequently used as an authoritative basis for further privately set standards. The Code of Conduct for Responsible Fisheries of the FAO and related instruments form the basis for private standard

---

[882] B. Simma/A. Heinemann, "Codes of Conduct" in: W. Korff et al. (eds.), Handbuch der Wirtschaftsethik, Band 2: Ethik wirtschaftlicher Ordnungen, 1999, 403-418 (416).

[883] Such as eco-labelling for sustainable fisheries, compare for a thorough assessment T. Cooper, "Picture this: promoting sustainable fisheries through eco-labeling and product certification", Ocean and coastal law journal 10 (2004), 1-49.

setting by fishermen or industry associations.[884] An outstanding example is the eco-labelling initiative of the Marine Stewardship Council – an environmental NGO working towards sustainable fishing practices which is modelled after the Forest Stewardship Council.[885] In a similar manner as the Forest Stewardship Council, the Marine Stewardship Council certifies organisations that grant eco-labels to particular fish products on the basis of its "Principles and Criteria for sustainable fishing". These principles and criteria are explicitly based on the FAO Code of Conduct for Responsible Fisheries.[886] By linking their instruments and policies to public standards adopted by over 180 states in the highest political body of the FAO, the Marine Stewardship Council strives to gain legitimacy for its own instruments. As a result, the norms of the CCRF have an indirect impact on the fisheries methods of a considerable part of the industry. About six percent of the world's total wild capture fisheries are now certified according to these standards, including 42 percent of the global wild salmon catch.

While such market mechanisms may be at times more effective than control by courts and administrative agencies,[887] their effectiveness depends – as with other reputation-based mechanisms – on the authority and standing of the instruments in the public conscience and on the efforts of NGOs.

II. "Implementation" efforts by private actors

The voluntary nature of international instruments does not necessarily mean that business actors remain passive. Since they typically try to avoid binding regulation until such regulation becomes inevitable,[888] or

---

[884] E.g. the Canadian Code of Conduct for Responsible Fisheries; the Code of Conduct for a Responsible Seafood Industry of the Australian Seafood Industry Council.

[885] MSC website, http://eng.msc.org/.

[886] Compare the website of the MSC at www.msc.org.

[887] B. Simma/A. Heinemann, "Codes of Conduct" in: W. Korff et al. (eds.), Handbuch der Wirtschaftsethik, Band 2: Ethik wirtschaftlicher Ordnungen, 1999, 403-418 (416).

[888] J.G. Ruggie, "Current Development: Business and Human Rights: The Evolving International Agenda", American Journal of International Law 101 (2007), 819-839 (822).

to deflect NGO criticism and complaints, business actors are often keen to demonstrate their compliance with and support for these instruments irrespective of their voluntary nature.[889]

A frequently employed tool used for this purpose is self-regulation and private standard-setting.[890] Oftentimes, such standards draw upon the international instruments of the kind analysed here. Again, the fact that these international instruments are directly addressed to private actors makes them particularly suited in this respect. Private standards can benefit from the particular legitimacy of *international* instruments which derives *inter alia* from the participation of states and international organisations.[891]

Industry and business associations are often particularly active in enhancing and enforcing the self-regulatory efforts of their membership. One explanation for this active role is the interest of industry and business associations in defending the voluntary option against societal pressures demanding binding regulation on both the international and domestic level.

In the area of corporate social responsibility, the OECD Guidelines have for example been incorporated into national codes of conduct by industry on several occasions.[892] Similarly, the Equator Principles – a set of standards established by a number of important international banks – use the Policy and Performance Standards on Social and Environmental Sustainability established by the International Finance Corporation of the World Bank[893] as a point of reference. The Equator Principles are now adhered to by a significant number of banks which are al-

---

[889] For an overview of private actor implementation see also U. Dieckert, Die Bedeutung unverbindlicher Entschließungen internationaler Organisationen für das innerstaatliche Recht der Bundesrepublik Deutschland, 1993, 140 et seq.

[890] J. Morrison/N. Roht-Arriaza, "Private and Quasi-Private Standard Setting" in: D. Bodansky/J. Brunnée/E. Hey (eds.), Oxford Handbook of International Environmental Law, 2007, 498-527; J. Friedrich, "Environment, Private Standard Setting", in: R. Wolfrum (ed.), Max Planck Encyclopedia of Public International Law, 2010, online at: www.mpepil.com.

[891] See on the bases of legitimacy of nonbinding international instruments Part 3, further below.

[892] E. Kocher, "Selbstverpflichtungen von Unternehmen zur sozialen Verantwortung: Erfahrungen mit sozialen Verhaltenskodizes in der transnationalen Produktion", Recht der Arbeit 1 (2004), 27-31 (27).

[893] These standards of the IFC are directly applied to all investment projects, see http://www.ifc.org/ifcext/sustainability.nsf/Content/EnvSocStandards.

together responsible for 80 per cent of global commercial project lending.[894]

The area of fisheries can also serve as an illustration. The Australian Seafood Industry Council for example established a Code of Conduct for a Responsible Seafood Industry which is based on the CCRF.[895] Similarly, the Federation of European Aquaculture Producers – an international organisation composed of 28 European National Aquaculture Associations – has adopted the European Aquaculture Code of Conduct which implements the CCRF as well as the FAO Technical Guidelines on Aquaculture.[896]

Provided that they are not only designed as window-dressing exercises, the advantage of these initiatives is the integration of the addressees, i.e. those affected by the rules, into the process. Such participation and integration may increase acceptance of the norms and again may trigger processes of persuasion and socialisation among the participants. For reasons which have been repeatedly mentioned in this study, institutionalisation can be generally assumed to support these processes if it establishes continuous dialogue and discourse.

The preference of industry for voluntary norms frequently leads to activities beyond compliance. Besides adopting self-regulatory norms that mirror the international instruments, industry and business associations can be powerful enforcement agents. The National Associations of Manufacturers of Agrochemical Products (GIFAP) for instance made adherence to the FAO Pesticide Code a condition for membership.[897] Moreover, the GIFAP even engaged in additional implementation activities such as capacity building, and thereby went beyond a mere

---

[894] Compare information at http://www.equator-principles.com/.

[895] Code of Conduct for a Responsible Seafood Industry, available at http://www.seafoodsite.com.au/sustainable/code.php.

[896] The code of conduct is available at http://www.feap.info/consumer/codes/feapintro_en.asp.

[897] B. Dinham, "The Success of a Voluntary Code in Reducing Pesticide Hazards in Developing Countries", Green Globe Yearbook of International Co-operation on Environment and Development 3 (1996), 29-36 (31); D.G. Victor, "'Learning by Doing' in the Nonbinding International Regime to Manage Trade in Hazardous Chemicals and Pesticides" in: D.G. Victor/K. Raustiala/E.B. Skolnikoff (eds.), The Implementation and Effectiveness of International Environmental Commitments: Theory and Practice, 1998, 221-281 (255); S.R. Ratner, "Business" in: D. Bodansky/J. Brunnée/E. Hey (eds.), The Oxford Handbook of International Environmental Law, 2007, 807-828 (823).

demonstration of compliance.[898] Thus, when the PIC procedure was adopted, it was added as one element of capacity building initiatives to the so-called "Safe Use Project" which was financed and conducted by the GIFAP.[899] However, this may not have been purely altruistic. The safe use of pesticides serves the interest of business to avoid accidents or negative health reports which are likely to discredit the use of pesticides altogether. Another case is even more surprising. By means of industry sponsored international capacity building projects, associations of pesticide producing companies such as CropLife aim to strengthen the capacity of farmers for integrated pest management.[900] This is noteworthy since integrated pest management is one method with which pesticide use can be diminished altogether. It therefore does not serve the immediate interest of the industry in securing their markets. Critics however point out that most firms are not focussed on these issues, or on raising awareness among the general population, but are rather interested in advocating their products.[901]

The example of the FAO Pesticide Code also serves to illustrate the potential of a nonbinding approach to commence a process of norm development and public discourse supported by business actors. Since industry is generally keen to support voluntary options, a nonbinding instrument could be used as a tool to draw these actors into a process of norm development from which it can only withdraw at considerable reputational costs. Thus, industry associations had favoured the nonbinding approach of the FAO Pesticide Code, and had therefore embraced the code firmly. At first, this only concerned the code without

---

[898] Generally S.R. Ratner, "Business" in: D. Bodansky/J. Brunnée/E. Hey (eds.), The Oxford Handbook of International Environmental Law, 2007, 807-828 (823).

[899] The "Safe Use Project" also included education, provision of protective clothing as well as the distribution of pesticide antidotes. Compare D.G. Victor, "'Learning by Doing' in the Nonbinding International Regime to Manage Trade in Hazardous Chemicals and Pesticides" in: D.G. Victor/K. Raustiala/E.B. Skolnikoff (eds.), The Implementation and Effectiveness of International Environmental Commitments: Theory and Practice, 1998, 221-281 (255).

[900] Information on the program is available at www.croplife.org (1 March 2010).

[901] M. Singh, "Report on India" in: FAO, 'Proceedings of the Asia Regional Workshop on the Implementation, Monitoring and Observance, International Code of Conduct on the Distribution and Use of Pesticides', 2005, RAP Publication 2005/29, available at www.fao.org/docrep/008/af340e/af340e0i.htm.

the PIC procedure. When the FAO members moved from resistance towards cooperation with UNEP, and began supporting the PIC concept in concert with powerful actors such as the United States, some kind of additional regulation seemed more and more inevitable. Consequently, the agrochemical industry also changed its strategy regarding PIC.[902] The original support of industry for the code of conduct hardly allowed for a rejection of the newly introduced PIC procedure if the industry did not want to risk considerable reputational costs.[903]

III. The potential and limits of the regulatory capacity regarding private actors

1. Potential

Even if most authors dealing with the OECD Guidelines stress their overall positive effects,[904] empirical evidence on the actual positive effects of this and similar instruments on the behaviour of private actors is slim. A look at the mechanisms and initiatives analysed in this chapter however suggests that international norms directed at private actors have some potential to change their behaviour. In certain constellations, nonbinding instruments may even outperform binding regimes in terms

---

902 R.L. Paarlberg, "Managing Pesticide Use in Developing Countries" in: P.M. Haas/R.O. Keohane/M.A. Levy (eds.), Institutions for the Earth: Sources of Effective International Environmental Protection, 1993, 309-350 (329).

903 R.L. Paarlberg, "Managing Pesticide Use in Developing Countries" in: P.M. Haas/R.O. Keohane/M.A. Levy (eds.), Institutions for the Earth: Sources of Effective International Environmental Protection, 1993, 309-350 (336-337).

904 H.W. Baade, "The Legal Effects of Codes of Conduct for Multinational Enterprises", German Yearbook of International Law 22 (1979), 11-52 (25-26); R. Blanpain, "Guidelines for Multinational Enterprises, for Ever?: The OECD Guidelines, 20 years later", The International Journal of Comparative Labour Law and Industrial Relations 14 (1998), 337-348 (348); A.A. Fatouros, "On the implementation of international codes of conduct: an analysis of future experience", American University Law Review 30 (1980-1981), 941-972 (959); P. Sanders, "Implementing International Codes of Conduct for Multinational Enterprises", American Journal of Comparative Law 30 (1982), 241-254 (244); T.W. Vogelaar, "The OECD Guidelines: Their Philosophy, History, Negotiation, Form, Legal Nature, Follow-Up Procedures and Review" in: N. Horn (ed.), Legal Problems of Codes of Conduct for Multinational Enterprises, 1980, 127-140 (135).

of compliance of corporations.[905] Their voluntary nature apparently does not keep international organisations from establishing institutionalised complaint procedures, membership systems or in some cases to publicly list non-compliant companies. Business actors concerned about reputation among consumers and business partners can hardly afford to ignore these instruments. Depending on the degree of institutionalisation, the initiatives may establish continuous interaction and dialogue between NGOs, trade unions and private business on the elaboration and implementation of norms. This may have the effect that companies are persuaded and internalise the values supported by the nonbinding instrument,[906] at least to the extent that such a change can be reconciled with a profitable business strategy in the long term. Even if economic considerations can be expected to remain paramount, companies may also eventually perceive a standard to be legitimate and comply out of a normative belief, in particular if they participated in its creation.[907]

Nonbinding instruments can also be used for labelling or naming and shaming by NGOs, and are useful tools for marketing purposes. Companies that wish to respond to societal expectations in order to deflect criticism or that wish to adapt their business strategy in order to improve their reputation are given guidance on how to do so. Once the established rules are internalised into organisational processes and decision-making, this by itself can have positive environmental repercussions.[908]

---

[905] D.G. Victor, "'Learning by Doing' in the Nonbinding International Regime to Manage Trade in Hazardous Chemicals and Pesticides" in: D.G. Victor/K. Raustiala/E.B. Skolnikoff (eds.), The Implementation and Effectiveness of International Environmental Commitments: Theory and Practice, 1998, 221-281 (254).

[906] S.D. Murphy, "Taking multinational corporate codes of conduct to the next level", Columbia journal of transnational law 43 (2005), 389-433 (423).

[907] I. Hurd, "Legitimacy and Authority in International Politics", International Organization 53 (1999), 379-408 (387); H. Keller, "Codes of Conduct and their Implementation: the Question of Legitimacy" in: R. Wolfrum/V. Röben (eds.), Legitimacy in International Law, 2008, 219-298 (275).

[908] R. Blanpain, "Guidelines for Multinational Enterprises, for Ever?: The OECD Guidelines, 20 years later", The International Journal of Comparative Labour Law and Industrial Relations 14 (1998), 337-348 (348).

To the extent that codes of conduct change the practice of companies, they contribute to new trade usages[909] and possibly to the emergence of a new *lex mercatoria*,[910] although this latter notion remains controversial.[911] Be it as it may, the OECD Guidelines and similar international codes of conduct have the potential to add a layer of ethical standards and values to this emerging body of norms.[912] The fact that multinational corporations are not accepted as subjects of international law does not present an obstacle to the emergence of a new *lex mercatoria* as it does for the emergence of customary public international law in this field.[913]

Altogether, nonbinding instruments can be important complements to other binding instruments in at least two ways. As analysed earlier,[914] they can initiate and support the emergence of widely accepted principles which could later find their way into treaty law. This is arguably already happening. Although still lacking concretisation, principles such as the precautionary principle or the requirement for environmental impact assessment appear in almost all of the public and private voluntary initiatives. Second, these approaches can complement more tradi-

---

[909] I. Seidl-Hohenveldern, "International Economic 'Soft Law'", Recueil des Cours de l'Académie de Droit International 163 (1979), 169-246 (212) (taking the fact that the outcome of the follow-up interpretation procedure of the OECD Guidelines is usually complied with by the respective enterprise, and that the result of the procedures are usually welcomed by the Business and Industry Advisory Committee to the OECD (BIAS) as a sign for the possibility of a gradual establishment of trade usages).

[910] B.-O. Bryde, Internationale Verhaltensregeln für Private – Völkerrechtliche und verfassungsrechtliche Aspekte, 1981, 34; N. Horn, "Codes of conduct for MNEs and Transnational Lex Mercatoria: An international Process of Learning and Law Making" in: N. Horn (ed.), Legal Problems of Codes of Conduct for Multinational Enterprises, 1980, 45-81 (59 et seq.).

[911] The literature is manifold. See only P. Zumbansen, "Lex mercatoria: Zum Geltungsanspruch transnationalen Rechts", Rabels Zeitschrift für ausländisches und internationales Privatrecht 67 (2003), 637-682.

[912] N. Horn, "Codes of conduct for MNEs and Transnational Lex Mercatoria: An international Process of Learning and Law Making" in: N. Horn (ed.), Legal Problems of Codes of Conduct for Multinational Enterprises, 1980, 45-81; E. Morgera, Corporate accountability in international environmental law, 2009, 115.

[913] B.-O. Bryde, Internationale Verhaltensregeln für Private – Völkerrechtliche und verfassungsrechtliche Aspekte, 1981, 34.

[914] See on this Part 2, at A.I.1., further above.

tional legal approaches. Voluntary approaches can be used to establish norms which go further than any existing law. The cooperative approach of instruments such as the OECD Guidelines enhances the acceptance of progressive norms.[915] Voluntary norms can be used for marketing purposes and are ideal indicators for companies that pursue pioneering long-term strategies. Simply by codifying in a widely accepted instrument what is socially recommended and expected from business actors, nonbinding instruments addressed to private actors can support and direct societal and market-related forces. In addition, international mediation and complaint procedures can complement national judicial proceedings by allowing non-state actors to bring complaints and relevant arguments which would otherwise not be possible due to their lack of standing or other procedural restraints in national law.[916] Private actors may consequently comply beyond the requirements of domestic or international law.[917] One concrete example in this respect is the FAO Pesticide Code. In this case, industry actors are engaged in capacity building activities in support of the overall objectives of the instrument, albeit arguably with a view to forestall binding regulation.

2. *Limits*

Despite the potential of nonbinding instruments as described above, this analysis has also already hinted at the limitations of these instruments and initiatives.

Generally speaking, nonbinding instruments cannot be alternatives but only complement international law and domestic law and its enforcement. In fact, in particular with respect to the regulation of private actors, they are inherently limited and can only play a supplementary role.

---

[915] Financial Times, "Letters to the Editor: Why business values OECD guidelines", 12 July 2005.

[916] In German administrative law, the right to bring a legal complaint presupposes a law that protects the individual interests of the claimant, compare Article 42 (2) of the German administrative procedural law ("Verwaltungsgerichtsordnung").

[917] J. Morrison/N. Roht-Arriaza, "Private and Quasi-Private Standard Setting" in: D. Bodansky/J. Brunnée/E. Hey (eds.), Oxford Handbook of International Environmental Law, 2007, 498-527 (523).

First, nonbinding instruments and compliance mechanisms which are primarily directed at private actors must be distinguished from those which primarily seek to influence states. The former are part of the ongoing attempt by international organisations to respond to the growing importance of these actors for social and environmental governance. The regulation of multinational enterprises but also of the fishing industry remains inadequate in the absence of effective international law and state-based enforcement. Directly to set international standards for these actors represents a departure from traditional international law which has been mostly addressed to and implemented by states. Given the current state of international legal doctrine where multinational enterprises are not considered subjects of international law, such regulatory endeavours must necessarily be conducted through nonbinding instruments. However, the mere fact of adoption of such norms with direct applicability to multinational enterprises by states at the international level reflects an increasing need to recognise the legal subjectivity of such actors.[918] This apparent need generally strengthens the arguments advanced by scholars for the recognition of a limited international legal personality for multinational enterprises.[919]

However, nonbinding instruments can hardly be a means effectively to regulate multinational enterprises without relying on national legislative implementation and enforcement. As could be seen, even highly institutionalised initiatives only encompass underdeveloped compliance mechanisms compared to both national and international binding law. Except for some rare occasions when a certificate or membership is actually withdrawn, sanctions or penalties for the violation of voluntary standards are largely absent. Where the possibility of complaints or sanctions exists, their functioning largely depends on sufficient public attention and NGO activity. The complaint mechanism for instance only works if supported by sufficient interest and capacity of the

---

[918] G. Dahm/J. Delbrück/R. Wolfrum, Völkerrecht, Band I/2, Der Staat und andere Völkerrechtssubjekte, Räume unter internationaler Verwaltung, 2002, 255; B.-O. Bryde, Internationale Verhaltensregeln für Private – Völkerrechtliche und verfassungsrechtliche Aspekte, 1981, 19.

[919] For the argument that such instruments addressing private actors reflect and support these tendencies and an overview of scholarship on the question of legal subjectivity see G. Dahm/J. Delbrück/R. Wolfrum, Völkerrecht, Band I/2, Der Staat und andere Völkerrechtssubjekte, Räume unter internationaler Verwaltung, 2002, 257; for an good overview of the discussion E. Morgera, Corporate accountability in international environmental law, 2009, 56-60; S. Hobe/O. Kimminich, Einführung in das Völkerrecht, 2004, 157-159.

NGOs and trade unions able to bring complaints and monitor behaviour.[920] Due to selective attention by the public and by NGOs, and due to differences in the reputational sensitivity of business actors, the potential of nonbinding standards will always remain limited to particular sectors, particular issues and particular time periods. Leaving aside for now the more fundamental legitimacy issue raised by NGOs performing public functions,[921] the unequal representation and power of NGOs in developed and developing countries may also lead to unequal implementation in developed states and developing states.

In sum, relying on activism will inevitably cause enforcement and monitoring to remain "piecemeal and inconsistent".[922] The need for enforceable legal rules which guarantee consistent and independent application, i.e. the development of international and national law, suddenly (re-)appears in the picture. Even the most advanced compliance review mechanism in the field of corporate responsibility, namely the one established by the OECD Guidelines, cannot guarantee consistency of application. Implementation by national governments is highly uneven and does not secure compliance.[923] Thus, not only does the composition of the National Contact Points vary greatly from state to state, but the NCPs also have wide discretion in deciding whether to admit complaints or whether to issue public statements. The analysis of the most frequently addressed NCPs – namely those in the United States, the Netherlands and France – indicates that the total number of cases which have been admitted is quite small, and implementation in areas outside of labour relations has not been substantial.[924] Therefore, in particular environmental NGOs demand better enforcement through states and binding international obligations as the only way to force states to implement and apply norms for corporations and thus to overcome the

---

[920] A.A. Fatouros, "On the implementation of international codes of conduct: an analysis of future experience", American University Law Review 30 (1980-1981), 941-972 (970).

[921] See for a detailed discussion in Part 3 on legitimacy.

[922] J.A. Zerk, Multinationals and Corporate Social Responsibility: Limitations and Opportunities in International Law, 2008, 243.

[923] J.G. Ruggie, "Current Development: Business and Human Rights: The Evolving International Agenda", American Journal of International Law 101 (2007), 819-839 (834).

[924] C.N. Franciose, "A critical assessment of the United States' implementation of the OECD guidelines for multinational enterprises", Boston College international and comparative law review 30 (2007), 223-236 (232).

existing discrepancies which exist between states and which allow multinational corporations to evade regulation.[925]

The fundamental problem with the regulatory strategy pursued by such initiatives is the reduced role of the state in regulation and enforcement. It is likely to be insufficient to change the behaviour of private actors whenever such a change cannot be justified economically.[926]

International codes of conduct directed at private actors thus do not represent sufficient alternatives to international and domestic national law and enforcement. This of course assumes that appeals to morality which contradict the underlying paradigm of profit maximisation are likely to remain without lasting effect.[927] Voluntary norms directed at business actors can therefore not be solely relied on for effective environmental governance, especially in areas that require substantial changes to the status quo, as is the case with climate policy for example.

International nonbinding instruments are not only inherently limited in their regulatory capacity to regulate private actors. They are also ill-equipped to replace public law as the legitimate provider of security and of a level playing field for business actors by means of broadly applicable and enforceable legal rules. Now, it is certainly true – as has been observed in the discussion on corporate responsibility – that other state-oriented conceptions are at least at the moment politically infeasible. As long as this is the case, voluntary approaches such as those of the OECD Guidelines can be of value. Perhaps equipped with better

---

[925] Financial Times, "OECD plea to raise corporate standards", 4 May 2005. In support of stronger rules is also A.A. Fatouros, "On the implementation of international codes of conduct: an analysis of future experience", American University Law Review 30 (1980-1981), 941-972 (950).

[926] The failure of the European car industry to live up to their voluntary commitments taken up in 1998 by 2008 illustrates this limitation of voluntary commitments; for the voluntary norm-replacing commitment see Commission Recommendation of 5 February 1999 on the reduction of CO2 emissions from passenger cars EU Commission Recommendation 1999/125/EG). The EU finally had to resort to binding legislation to force industry to develop cars with lower emissions, compare Regulation (EC) No 443/2009 of the European Parliament and of the Council of 23 April 2009 setting emission performance standards for new passenger cars as part of the Community's integrated approach to reduce CO 2 emissions from light-duty vehicles.

[927] J. Dine, "Multinational enterprises: international codes and the challenge of sustainable development", Non-State actors and international law 1 (2001), 81-106 (106).

monitoring mechanisms, international codes of conduct directed at multinational corporations may be the only solution for the time being.[928]

But one should not disregard the danger that lies in putting too much emphasis and resources on developing voluntary international standards directed at private corporations if this means that other efforts for binding international rules are put off or pressure to regulate is lowered. Any long term strategy should not neglect the need to reinforce the regulatory capacity of both host and home states of multinational corporations and should aim to provide harmonisation of domestic law through international law.[929] For the international level, this means that efforts to develop an international treaty defining duties for states on how to treat multinational corporations as well as minimum standards for multinational corporations have not become superfluous.[930] So far, the law on the responsibility of multinational corporations for their activities has developed disparately through national legal systems and practice, driven largely by judicial decisions under the Alien Tort Claims Act which opens jurisdiction of US Courts for torts that violate international law.[931] An international treaty defining minimum standards for multinational corporations could prevent the ongoing disparate and largely independent domestic development of norms on the responsibility of multinational corporations.[932] It could fill the gaps at the

---

[928] S.D. Murphy, "Taking multinational corporate codes of conduct to the next level", Columbia journal of transnational law 43 (2005), 389-433 (423).

[929] This is also one of the conclusion drawn by John Gerard Ruggie after having comprehensively studied the issue of corporate responsibility and voluntary standard-setting in the field of human rights, compare J.G. Ruggie, "Current Development: Business and Human Rights: The Evolving International Agenda", American Journal of International Law 101 (2007), 819-839 (839).

[930] It is beyond the scope of this dissertation to discuss the further reaching proposals which call for binding treaty norms for multinationals, compare instead J.A. Zerk, Multinationals and Corporate Social Responsibility: Limitations and Opportunities in International Law, 2008, 297-298.

[931] See on this development A. Seibert-Fohr/R. Wolfrum, "Die einzelstaatliche Durchsetzung völkerrechtlicher Mindeststandards gegenüber transnationalen Unternehmen", Archiv des Völkerrechts 43 (2005), 153-186.

[932] A. Seibert-Fohr/R. Wolfrum, "Die einzelstaatliche Durchsetzung völkerrechtlicher Mindeststandards gegenüber transnationalen Unternehmen", Archiv des Völkerrechts 43 (2005), 153-186 (185).

international level which NGOs try to fill by resorting to national law procedures where they exist.

Part 3

The legitimacy of nonbinding instruments

A. Introduction

The Common Rules of Procedure of the German Federal Ministries explicitly provide that each ministry should always verify whether the negotiation of a treaty is unavoidable or whether the pursued purpose can also be achieved by an arrangement below the level of binding international law.[1] This Part considers whether specific legitimacy challenges arise from increased reliance on nonbinding instruments.

The question of legitimacy has recently gained heightened interest in scholarship of international law.[2] The perceived need for improved or new legitimation structures stems largely from the perception that international law and international institutions are undergoing a funda-

---

1 Compare § 72 para. 1 of the Gemeinsame Geschäftsordnung der Bundesministerien (Common Rules of Procedure of the Federal Ministries) of the German Government, available at www.verwaltung-innovativ.de.

2 See only the contributions in R. Wolfrum/V. Röben (eds.), Legitimacy in International Law, 2008; the research undertaken by the Global Administrative Law Project at the New York University, available through www.iilj.org/GAL; H. Charlesworth/J.-M. Coicaud (eds.), Fault Lines of International Legitimacy, 2010; J.-M. Coicaud/V. Heiskanen (eds.), The legitimacy of international organizations, 2001; R. Wolfrum, "Legitimacy of International Law and the Exercise of Administrative Functions: The Example of the International Seabed Authority, the International Maritime Organization (IMO) and International Fisheries Organizations" in: A. von Bogdandy/R. Wolfrum/J. von Bernstorff/P. Dann/M. Goldmann (eds.), The Exercise of Public Authority by International Institutions: Advancing International Institutional Law 2010, 917-940; D. Bodansky, "Legitimacy" in: D. Bodansky/J. Brunnée/E. Hey (eds.), Oxford Handbook of International Environmental Law, 2007, 704-723 (715 et seq.); J.H.H. Weiler, "The geology of international law – governance, democracy and legitimacy", Zeitschrift für ausländisches öffentliches Recht und Völkerrecht 64 (2004), 547-562; M. Eifert, "Legitimationsstrukturen internationaler Verwaltung" in: H.-H. Trute/T. Groß/H.C. Röhl/C. Möllers (eds.), Allgemeines Verwaltungsrecht – zur Tragfähigkeit eines Konzepts, 2007, 307-331; I. Clark, International Legitimacy and World Society, 2007.

mental shift towards "global governance"[3] or a "global administrative space".[4] With the ongoing development of a body of international law of cooperation that goes beyond mere coordination, the issues addressed by international law and international institutions increasingly expand into almost all domains of social and economic life formerly within the exclusive realm of the nation state. International law and institutions therefore increasingly have a direct effect on private actors which previously were only subjected to regulation of the state.[5] Parallel to these developments, the procedures of law making and the actors that engage in law making and implementation processes appear to be changing.[6] International institutions also increasingly take on legislative, administrative and adjudicatory functions.[7]

So what about nonbinding instruments used by international institutions? A specific analysis of legitimacy in the context of nonbinding instruments is required because discussions on legitimacy often do not differentiate between binding and nonbinding instruments, even though important differences exist between them. The differences exist both in terms of the functions and steering power as well as in terms of the (legitimising) procedures and mechanisms that apply to binding but not to nonbinding instruments. Therefore, a more specific perspective is needed that identifies the particular challenges raised by nonbinding instruments but also the different needs for legitimation triggered by their usage. Without claiming to treat the subject comprehensively, the study will suggest some responses to the specific challenges arising from the use of nonbinding norms.

---

[3] The term is discussed and used as a conceptual basis of legal analysis by C. Tietje, Internationalisiertes Verwaltungshandeln, 2001, 164-170 and 278-287.

[4] B. Kingsbury/N. Krisch/R. Stewart, "The Emergence of Global Administrative Law", Law & Contemporary Problems (2004-2005), 15-62 (18).

[5] W. Friedmann, The Changing Structure of International Law, 1964, 60 et seq.; C. Tomuschat, "International law: ensuring the survival of mankind on the eve of a new century", Recueil des cours de l'Académie de Droit International de La Haye 281 (2001), 13-438 (70).

[6] R. Wolfrum, "Legitimacy in International Law from a Legal Perspective: Some Introductory Considerations" in: R. Wolfrum/V. Röben (eds.), Legitimacy in International Law, 2008, 1-24 (10).

[7] J.E. Alvarez, "International Organizations: Then and Now", American Journal of International Law 100 (2006), 324-347.

I. Notion of legitimacy

The notion of legitimacy refers to the justification of the exercise of authority.[8] As indicated, the question of legitimacy thus goes beyond legality. An act may be legal but nevertheless be illegitimate if it cannot bring to bear some more fundamental reason why it should be followed or accepted by others. Even if international cooperation is legal, legitimacy issues can still arise.[9]

In assessing whether authority is justified and thus legitimate, a so-called objective or normative perspective and a subjective perspective are often distinguished.[10] What is referred to as objective or normative legitimacy pertains to the question whether a particular rule or decision is based on some generally agreed norm or procedure which justifies the rule or decision.[11] A reliable long-term legitimation of authority cannot disregard the so-called subjective dimension, namely the need for a belief or perception of the addressees that the exercise of authority

---

[8] Similar D. Bodansky, "Legitimacy" in: D. Bodansky/J. Brunnée/E. Hey (eds.), Oxford Handbook of International Environmental Law, 2007, 704-723 (705); R. Wolfrum, "Legitimacy in International Law from a Legal Perspective: Some Introductory Considerations" in: R. Wolfrum/V. Röben (eds.), Legitimacy in International Law, 2008, 1-24 (6); J. Delbrück, "Exercising Public Authority Beyond the State: Transnational Democracy and/or Alternative Legitimation Strategies?", Indiana Journal of Global Legal Studies 10 (2003), 29-43 (31-32).

[9] C. Möllers, "Transnationale Behördenkooperation: Verfassungs- und völkerrechtliche Probleme transnationaler administrativer Standardsetzung", Zeitschrift für ausländisches öffentliches Recht und Völkerrecht 65 (2005), 351-389 (378).

[10] The distinction is often made in legal scholarship, compare e.g. D. Bodansky, "The Legitimacy of International Governance: A Coming Challenge for International Environmental Law?", American Journal of International Law 93 (1999), 596-624 (601); A. Buchanan/R.O. Keohane, "The Legitimacy of Global Governance Institutions" in: R. Wolfrum/V. Röben (eds.), Legitimacy in International Law, 2008, 25-62 (25); E. Hey, "Sustainable development, normative development and the legitimacy of decision-making", Netherlands yearbook of international law 34 (2004), 3-53 (13).

[11] D. Bodansky, "Legitimacy" in: D. Bodansky/J. Brunnée/E. Hey (eds.), Oxford Handbook of International Environmental Law, 2007, 704-723 (709); similar M. Zürn, "Global Governance and Legitimacy Problems", Government and Opposition 39 (2004), 260-287 (260).

is indeed justified.[12] As a legal study, the following analysis is mostly concerned with the conditions required to achieve objective legitimacy, since these are often of a legal nature, rather than the sociological conditions under which an entity is believed to be legitimate.

Legitimacy as a precondition for the exercise of authority is from a functional perspective needed to provide stability to the way the international system is organised, including the stability to manage necessary changes.[13] It is also an important compliance-enhancing factor, and therefore a precondition for the effectiveness of a particular norm or institution.[14] Compliance by an actor with a rule rests not primarily on police enforcement or mere calculation of interests, but on the internal sense of obligation to obey the rule based upon "the normative belief by an actor that a rule or institution ought to be obeyed."[15] Legitimate norms thus exercise a "compliance pull" on those to whom they are addressed.[16] In a voluntarist system such as international law where coercion is the exception rather than the rule, legitimacy therefore gains particular importance.[17] This is particularly true in the case of nonbinding

---

[12] This has been highlighted in detail by M. Weber, Wirtschaft und Gesellschaft, 2002, 122 et seq.; Franck also stresses the perception of a rule as legitimate, compare T.M. Franck, "Legitimacy in the international system", American Journal of International Law 82 (1988), 705-759 (706); for a subjective view see also I. Hurd, "Legitimacy and Authority in International Politics", International Organization 53 (1999), 379-408 (381).

[13] J.-M. Coicaud, "The Evolution of the International Order and Fault Lines of International Legitimacy" in: H. Charlesworth/J.-M. Coicaud (eds.), Fault Lines of International Legitimacy, 2010, 87-114 (87).

[14] D. Bodansky, "The Concept of Legitimacy in International Law" in: R. Wolfrum/V. Röben (eds.), Legitimacy in International Law, 2008, 309-317 (310); A.C. Kiss/D. Shelton, International environmental law, 2004, 98.

[15] I. Hurd, "Legitimacy and Authority in International Politics", International Organization 53 (1999), 379-408 (381 and 387).

[16] T.M. Franck, "Legitimacy in the international system", American Journal of International Law 82 (1988), 705-759 (712); T.M. Franck, "The Power of Legitimacy and the Legitimacy of Power: International Law in an Age of Power Disequilibrium", American Journal of International Law 100 (2006), 88-106 (93); critical of this approach due to the lack of explanations why states care about legitimacy is A.T. Guzman, "A Compliance-Based Theory Of International Law", California Law Review 90 (2002), 1823-1887 (1834-1835).

[17] T.M. Franck, "Legitimacy in the international system", American Journal of International Law 82 (1988), 705-759 (711); I. Hurd, "Legitimacy and Au-

or voluntary instruments which lack even those enforcement measures available for binding norms. Legitimacy is thus particularly relevant for the impact of nonbinding instruments.

Is there any added value to a legitimacy analysis over an analysis of legality? An analysis in terms of legality does not seem to suffice for identifying the challenges arising from new activities of international actors. Only very rarely are the activities of international institutions illegal, but they nevertheless often face legitimacy challenges. The reason is simple. Legality can only assess such activities in terms of existing international or institutional law. It therefore has inherent limits when institution law has not yet developed, as is necessarily the case when international institutions are undergoing a fundamental change. Legitimacy considers the foundations of the system of law and whether it is necessary to adapt law and institutions to new developments. Legitimacy assessments are therefore suited to identify the challenges that arise when basic premises of a particular legal system change.[18]

II. The legitimacy question in the context of nonbinding norms

It is not at all self-evident that the question of legitimacy should be raised in the context of nonbinding instruments. Legislative, administrative or enforcement activities of international institutions with binding effect clearly constitute an exercise of authority. But where instruments do not legally oblige a state or a private actor to act in a particular manner, the addressees of instruments which are (expressly) nonbinding or "voluntary" formally retain their full freedom of action. Or more concretely with respect to states, the instruments are not an act of the exercise of sovereign power, since states have in these cases deliberately refrained from transferring any part of their sovereignty to the international level. Similarly, if private actors follow international recommendations and other voluntary norms, they do so freely, not because they are subject to the authority of the issuing organisation. Arguably, where the actions of states and – directly or indirectly – private actors are not predetermined from the international level, an issue of le-

---

thority in International Politics", International Organization 53 (1999), 379-408 (401).

[18] Similarly J.-M. Coicaud, "The Evolution of the International Order and Fault Lines of International Legitimacy" in: H. Charlesworth/J.-M. Coicaud (eds.), Fault Lines of International Legitimacy, 2010, 87-114 (96-97).

gitimacy could not arise. Or in other words, could one not say that nonbinding instruments of international institutions could not even theoretically raise any legitimacy concerns due to their voluntary character?

Even though one must distinguish between binding and nonbinding prescriptions and administration of international institutions, the question of legitimacy is not *per se* obsolete in the context of nonbinding norms. Two considerations are important in this context. The first is the need for effective instruments. Legitimacy is important for to enhance compliance with the nonbinding instrument. Regardless of the potential of a nonbinding approach in a given area, i.e. whether it is a superior or inferior or a supplementary tool, the ability of the instruments to develop their full potential is often directly related to their ability to induce compliance. As explained, nonbinding instruments are particularly dependent on legitimacy to be influential upon behaviour. The legitimacy issue must therefore be raised in the context of nonbinding norms from a functional, compliance oriented perspective.

Secondly, one must not ignore the legal and factual consequences of nonbinding instruments.[19] As shown above, nonbinding norms play important roles in the generation of law and as complements to treaty law by concretising binding norms through direct application or through references. The importance of nonbinding instruments in the development of international law and as supplements to existing international law varies, but can no longer be denied.[20] Even nonbinding acts of international institutions must therefore meet some kind of legitimacy test.[21] While states can object to these developments by not adopting or not implementing a certain nonbinding instrument, the im-

---

[19] J. Delbrück, "Exercising Public Authority Beyond the State: Transnational Democracy and/or Alternative Legitimation Strategies?", Indiana Journal of Global Legal Studies 10 (2003), 29-43 (35).

[20] See above in Part 2, at A.I.1.; compare also C. Chinkin, "Normative Development in the International Legal System" in: D. Shelton (ed.), Commitment and Compliance: the Role of Non-binding Norms in the International Legal System, 2000, 21-42; E. Hey, "International Institutions" in: D. Bodansky/ J. Brunnée/E. Hey (eds.), Oxford Handbook of International Environmental Law, 2007, 749-769 (755 et seq.).

[21] E. Hey, "International Institutions" in: D. Bodansky/J. Brunnée/E. Hey (eds.), Oxford Handbook of International Environmental Law, 2007, 749-769 (757-758); R. Wolfrum, "Introduction" in: R. Wolfrum/V. Röben (eds.), Developments of International Law in Treaty Making, 2005, 1-13 (5 and 6).

pact on the normative development of new or existing customary and treaty law as described above can hardly be influenced by any individual objecting state.

It has also been observed that international institutions use nonbinding instruments to prescribe norms that guide and constrain states as well private actors. In exercising such a prescriptive function, international institutions exercise public authority, even though it comes in a different form than in the case of binding instruments. As this study has shown,[22] international institutions may at times be as successful in steering the behaviour of addressees through nonbinding norms as they are with binding instruments.[23] Even though the impact may vary in degree and from one instrument to another, many of the analysed instruments had a considerable impact on further regulatory and administrative measures as well as on the behaviour of private actors. International institutions use nonbinding instruments to constrain and predetermine the choices of states, domestic policy makers or administrators as well as those of private actors, either directly or indirectly through states. This effect is often not simply coincidental, but is directly related to the compliance-enhancing means and procedures established by international organisations.[24] In sum, the exercise of authority can take different forms, and if one takes the legal and factual effects of nonbinding instruments seriously, nonbinding instruments may constitute one of them.[25] In particular as these activities interfere with formerly domestic issues – as is the case for most environmental policies[26] – nonbinding activities of international organisations should not be per se excluded from legitimacy questions but to the contrary pose specific legitimacy concerns.[27]

---

[22] J.E. Alvarez, International Organizations as Law-makers, 2005, 217 et seq; D. Shelton (ed.), Commitment and Compliance: The Role of Non-Binding Norms in the International Legal System, 2000.

[23] A. Verdross/B. Simma, Universelles Völkerrecht: Theorie und Praxis, 1984, § 656.

[24] See above, in Part 2, at B.I.4. and 5a).

[25] D. Bodansky, "The Concept of Legitimacy in International Law" in: R. Wolfrum/V. Röben (eds.), Legitimacy in International Law, 2008, 309-317 (312).

[26] M. Zürn, "Global Governance and Legitimacy Problems", Government and Opposition 39 (2004), 260-287 (277).

[27] Similarly S. Oeter, "The openness of international organisations to transnational rule-making" in: O. Dilling/M. Herberg/G. Winter (eds.), Transna-

Realising that authority can be exercised through nonbinding instruments does not however mean that no distinction should be made between binding and nonbinding norms with respect to legitimacy. As the exercise of authority can be a matter of degree, so is the need for justification.[28] As discussed previously,[29] the binding nature of an instrument is one factor which increases its authority, and therefore that of the institution.[30] Even if they may be as effective as binding norms in inducing compliance, the authority and therefore the pressure to comply exerted by nonbinding norms is generally lower. In addition to the nonbinding form, the study has revealed other factors which can usually be associated with nonbinding instruments which limit their general regulatory potential if compared to treaty law.[31] Everything else being equal, the need for justification is thus lower than for binding norms. [32]

---

tional administrative rule making: performance, legal effects and legitimacy, 2011, 235-252 (249); this conclusion is also drawn by J. Delbrück, "Exercising Public Authority Beyond the State: Transnational Democracy and/or Alternative Legitimation Strategies?", Indiana Journal of Global Legal Studies 10 (2003), 29-43 (35); E. Hey, "International Institutions" in: D. Bodansky/J. Brunnée/E. Hey (eds.), Oxford Handbook of International Environmental Law, 2007, 749-769 (758); P. Muchlinski, "Human Rights, social responsibility and the regulation of international business: The development of international standards by intergovernmental organisations", Non-State Actors and International Law 3 (2003), 123-152 (146); A. von Bogdandy, "Lawmaking by International Organisations: Some Thoughts on Non-Binding Instruments and Democratic Legitimacy" in: R. Wolfrum/V. Röben (eds.), Developments of International Law in Treaty Making, 2005, 171-182 (173); similarly apparently also Alvarez who considers nonbinding instruments to be part of the law making function of international organizations to which he connects his legitimacy critique, see J.E. Alvarez, International Organizations as Law-makers, 2005.

[28] D. Bodansky, "The Concept of Legitimacy in International Law" in: R. Wolfrum/V. Röben (eds.), Legitimacy in International Law, 2008, 309-317 (316).

[29] Compare Part 2, at B.I.1., further above.

[30] D. Bodansky, "The Concept of Legitimacy in International Law" in: R. Wolfrum/V. Röben (eds.), Legitimacy in International Law, 2008, 309-317 (316).

[31] Compare in particular Part 2, at B.I. on compliance and implementation, further above.

[32] D.C. Esty, "Good Governance at the Supranational Scale: Globalizing Administrative Law", Yale Law Journal 115 (2006), 1490-1562 (1538); D. Bodansky, "The Concept of Legitimacy in International Law" in: R. Wolfrum/V. Röben (eds.), Legitimacy in International Law, 2008, 309-317 (316).

B. Legitimacy challenges

The point of departure for this analysis is the traditional conception of legitimacy of international law. The objective is to outline the traditional bases of legitimacy in order to assess whether the existing structures already suffice to legitimize the described nonbinding activities of international institutions, or whether the increasing reliance on nonbinding instruments represents certain challenges which must be addressed.

Furthermore, the analysis is oriented at the ideal of parliamentary democracies. In particular, for the national level analysis, it takes the example of Germany – as a country with a constitutionally secured openness to international law and cooperation – as a reference.[33] It is not the intention to imply that the issue of legitimacy could not be viewed differently from the standpoint of a different political and legal system. Nevertheless it is hoped that some of the general lines of argument are – possibly with adaptations – transferable to such other legal systems.

I. Traditional bases of legitimacy: consent and domestic implementation process

Rooted deeply in the concept of sovereignty of states but not democracy,[34] the legitimacy of international law is traditionally secured through the idea that states must consent to rules in order to be bound.[35] Binding treaty norms usually have to be ratified in order to

---

[33] The openness to international cooperation is enshrined in the preamble and Articles 24 to 26 of the German Basic Law.

[34] On the distinct traditional conceptions of the national and international legal systems instructively J.H.H. Weiler, "The geology of international law – governance, democracy and legitimacy", Zeitschrift für ausländisches öffentliches Recht und Völkerrecht 64 (2004), 547-562 (547-548); on the concept of sovereignty and its value today S. Oeter, "Souveränität – ein überholtes Konzept?", Tradition und Weltoffenheit des Rechts (2002), 259-290 (286-287).

[35] D. Bodansky, "The Legitimacy of International Governance: A Coming Challenge for International Environmental Law?", American Journal of International Law 93 (1999), 596-624 (604); V. Röben, Außenverfassungsrecht: eine Untersuchung zur auswärtigen Gewalt des offenen Staates, 2007, 42; R. Wolfrum, "Legitimacy in International Law from a Legal Perspective: Some Introductory Considerations" in: R. Wolfrum/V. Röben (eds.), Legitimacy in

bind the respective state. The traditional conception of international law thus contrasts with conceptions of democratic legitimacy and rule of law at the national level where the majority of a collectivity governs individuals even against their will.[36] Domestic procedures based on domestic constitutional law ensure that the implementation of international law vis-à-vis private actors is legitimated.[37] For democratic states, theses procedures ensure that acts of authority vis-à-vis individuals can be traced back to a decision of the electorate.

International organisations or treaty organs such as Conferences of the Parties, and their acts, are creations of states and therefore in principle derive their legitimacy from the consent of states. With respect to the legitimation of institutions through consent, it is helpful to distinguish two forms of consent: on the one hand general consent to a treaty system or to the constitution of an organisation, and on the other hand the specific consent to particular obligations or decisions, as for instance in the form of ratification of a treaty amendment, of a change of an annex or the approval of a resolution.[38] The constitutive treaty to which states have consented (through general consent) ideally delineates the legal boundaries for subsequent institutional activity to which states give their specific consent. The concept of legality and the principle of attribution of competencies serve to connect the subsequent activities to the original basis in (general) state consent.[39] Legality refers to the condition of being in accordance with law or principle, and the principle of attribution of competencies means that the powers of an organisation

---

International Law, 2008, 1-24 (7); for a discussion of consent theory and its shortcoming in meeting the realities of social organization compare A.E. Buchanan, Justice, legitimacy, and self-determination: moral foundations for international law, 2003, 151.

[36] J.H.H. Weiler, "The geology of international law – governance, democracy and legitimacy", Zeitschrift für ausländisches öffentliches Recht und Völkerrecht 64 (2004), 547-562 (548).

[37] Compare R. Wolfrum, "Legitimacy in International Law from a Legal Perspective: Some Introductory Considerations" in: R. Wolfrum/V. Röben (eds.), Legitimacy in International Law, 2008, 1-24 (7).

[38] This distinction is also made by D. Bodansky, "The Legitimacy of International Governance: A Coming Challenge for International Environmental Law?", American Journal of International Law 93 (1999), 596-624 (604).

[39] D. Bodansky, "The Concept of Legitimacy in International Law" in: R. Wolfrum/V. Röben (eds.), Legitimacy in International Law, 2008, 309-317 (311).

are limited to those attributed to them by states, i.e. that they always need a legal basis for their activities.[40] If they exceed these limitations or act without competence, i.e. when their acts cannot be based on explicit or implied powers[41] or on customary powers, an act of the organisation is void for being *ultra vires*.[42]

A further safeguard of legitimacy is the conception that the international institutions are controlled by a plenary organ that in turn remains under the control of states. Furthermore, decision-making procedures typically requiring specific consent to the adoption of instruments ensure that the further development of international organisations cannot be undertaken against the will of individual states.[43] In many areas of international law, and in particular in the environmental field, this is the point where more flexible forms of decision-making have been developed. One such form is decision-making through consensus which is mostly used in international treaty regimes and institutions today. Not having to vote positively for a particular decision allows states to let less significant decisions pass even though they may not have the position to support it. It is thus more flexible than unanimity, but still secures the rights of the minority.[44] Other forms such as contracting-in and opt-out procedures or, in exceptional cases of such as the Montreal Protocol, the possibility to amend Annexes to the treaty through two-thirds majority decision[45] more clearly deviate from the ideal of sover-

---

[40] International organizations do not have a Kompetenz-Kompetenz, see for details H.G. Schermers/N.M. Blokker, International Institutional Law: Unity within diversity, 2003, § 209.

[41] Not all powers of an international organization can be laid down in the founding documents, since it must be able to respond to developments that could not have been foreseen, compare on implied powers H.G. Schermers/ N.M. Blokker, International Institutional Law: Unity within diversity, 2003, § 232.

[42] H.G. Schermers/N.M. Blokker, International Institutional Law: Unity within diversity, 2003, § 206 et seq.

[43] See for details on the distinction of specific and general consent D. Bodansky, "The Legitimacy of International Governance: A Coming Challenge for International Environmental Law?", American Journal of International Law 93 (1999), 596-624 (604).

[44] On this R. Wolfrum, "Konsens im Völkerrecht" in: H. Hattenhauer (ed.), Mehrheitsprinzip, Konsens und Verfassung, 1986, 79-91 (87).

[45] Article 2.9 c) Montreal Protocol on Substances that Deplete the Ozone Layer (16 September 1987), 26 ILM 1550 (1987). On the various possibilities of secondary law-making by international organizations, including rare legislative

eign equality for the sake of flexibility. Apart from these exceptions, however, only nonbinding instruments can generally be adopted by majority decision-making, even though consensus remains the rule in practice. The specific consent to and control of international activities by state representatives in this ideal model also ensures domestic balancing of the interests of various parts of the government. Foreign ministries or other leading ministries hereby coordinate the positions to be taken with other ministries.

By taking this conception as a point of departure, it is acknowledged that this model has certain merits and is not entirely outdated. This is not to ignore the fact that the concept of consent-based legitimation is increasingly subject to challenge. In particular, consent as the main basis of legitimacy at the international level becomes problematic where international law more and more resembles domestic law.[46] Consent is not sufficient if prescriptions of international institutions are indirectly or directly applicable to private actors such as private individuals and organisations, but must then be balanced by the rule of law and judicial control.[47] As aptly described by Joseph Weiler, in a system where the international and the internal are blurred, the conflation of government and state which is implicit in the traditional concept of consent is increasingly untenable as it unduly empowers the executive over other

---

acts, opt-out procedures in technical coordination and the more common contracting-out procedures which safeguard consent of states J.D. Aston, Sekundärgesetzgebung internationaler Organisationen zwischen mitgliedstaatlicher Souveränität und Gemeinschaftsdisziplin, 2005, 166-175; for an assessment of flexibilisation through COP decision-making in Multilateral Environmental Agreements see J. Brunnée, "COPing with Consent: Law-Making under Multilateral Environmental Agreements", Leiden Journal of International Law 15 (2002), 1-52.

[46] D. Bodansky, "The Legitimacy of International Governance: A Coming Challenge for International Environmental Law?", American Journal of International Law 93 (1999), 596-624 (606).

[47] R. Wolfrum, "Legitimacy of International Law and the Exercise of Administrative Functions: The Example of the International Seabed Authority, the International Maritime Organization (IMO) and International Fisheries Organizations" in: A. von Bogdandy/R. Wolfrum/J. von Bernstorff/P. Dann/M. Goldmann (eds.), The Exercise of Public Authority by International Institutions: Advancing International Institutional Law 2010, 917-940 (924).

domestic political constituencies.[48] And further, consent is often a fiction in an international political system where most states cannot afford to stay out, and therefore not consenting ceases to be a real option.[49]

But to take this model as a point of departure does not mean that it is considered ideal, but that it still serves as a valid background for an analysis of challenges because it shows where traditional legitimation structures may be insufficient in the case of nonbinding instruments. Contrasting new developments with the traditional model can help to understand and identify areas where one may have to think about improved legitimation structures. In addition, in spite the existing challenges to traditional conceptions, taking the traditional model as a point of departure pays tribute to the fact that the nation state continues to be the most important actor in international relations. General and specific consent of states at the international level and legitimation of domestic implementation through procedures of domestic law therefore continue to be of importance as a minimum requirement for the legitimacy of international decision making,[50] in particular since adequate alternative strategies to legitimise the work of international institutions are still wanting. The formation of political will within states is still highly relevant for the adherence of a collectivity to rules made at least in part by others.[51]

---

[48] J.H.H. Weiler, "The geology of international law – governance, democracy and legitimacy", Zeitschrift für ausländisches öffentliches Recht und Völkerrecht 64 (2004), 547-562 (558).

[49] J.H.H. Weiler, "The geology of international law – governance, democracy and legitimacy", Zeitschrift für ausländisches öffentliches Recht und Völkerrecht 64 (2004), 547-562 (557-558).

[50] V. Röben, Außenverfassungsrecht: eine Untersuchung zur auswärtigen Gewalt des offenen Staates, 2007, 42; similarly R. Wolfrum, "Legitimacy in International Law from a Legal Perspective: Some Introductory Considerations" in: R. Wolfrum/V. Röben (eds.), Legitimacy in International Law, 2008, 1-24 (7).

[51] S. Oeter, "Souveränität – ein überholtes Konzept?", Tradition und Weltoffenheit des Rechts (2002), 259-290 (290).

II. International level challenges

1. The gradual attenuation of original consent

From the traditional perspective described in the last section, the original consent to the constitutive treaty (which outlines the mandate for subsequent international activities) and associated domestic ratification procedures is an important means of legitimation of the activities of international organisations and similar institutions.

This chain of legitimacy from the national to the international level established at least in part by the general consent of states to the base treaty is attenuated when the institutional bodies develop their mandates over time.

Thus, international organisations tend to expand their field of competence with a view to addressing what is being perceived as a functional necessity and a response to changing circumstances.[52] Constitutions or charters of international organisations often cannot, or only rather poorly, account for such "mission creep", which they usually experience over time both with regard to their objectives and the competencies of their organs.[53] As seen in this study, such a trend can be observed in particular with respect to nonbinding instruments. Sustainable development and environmental protection have become central principles of many nonbinding instruments of the OECD or the World Bank. However, the Charter of the World Bank for instance does not contain any provision on environmental protection, nor does it foresee a procedure for the adoption of environmental policies. Nonetheless the World Bank adopted environmental policies to guide the approval of loans and credits.[54] A further illustrative example is that of the OECD Guidelines. While environmental objectives are not mentioned in the OECD Convention and did not play a role in the first version of the OECD Guide-

---

[52] This is a recurring theme in the insightful study by J.E. Alvarez, International Organizations as Law-makers, 2005; for a similar assessment see also E. Hey, "International Institutions" in: D. Bodansky/J. Brunnée/E. Hey (eds.), Oxford Handbook of International Environmental Law, 2007, 749-769 (758).

[53] J.E. Alvarez, International Organizations as Law-makers, 2005, 257.

[54] This is also emphasized by D.A. Wirth, "Compliance with Non-Binding Norms of Trade and Finance" in: D. Shelton (ed.), Commitment and Compliance: The Role of Non-Binding Norms in the International Legal System, 2000, 330-344 (333); J.E. Alvarez, International Organizations as Law-makers, 2005, 235-236.

lines of 1976, an environmental chapter was introduced in 1991 and the revision of 2000 saw the emergence of sustainable development as a guiding principle.

The expansion of mandates can also be observed in institutional developments and in the creation of new kinds of instruments. Due to their embeddedness in traditional international law doctrine, the founding documents of international institutions never provide for instruments that directly address non-state actors. As analysed above,[55] the practice of the FAO to adopt numerous codes of conduct which are not only addressed to states but also to private actors in effect expands the nomenclature of the FAO Constitution, because that only foresees instruments that are addressed to states.[56] And similarly, even though the joint administration of the PIC procedure by UNEP and the FAO can be based on a general substantive mandate of the founding documents of both organisations, neither of their constitutional documents foresees such institutional collaboration.[57]

Now, this expansion of the original mandates does not necessarily render these activities ultra vires and therefore void. Although the doctrine of attributed powers generally ensures that international institutions always need a legal basis for their activities, custom and the implied powers doctrine allow for organisations to develop further in ways not foreseen when they were created.[58] The basis for custom is the implicit consent of states to new powers subsequent to the adoption of the founding document.[59] In the case of continuous cooperation of states and their continuous approval of the activities through political bodies, the powers of the organisation may gradually expand. Whatever is necessary for the organisation to perform its explicitly provided functions is generally considered as authorised by an implied power.[60]

---

55 See the analysis of the FAO CCRF undertaken above, in Part 1, at B.I.

56 FAO Constitution, Article IV.

57 This is emphasized by J.E. Alvarez, International Organizations as Lawmakers, 2005, 236.

58 H.G. Schermers/N.M. Blokker, International Institutional Law: Unity within diversity, 2003, § 232.

59 H.G. Schermers/N.M. Blokker, International Institutional Law: Unity within diversity, 2003, § 232.

60 H.G. Schermers/N.M. Blokker, International Institutional Law: Unity within diversity, 2003, §§ 232 and 233.

Expansive tendencies have been facilitated by a wide interpretation of implied powers. The basis for such a liberal approach towards implied powers was laid down by the ICJ in the *Certain Expenses case* when it opined that, "when the Organisation takes action which warrants the assertion that it was appropriate for the fulfilment of one of the stated purposes of the United Nations, the assumption is that such an action is not ultra vires the Organisation".[61] However, the implied powers doctrine, if applied expansively, weakens the utility of the *ultra vires* doctrine – and therefore of the principle of attribution of competence and legality – to serve as a constraining device that ensures the linkage between the original consent and the democratic process of ratification and later activities. The more recent advisory opinion of the ICJ on the Legality of the Use by a State of Nuclear Weapons in Armed Conflict where the Court found that the mandate of the World Health Organization could not be interpreted to include the competence to deal with the question of the legality of Nuclear Weapons[62] could be seen as an example of a more restrictive approach of the ICJ in the interpretation of the mandates of organisations.[63] Leaving aside the question whether or not this is indicative of a general trend of international and national jurisprudence and scholarship that views international organisations more critically, as suggested by Jan Klabbers,[64] it seems clear that the *ultra vires* doctrine would need considerable refinement and additional force in order to strengthen the principles of legality and attribution of competences. And only if these are strengthened they can secure the link to formal state consent which remains one of the bases for the legitimacy of dynamic institutional developments.

2. *Flexible decision making procedures*

The chain of legitimacy from the state level to the level of international decision making identified above as one of the cornerstones of tradi-

---

[61] ICJ, Certain Expenses of the United Nations (Article 17, Paragraph 2, of the Charter), Advisory Opinion of 20 July 1962, ICJ Reports 1962, p. 151, 168.

[62] ICJ, Legality of the Use by a State of Nuclear Weapons in Armed Conflict, Advisory Opinion of 8 July 1996, ICJ Reports 1996, p. 66, para. 21.

[63] Compare J. Klabbers, An Introduction to International Institutional Law, 2002, 80.

[64] J. Klabbers, "The changing image of international organizations" in: J.-M. Coicaud/V. Heiskanen (eds.), The Legitimacy of International Organizations, 2001, 221-255 (236 et seq.).

tional legitimacy conceptions is weakened in the case of majority decision making. Majority decision making increases the possibility that decisions are taken even against the will of an individual state. While binding instruments – notwithstanding some exceptional cases[65] – usually require the formal consent of every state in order for them to be legally binding for that particular state to be bound by them, nonbinding instruments can be adopted without positive consent. They could even at least from a formal point of view be decided upon by two-third majorities at the highest and even simple majorities for lower level bodies, a fact that increases the pressure on reluctant states.[66] Even if in practice consensus decision making prevails, consensus procedures increases the pressure on individual states to join the majority at least if they are isolated. And after the instrument is adopted, a state can of course choose not to implement the measures, but it cannot as in the case of binding treaty law and secondary law opt out or choose not to ratify. Furthermore, a state that is not in clear support of the instrument can hardly prevent that the adopted instrument gains impact on international norm development, state practice and that it may be used by other states or domestic constituencies to raise pressure on that state.

Moreover, as could be seen in the case of the FAO but also the OECD, the nonbinding instruments are often even addressed to non-members and private actors of non-participating states. These deviations from consensus carry the danger that addressees are subjected to norms and decisions which their representatives have not approved.[67] This is demonstrated by the way in which corporations of non-adherent states to the OECD Guidelines, i.e. states which did not officially sign up to the Guidelines, were listed by the expert group of the Security Council irrespective of the consent of their home states to the Guidelines.[68]

---

[65] As Article 2 (9) c) of the Montreal Protocol.

[66] See e.g. FAO Constitution, Article IV (3), which provides for the possibility to make recommendations by a 2/3 majority. See on the various institutions already the Part 1 section A, further above.

[67] C. Kirchner, "Bedingungen interstaatlicher Institutionalisierung von wirtschaftlichen Prozessen" in: W. Hauff (ed.), Handbuch der Wirtschaftsethik, Band 2: Ethik wirtschaftlicher Ordnungen, 1999, 376-389.

[68] See on this in Part 2, at A.II.2 b) (2).

As mentioned, except for rare cases the main instruments are usually adopted by consensus or unanimity of Members of an organisation.[69] In the case of the OECD Guidelines, its application to non-members is contingent on an act of "adherence", i.e. an act of approval by the respective state. However, lower level bodies may be less reluctant to take decisions by majority vote.[70] Even when decisions are in practice taken by consensus, the mere availability of procedures for majority voting is also of political significance, because it increases the pressure on states to join the consensus to avoid being outvoted. For example, the possibility of a majority decision has increased the pressure on the minority of developed member states to agree to the adoption of the PIC clause at the FAO Conference and the UNEP Council.[71]

3. Informalisation and the growing influence of the executive and experts

The traditional model described above starts from the assumption that the political state-controlled decision-making bodies of the organisation fully control the activities of the organisation. However, this traditional model increasingly comes under pressure.[72] Informalisation and the ongoing shift of norm-making authority to executive specialists and experts represent a challenge for traditional legitimation structures.[73]

---

[69] One exception was the abstention of Turkey in the adoption of the OECD Guidelines, see above in the case study on the OECD Guidelines, in Part 1, at B.III.

[70] See for instance the decision of the Committee of Fisheries of the FAO to mandate the secretariat of the FAO with the development of a standard for marine protected areas even against the expressly stated wish of one member state, compare the Report of the Twenty-Sixth Session of the Committee on Fisheries, 7-11 March 2005, para. 103, available at: ftp://ftp.fao.org/docrep/fao/008/a0008e/a0008e00.pdf.

[71] For details see the case study in Part 1, at B.II., further above.

[72] M. Eifert, "Legitimationsstrukturen internationaler Verwaltung" in: H.-H. Trute/T. Groß/H.C. Röhl/C. Möllers (eds.), Allgemeines Verwaltungsrecht – zur Tragfähigkeit eines Konzepts, 2007, 307-331 (315).

[73] Similarly S. Oeter, "The openness of international organisations to transnational rule-making" in: O. Dilling/M. Herberg/G. Winter (eds.), Transnational administrative rule making: performance, legal effects and legitimacy, 2011, 235-252, 249.

In today's international institutions, organs and bodies on an organisationally subordinated level of international institutions and in particular secretariats and expert bodies are gaining significance and autonomy.[74] They increasingly act as autonomous actors in the pursuit of interests that may deviate to some extent from those of states.[75] One reason for this development is that a number of significant institutional activities are not designed, elaborated and decided upon at the highest political levels, but at lower political levels or by civil servants of the secretariats.

This is also and in particular true for nonbinding instruments. Informalisation through nonbinding instruments plays a significant part in this development. Organisationally subordinated bodies, secretariats and expert bodies gain particular influence through the use and impact of nonbinding instruments.[76] It is a defining feature of the nonbinding instruments analysed in this study that they are not one-time endeavours. Rather, their impact largely derives from their long-term influence brought about by institutionalised processes of continuous norm production, norm concretisation and the management of norm implementation and compliance assistance.[77] It is nothing unusual that decision making is delegated to lower administrative bodies and experts. Decision-making in plenary organs is often too unproductive and inflexible for these tasks.[78] However, any act of delegation implies an extension of the chain of legitimacy and almost inevitably entails a loss of control by states representatives. The collectivity of states, often represented by the plenary body, loses control to the extent to which the mandated body is given the ability to act autonomously. However, the degree of the autonomy depends on a variety of factors. Among them are the degree of specificity and formalisation of the mandate and the availability of oversight mechanisms. A mandate of vague and general character increases the autonomy of the authorised body. Finally, the composition

[74] M. Eifert, "Legitimationsstrukturen internationaler Verwaltung" in: H.-H. Trute/T. Groß/H.C. Röhl/C. Möllers (eds.), Allgemeines Verwaltungsrecht – zur Tragfähigkeit eines Konzepts, 2007, 307-331 (315).

[75] J.E. Alvarez, International Organizations as Law-makers, 2005, 585.

[76] Similarly M. Eifert, "Legitimationsstrukturen internationaler Verwaltung" in: H.-H. Trute/T. Groß/H.C. Röhl/C. Möllers (eds.), Allgemeines Verwaltungsrecht – zur Tragfähigkeit eines Konzepts, 2007, 307-331 (315).

[77] See already the compliance analysis in Part 2, at B.I., further above.

[78] FAO, The Challenge of Renewal, Independent External Evaluation of the Food and Agriculture Organization, Working Draft (2007), available at http://www.fao.org/unfao/bodies/IEE-Working-Draft-Report/K0489E. pdf.

of the body, and in particular whether it is composed of government representatives or of civil servants or other experts, is of significance for the degree of influence of states.

a) Broad mandates and lack of institutional law

It has been observed that the delegation of norm development or norm implementation activities is generally based on broad and unspecific mandates and only rarely accompanied by any instrument-specific procedural or substantive rules. It is true that there usually exist rules of procedures and terms of reference outlining such things as voting procedures, composition and elections of committees. However, at least with respect to nonbinding instruments and related activities, one searches in vain for specific guidance on how these activities should be exercised. The broad and unspecific mandates given to subordinate bodies or secretariats of the FAO or of UNEP have provided them with wide discretion in how to conduct their activities. They have made use of this discretion both in terms of substantive decision making and procedural questions.[79] A particularly illustrative example in this respect is the informal delegation of decision-making authority to the Joint Group of Experts under the voluntary PIC system.[80] No specific mandate existed for the secretariat to assemble such an expert group, and no particular predetermined decision-making procedure was provided for by either the political bodies or the secretariat. Instead, the Joint Expert Group adopted its own procedures, often acted on an ad hoc basis, and used input from industry, NGOs or governments as it saw fit. The only real exception is the procedural guidance adopted by the OECD Council for the activities of the Investment Committee and the National Contact Points. This may be because the implementation procedures of the OECD Guidelines directly implicate private actors which are represented at the OECD by the Business and Industry Advisory Committee.

79 See on this already Part 1, on Instruments, further above.

80 See the detailed analysis in the case study of the PIC system, in Part 1, at B.II.3, further above.

b) Weak oversight and control

Another factor which increases the potential for autonomous decision making by subsidiary bodies is the general lack of substantive oversight and control within the organisation. Most organisations simply rely on internal reporting mechanisms and budgeting as the main mechanism to control lower level bodies. An accountability deficit results in particular with respect to the secretariats that are made up of international civil servants who are not accountable to states.[81]

c) Delegation to subsidiary decision-making bodies

In subsidiary organisational bodies – as for example the Committee for Fisheries (COFI) and its subsidiaries or the Investment Committee (IC) of the OECD and subsidiaries – state representatives remain responsible for decision-making. But independence from political oversight is not only a question of whether a body is composed of representatives of governments or not.[82] The degree of independence of a body from control of higher political levels also depends on the individual members of the body. The FAO's COFI and the Investment Committee of the OECD are largely made up of representatives of specialised ministries. Representatives of specialised ministries may pursue an agenda and have an "agency mindset" that is different from other parts of their governments or the foreign ministries.[83] Once agents with a common agenda or mindset come together, their continuous and often more frequent interaction is likely to produce results which may often differ from those of higher level bodies composed of diplomats at the highest political level.[84] Through their international cooperation, networks of specialised agents may be able to insulate themselves against criticism or influence by other parts of the government. The ensuing constraint on

81 The insufficient oversight of the FAO Council over the secretariat has been highlighted in the independent expert evaluation of the FAO, The Challenge of Renewal, Independent External Evaluation of the Food and Agriculture Organization, Working Draft (2007), available at http://www.fao.org/unfao/bodies/IEE-Working-Draft-Report/K0489E.pdf.

82 J.E. Alvarez, International Organizations as Law-makers, 2005, 247.

83 K.W. Abbott/D. Snidal, "Hard and Soft Law in International Governance", International Organization 54 (2000), 421-456 (453).

84 Similarly J.E. Alvarez, International Organizations as Law-makers, 2005, 247; A.-M. Slaughter, A New World Order, 2004.

proper balancing of interests at the national and international level may be more pronounced in the case of nonbinding instruments where control on the part of higher political bodies is likely to be more lenient.

d) Delegation to secretariats

The delegation of norm elaboration and implementation activities to civil servants working in particular in secretariats creates areas of executive discretion in both norm creation and norm implementation activities. Of course, secretariats are not acting independently from political bodies. However, tight control and oversight by political bodies is rare.[85] The significance of all of these activities and the lack of oversight stands in contrast to the limited procedural or substantive guidance from political organs.

For example, the examples of the OECD Guidelines and the FAO Code of Conduct for Responsible Fisheries (CCRF) illustrate the extent to which secretariats – on the basis of broad unspecific mandates – provide for concretisations and commentaries of the main instruments. They enact model legislation and best practices which they disseminate and promote, often through their advisory role in developing countries. As has been analysed in this study, the impact of these activities is not negligible. Secretariats are instrumental in administering and conducting international implementation efforts and follow-up processes. Again acting on a broad mandate unrestricted by any procedural or substantive institutional law, secretariats often decide on such issues as the granting of compliance assistance. As illustrated by the cooperation of the UNEP and FAO secretariats on the voluntary PIC regime or the cooperation of FAO with CITES and the WTO, they also conduct inter-institutional cooperation.

e) Delegation to experts

The structural particularities of environmental issues, namely high levels of scientific uncertainty and problem complexity, are only some of

[85] The tight governmental control of secretariats of international organizations or within treaty regimes is identified as a myth by E. Brown Weiss, "Understanding compliance with international environmental agreements: the Baker's dozen myths", University of Richmond law review 32 (1999), 1555-1589 (1570).

the factors which render expertise essential for solving environmental problems. Expert bodies and experts working for secretariats of international institutions not only provide important informational input for modern environmental institutions. As could be seen in the case studies, they also conduct some of the norm development and centralised implementation activities. As a result of the influence of expertise at all levels of decision making, the general understanding of the issues in the organisation and of political bodies is shaped according to the conceptions of these so-called epistemic communities which often share a common mindset.[86]

Expert groups, important as they are, often enjoy considerable autonomy from political control. This is often the result of their particular authority stemming from their expertise. Even when their work is subject to controlling decisions of a political body, the proposals of experts are often of such a complex nature that they are not always or not in detail scrutinised by political bodies such as the plenary.[87] The most striking example of autonomous expert decision making was provided for by the activities of the FAO/UNEP Joint Expert Group. Only subject to internal reporting mechanisms, it basically administered the voluntary PIC procedure and took all important decisions. It not only approved the decision guidance documents but also decided which pesticides and chemicals should be included in the system: decisions which were politically and economically significant and contested. Without being mandated to do so, the expert group also changed the details of the PIC procedure, for example by broadening the criteria for determining which substances should be included in the PIC system.

Although it is difficult to generalise, the transformation of the PIC system into a binding legal regime suggests that expert discretion may of-

---

[86] P. Haas, "Epistemic Communities" in: D. Bodansky/J. Brunnée/E. Hey (eds.), The Oxford Handbook of International Environmental Law, 2007, 791-806; J. von Bernstorff, "Procedures of Decision-Making and the Role of Law in International Organizations" in: A. von Bogdandy/R. Wolfrum/J. von Bernstorff/P. Dann/M. Goldmann (eds.), The Exercise of Public Authority by International Institutions: Advancing International Institutional Law, 2010, 777-806 (788).

[87] J. von Bernstorff, "Procedures of Decision-Making and the Role of Law in International Organizations" in: A. von Bogdandy/R. Wolfrum/J. von Bernstorff/P. Dann/M. Goldmann (eds.), The Exercise of Public Authority by International Institutions: Advancing International Institutional Law, 2010, 777-806 (787) ("rubber-stamping role" of national representatives).

ten be greater if the decisions taken are of nonbinding nature. In contrast to the voluntary PIC system, the PIC Convention provides for a clear mandate and functions of a joint secretariat[88] as well as a clear mandate for the establishment of an expert body – the Chemical Review Committee – by the Conference of the Parties.[89] Unlike the voluntary instruments, rules on the composition and voting procedures of this so-called Chemical Review Committees are also included.[90] Most importantly perhaps, the Convention clarifies the competencies and functions of this Committee and even contains detailed criteria for the listing of substances to be considered by the Committee.[91] This formalisation through procedural law has certainly improved the legitimacy of the procedure. On the other hand, this development also has repercussions for the flexibility of the system. In particular administration by the expert body, under the Convention called Chemical Review Committee, is curtailed. Even though some flexibility is maintained through the two-thirds majority voting of the Chemical Review Committee,[92] it cannot autonomously decide on new definitions, new criteria or the inclusion of new categories of substances as did the Joint Expert Group. Above all, the Chemical Review Committee's decisions are now subjected to the political control of the Conference of the Parties. While none of the operational decisions were made by a political body under the voluntary PIC system, the Conference of the Parties now ultimately decides on the listing or de-listing of chemicals on the basis of the recommendations of the Chemical Review Committee by consensus.[93] Consequently, the listing of substances has now become more politicised. As could be expected, Parties to the PIC Convention now struggle over each listing. The consensus rule has also raised the prospects for a manufacturer to gain an economic advantage from lobbying its government to veto the inclusion of particular substances into the PIC system.[94] As a consequence, the first four Conferences of the Parties,

---

[88] PIC Convention, Article 19.

[89] PIC Convention, Article 18.6.

[90] PIC Convention, Article 18.6.

[91] Compare PIC Convention, Articles 5.6, 6.5, 7.1 and 9.2 as well as Annexes II and IV, Part 3.

[92] PIC Convention, Article 18.6 (c).

[93] PIC Convention, Articles 7.3 and 9.3.

[94] N.S. Zahedi, "Implementing the Rotterdam Convention: the challenges of transforming aspirational goals into effective controls on hazardous pesticide

despite a recommendation to this effect by the expert group, could not agree on listing the chemical chrysotile asbestos – a hazardous chemical with potentially harmful effects on human health and the environment which is banned in the European Union but still produced in other countries such as Canada.[95] The opposition of some exporting countries, including in particular Canada, has impeded this. This issue has added to increasing concerns that the decision making procedures of the Convention render the Convention ineffective.[96] In other words, the additional formalisation and stronger political control in the case of the PIC Convention has come at the cost of reduced flexibility and possible ineffectiveness.

III. National level challenges

1. Deparliamentarisation

Domestic procedures at least in democratic countries help to ensure that regulatory decisions with implications for private actors are being taken with adequate parliamentary participation. This model builds on the understanding that while the executive conducts foreign affairs, it is the legislature which is responsible for implementation.

The growing economic and ecological interdependence of states and the ensuing need for decision-making at the international level in areas formerly within the realm of states is accompanied by an expansion of the powers of the executive and parallel deparliamentarisation.[97] The

exports to developing countries", Georgetown international environmental law review 11 (1999), 707-739 (727).

95 Report of the Conference of the Parties to the Rotterdam Convention on the Prior Informed Consent Procedure for Certain Hazardous Chemicals and Pesticides in International Trade on the work of its third meeting, Doc. UNEP/FAO/RC/COP.3/26 of 10 November 2006, Decision RC-3/3.

96 International Institute for Sustainable Development, Earth Negotiations Bulletin, Vol. 15 No. 168, at 10, available at http://www.iisd.ca/download/pdf/enb15168e.pdf.

97 R. Wolfrum, "Legitimacy in International Law from a Legal Perspective: Some Introductory Considerations" in: R. Wolfrum/V. Röben (eds.), Legitimacy in International Law, 2008, 1-24 (19-20); from a US perspective C.A. Bradley, "International delegations, the structural constitution, and non-self-execution", Stanford law review 55 (2003), 1557-1596; for a detailed assessment of various forms of deparliamentarisation from a German constitutional law

outsourcing of decision making with implications for private actors includes a "factual loss of power" for parliamentarians who do not participate in the formulation of the norms, and afterwards only have the chance to ratify the result, or – as is sometimes the case for nonbinding instruments – not even that.[98]

The move to nonbinding instruments is part and parcel of this trend to exclude parliaments from international norm development, and poses particular challenges. As Christian Tomuschat observed already in 1978, the increasing reliance on nonbinding instruments increases the weight of the executive to the detriment of parliamentary participation and control.[99] Nonbinding instruments contribute to the internationalisation of domestic affairs, but do not trigger ratification procedures,[100] and they may at times be implemented in the domestic legal order without a specific legislative act. Some consequently fear the possible misuse of nonbinding instruments which circumvent parliamentary participation but nevertheless influence international and domestic lawmaking and administrative implementation.[101] Even though such devel-

---

perspective W. Kahl, "Parlamentarische Steuerung der internationalen Verwaltungsvorgänge" in: H.-H. Trute/T. Groß/H.C. Röhl/C. Möllers (eds.), Allgemeines Verwaltungsrecht – Zur Tragfähigkeit eines Konzepts, 2008, 71-106; T. Puhl, "Entparlamentarisierung und Auslagerung staatlicher Entscheidungsverantwortung" in: J. Isensee/P. Kirchhof (eds.), Handbuch des Staatsrechts der Bundesrepublik Deutschland, Band III, 2005, 639-682 (79-82).

98 T. Puhl, "Entparlamentarisierung und Auslagerung staatlicher Entscheidungsverantwortung" in: J. Isensee/P. Kirchhof (eds.), Handbuch des Staatsrechts der Bundesrepublik Deutschland, Band III, 2005, 639-682 (641).

99 C. Tomuschat, "Der Verfassungsstaat im Geflecht der internationalen Beziehungen", Veröffentlichungen der Vereinigung der Deutschen Staatsrechtslehrer 36 (1978), 7-63 (32-34); compare also B.-O. Bryde, Internationale Verhaltensregeln für Private – Völkerrechtliche und verfassungsrechtliche Aspekte, 1981, 36.

100 In Germany, for instance, only binding treaty law is subject to ratification requirements, Article 59 (2) of the Basic Law does not apply to nonbinding norms irrespective of their factual significance, see R. Streinz, "Art. 59 [Völkerrechtliche Vertretungsmacht]" in: M. Sachs (ed.), Grundgesetz: Kommentar, 2009, (mn. 40); S. Kadelbach, "Die parlamentarische Kontrolle des Regierungshandelns bei der Beschlußfassung in internationalen Organisationen" in: R. Geiger (ed.), Neuere Probleme der parlamentarischen Legitimation im Bereich der auswärtigen Gewalt, 2003, 41-57 (50).

101 F. Orrego Vicuña, "In Memory of Triepel and Anzilotti: The Use and Abuse of Non-Conventional Lawmaking" in: R. Wolfrum/V. Röben (eds.), De-

opments can be justified as being legal under the respective constitutional law,[102] the shift to the executive may in certain situations require measures to adjust traditional legitimation processes at the domestic level or to adopt additional ones at the international level.

In terms of parliamentary legitimation, one could of course always point to the adoption and ratification of the constitutive treaty of the respective international organisation or treaty. However, the legitimating role of domestic ratification is limited for several reasons.[103] One is the fact that often parliaments only have the choice to approve or disapprove a fully negotiated treaty. In reality, given the international political situation, parliamentarians are often under great pressure to approve. Furthermore, and more importantly, international treaty regimes and international institutions today usually work on the basis of framework treaties, and actual decision-making is thus often far removed from the original consent. And thirdly, one cannot even assume that parliaments have approved the constitutive documents of international organisations. In German scholarship, there is some insecurity as to when the accession to an international organisation must be ratified, i.e. when it constitutes a treaty that regulates the political relations of the federal state under Article 59 (2) of the German Basic Law.[104] The accession of Germany to the FAO, for example, was in contrast to the

---

velopments of International Law in Treaty Making, 2005, 497-506; similarly also K. Hailbronner, "Völkerrechtliche und staatsrechtliche Überlegungen zu Verhaltenskodizes für transnationale Unternehmen" in: I. v. Münch (ed.), Staatsrecht – Völkerrecht – Europarecht: Festschrift für Hans-Jürgen Schlochauer, 1981, 329-362 (357); C. Tomuschat, "Der Verfassungsstaat im Geflecht der internationalen Beziehungen", Veröffentlichungen der Vereinigung der Deutschen Staatsrechtslehrer 36 (1978), 7-63 (34).

[102] In Germany this development is arguably justifiable under the German Basic Law with its openness to international cooperation which is enshrined in its preamble and Articles 24 to 26; for this interpretation see B.-O. Bryde, Internationale Verhaltensregeln für Private – Völkerrechtliche und verfassungsrechtliche Aspekte, 1981, 36.

[103] On the problems of parliamentarian control through domestic ratification see also W. Kahl, "Parlamentarische Steuerung der internationalen Verwaltungsvorgänge" in: H.-H. Trute/T. Groß/H. C. Röhl/C. Möllers (eds.), Allgemeines Verwaltungsrecht – Zur Tragfähigkeit eines Konzepts, 2008, 71-106 (79 et seq.).

[104] See instead of many R. Wolfrum, "Kontrolle der auswärtigen Gewalt", Veröffentlichungen der Vereinigung der Deutschen Staatsrechtslehrer 56 (1997), 38-66 (51).

UN, NATO or the WTO undertaken without the approval of the German Bundestag.[105]

The nature and seriousness of the challenge of deparliamentarisation depends on whether and to what extent the parliament may be involved in the implementation of the respective instruments. It thus appears necessary to differentiate between the various ways in which nonbinding instruments enter the national system as analysed in detail above, namely implementation without a specific legislative act and implementation though specific legislation.

a) Implementation without a specific legislative act

National legal systems provide avenues for the implementation of nonbinding instruments without any specific legislative implementing act, for example through internal administrative guidelines.[106] However, the administrative implementation of the international instruments without specific legislative basis is – at least in the German legal order – only possible on the basis of a general legal authorisation to do so. General legal authorisation refers to the mandate for the administration to act, as for example to give export credits or to require impact assessments. Specific legislation refers to the specific legislative approval of a particular international instrument. The requirement that the administration must always act on the basis at least of a general legal authorisation is a cornerstone of a democratic and rule of law based legal system. It renders illegal any administrative or judicial action which is solely based on the considerations of international nonbinding norms, so that international they cannot be the exclusive basis for executive decisions.[107]

---

[105] S. Kadelbach, "Die parlamentarische Kontrolle des Regierungshandelns bei der Beschlußfassung in internationalen Organisationen" in: R. Geiger (ed.), Neuere Probleme der parlamentarischen Legitimation im Bereich der auswärtigen Gewalt, 2003, 41-57 (43).

[106] Compare the analysis of national implementation, in Part 2, at B.II., further above.

[107] See already B.-O. Bryde, Internationale Verhaltensregeln für Private – Völkerrechtliche und verfassungsrechtliche Aspekte, 1981, 39; K. Hailbronner, "Völkerrechtliche und staatsrechtliche Überlegungen zu Verhaltenskodizes für transnationale Unternehmen" in: I. v. Münch (ed.), Staatsrecht – Völkerrecht – Europarecht: Festschrift für Hans-Jürgen Schlochauer, 1981, 329-362 (355 and 362).

Even though such a general authorisation to act exists, parliaments remain left out with respect to the specific issue at hand. Without a signalling event such as ratification, debate on the specific issue becomes less likely.[108] Depending on the leeway given to administrators, nonbinding instruments may thus be adopted internationally and implemented nationally without any parliamentary interference. Opting for nonbinding instruments instead of a treaty can insulate the policy from domestic political scrutiny which would at least theoretically be possible if ratification and legislative implementation were required.[109]

Some commentators consequently warn against the tendency to allow administrative implementation on the basis of general clauses.[110] The lack of specific parliamentary participation reduces opportunities for the public and the parliamentary opposition to demand justification for the specific international decision making or to publicly debate particular developments. Consequently there is a need for enhanced participation of parliaments and for transparency in international negotiations even in the context of nonbinding instruments.

b) Implementation by means of a specific legislative act

In most cases, effective implementation of a particular nonbinding instrument requires a legislative act at least under German law. In these cases, the national parliament is specifically involved.[111] Nevertheless, legitimacy problems related to deparliamentarisation may arise here as well.

---

[108] K. Raustiala, "Form and Substance in International Agreements", American Journal of International Law 99 (2005), 581-614 (597).

[109] Similarly e.g. H. Hillgenberg, "A Fresh Look at Soft Law", European Journal of International Law 10 (1999), 499-515 (504).

[110] C. Engel, Völkerrecht als Tatbestandsmerkmal deutscher Normen, 1989, 252 who differentiates between general clauses and administrative discretion, and sees no limitations for the latter; K. Hailbronner, "Völkerrechtliche und staatsrechtliche Überlegungen zu Verhaltenskodizes für transnationale Unternehmen" in: I. v. Münch (ed.), Staatsrecht – Völkerrecht – Europarecht: Festschrift für Hans-Jürgen Schlochauer, 1981, 329-362 (357)

[111] See on this Part 2, at B.II.1., further above.

(1) References in legislative acts to nonbinding instruments (dynamic references)

References in national legislation to specific nonbinding instruments may also pose problems of democratic legitimation.[112] For example, references to a specific instrument in its most recent form, which in German legal doctrine is often described as a "dynamic reference", relocates legislative competencies by allowing regulation and law making to be conducted outside of the realm of influence of the domestic legislative branch.[113] The national legal order is directly opened to the influence of representatives of the governing branch of all member states of an international institution.[114] Of course, one should not forget that the legislative approval of the constitutive treaty in combination with the legislative approval of the dynamic reference results in some parliamentary participation.[115] In the case of nonbinding instruments which can usually be adopted by two-thirds or simple majority vote,[116] the parliament however factually surrenders its influence on decision-making for the future. In the case nonbinding instruments are given effect domestically by way of dynamic references, additional legitimating mechanisms appear to be necessary.[117]

(2) "Programmed" legislation

The implementation of international nonbinding instruments through a specific legislative act also gives rise to legitimacy challenges in the case of what can be referred to as "programmed legislation". "Programmed

---

112 C. Engel, Völkerrecht als Tatbestandsmerkmal deutscher Normen, 1989, 41; for an overview of the discussion on this issue consider in German constitutional law scholarship, see instead of many C. Tietje, Internationalisiertes Verwaltungshandeln, 2001, 600 et seq.

113 Bundesverfassungsgericht, BVerfGE 42, 285, 312 (which sees this as a problem where basic freedoms are affected; the dynamic reference is then interpreted as a static one); see also more recently Bundesverfassungsgericht, BVerfG, 2 BvR 2408/06.

114 C. Tietje, Internationalisiertes Verwaltungshandeln, 2001, 601 and 603-616.

115 C. Tietje, Internationalisiertes Verwaltungshandeln, 2001, 210-212.

116 Instead of many, compare the FAO.

117 B.-O. Bryde, Internationale Verhaltensregeln für Private – Völkerrechtliche und verfassungsrechtliche Aspekte, 1981, 37.

legislation" is here used to describe those instances where the international nonbinding norms serve as a blueprint that largely predetermines the content of domestic implementing legislation. Contrary to traditional treaty law which outlines the objectives but left the means of implementation to the discretion of states, one of the characteristics of many more recent (binding and nonbinding) instruments is that they entail detailed prescriptions of actions to be taken.[118] The prescription of particular implementation measures reduces the leeway of parliaments if they seek to implement the respective instrument. Much of the substance of the law is *de facto* predetermined by the international discourse and debate.

In fact, the entire exercise of adopting an international instrument outlining norms of behaviour for a specific issue area only makes sense if the instrument can be implemented without calling into question its content, process of elaboration and the balance of interests achieved in the negotiations. It is in the interest of all actors participating in the international initiative that implementation is comprehensive and the package not reopened, so as not to upset the balance of different economic, environmental and social interests reached in international negotiations. In particular developing countries, due to limited resources, often need to rely on the international binding but also nonbinding instruments as a reliable and legitimate source of norms.[119] Thus legislative acts that implement nonbinding instruments are often not genuinely homemade but amount to "rubberstamped regulation worked out at the level of [International Organisations]".[120]

---

118 Compare the analysis of the characteristics of the instruments in Part 1, at C, further above.

119 Recent examples for nonbinding instruments that are intended specifically to serve as guidance instruments for national legislation are UNEP Guidelines for the development of national legislation on liability, response action and compensation for damage caused by activities dangerous to the environment and the UNEP Guidelines for the development of national legislation on access to information, public participation and access to justice in environmental matters, both adopted at the 11th Special Session of the UNEP Governing Council in 2010, see UN Doc. UNEP/GCSS.XI/L.5 (2010).

120 J. Delbrück, "Exercising Public Authority Beyond the State: Transnational Democracy and/or Alternative Legitimation Strategies?", Indiana Journal of Global Legal Studies 10 (2003), 29-43 (35-36); C. Tietje, Internationalisiertes Verwaltungshandeln, 2001, 288-487; similarly also T. Stein, "Demokratische Legitimierung auf supranationaler und internationaler Ebene", Zeitschrift für ausländisches öffentliches Recht und Völkerrecht 64 (2004), 563-570 (566).

Given the apparent functional need for a comprehensive treatment of certain issues at the international level, it is in the interest of all participants in the international process to develop procedures which remedy the lack of domestic parliamentary participation at the negotiation stage. In other words, ways must be found to improve the possibilities for the parliament to influence the activities of the government at the international level even for nonbinding instruments.

2. *Internationalisation and public discourse*

To the extent that decision-making shifts to international fora and thus parliamentary participation is weakened, the distance between the national public discourse and international decision making is increased. This poses some risk. Public discourse fulfils an essential role in a democratic system. Constitutionally safeguarded public discourse and debate in the public realm close the gap between the citizens of a state and the governing organs.[121] Furthermore, the general public or so-called civil society is instrumental in making decision makers sensitive to otherwise neglected or ignored issues.[122] By shifting debate and discourse to international fora, processes of internationalisation erode these mechanisms.

In the absence of a strong global public at the international level,[123] international public discourse cannot take up this function. Mechanisms which serve to improve the access and participation of the national public in international processes then become necessary to maintain the legitimate function of the national public in debating international issues, including the international activities and behaviour of executives, and in generally stimulating debate on such issues in parliament.[124] One way to

---

121 R. Herzog, "Art. 20 GG" in: T. Maunz/G. Dürig (eds.), Grundgesetz: Kommentar, 2009, (mn. 69-73).

122 J. Habermas, Faktizität und Geltung: Beiträge zur Diskurstheorie des Rechts und des demokratischen Rechtsstaats, 1998, 460-463.

123 For the distinction of strong and weak public see J. Habermas, Faktizität und Geltung: Beiträge zur Diskurstheorie des Rechts und des demokratischen Rechtsstaats, 1998; the emerging global public is at best a weak one which does not meet the exigencies of a functioning public of domestic systems, compare H. Brunkhorst, Solidarität: von der Bürgerfreundschaft zur globalen Rechtsgenossenschaft, 2002, 187.

124 J. von Bernstorff, "Zivilgesellschaftliche Partizipation in Internationalen Organisationen: Form globaler Demokratie oder Baustein westlicher Experten-

stimulate such processes is to introduce procedures for access to information and public participation in international institutional processes, as discussed below.[125]

3. *Sectoralisation*

Where administrative officials cooperate without involving foreign ministries or other parts of the government, and possibly without any parliamentary participation, this may endanger the balancing of interests at the domestic level which provides a necessary counterbalance to the sectoralisation and fragmentation at the international level.[126] The development of nonbinding instruments does not trigger procedures of participation of other ministries to the extent that treaty instruments do. In the case of the development of treaty law, the Federal Ministry of Germany which is responsible for the area dealt with by the treaty is required to consult with other Ministries in much the same way as they consult in preparing legislation. In addition, the Common Rules of Procedure of the Federal Ministries obligate the respective Ministry to inform the Foreign Ministry and obtain its approval.[127] Such rules do not exist for nonbinding instruments.

Certainly, such balancing is often nevertheless done in practice, but there is no obligation to do so. As a result, the use of nonbinding instruments may enable a single ministry to circumvent parliamentary ratification procedures, but also – at least theoretically – provide an opportunity to avoid coordinating their position with that of other minis-

---

herrschaft?" in: H. Brunkhorst (ed.), Demokratie in der Weltgesellschaft, 2009, 277-302 (296 et seq.).

125 See section C.III.3. of this Part.

126 On the issue of transnational cooperation and its implications C. Möllers, "Transnationale Behördenkooperation: Verfassungs- und völkerrechtliche Probleme transnationaler administrativer Standardsetzung", Zeitschrift für ausländisches öffentliches Recht und Völkerrecht 65 (2005), 351-389 (379-380).

127 § 72 (2) of the Gemeinsame Geschäftsordnung der Bundesministerien (Common Rules of Procedure of the Federal Ministries of the German Government) reads: "Vor Aufnahme von Verhandlungen und Teilnahme an Konferenzen über völkerrechtliche Übereinkünfte mit auswärtigen Staaten, ihren Organen und mit internationalen Organisationen hat das federführende Bundesministerium das Auswärtige Amt rechtzeitig zu unterrichten und seine Zustimmung einzuholen, soweit keine besondere Regelung getroffen wurde." The rules of procedure are available at www.verwaltung-innovativ.de.

tries. Consequently, the turn to nonbinding instruments and the avoidance of binding treaty instruments which is explicitly recommended in the Common Rules of Procedures of the Federal Ministries in Germany[128] also means less certainty that different interests will be weighed against each other at the national level before international norms are adopted. As indicated, the issue is particularly problematic where ministerial bureaucrats of a lower political level cooperate transnationally. Where higher level bodies of international organisations participate in the processes, participation of foreign ministries or the governmental level besides ministerial bureaucrats is most often secured.

4. Challenges arising from addressing private actors

Nonbinding instruments which largely build on state implementation must be distinguished from those which almost exclusively aim at regulating private actors, such as codes of conduct. Comparable to national administrative law, many of these instruments indeed directly impact private actors.[129] In this case, the traditional separation of the international and the national spheres become blurred. The traditional concept that legitimation of acts concerning private actors derives from domestic implementation is therefore eroded.

In some cases domestic administrations are involved in the implementation. An example is the voluntary PIC system. As illustrated by the Designated National Authorities established by the voluntary PIC system or in the National Contact Points of the compliance review procedures of the OECD Guidelines, international organisations sometimes resort to national administrations to administer international procedures. Nonbinding international norms are then directly applied by domestic administrators taking part in international procedures. These cases deviate considerably from the traditional implementation model.

The diminished role of domestic substantive legislation and procedural administrative law also means that law plays a more limited role in guaranteeing legitimacy. Where international instruments remain entirely voluntary without any implications for private actors in the case of non-compliance, this may not be problematic. However, when the international instruments are supported by specific review mechanisms which cannot be ignored by private actors, as in the case of the OECD

---

128 *Ibid.*, § 72 (1).

129 J.E. Alvarez, International Organizations as Law-makers, 2005, 245.

Guidelines, additional forms of legitimation must be sought. Some form of international procedural law that ensures the balancing of interests, participation of stakeholders as well as minimum of due process for private addressees becomes necessary. The Procedural Guidance of the OECD Council to the Investment Committee and the National Contact Points which includes a review process, a nascent right to be heard as well as procedures for the protection of confidential information, appears to be a development in this direction.[130]

C. Addressing the challenges

I. General approach

In searching for new and improved legitimation strategies to address the challenges outlined above several paths can be pursued. In the following, I will distinguish output from input as well as international and domestic level approaches.

One implicit assumption in the following selection of possible improvements is that new forms of international cooperation through international institutions can be legitimated without necessarily requiring a global parliament or global institutions governed by popularly elected representatives.[131] Such conceptions do not appear as an adequate model for assessing international institutions today. Apart from their highly questionable political feasibility and desirability,[132] they face the prob-

---

[130] See for details the analysis of the specific instances procedure of the OECD in Part 2, at C.I.1.

[131] This is proposed as the long-term objective in the conception of cosmopolitan model of global law and democracy, compare e.g. D. Held, Democracy and the global order: from the modern state to cosmopolitan governance, 1995, 278; for a proposal of a global parliament R.A. Falk/A. Strauss, "On the creation of a global peoples assembly: legitimacy and the power of popular sovereignty", Stanford journal of international law 36 (2000), 191-220 (191).

[132] It is questionable whether the domestic model of democracy could be employed as a tool for better legitimation of international institutions and governance, see only J.H.H. Weiler, "The geology of international law – governance, democracy and legitimacy", Zeitschrift für ausländisches öffentliches Recht und Völkerrecht 64 (2004), 547-562 (548); on the various existing conceptions of how to square democracy with governance and globalisation A. von Bogdandy, "Demokratie, Globalisierung, Zukunft des Völkerrechts – eine

lem that democracy depends on the existence of a global public or *demos*, i.e. a collective sense of community or identity. A global *demos* is neither existing nor perceivable at the international level in the near future and consequently, other alternative conceptions of legitimation must be found.[133]

Similarly inadequate are approaches that only accept national democratic states as being able to exercise authority legitimately, and are consequently sceptical towards international governance.[134] This concept of sovereignty that exclusively rests with the nation state is however outdated in times where growing interdependence forces states to cooperate internationally for the sake of maintaining a certain minimum of independence and the ability to pursue their interests.[135] Besides, this latter model hardly appears as a viable option for solving today's pressing problems, such as those relating to the environment, which require international cooperation.[136]

Bestandsaufnahme", Zeitschrift für ausländisches öffentliches Recht und Völkerrecht 63 (2003), 853-877.

133 Similarly A. Buchanan/R.O. Keohane, "The Legitimacy of Global Governance Institutions" in: R. Wolfrum/V. Röben (eds.), Legitimacy in International Law, 2008, 25-62 (39); D. Bodansky, "Legitimacy" in: D. Bodansky/J. Brunnée/E. Hey (eds.), Oxford Handbook of International Environmental Law, 2007, 704-723 (716); J. Delbrück, "Exercising Public Authority Beyond the State: Transnational Democracy and/or Alternative Legitimation Strategies?", Indiana Journal of Global Legal Studies 10 (2003), 29-43 (40).

134 E.-W. Böckenförde, "Die Zukunft politischer Autonomie: Demokratie und Staatlichkeit im Zeichen von Globalisierung, Europäisierung und Individualisierung" in: E.-W. Böckenförde (ed.), Staat, Nation, Europa: Studien zur Staatslehre, Verfassungstheorie und Rechtsphilosophie, 1999, 103-126 (123-125); J.A. Rabkin, Law without nations?: Why constitutional government requires sovereign states, 2005, in particular 233 et seq.

135 G. Dahm/J. Delbrück/R. Wolfrum, Völkerrecht, Band I/1, Die Grundlagen. Die Völkerrechtssubjekte, 1989, 218 et seq.; on the concept and role of a sovereignty as a changing but still necessary concept to describe the central role of states in the development and legitimation of international law, compare S. Oeter, "Souveränität – ein überholtes Konzept?", Tradition und Weltoffenheit des Rechts (2002), 259-290 (290).

136 Similar D. Bodansky, "Legitimacy" in: D. Bodansky/J. Brunnée/E. Hey (eds.), Oxford Handbook of International Environmental Law, 2007, 704-723 (721); E. Hey, "International Institutions" in: D. Bodansky/J. Brunnée/E. Hey (eds.), Oxford Handbook of International Environmental Law, 2007, 749-769 (759 and 764).

In sum, therefore, the underlying assumptions of the following suggestions are that international governance and law are necessary, that a retreat to governance only through nation states is neither desirable nor feasible, and that international institutions can be legitimated in ways that do not need to reproduce the model of democratic nation states at the international level.

1. Domestic and international approaches

One possible strategy could be to improve the chain of legitimation at the domestic level from the representatives of the people to the executive engaged in international negotiations and decision making.[137] Another strategy often proposed is to strive for legitimation of international processes, for example by identifying and developing administrative law procedures which ensure transparency, minimum participatory requirements and fairness.[138] Both remedial strategies respond to different aspects of the challenge, which concern both the national (e.g. deparliamentarisation) and the international level (e.g. erosion of consent and dynamic regimes). Just as the challenges are not restricted to one level, so should the strategies for improvement not be seen as exclusive. A national level response could improve the parliamentary influence and control of the international activities of government agents engaged in international decision making.[139] As a consequence of the challenges described above, i.e. the diminishing influence of national representatives on the international level, the internationalisation of discourse and deparliamentarisation in implementation, additional solutions must be

---

[137] In this sense, albeit for the context of treaty making, R. Wolfrum, "Kontrolle der auswärtigen Gewalt", Veröffentlichungen der Vereinigung der Deutschen Staatsrechtslehrer 56 (1997), 38-66 (43 et seq.); in the broader context of international institutions and their activities R. Wolfrum, "Legitimacy in International Law from a Legal Perspective: Some Introductory Considerations" in: R. Wolfrum/V. Röben (eds.), Legitimacy in International Law, 2008, 1-24 (22-23).

[138] S. Cassese, "Il diritto amministrativo globale: una introduzione", Rivista trimestrale di diritto pubblico 55 (2005), 331-357.

[139] R. Wolfrum, "Kontrolle der auswärtigen Gewalt", Veröffentlichungen der Vereinigung der Deutschen Staatsrechtslehrer 56 (1997), 38-66 (43 et seq.); W. Kahl, "Parlamentarische Steuerung der internationalen Verwaltungsvorgänge" in: H.-H. Trute/T. Groß/H.C. Röhl/C. Möllers (eds.), Allgemeines Verwaltungsrecht – Zur Tragfähigkeit eines Konzepts, 2008, 71-106.

found at the international level. Therefore, both national and international strategies must be sought in a complementary way.[140]

2. *Input and output oriented approaches*

The following prescriptions will draw on both input-oriented and output-oriented approaches.[141] As in nation states where both forms of legitimacy coexist, both output and input legitimacy are necessary elements of long-term legitimation at the international level.[142]

The concept of input legitimacy emphasizes that political decisions are legitimate if they represent the will of the people, i.e. if they can be derived in some way from the preferences of the members of a particular community.[143] Democratic legitimacy must thus be based on processes of democratic will formation that ultimately influence and control decision making.

As a complementary but distinct concept, output oriented legitimacy emphasises that the exercise of authority is legitimate when it contributes to solving efficiently problems that require collective solutions.[144] Factors such as efficiency and the input of expertise in law making and

---

[140] This idea of combining various (national and international level) approaches as an answer to the challenges is proposed e.g. by R. Wolfrum, "Legitimacy in International Law from a Legal Perspective: Some Introductory Considerations" in: R. Wolfrum/V. Röben (eds.), Legitimacy in International Law, 2008, 1-24 (22); similar also J. Delbrück, "Exercising Public Authority Beyond the State: Transnational Democracy and/or Alternative Legitimation Strategies?", Indiana Journal of Global Legal Studies 10 (2003), 29-43 (43); for a proposal of a complex legitimacy standard that integrates various political and legal elements consider A. Buchanan/R.O. Keohane, "The Legitimacy of Global Governance Institutions" in: R. Wolfrum/V. Röben (eds.), Legitimacy in International Law, 2008, 25-62.

[141] R. Wolfrum, "Legitimacy in International Law from a Legal Perspective: Some Introductory Considerations" in: R. Wolfrum/V. Röben (eds.), Legitimacy in International Law, 2008, 1-24 (6); D. Bodansky, "The Legitimacy of International Governance: A Coming Challenge for International Environmental Law?", American Journal of International Law 93 (1999), 596-624 (612).

[142] F.W. Scharpf, Regieren in Europa: effektiv und demokratisch?, 1999, 21.

[143] F.W. Scharpf, Regieren in Europa: effektiv und demokratisch?, 1999, 16.

[144] F.W. Scharpf, Regieren in Europa: effektiv und demokratisch?, 1999, 16 and 20 et seq.; D.C. Esty, "Good Governance at the Supranational Scale: Globalizing Administrative Law", Yale Law Journal 115 (2006), 1490-1562 (1517).

implementation are often considered to improve the substantive results and thus the output legitimacy of an institution or a legal rule.[145] Even if one leaves aside the difficulties in empirically relating regime design to output,[146] the exercise of authority cannot be legitimated solely on the basis of its functionality and superior efficiency.

Nonetheless a lack of effectiveness has a delegitimising effect as it can erode the legitimacy of an institution over time.[147] Consequently, expertise – as an important element through which output legitimacy is secured – is one factor that contributes to legitimate institutional activities.[148] The expertise of NGOs can play an important role in this respect.[149] Effectiveness may also be enhanced if decisions are taken by smaller non-plenary bodies, by majority votes and when administrative support is provided by the secretariat.

---

145 On expertise D. Bodansky, "The Legitimacy of International Governance: A Coming Challenge for International Environmental Law?", American Journal of International Law 93 (1999), 596-624 (619-620).

146 For a noteworthy empirical research project that attempted to assess the effectiveness of environmental regimes through an international regime database, see H. Breitmeier/O.R. Young/M. Zürn, "The International Regimes Database: Architecture, Key Findings, and Implications for the Study of Environmental Regimes" in: K. Jacob/F. Biermann/P.-O. Busch/P.H. Feindt (eds.), Politik und Umwelt, 2007, 41-59; compare for an assessment of the findings in terms of legitimacy H. Breitmeier, The legitimacy of international regimes, 2008.

147 Referring to efficiency but similar J. Delbrück, "Exercising Public Authority Beyond the State: Transnational Democracy and/or Alternative Legitimation Strategies?", Indiana Journal of Global Legal Studies 10 (2003), 29-43 (42); similarly R. Wolfrum, "Legitimacy in International Law from a Legal Perspective: Some Introductory Considerations" in: R. Wolfrum/V. Röben (eds.), Legitimacy in International Law, 2008, 1-24 (7).

148 J. Delbrück, "Exercising Public Authority Beyond the State: Transnational Democracy and/or Alternative Legitimation Strategies?", Indiana Journal of Global Legal Studies 10 (2003), 29-43 (43); R. Wolfrum, "Legitimacy in International Law from a Legal Perspective: Some Introductory Considerations" in: R. Wolfrum/V. Röben (eds.), Legitimacy in International Law, 2008, 1-24 (24); D. Bodansky, "Legitimacy" in: D. Bodansky/J. Brunnée/E. Hey (eds.), Oxford Handbook of International Environmental Law, 2007, 704-723 (719); V. Röben, Außenverfassungsrecht: eine Untersuchung zur auswärtigen Gewalt des offenen Staates, 2007, 19.

149 J. Delbrück, "Exercising Public Authority Beyond the State: Transnational Democracy and/or Alternative Legitimation Strategies?", Indiana Journal of Global Legal Studies 10 (2003), 29-43 (43).

The need for effectiveness means that the legitimacy challenges of decision making by international institutions cannot be solved by simply delegating decision-making powers to high-level plenary organs. This would seriously impede effectiveness. Procedural law is therefore proposed as a way to legitimate decisions by providing for sufficient access and representation of various actors, including experts, NGOs and civil servants. Adequate access and control procedures based on procedural law could address the input-related problems of the shift to the executive as well as the increased distance between national public discourse and international decision-making.

3. Formalisation through procedural law

It is one of the basic assumptions of this study that procedural law can enhance the legitimacy of international institutions and the use of non-binding instruments. Procedural legitimacy in simple terms refers to the conception that fair and adequate procedures based on due process can enhance the legitimacy of the norms and authoritative actions deriving from this process.[150] How is this achieved? Procedural rules structure decision making by pre-determining whose view should be taken into account, which aspects to consider and which to exclude, and how different aspects and contributions should affect the decision. In this sense, procedures reduce the complexity of decision-making.[151] At the same time, procedures provide reasons why the outcome of the decision making process becomes acceptable and therefore legitimate.[152] This focus on the subjective dimension is highlighted by Thomas Franck when he states that "legitimacy is a property of a rule or rule-making institution which itself exerts a pull toward compliance on those addressed

---

[150] Similarly N. Luhmann, Legitimation durch Verfahren, 1989, 25; R. Wolfrum, "Legitimacy in International Law from a Legal Perspective: Some Introductory Considerations" in: R. Wolfrum/V. Röben (eds.), Legitimacy in International Law, 2008, 1-24 (6); T.M. Franck, Fairness in international law and institutions, 1995, 7; E. Hey, "International Institutions" in: D. Bodansky/ J. Brunnée/E. Hey (eds.), Oxford Handbook of International Environmental Law, 2007, 749-769 (767); D.A. Wirth, "Reexamining Decision-Making Processes in International Environmental Law", Iowa Law Review 79 (1994), 769-802 (798).

[151] N. Luhmann, Legitimation durch Verfahren, 1989, 11-26.

[152] T.M. Franck, Fairness in international law and institutions, 1995, 7; N. Luhmann, Legitimation durch Verfahren, 1989, 25-26.

normatively because those addressed believe that the rule or institution has come into being and operates in accordance with generally accepted principles of right process."[153] In other words, formalised pre-determined procedures can make arbitrary decision making legitimate.[154] The deformalisation which is often implicit in the use of non-binding instruments does the opposite, it reduces legitimacy.[155] From the perspective of distributive justice and fairness, a lack of procedures also contributes to illegitimacy because formal procedures tend to protect the weaker actor. Procedural law on the international level may therefore be a prerequisite for fairness by allowing access for weak and strong actors, developing and developed states, alike.[156] Procedural law is thus both enabling and legitimising. This renders procedural law an ideal candidate for legitimation of decision making not only on the national, but also on the international plane.[157]

II. National level improvements

1. Procedures for improved parliamentary participation

The main basis for democratic legitimacy remains to date the national level. The original conception of the relationship of executive and legislature must consequently be adapted to new developments, at least to the extent that regulation and norm setting with direct implications for

---

[153] T.M. Franck, The power of legitimacy among nations, 1990, 24.

[154] N. Luhmann, Legitimation durch Verfahren, 1989, 25-26; with a similar consequence, Thomas Franck stresses "right process", i.e. the rooting of a legal system in a framework of formal requirements about how rules are made, interpreted, and applied, as a basis for legitimacy and ultimately for fairness, because it accommodates a popular belief that a system of rules based on formal procedure is also fair, compare T.M. Franck, Fairness in international law and institutions, 1995, 7.

[155] J. Klabbers, "The changing image of international organizations" in: J.-M. Coicaud/V. Heiskanen (eds.), The Legitimacy of International Organizations, 2001, 221-255 (237).

[156] E. Hey, "International Institutions" in: D. Bodansky/J. Brunnée/E. Hey (eds.), Oxford Handbook of International Environmental Law, 2007, 749-769 (758-759 and 767).

[157] R. Wolfrum, "Legitimacy in International Law from a Legal Perspective: Some Introductory Considerations" in: R. Wolfrum/V. Röben (eds.), Legitimacy in International Law, 2008, 1-24 (23).

domestic policy and private actors takes place at the international level.[158] The issue becomes even more pressing because nonbinding instruments provide for a possibility to avoid parliamentary approval through the wide discretion enjoyed by the executive and administration at the international and the national level.[159] Just as proposed for treaty law,[160] these developments therefore require a strengthening of the influence of parliaments on international decision making, especially before an instrument is adopted.[161] Increased internationalisation

---

[158] S. Kadelbach, "Die parlamentarische Kontrolle des Regierungshandelns bei der Beschlußfassung in internationalen Organisationen" in: R. Geiger (ed.), Neuere Probleme der parlamentarischen Legitimation im Bereich der auswärtigen Gewalt, 2003, 41-57 (53); similarly from a US perspective C.A. Bradley, "International delegations, the structural constitution, and non-self-execution", Stanford law review 55 (2003), 1557-1596.

[159] This fear is also shared by F. Orrego Vicuña, "In Memory of Triepel and Anzilotti: The Use and Abuse of Non-Conventional Lawmaking" in: R. Wolfrum/V. Röben (eds.), Developments of International Law in Treaty Making, 2005, 497-506 (497-506).

[160] W. Kahl, "Parlamentarische Steuerung der internationalen Verwaltungsvorgänge" in: H.-H. Trute/T. Groß/H.C. Röhl/C. Möllers (eds.), Allgemeines Verwaltungsrecht – Zur Tragfähigkeit eines Konzepts, 2008, 71-106 (85 et seq.); R. Wolfrum, "Kontrolle der auswärtigen Gewalt", Veröffentlichungen der Vereinigung der Deutschen Staatsrechtslehrer 56 (1997), 38-66 (43 et seq.); C. Möllers, Gewaltengliederung: Legitimation und Dogmatik im nationalen und internationalen Rechtsvergleich, 2005, 373-374; T. Puhl, "Entparlamentarisierung und Auslagerung staatlicher Entscheidungsverantwortung" in: J. Isensee/P. Kirchhof (eds.), Handbuch des Staatsrechts der Bundesrepublik Deutschland, Band III, 2005, 639-682 (mn. 3).

[161] F. Orrego Vicuña, "In Memory of Triepel and Anzilotti: The Use and Abuse of Non-Conventional Lawmaking" in: R. Wolfrum/V. Röben (eds.), Developments of International Law in Treaty Making, 2005, 497-506; B.-O. Bryde, Internationale Verhaltensregeln für Private – Völkerrechtliche und verfassungsrechtliche Aspekte, 1981, 37; E. Schmidt-Aßmann, "Die Herausforderung der Verwaltungsrechtswissenschaft durch die Internationalisierung der Verwaltungsbeziehungen", Der Staat 3 (2006), 315-338 (332) (not only but apparently also on nonbinding instruments); C. Tomuschat, "Der Verfassungsstaat im Geflecht der internationalen Beziehungen", Veröffentlichungen der Vereinigung der Deutschen Staatsrechtslehrer 36 (1978), 7-63 (36); T. Stein, "Demokratische Legitimierung auf supranationaler und internationaler Ebene", Zeitschrift für ausländisches öffentliches Recht und Völkerrecht 64 (2004), 563-570 (566); S. Kadelbach, "Die parlamentarische Kontrolle des Regierungshandelns bei der Beschlußfassung in internationalen Organisationen" in: R. Geiger

requires relinking the processes of norm creation to domestic parliaments in order to provide international decisions with domestic democratic legitimation.[162] However, one must be careful not to put into question that the executive should be leading foreign relations, as foreseen for instance by the German Basic Law. The predominance of the executive in foreign relations is necessary to provide the flexibility and autonomy needed to conduct international relations effectively, and must therefore be taken into account when thinking about possible improvements as done in the following paragraphs.[163]

Several means can be thought of to balance out the trend of deparliamentarisation at the national level: First of all, international organisations should not be established without approval by parliament as sometimes is the case at least in Germany with respect to the FAO, the WHO or UNESCO.[164] Furthermore, in German constitutional law it has been suggested that the constitutional requirement of ratification and parliamentary participation should be extended to nonbinding instruments with implications for private actors.[165] Even though the provision of the German Basic Law that requires parliamentary approval for treaty making and treaty implementation is only directly applicable

(ed.), Neuere Probleme der parlamentarischen Legitimation im Bereich der auswärtigen Gewalt, 2003, 41-57 (53).

162 V. Röben, Außenverfassungsrecht: eine Untersuchung zur auswärtigen Gewalt des offenen Staates, 2007, 508 et seq.; from a US perspective, albeit only for treaty law, this is also argued by C.A. Bradley, "International delegations, the structural constitution, and non-self-execution", Stanford law review 55 (2003), 1557-1596; S. Kadelbach, "Die parlamentarische Kontrolle des Regierungshandelns bei der Beschlußfassung in internationalen Organisationen" in: R. Geiger (ed.), Neuere Probleme der parlamentarischen Legitimation im Bereich der auswärtigen Gewalt, 2003, 41-57 (51 et seq.).

163 W. Kahl, "Parlamentarische Steuerung der internationalen Verwaltungsvorgänge" in: H.-H. Trute/T. Groß/H.C. Röhl/C. Möllers (eds.), Allgemeines Verwaltungsrecht – Zur Tragfähigkeit eines Konzepts, 2008, 71-106 (88-89).

164 S. Kadelbach, "Die parlamentarische Kontrolle des Regierungshandelns bei der Beschlußfassung in internationalen Organisationen" in: R. Geiger (ed.), Neuere Probleme der parlamentarischen Legitimation im Bereich der auswärtigen Gewalt, 2003, 41-57 (46).

165 C. Engel, Völkerrecht als Tatbestandsmerkmal deutscher Normen, 1989, 224 et seq.; I. Pernice, "Gesetzesvorbehalt für besondere Verträge (Art. 59 II 1 GG)" in: H. Dreier (ed.), Grundgesetz-Kommentar Band II, 2006, (mn. 46).

to treaty law,[166] it has been argued in German legal scholarship that this provision should be applied to non-binding instruments at least in those cases where an international treaty references these instruments.[167]

However, the importance of parliamentary approval as required for treaty law in Article 59 paragraph 2 of the German Basic Law should not be overestimated. Even in the case of treaty law, the options of the parliament are in practice generally limited to either accepting, accepting with reservations (an option which is frequently excluded by the treaty) or rejecting the treaty, but it can hardly subsequently demand changes to the treaty. The German Parliament for example hardly ever questions the results of international negotiations.[168] Of course, the requirement of parliamentary approval also has a preventive function which may be even more important than approval itself: because of the theoretical possibility of rejection, the executive can hardly present an unacceptable legal instrument to parliament.[169]

Nonetheless, the constitutional requirement of parliamentary approval should not be simply expanded, for Germany as an expansive application of Article 59 II of the German Basic Law.[170] This would not only neglect the differences between treaty norms and nonbinding instruments, but such a solution would also be too formalistic and not go to

---

[166] O. Rojahn, "Artikel 59 II" in: I. v. Münch/P. Kunig (eds.), Grundgesetz-Kommentar (Art. 20 bis Art. 69), 2001, 1250 et seq. (mn. 3 and 45b) (with further references).

[167] I. Pernice, "Gesetzesvorbehalt für besondere Verträge (Art. 59 II 1 GG)" in: H. Dreier (ed.), Grundgesetz-Kommentar Band II, 2006, (mn. 45).

[168] B.-O. Bryde, Internationale Verhaltensregeln für Private – Völkerrechtliche und verfassungsrechtliche Aspekte, 1981, 36; C. Tomuschat, "Der Verfassungsstaat im Geflecht der internationalen Beziehungen", Veröffentlichungen der Vereinigung der Deutschen Staatsrechtslehrer 36 (1978), 7-63 (28 ff.).

[169] U. Dieckert, Die Bedeutung unverbindlicher Entschließungen internationaler Organisationen für das innerstaatliche Recht der Bundesrepublik Deutschland, 1993, 163; C. Tomuschat, "Der Verfassungsstaat im Geflecht der internationalen Beziehungen", Veröffentlichungen der Vereinigung der Deutschen Staatsrechtslehrer 36 (1978), 7-63 (31).

[170] For an expansion of domestic ratification requirements at least for those nonbinding instruments which are referenced through dynamic references in treaties, W. Kahl, "Parlamentarische Steuerung der internationalen Verwaltungsvorgänge" in: H.-H. Trute/T. Groß/H.C. Röhl/C. Möllers (eds.), Allgemeines Verwaltungsrecht – Zur Tragfähigkeit eines Konzepts, 2008, 71-106 (87).

the heart of the matter.[171] It is not far-fetched to assume that the instruments have proven to be effective exactly because no intermediate legislative act was required.[172] To subject treaty and non treaty instruments to the same processes would disregard the apparent need for some more flexible informal form of cooperation at the international level. In order to maintain the functionality of the instruments, and therefore their output legitimacy, the need for control and input by national parliaments should be met by flexible approaches.[173]

It appears more promising to improve participation of parliament by allowing some possibility of parliamentary participation *before* the adoption of instruments at the international level, preferably already at the early stages of the negotiation processes. In Germany, the Bundestag disposes of important possibilities for controlling the executive in the conduct of international relations apart from ratification.[174]

As for treaty law, such forms of participation and control should also be considered and used in the context of the development of important nonbinding instruments.[175] One of these means is the exercise of parliamentary control over the activities of the executive government. Thus, the Bundestag the right to ask the Government (Bundesregierung) to report and inform the Bundestag on international devel-

---

171 C. Tomuschat, "Der Verfassungsstaat im Geflecht der internationalen Beziehungen", Veröffentlichungen der Vereinigung der Deutschen Staatsrechtslehrer 36 (1978), 7-63 (35); U. Dieckert, Die Bedeutung unverbindlicher Entschließungen internationaler Organisationen für das innerstaatliche Recht der Bundesrepublik Deutschland, 1993, 67 et seq.

172 J.E. Alvarez, International Organizations as Law-makers, 2005, 245.

173 B.-O. Bryde, Internationale Verhaltensregeln für Private – Völkerrechtliche und verfassungsrechtliche Aspekte, 1981, 37; C. Tomuschat, "Der Verfassungsstaat im Geflecht der internationalen Beziehungen", Veröffentlichungen der Vereinigung der Deutschen Staatsrechtslehrer 36 (1978), 7-63 (35).

174 This is stressed by C. Tomuschat, "Der Verfassungsstaat im Geflecht der internationalen Beziehungen", Veröffentlichungen der Vereinigung der Deutschen Staatsrechtslehrer 36 (1978), 7-63 (34-36); R. Wolfrum, "Kontrolle der auswärtigen Gewalt", Veröffentlichungen der Vereinigung der Deutschen Staatsrechtslehrer 56 (1997), 38-66 (43); V. Röben, Außenverfassungsrecht: eine Untersuchung zur auswärtigen Gewalt des offenen Staates, 2007, 512.

175 In this sense apparently also S. Kadelbach, "Die parlamentarische Kontrolle des Regierungshandelns bei der Beschlußfassung in internationalen Organisationen" in: R. Geiger (ed.), Neuere Probleme der parlamentarischen Legitimation im Bereich der auswärtigen Gewalt, 2003, 41-57 (50).

opments and discuss these issues publicly.[176] The German Basic Law also allows the German Bundestag to issue resolutions binding on the executive regarding specific issues in foreign affairs beyond the constitutionally provided forms.[177] Even though this possibility has so far not been accepted by the Bundesverfassungsgericht,[178] even the existing rights potentially allow the parliament to somewhat participate in and control the executive's foreign policy. They should therefore not be underestimated.[179] Further, if the implementation of international instruments requires financial resources, the parliament can use its spending power as a means of control. So far this has rarely been used in practice but could be used more.[180] A primary institutional mechanism of control of the activities of the executive which could also be used more effectively for nonbinding instruments are the parliamentary Committees and their right to information. Parliamentary Committees, in Germany for instance the Foreign Affairs Committee and issue-specific Committees such as the Committee on the Environment, Nature Conservation and Nuclear Safety, have under the Rules of Procedure of the

---

176 This is stressed e.g. by B.-O. Bryde, Internationale Verhaltensregeln für Private – Völkerrechtliche und verfassungsrechtliche Aspekte, 1981, 37; S. Kadelbach, "Die parlamentarische Kontrolle des Regierungshandelns bei der Beschlußfassung in internationalen Organisationen" in: R. Geiger (ed.), Neuere Probleme der parlamentarischen Legitimation im Bereich der auswärtigen Gewalt, 2003, 41-57 (52-53).

177 T. Puhl, "Entparlamentarisierung und Auslagerung staatlicher Entscheidungsverantwortung" in: J. Isensee/P. Kirchhof (eds.), Handbuch des Staatsrechts der Bundesrepublik Deutschland, Band III, 2005, 639-682 (mn. 7); B.-O. Bryde, Internationale Verhaltensregeln für Private – Völkerrechtliche und verfassungsrechtliche Aspekte, 1981, 37.

178 The Bundesverfassungsgericht restrictively reserves foreign affairs matters for the executive except in matters expressly foreseen in the German Basic Law; compare Bundesverfassungsgericht, BVerfGE 68, 1 (68 et seq.); confirmatory BVerfGE 104, 151 (207).

179 U. Dieckert, Die Bedeutung unverbindlicher Entschließungen internationaler Organisationen für das innerstaatliche Recht der Bundesrepublik Deutschland, 1993, 164; C. Tomuschat, "Der Verfassungsstaat im Geflecht der internationalen Beziehungen", Veröffentlichungen der Vereinigung der Deutschen Staatsrechtslehrer 36 (1978), 7-63 (36).

180 S. Kadelbach, "Die parlamentarische Kontrolle des Regierungshandelns bei der Beschlußfassung in internationalen Organisationen" in: R. Geiger (ed.), Neuere Probleme der parlamentarischen Legitimation im Bereich der auswärtigen Gewalt, 2003, 41-57 (51).

Bundestag the right to information, such as the right to summon a member of the Federal Government.[181] Generally speaking, the rights are not restricted to matters which are subject to domestic ratification procedures.[182] Certain requirements have to be met so that these rights may be effectively exercised by parliamentarians.[183] As already done in the process of European integration for matters of European affairs and most recently due to the entry into force of the Lisbon Treaty,[184] these requirements and the mentioned rights should be adjusted and strengthened over time to the growing importance of international institutions other than the EU.

2. *Avoidance of dynamic references*

Parliaments could counteract the shift to the executive by avoiding (dynamic) references in legislation to international nonbinding instruments at least where these instrument are designed to be continuously adapted. At a minimum, the referenced instruments and their amendments should be published in the official journal in order to secure legal clarity and certainty for those affected.[185] Furthermore, the referenced instruments should not become the basis for penal or other direct sanctions. The one clear example of a dynamic reference found in the context of this study – the reference in the German Plant Protection Act to the FAO Pesticide Code – shows how the legislature was careful not to

---

[181] Article 68 of the Rules of Procedure of the German Bundestag.

[182] The rights of these Committees cannot be reduced to areas that fall under Article 59 (2) German Basic Law, compare for the Foreign Affairs Committee W. Heun, "Artikel 45a" in: H. Dreier (ed.), Grundgesetz-Kommentar, Bd. 2, 2006, (mn. 5).

[183] S. Kadelbach, "Die parlamentarische Kontrolle des Regierungshandelns bei der Beschlußfassung in internationalen Organisationen" in: R. Geiger (ed.), Neuere Probleme der parlamentarischen Legitimation im Bereich der auswärtigen Gewalt, 2003, 41-57 (54).

[184] Compare in particular the Judgment of the Bundesverfassungsgericht in which it annulled the national law accompanying the approval of the Lisbon Treaty for lack of sufficient participation rights of the Parliament, compare Bundesverfassungsgericht, Judgement of 30 June 2009, 2 BvE 2/08; see also Article 93 of the Rules of Procedure of the German Bundestag.

[185] Bundesverfassungsgericht, BVerfGE 47, 285, 311.

attach penal sanctions to non-compliance with the norm that entailed the dynamic reference to the code of conduct.[186]

3. Improved intra-governmental consultation and approval

It has been observed that the use of nonbinding instruments may risk circumventing a proper balancing of interests, in particular in those cases where specialised ministerial bureaucrats cooperate at the international level but where oversight is slim. One desideratum for the national level is therefore to ensure that some form of consultation requirements between specialised ministries for nonbinding activities exist as in the case of treaty negotiations. Consultations between ministries should not only include the preparation of nonbinding instruments, but also provide for mechanisms of consultation and information about developments at the subsidiary bodies that often determine the substance of norms. This must be secured through procedural law. For Germany, a first step could be a change in the Common Rules of Procedure of the Federal Ministries (*Gemeinsame Geschäftsordnung der Bundesministerien)* which already contain such a mechanism for treaty law preparations.[187]

III. International level improvements

The following suggestions largely pursue two objectives. Following the above analysis, the first objective is to improve the chain of legitimacy from legitimated state representatives to international decision making. Improving this chain not only requires strengthening delegation from national decision-makers to the international institution, but also delegation *within* the international institution. In addition to possible improvements at the national level as mentioned above, further means to be discussed in this section include strengthening the principle of dele-

---

[186] See for details the analysis above in Part 2, at B.II.1.

[187] Compare § 72 (2) and (3) in conjunction with §§ 45, 46, 49, 62 of the Common Rules of Procedures of the Federal Ministries, in German available at www.verwaltung-innovativ.de; this is also proposed by U. Dieckert, Die Bedeutung unverbindlicher Entschließungen internationaler Organisationen für das innerstaatliche Recht der Bundesrepublik Deutschland, 1993, 245 et seq.

gated competences and improving procedural law for intra-institutional delegation.

The second objective is to re-link public debate at the national level to the international processes, and vice versa. To the extent that internationalisation impedes the public debate and contestability of decisions, procedures are needed at the international level which can improve a linkage to such public debate. To be clear, I do not intend to suggest that these improvements solve all the legitimacy problems of international institutions. Rather, they appear as minimum improvements if one takes the traditional state-centred model outlined above as a point of departure. As minimum requirements, the suggestions take into account that the legitimacy threshold for nonbinding instruments may be lower than for binding ones.

1. Strengthening the principle of legality

One means to strengthen the chain of delegation from the national level to international level decision making is to ensure that activities of international institutions remain within the limits of the legal framework provided.[188] Possible avenues to do so could be to limit the argument of implied powers, possibly by further refining and developing the *ultra vires* doctrine.[189] The fundamental concern that international institutional activities should remain within the limits of their attributed competencies also lies behind suggestions by the International Law Association to improve accountability and legitimacy of international institutions through the principles of constitutionality and institutional balance. The principle of constitutionality in the ILA report on accountability of international organisations refers to the legal obligation of the international organisation "to carry out its functions and exercise its powers in accordance with the rules of the IO"; institutional balance

---

[188] The significance of the legality principle and rule of law applied to international institutions is e.g. stressed by R. Wolfrum, "Legitimacy in International Law from a Legal Perspective: Some Introductory Considerations" in: R. Wolfrum/V. Röben (eds.), Legitimacy in International Law, 2008, 1-24 (23); E. Hey, "Sustainable development, normative development and the legitimacy of decision-making", Netherlands yearbook of international law 34 (2004), 3-53 (24).

[189] E. Schmidt-Aßmann, "Die Herausforderung der Verwaltungsrechtswissenschaft durch die Internationalisierung der Verwaltungsbeziehungen", Der Staat 3 (2006), 315-338 (329).

means that "organs of an international Organisation cannot overstep the institutional restraints laid down in the constituent instrument determining how they exercise their powers."[190]

It is beyond the scope of this study to develop a full-blown conception of how to achieve this goal. Strengthening the legality principle of course requires sufficiently precise institutional law on mandates and terms of reference. One central element of such a strategy is and should be improved judicial review. Apart from that it should suffice to point out that endeavours to refine the delegated competences doctrine should not ignore nonbinding activities of international institutions. In addition, it appears necessary to establish institutional law and procedures even for nonbinding activities. Procedural law must be developed that, even for nonbinding processes, clearly defines mandates, modes of decision-making, the inclusion and use of information and the scope of participation, as this is a prerequisite both for defining the responsibilities of decision makers and for assessing legality of acts through judicial review.

2. *Improved procedural law to define intra-institutional delegation*

As in national legal systems, discretionary leeway for civil servants and expert groups is often necessary to ensure the functioning of institutions. There is a need for expert input and flexible management that is not constrained by constant political considerations. Furthermore, administrative discretion is necessary since legal rules cannot address all possible contingencies beforehand.[191] Flexible decision-making in subsidiary or expert bodies facilitates the translation of basic principles into concrete actionable measures as well as their continuous revision to adapt them to newly arising circumstances.[192]

As in national legal systems, however, administrative or expert discretion cannot be unlimited but requires a strong legal framework which

---

[190] International Law Association, Final Conference Report Berlin (2004) entitled "Accountability of International Organisations" (2004), at 13; available at http://www.ila-hq.org/en/committees/index.cfm/cid/9.

[191] E. Benvenisti, "Public Choice and Global Administrative Law: Who's Afraid of Executive Discretion?", IILJ Working Paper 2004/3 (Global Administrative Law Series), available at: www.iilj.org.

[192] Compare the cases of the FAO CCRF and the PIC procedure, discussed above in Part 1, at B.I. and II.

legitimates their activities. As with national administrative systems, it matters how and to whom norm development and decision-making are delegated.[193] Where expert bodies or civil servants take influential decisions, it is paramount not only that they are well qualified, but also that they operate transparently to ensure a minimum of public accountability.[194] This demand acknowledges that decision-making by expert bodies is not a purely technical and thus apolitical question. Technical issues, as for example the decision whether a certain pesticide should be included in the list or not, are often entangled with important economic and political interests.[195]

In accordance with the desideratum just mentioned, numerous examples in this and other studies point to the emergence of an ever complex body of procedural institutional law for international institutions. One remarkable example for evolving procedures in the realm of voluntary instruments is the detailed procedural guidance of the OECD for the complaint procedure under the OECD Guidelines.[196] These detailed procedural rules provide a comprehensive legal framework for the Investment Committee and the National Contact Points (NCPs). They establish core criteria for the operation of NCPs.[197] They also contain some rudimentary guidance on the composition of NCPs, even if states maintain considerable discretionary leeway. The discrepancies between the composition of NCPs in various countries resulting from this freedom has been a point of contention and subject to criticism by NGOs ever since. This illustrates how procedural rules may also contribute to constructive and more differentiated critique.

---

[193] J.E. Alvarez, International Organizations as Law-makers, 2005, 245-246.

[194] P.H. Sand, Transnational Environmental Law: Lessons in Global Change, 1999, 57.

[195] Critical of approaches which consider some issue areas to be guided by technical rationalities without any political interferences W. Mattli/T. Büthe, "Setting International Standards: Technological Rationality or Primacy of Power", World Politics 56 (2003), 1-42; D.M. Leive, International regulatory regimes: case studies in health, meteorology, and food, 1976, 583; J.E. Alvarez, International Organizations as Law-makers, 2005, 252.

[196] For details see the analysis in Part 2, at C.I.1, further above.

[197] These are visibility, accessibility, transparency and accountability, compare the OECD Decision C(2000)96, Procedural Guidance, Part I.

3. Procedures of access to information and public participation

a) Stakeholder and NGO participation as a general remedy?

The improved participation in and access to international institutions by stakeholders and NGOs is often proposed as a remedy through which the perceived legitimacy challenges can be addressed or at least reduced.[198] This approach is based on expansive notions of democracy which focus on self-determination of affected individuals.[199] Legitimacy in this view not only derives from discourse between state representatives, but of all those potentially affected by the norms, including citizens and NGOs. NGOs can lend a voice to affected stakeholders or defend the interests of future generations across geographical boundaries.[200]

---

[198] B.-O. Bryde, "Konstitutionalisierung des Völkerechts und Internationalisierung des Verfassungsrechts", Der Staat 42 (2003), 61-75 (65-66); B. Boutros Ghali, An agenda for democratization, 1996, 25 et seq.; similarly R. Khan, "The Anti-Globalization Protests: Side-show of Global Governance or Law-making on the Streets?", Zeitschrift für ausländisches öffentliches Recht und Völkerrecht 61 (2001), 323-355; stressing the role of NGOs as representatives of world public opinion is D. Thürer, "The emergence of non-governmental organizations and transnational enterprises in international law and the changing role of the state", Non-state actors as new subjects of international law (1999), 37-58 (46); F.W. Stoecker, NGOs und die UNO: die Einbindung von Nichtregierungsorganisationen (NGOs) in die Strukturen der Vereinten Nationen, 2000, 99-121; J. Delbrück, "Exercising Public Authority Beyond the State: Transnational Democracy and/or Alternative Legitimation Strategies?", Indiana Journal of Global Legal Studies 10 (2003), 29-43 (40); H. Brunkhorst, Solidarität: von der Bürgerfreundschaft zur globalen Rechtsgenossenschaft, 2002, 213 et seq.; F. Müller, Demokratie zwischen Staatsrecht und Weltrecht: nationale, staatlose und globale Formen menschenrechtsgestützter Demokratisierung, 2003, 38 et seq.; J. Ebbesson, "Public Participation" in: D. Bodansky/J. Brunnée/E. Hey (eds.), Oxford Handbook of International Environmental Law 2007, 681-703 (687); A.C. Kiss/D. Shelton, International environmental law, 2004, 98.

[199] For assessments of the underlying premises of these conceptions compare e.g. J. Ebbesson, "Public Participation" in: D. Bodansky/J. Brunnée/E. Hey (eds.), Oxford Handbook of International Environmental Law 2007, 681-703 (687); A. von Bogdandy, "Demokratie, Globalisierung, Zukunft des Völkerrechts – eine Bestandsaufnahme", Zeitschrift für ausländisches öffentliches Recht und Völkerrecht 63 (2003), 853-877 (903-904).

[200] S. Riedinger, Die Rolle nichtstaatlicher Organisationen bei der Entwicklung und Durchsetzung internationalen Umweltrechts, 2001, 285-289; on the

NGO participation does not however directly contribute to an additional legitimation of international institutions. Any conception that aims to base the legitimacy of international institutions directly on the participation of non-governmental actors faces serious problems of representation and power asymmetries.[201] NGOs cannot claim to have a democratic mandate and do not necessarily have an internal democratic structure. This is not to say that the interests of NGOs may not often coincide with concerns of stakeholders or public interests in the environment. But there is no mechanism for ensuring that this is the case, and most of the time the degree of their influence can hardly be justified by the number of stakeholders they represent.[202] Even though they demand accountability from states, they face problems of accountability themselves.[203] In extreme cases, they may only represent the interests of a few or even single powerful individuals, or what those individuals perceive to be the common interest. Additionally, the geographical and cultural imbalances among NGOs, and in particular the dominance of

---

potential contributions of the "world society" to international legitimacy see I. Clark, International Legitimacy and World Society, 2007.

[201] J. Gupta, "The role of non-state actors in international environmental affairs", Zeitschrift für ausländisches öffentliches Recht und Völkerrecht 63 (2003), 459-486 (485); critical of conceptions that assign such a role to NGOs are R. Wolfrum, "American-European Dialogue: Different Perceptions of International Law – Introduction", Zeitschrift für ausländisches öffentliches Recht und Völkerrecht 64 (2001), 255-262 (262); T. Stein, "Demokratische Legitimierung auf supranationaler und internationaler Ebene", Zeitschrift für ausländisches öffentliches Recht und Völkerrecht 64 (2004), 563-570 (564); S. Kadelbach, "Die parlamentarische Kontrolle des Regierungshandelns bei der Beschlußfassung in internationalen Organisationen" in: R. Geiger (ed.), Neuere Probleme der parlamentarischen Legitimation im Bereich der auswärtigen Gewalt, 2003, 41-57 (55).

[202] D. Bodansky, "Legitimacy" in: D. Bodansky/J. Brunnée/E. Hey (eds.), Oxford Handbook of International Environmental Law, 2007, 704-723 (718); J. Gupta, "The role of non-state actors in international environmental affairs", Zeitschrift für ausländisches öffentliches Recht und Völkerrecht 63 (2003), 459-486 (485).

[203] J. Gupta, "The role of non-state actors in international environmental affairs", Zeitschrift für ausländisches öffentliches Recht und Völkerrecht 63 (2003), 459-486 (485).

western and northern NGOs, also impede that NGOs in their entirety could be seen to represent a "global civil society" or "world society"[204].

In light of these difficulties, generalised arguments in favour of or against NGO participation must give way to a more nuanced view. NGO participation does not generally enhance the legitimacy of international institutions,[205] but their specific contributions must be taken into account when thinking about strategies for improvements of institutional law. In some specific ways, however, NGOs and stakeholder participation can make a contribution, as discussed in the following.

b) Public participation, access to information and legitimacy

Procedures granting access to information and public participation on the international level can help to address some of the legitimacy challenges outlined earlier. Access to information refers to the right to obtain environmental information from decision makers.[206] Procedures providing for access to information and public participation are in two distinct ways relevant for the challenges outlined above.

First, access to information and public participation by non-governmental organisations[207] and stakeholders in general terms help to contribute to effectiveness and therefore improve the output of institutions. This general perception is reflected in institutional practice,[208] in-

---

[204] This term is promulgated by I. Clark, International Legitimacy and World Society, 2007, 193.

[205] With this clarity also J. von Bernstorff, "Zivilgesellschaftliche Partizipation in Internationalen Organisationen: Form globaler Demokratie oder Baustein westlicher Expertenherrschaft?" in: H. Brunkhorst (ed.), Demokratie in der Weltgesellschaft, 2009, 277-302 (296).

[206] Compare Article 3 of the Convention on Access to Information, Public Participation in Decision-Making and Access to Justice in Environmental Matters (28 June 1998), 2161 UNTS 447, 38 ILM 517 (1999).

[207] For the attempt of a definition of NGOs see J.A. Fuentes Véliz, "L'evolution du role des organisations non gouvernementales dans le droit de l'environnement", Revue Européenne de droit de l'Environnement 4 (2007), 401-430 (5-8).

[208] Consider e.g. the inclusion of NGOs in the reporting exercise of the FAO and the norm elaboration process of the OECD Guidelines.

ternational treaty making[209] and various nonbinding instruments.[210] One important element in this assumption is improved acceptance. The provision of information to stakeholders conveys the message that their concerns are being taken into account, and thus contributes to increasing their acceptance of environmental measures.[211] The notice-and-comment procedure of the OECD Guidelines' review, for instance, has made an important contribution to the success of the review by securing support of the various stakeholders.[212]

Second, the expansion of public, but in particular of NGO participation carries the potential to improve the quality of decision-making and increase the effectiveness of norm implementation.[213] NGOs contribute to the effectiveness of international institutions through the provision

---

209 See in particular the Aarhus Convention on Access to Information, Public Participation in Decision-Making and Access to Justice in Environmental Matters (28 June 1998), 2161 UNTS 447, 38 ILM 517 (1999).

210 Agenda 21 (1992), paras. 23.2., 8.18., 27.9, 27.10, 27.13; WSSD Johannesburg Plan of Implementation (2002), para. 119. The Meeting of State Parties to the Aarhus Convention in the so-called Almaty Guidelines, para. 12, considers that "[p]roviding international access opportunities in environmental matters, and establishing and strengthening procedures that enable the taking of these opportunities, generally improves the quality of decision-making and the implementation of decisions."

211 M. Eifert, "Umweltinformation als Regelungsinstrument", Die öffentliche Verwaltung 47 (1994), 544-552 (545 et seq.); M. Butt, Die Ausweitung des Rechts auf Umweltinformation durch die Aarhus-Konvention, 2001, 16.

212 For this positive assessment, see e.g. C. Wilkie, "Enhancing Global Governance: Corporate Social Responsibility and the International Trade and Investment Framework" in: J.J. Kirton/M.J. Trebilcock (eds.), Hard Choices, Soft Law: Voluntary Standards in Global Trade, Environment and Social Governance, 2004, 288-322 (296).

213 S. Riedinger, Die Rolle nichtstaatlicher Organisationen bei der Entwicklung und Durchsetzung internationalen Umweltrechts, 2001, 39 et seq.; K. Raustiala, "States, NGOs, and International Environmental Institutions", International Studies Quarterly 41 (1997), 719-740; M. Bothe, "The Evaluation of Enforcement Mechanisms in International Environmental Law: An Overview" in: R. Wolfrum (ed.), Enforcing environmental standards: economic mechanisms as viable means?, 1996, 13-38; J. Ebbesson, "Public Participation" in: D. Bodansky/J. Brunnée/E. Hey (eds.), Oxford Handbook of International Environmental Law 2007, 681-703 (688); see also the contributions in P.-M. Dupuy/L. Vierucci (eds.), NGOs in international law: efficiency in flexibility?, 2008.

of expertise to norm making processes.[214] NGOs are often instrumental and indispensable for the monitoring of implementation and for capacity-building programmes.[215] By improving the effectiveness of institutions without formally participating in decision making, NGOs can contribute to output legitimacy,[216] and thereby help to prevent institutions deparliamentarisation and internationalisation. The longer and necessarily more attenuated chain of legitimacy from the national to the international level can at least partly be responded to by relinking national public discourse to decision-making at the international level. Procedures allowing for public participation and access to information could serve this purpose. By relinking international activities and negotiations to the national public, these procedures could help stimulating public debates within states as well as, as a consequence thereof, in national parliaments. This can further stimulate the interest of the national Parliaments, through Committees as mentioned earlier, in international developments and furthermore help to scrutinise the performance of the government in international negotiations.[217] Access to information

[214] The importance of expertise of NGOs for the work of international organizations is generally recognized, consider for instance the analyses of E. Riedel, "The Development of International Law: Alternatives to Treaty-Making? International Organizations and Non-State Actors" in: R. Wolfrum/V. Röben (eds.), Developments of International Law in Treaty Making, 2005, 301-318 (304 et seq.); V. Röben, "Proliferation of Actors" in: R. Wolfrum/V. Röben (eds.), Developments of International Law in Treaty Making, 2005, 511-542 (526); J. Ebbesson, "Public Participation" in: D. Bodansky/J. Brunnée/E. Hey (eds.), Oxford Handbook of International Environmental Law 2007, 681-703 (687); A.-K. Lindblom, Non-governmental organisations in international law, 2005, 29; D. Thürer, "The emergence of non-governmental organizations and transnational enterprises in international law and the changing role of the state", Non-state actors as new subjects of international law (1999), 37-58 (44).

[215] J. Delbrück, "Exercising Public Authority Beyond the State: Transnational Democracy and/or Alternative Legitimation Strategies?", Indiana Journal of Global Legal Studies 10 (2003), 29-43 (42).

[216] K. Raustiala, "The "participatory revolution" in international environmental law", The Harvard environmental law review 21 (1997), 537-586 (583).

[217] In this sense also J. von Bernstorff, "Zivilgesellschaftliche Partizipation in Internationalen Organisationen: Form globaler Demokratie oder Baustein westlicher Expertenherrschaft?" in: H. Brunkhorst (ed.), Demokratie in der Weltgesellschaft, 2009, 277-302 (296-298).

and participation can increase the transparency of decision making[218] and provide interested individuals or groups with the possibility to remain informed about particular developments. Where NGOs are able to contribute in that process, they can in this narrow sense contribute to the legitimacy of international activities. NGOs have thus a role to play as intermediaries in this linkage between the national political process and the international decision making on the part of executives.[219] The organisational form of NGOs is often the only form through which stakeholders may take part in international discourse, for example as observers.[220] And frequently only NGOs have the resources necessary to monitor closely international negotiations and thereby help to improve their transparency.[221] The International Institute for Sustainable Development, for instance, is specifically dedicated to the task of monitoring and reporting through various online publications such as the Earth Negotiation Bulletin.[222]

Numerous documents exist which include concrete suggestions how the objective of enhanced public participation and access to information could be implemented at the international level. Two of them should be highlighted here. The first stems from the International Law Association (ILA) which has produced a number of recommendations on the accountability of international organisations. These recommendations stress in particular the concept of better access to information. The ILA Report recommends that international organisations adhere to the prin-

---

[218] M. Butt, Die Ausweitung des Rechts auf Umweltinformation durch die Aarhus-Konvention, 2001, 13-14; M. Eifert, "Umweltinformation als Regelungsinstrument", Die öffentliche Verwaltung 47 (1994), 544-552 (545).

[219] J. von Bernstorff, "Zivilgesellschaftliche Partizipation in Internationalen Organisationen: Form globaler Demokratie oder Baustein westlicher Expertenherrschaft?" in: H. Brunkhorst (ed.), Demokratie in der Weltgesellschaft, 2009, 277-302 (296); a similar role of NGOs in the context of treaty development is stressed by M. Schroeder, Die Koordinierung der internationalen Bemühungen zum Schutz der Umwelt, 2005, 149.

[220] A.-K. Lindblom, Non-governmental organisations in international law, 2005, 524.

[221] J. Gupta, "The role of non-state actors in international environmental affairs", Zeitschrift für ausländisches öffentliches Recht und Völkerrecht 63 (2003), 459-486 (485); K. Raustiala, "The "participatory revolution" in international environmental law", The Harvard environmental law review 21 (1997), 537-586 (563 and 586).

[222] The Earth Negotiation Bulletin is available at www.iisd.ca.

ciples of transparency and access to information by adopting all decisions in a public vote and by opening non-plenary meetings to the public.[223] Indeed, opening the process in this manner could enhance the possibility of parliaments and the national publics following and eventually contesting the decision-making of their representatives before a final document or decision is produced.

The second document, the Almaty Guidelines adopted by the Meeting of the Parties to the Aarhus Convention in 2005,[224] recommends that the principles of the 1998 Convention on Access to Information, Public Participation in Decision-making and Access to Justice in Environmental Matters (Aarhus Convention), in particular access to information and public participation, also be applied at the international level. The Guidelines concretise the objectives outlined in the Aarhus Convention, which obliges states to promote transparency and participation in "international environmental decision-making processes and within the framework of international organisations in matters relating to the environment".[225]

The transposition of the principle allowing for access to information to the international level should result in a presumption in favour of disclosure of information except for limited, clearly defined exceptions.[226] One of these exceptions should be the protection of confidential business information. The Guidelines follow this objective by requiring full and active disclosure of information to interested sections of the public,

---

223 International Law Association, Final Conference Report Berlin (2004) entitled "Accountability of International Organisations" (2004), at 229.

224 United Nations Economic Commission for Europe, Second Meeting of the Parties to the Convention on Access to Information, Public Participation in Decision-making and Access to Justice in Environmental Matters, Decision II/4 entitled "Promoting the Application of the Principles of the Aarhus Convention in International Forums", ECE/MP.PP/2005/2/Add.5 (20 June 2005) (hereinafter Almaty Guidelines).

225 Aarhus Convention, Article 3 (7).

226 S. Oberthür/M. Buch/S. Müller/S. Pfahl/R.G. Tarasofsky/J. Werksman/A. Palmer, Participation of non-governmental organisations in international environmental co-operation: legal basis and practical experience, 2002, 216; C. Saladin/B. VanDyke, "Implementing the principles of the Public Participation Convention in international organizations", Yearbook of human rights & environment 1 (2001), 1-96 (8).

including NGOs.[227] This includes free access to documents and meetings in all phases of decision-making.

With respect to public participation, the Aarhus Convention distinguishes between specific activities, plans and policies and executive regulations. This general idea to differentiate according to types of instruments and activities should also be followed on the international level. For example, one should differentiate between institutional activities with direct impact on private actors on the one hand and general elaboration and adoption of standards and norms on the other. In the former case, direct access of individual stakeholders and affected citizens, possibly represented by NGOs, should be allowed. In the case of general guidelines, the emphasis should be on observer access at an early stage in the negotiations, notice-and comments procedures and requirements to take the results of such consultations into account. In this sense, the Almaty Guidelines promote access for observers, the opening of advisory committees to stakeholders and general calls for comments.[228] They also stress that the results of public participation have to be taken into account by the decision-makers.[229] As with the Aarhus Convention, the Almaty Guidelines include a suggestion for access of "public interest organisations" in addition to affected members of the public.[230] In promoting general access of NGOs they are however too general and should be improved by taking into account an individual NGO's particular capabilities.

Some of the procedural requirements delineated in the Almaty Guidelines and supported by the ILA recommendations are already somewhat mirrored in institutional law analysed in this study. In particular the OECD increasingly reflects this approach. The procedural guidance issued by the OECD Council calls upon the National Contact Points to pursue a proactive approach regarding access to information. NCPs must inform and promote the guidelines, respond to enquiries issued by particular stakeholders, including business community, NGOs and the public.[231] Further, the web-based notice-and-comment procedures undertaken during the review of the OECD Guidelines in 2000 and the

[227] Almaty Guidelines, at IV.

[228] Almaty Guidelines, para. 33.

[229] Almaty Guidelines, para. V.37.

[230] Almaty Guidelines, para. 30 b).

[231] OECD Council Decision C(2000)96, Procedural Guidance, Part B.

updating process in 2010/2011 were a step in the right direction.[232] However, the application of the notice-and-comment procedures was not provided for in procedural law but developed on an ad hoc basis.

c) Procedural institutional law

As the above analysis shows, public participation is not *per se* an adequate tool to enhance legitimacy. Rather, the legitimating force of public participation depends on who participates, and whether such participation is in accordance with the purposes of achieving effectiveness, transparency and linkages between national public discourse and international decision making. Access of NGOs in particular is not *per se* a means for improving legitimacy, but a differentiation analysis is required to ensure that the participatory mechanisms and the participants actually remain within the purposes of such participation. One has to differentiate clearly between NGOs which can help to activate national public discourse and NGOs that do not have that capacity. Nationally active NGOs generally have better opportunities to influence and stimulate debate in their respective national public sphere.[233] NGOs which do not strive to contribute to national debates but rather attempt to push through special interests do not improve but rather erode the legitimacy of international institutions. Thus, the idea of giving a legitimate role to those NGOs that contribute to a linkage of international negotiations and national public discourse can also contribute, if translated into institutional procedural law, to a controlling function at the international level.[234]

Procedural law can provide for the necessary differentiation between those actors which can contribute in legitimate ways and those that

[232] See already the discussion above, in Part 1 at B.III.5.; compare also J. Salzman, "Decentralized Administrative Law in the Organization for Economic Cooperation and Development", Law and Contemporary Problems 68 (2005), 189-224 (217).

[233] J. von Bernstorff, "Zivilgesellschaftliche Partizipation in Internationalen Organisationen: Form globaler Demokratie oder Baustein westlicher Expertenherrschaft?" in: H. Brunkhorst (ed.), Demokratie in der Weltgesellschaft, 2009, 277-302 (297).

[234] Similar J. von Bernstorff, "Zivilgesellschaftliche Partizipation in Internationalen Organisationen: Form globaler Demokratie oder Baustein westlicher Expertenherrschaft?" in: H. Brunkhorst (ed.), Demokratie in der Weltgesellschaft, 2009, 277-302.

cannot. Procedural law can also keep the positive contributions of the mechanisms from eroding other legitimising factors which are mediated through state representatives. Determining the procedural rights of NGOs not only counterbalances tendencies of informal and unregulated influence of environmental and industry NGOs.[235] Seen from the perspective of NGOs, only pre-determined procedural law can make sure that their participation is not dependent on the good will of institutions or specific political constellations.[236] Unregulated and undifferentiated access carries the danger of lobbyism and can blur the lines between public interests represented by state representatives and private interest groups. As only government representatives can be held accountable to their constituencies through political mechanisms, this would create new problems for the legitimacy of these institutions rather than solving the existing ones.[237]

One necessary means which is already widely used for the purpose of securing legimate NGO participation and fostering that of international institutions is that of pre-determined accreditation procedures.[238] Through such procedures, the legitimacy of NGO participation, and consequently the input and output legitimacy of the activities of international organisations can be improved.[239] In light of the analysis undertaken here, these accreditation procedures should take into account the expertise of NGOs, along with their position at the national level. The institutional practice of the United Nations regarding nongovernmental organisations is one example of already-existing institutional law that takes into account some of the above considerations. The practice is defined by the UN Charter and four resolutions of the

---

[235] In this sense also S. Kadelbach, "Die parlamentarische Kontrolle des Regierungshandelns bei der Beschlußfassung in internationalen Organisationen" in: R. Geiger (ed.), Neuere Probleme der parlamentarischen Legitimation im Bereich der auswärtigen Gewalt, 2003, 41-57 (56).

[236] S. Oberthür/M. Buch/S. Müller/S. Pfahl/R.G. Tarasofsky/J. Werksman/A. Palmer, Participation of non-governmental organisations in international environmental co-operation: legal basis and practical experience, 2002, 225-227

[237] A.-M. Slaughter, A New World Order, 2004, 10.

[238] Similarly D.B. Reiser/C.R. Kelly, "Linking NGO accountability and the legitimacy of global governance", Brooklyn Journal of International Law 36 (2011), 1011-1073.

[239] D.B. Reiser/C.R. Kelly, "Linking NGO accountability and the legitimacy of global governance", Brooklyn Journal of International Law 36 (2011), 1011-1073, 1068.

Economic and Social Council of the United Nations[240] as well as corresponding institutional law within the specialised organisations.[241] These instruments establish a system of quality control for the access of NGOs. Such a mechanism ensures that NGO activity does not become a legitimacy issue in itself.[242] Following the fundamental distinction between participation and consultation established by the UN Charter in Articles 69 to 71, NGOs are only accorded a role as consultants without any direct role in the inter-governmental deliberations.

Further, the procedures – albeit in a rudimentary manner – take into account the purposes outlined above, and thus distinguish between different NGOs. NGOs must demonstrate particular expertise and represent public opinion on a certain matter. Thus, according to the ECOSOC Resolutions, "decisions on arrangements for consultation should be guided by the principle that consultative arrangements are to be made, on the one hand, for the purpose of enabling the Council or one of its bodies to secure expert information or advice from organisations having special competence in the subjects for which consultative arrangements are made, and, on the other hand, to enable international, regional, subregional and national organisations that represent important elements of public opinion to express their views."[243] Moreover, the granting of any kind of status is, among other criteria, subject to the degree and nature of representativeness of a particular organisation. At ECOSOC, the highest "general consultative status" which allows NGOs to be an observer is only attributed to an NGO which "is broadly representative of major segments of society in a large number

---

240 ECOSOC Resolutions 288B/X (1950), 3/II (1964), 1296/XLIV (1969) and 1996/31 (1996).

241 Compare only the procedures and guidelines adopted at the FAO on NGO participation, see Basic texts of the Food and Agriculture Organization of the United Nations, Volumes I and II, 2006, FAO Policy Concerning Relations with International Non-Governmental Organizations, available at http://www.fao.org/docrep/009/j8038e/j8038e00.htm; for details on various types of accreditation procedures and conditions, see D.B. Reiser/C.R. Kelly, "Linking NGO accountability and the legitimacy of global governance", Brooklyn Journal of International Law 36 (2011), 1011-1073, 1052 et seq.

242 E. Riedel, "The Development of International Law: Alternatives to Treaty-Making? International Organizations and Non-State Actors" in: R. Wolfrum/V. Röben (eds.), Developments of International Law in Treaty Making, 2005, 301-318 (318).

243 ECOSOC Resolution 1996/31 (1996), para. 20.

of countries in different regions of the world".[244] This could be interpreted as an acknowledgement that an organisation must somehow be embedded within the national societies, but remains vague in this respect.

Contrary to ECOSOC, the embeddedness at the national level is not a criterion in the FAO. The requirements that apply to both general and special consultative status relates only to the international structure and scope of the NGO.[245] Thus, there still remains need for further refinement. In particular, accreditation procedures should specify clearly that NGOs are privileged if they function as a transmitter between the public discourse at the national level and the international level discourse. A step in that direction is the new criteria of the ECOSOC, which explicitly put national NGOs on an (almost) equal footing with international NGOs, even though their home state has to be consulted before granting of the status.[246]

4. Independent (judicial) review

Improving legitimacy through institutional law and procedures ultimately also hinges on the availability of some kind of independent review, preferably by judicial institutions.[247] The judiciary has a role to play in securing the legality principle, in the development of principles of international institutional law, and as the provider of due process and review of the activities of international institutions.

Meaningful judicial control of international institutions, although not yet fully developed, is not a utopian ideal. Thus, in particular national courts increasingly take on the task to control international organisa-

---

[244] ECOSOC Resolution 1996/31 (1996), para. 22.

[245] Basic texts of the Food and Agriculture Organization of the United Nations, Volumes I and II, 2006, FAO Policy Concerning Relations with International Non-Governmental Organizations, paras 6-8, the procedures are available at http://www.fao.org/docrep/009/j8038e/j8038e00.htm.

[246] ECOSOC Res. 1996/31 (1996), paras 5 and 8.

[247] R. Wolfrum, "Legitimacy in International Law from a Legal Perspective: Some Introductory Considerations" in: R. Wolfrum/V. Röben (eds.), Legitimacy in International Law, 2008, 1-24 (23); detailed and insightful on possible judicial remedies for states and non-state actors against international organizations as well as the role of the International Court of Justice K. Wellens, Remedies against international organisations, 2002.

tions, and even engage in transnational judicial cooperation and exchange of information to do so.[248] They thereby control indirectly the growing influence of the executive branch without damaging the effectiveness in the conduct of foreign policy. Overall, national courts appear less willing to accept absolute immunities but stress functional immunities. If taken seriously, functional immunities do not always require a high level of protection, in particular not in those cases where the institution engages in private matters and is not only concerned with its own internal affairs.[249]

In the context of nonbinding instruments, independent and ideally judicial review is particularly important in respect of two types of instruments and activities: instruments and associated procedures which have direct effects on private actors on the one hand, and instruments defining norms of conduct which are referenced by treaty law on the other.

It is fairly obvious that independent judicial review is needed whenever and to the extent that private actors are directly affected by decisions taken at the international level, and where no other system that safeguards due process rights exists, as in the case of terrorist listings by the Security Council.[250] If one looks at nonbinding instruments, the need is less obvious. Indeed, one could be tempted to suggest that review does not play a role due to the less intrusive nature of these instruments. Although less intrusive, nonbinding instruments can at times also have significant direct implications for private actors. Examples are the formally nonbinding operational procedures and safeguards of the World Bank and other multilateral financial institutions. Where national procedures guaranteeing due process or judicial review are not available, international institutional law and ideally national and/or international judicial review of such law must be available to safeguard due process rights and the rule of law. One can already observe the tentative beginning towards stronger review mechanisms in international institutional

---

248 E. Benvenisti/G.W. Downs, "National Courts, Domestic Democracy, and the Evolution of International Law", European Journal of International Law 20 (2009), 59-72 (68).

249 A. Reinisch, International organizations before national courts, 2000, 391-393.

250 European Court of Justice, Kadi, 2005 E.C.R. II-3649, 214; Yusuf & Al Barakaat, 2005 E.C.R. II-3533, 265; for an analysis see e.g. P. Hilpold, "EU Law and UN Law in Conflict: The Kadi Case", Max Planck United Nations Yearbook 13 (2009), 141-181.

law in that respect. A prominent example is the establishment of the World Bank Inspection Panel whose responsibilities include reviewing the application of the (legally nonbinding) Operational Guidelines of the World Bank by giving affected persons and groups the right to bring a complaint.[251] But also the review procedure under the OECD Guidelines can be seen as part of that development. As described above, advisory bodies or states can ask for a review of the activities of the National Contact Points, thereby giving these actors a kind of appeal possibility under which international institutional law can be scrutinized. The procedure specifically provides that affected corporations have a right to be heard in front of the Investment Committee. However, the mechanism is still largely political in nature and lacks independence.

Second, it has been seen in this study that treaty instruments frequently and perhaps increasingly so reference nonbinding instruments. In these cases, the instrument is hardened and gains authority with direct (legally binding) implications for states and, indirectly, for private actors. At least from a legitimacy perspective, these cases therefore warrant particular scrutiny. Dispute settlement bodies can play a twofold role here. On the one hand, dispute settlement bodies, for instance under UNCLOS or the WTO, can play a traditional judicial role when determining whether the national measures such as SPS measures or laws to prevent pollution from ships are in conformity with international rules and standards.[252] The relevant rules and standards are then prescribed at the international level, implemented at national level and this application is controlled by the international judiciary.[253]

---

[251] On this see A. Naudé Fourie, The World Bank Inspection Panel and quasi-judicial oversight: in search of the 'judicial spirit' in public international law 2009.

[252] With respect to the IMO and UNCLOS this is stressed by R. Wolfrum, "Legitimacy of International Law and the Exercise of Administrative Functions: The Example of the International Seabed Authority, the International Maritime Organization (IMO) and International Fisheries Organizations" in: A. von Bogdandy/R. Wolfrum/J. von Bernstorff/P. Dann/M. Goldmann (eds.), The Exercise of Public Authority by International Institutions: Advancing International Institutional Law 2010, 917-940 (936).

[253] Similarly R. Wolfrum, "Legitimacy of International Law and the Exercise of Administrative Functions: The Example of the International Seabed Authority, the International Maritime Organization (IMO) and International Fisheries Organizations" in: A. von Bogdandy/R. Wolfrum/J. von Bernstorff/P. Dann/M. Goldmann (eds.), The Exercise of Public Authority by Interna-

On the other hand, dispute settlement bodies, having competence to interpret the provisions of the respective treaty, also have the authority to interpret the provision entailing the references. Consequently, they may in principle also assess to which extent a particular instrument plays a role for determining the binding rule in question. This "gatekeeper role"[254] is legally possible due to the often generally worded references which require further judicial concretisation, but of course also depend on adequate possibilities of states to bring such cases.[255] In being authorised to interpret these provisions, the dispute settlement bodies are also mandated to thus decide what qualifies as an "international standard" under the SPS and TBT Agreement or what is a "generally recommended" or a "generally accepted" international standard under UNCLOS. The same is true for the interpretation of Article XX GATT, and in particular its chapeau.[256]

Within the limits defined by the wording of the references, the dispute settlement bodies could thus scrutinise the authority claims of these standards before providing them additional authority under the respective treaty. To some extent, this is already done by the WTO Appellate Body.[257] To be sure, the argument is not to give the dispute settlement bodies the power to judge beyond the boundaries of the treaty for which they have jurisdiction, but merely to assess whether any additional authority can be given to these instruments which they do not al-

---

tional Institutions: Advancing International Institutional Law 2010, 917-940 (936).

254 J. Scott, "International Trade and Environmental Governance: Relating Rules (and Standards) in the EU and the WTO", European Journal of International Law 15 (2004), 307-354 (311).

255 On the problem of standing in the case of fisheries law which depends on the interpretation of Article 297 (3) UNCLOS, see R. Wolfrum, "The Role of the International Tribunal for the Law of the Sea" in: M.H. Nordquist/J.N. Moore (eds.), Current Fisheries Issues and the Food and Agriculture Organization of the United Nations, 2000, 369-385 (370).

256 See for details on the interpretation of references and Article XX GATT the analysis in Part 2, at A.I.2, further above.

257 J. Scott, "International Trade and Environmental Governance: Relating Rules (and Standards) in the EU and the WTO", European Journal of International Law 15 (2004), 307-354.

ready have.[258] One could thus in the long term imagine dispute settlement bodies establishing criteria for assessing whether and to which extent a particular instrument is accepted under the references of the treaty in question. With this gatekeeper function, dispute settlement bodies could at the same time foster inter-institutional cooperation, stimulate the development of institutional law within institutions, and provide a safeguard against influences of illegitimate institutions and standards.[259]

D. Conclusion and outlook

International cooperation requires flexible means of cooperation that transcend traditional forms of law making, such as dynamic treaty regimes or international cooperation through nonbinding instruments.

These flexible forms of decision making present challenges to the traditional safeguards designed to ensure legitimacy. Clearly, nonbinding instruments such as the ones analysed in this study do not raise the kind of legitimacy problems faced by international institutions acting as quasi-legislators. Cases of international legislation or international administration where international institutions take binding decisions with implications for individuals pose more significant legitimacy problems.[260] In contrast, the adoption of nonbinding instruments does not

258 J. Scott, "International Trade and Environmental Governance: Relating Rules (and Standards) in the EU and the WTO", European Journal of International Law 15 (2004), 307-354 (349).

259 J. Scott, "International Trade and Environmental Governance: Relating Rules (and Standards) in the EU and the WTO", European Journal of International Law 15 (2004), 307-354 (310).

260 See for example the sanctions against individuals in resolutions on Sudan, compare e.g. Security Council Resolutions S/RES/1591 (2005) of 29 March 2005 S/RES/1672 (2006) of 25 April 2006 and in resolutions regarding suspected terrorists which are triggered through decisions of the Sanctions Committee established through Resolutions S/RES/1267 (1999) of 15 October 1999 and specified by Resolution S/RES/1333 (2000) of 19 December 2000 (the enforcement of this procedure through EU Council Regulation (EC) No 881/2002 of 27 May 2002 has been annulled for its infringement with basic human rights, such as the right to a judicial remedy and right to property, by the European Court of Justice. Compare ECJ, Kadi v Council and Commission, Judgment of the Court (Grand Chamber) of 3 September 2008, C-315/01 and ECJ, Yusuf and Al Barakaat International Foundation v Council and Commis-

directly create a binding obligation for states. The international institutions in these cases do not "legislate": even if their activities may contribute or give rise to normative developments, they cannot predetermine national implementation in the same way as binding law. Distinguishing the rather clear-cut cases of dynamic treaty development through binding law is thus warranted, not least in order to foreclose undifferentiated critique of the legitimacy of all activities of international institutions.

However, the issuing of nonbinding instruments by international institutions still has legal and factual compliance-inducing effects on states as well as direct or indirect implications for private actors. Although of lesser concern than for binding instruments, the erosion of traditional legitimacy safeguards that comes with these nonbinding forms of cooperation must therefore be addressed.

This study has attempted to identify some of the legitimacy challenges of nonbinding norms. It has therefore set aside the larger questions of how to conceptualise international governance, in order pragmatically to pinpoint some challenges that must and can be addressed without questioning or reinventing the entire system.

The challenges of the nonbinding activities of international institutions are not directly induced by the one-time adoption of obligations, but result from the more subtle processes of norm development and implementation. The first one is increasing internationalisation and corresponding deparliamentarisation. An increasing recourse to nonbinding instruments by international institutions internationalises issues formerly situated in the realm of nation states. Although nonbinding norms do not predetermine the national implementation decision in the same way as concrete binding norms, this internationalisation shifts responsibility away from parliaments to the executive. Deparliamentarisation and internationalisation further reduce the ability of the parliamentary opposition and the general public to challenge and discuss ongoing activities now happening at the international level.

Second, the founding treaty of international institutions proves to be of limited value to link the original consent to the subsequent institutional activities. It could therefore only to a limited extent legitimise nonbind-

sion, Case C-306/01 of the European Court of Justice in 2008, compare, Joined Cases C-402/05 P and C-415/05). On the legislative function of the Security Council in general e.g. L.M. Hinojosa Martínez, "The legislative role of the Security Council in its fight against terrorism: legal, political and practical limits", The international and comparative law quarterly (2008), 333-359.

ing institutional activities. The nonbinding form of action facilitates a dynamic development within the organisation. Most importantly perhaps, the traditional premise that the international organisations or treaty regimes could be controlled through the plenary, and therefore be legitimated through a chain of legitimacy from the national to the international level, is eroded by the increasing importance and autonomy of subsidiary organs, expert bodies and international civil servants which often act without specific institutional law that delineates and conditions their actions.

Some suggestions have been made in this study of how to respond to these challenges. In sum, these are directed at improving the participation and control by national parliaments; improving the chain of delegation by strengthening the legality principle and specific intra-institutional law; and strengthening the linkage between national public discourse and international decision making through mechanisms of access to information and public participation. The further development of procedural institutional law appears to be a promising avenue to balance needed flexibility with a minimum of legal formalism. Procedural law can turn arbitrariness into a legitimate exercise of power, and thereby strengthen chains of delegation and secure mechanisms of participation while determining those actors that should be allowed access.

In applying these suggestions, the challenge remains is to find the right balance between flexibility and stability, between delegation to experts and political control, between confidential informal negotiations and public accessibility and accountability. This difficult task is likely to require tailor-cut solutions for each regime and instrument.

Part 4

Concluding summary

1. Many of today's activities of international institutions – be it traditional international organisations or treaty bodies of Multilateral Environmental Agreements – involve the use of some kind of nonbinding instrument. In fact, the maturing international system is a complex system of interrelated nonbinding and binding instruments. This applies in particular to environmental law, where the need to protect common resources, the diversification of actors, the speed of technological advancement and the high level of uncertainty pose particular challenges to traditional treaty law.

2. As the survey and the case studies in the Part 1 of this study demonstrate, nonbinding instruments are not only widely used, but they are also highly diverse. In a first systematisation, various categories of nonbinding instruments were identified: memoranda of understanding; declarations of principles; international action plans; international recommendations such as guidelines and codes of conduct; operational procedures and safeguards; and technical standards. Each of these categories differs from another either in respect of adopting actors, norm characteristics or addressees, but they also often overlap in practice. [see Part 1, A]

3. The case studies and the examples in this study show that different nonbinding instruments are employed by international institutions to address a variety of regulatory challenges, and that the institutions establish a variety of different institutional means to enhance implementation and compliance. [Part 1, B]

a) The first study on the FAO's Code of Conduct for Responsible Fisheries and related nonbinding fisheries instruments adopted by the FAO illustrates the capacity of a nonbinding instrument to establish and shape a broad and integrative international discourse as well as norm making processes at the international, but also at regional and domestic levels. The FAO contributed to this effect through the employment of a variety of soft compliance-enhancing means such as reporting and capacity building, and strategically attempted to engage non-state actors and regional fisheries organisations in this endeavour. Despite these efforts, however, the adoption and promotion of the Code of Conduct so far if at all had only insufficient effect on unsustainable fishing management practices. Implementation and compliance lacks behind despite normative changes in policy and legal frameworks. This fact highlights the limitations of nonbinding instruments in areas where the objective is to protect a common good and achieve sustainable use despite countervailing short-term oriented economic interests. Nonbinding instruments appear to be clearly limited and cannot replace treaty law with respect to implementation and compliance control. [Part 1, B.I.]

b) A successful use of nonbinding instruments can be found in the area of hazardous substances regulation. UNEP and the FAO jointly introduced the prior informed consent mechanism through nonbinding instruments. It was broadly applied within a comparatively short period of time. In a multi-level system, the procedure was implemented through a network of national focal points, a joint expert body of the FAO and UNEP at the international level and corresponding national (or European) legislation. The voluntary form and the corresponding flexible institutionalisation proved to be critical elements in facilitating the learning process which proved to be essential in the initial phase of establishing the procedure. The later shift to the binding PIC Convention, modelled on the norms and mechanisms as they had developed under the voluntary system, demonstrates the need for legal certainty with respect to such procedural rules over time. The treaty format reduced flexibility by linking listing decisions more closely to intergovernmental decisions. While increasing political input, this particular change has been a drawback for the effective functioning of the mechanism. [Part 1, B.II.]

c) The case study on the OECD Guidelines for Multinational Enterprises shows how international institutions employ nonbinding instruments to directly address non-state actors such as corporations in order to improve corporate social and environmental responsibility. It also shows the remarkable institutionalisation at the international and national level that may accompany these attempts. The OECD established a complaint procedure under which NGOs or even private persons can bring complaints about the non-compliance of corporations with the OECD Guidelines. Even though difficult to prove empirically, the reformed international complaint mechanism of the OECD appears to be having an effect on corporations. [Part 1, B.III.]

4. From the survey of instruments and the case studies, one can draw a set of cross-cutting parameters which describe characteristics of nonbinding instruments with an effect on the utility, effectiveness and legitimacy of any particular instrument. The way in which these characteristics play a role was demonstrated throughout Parts 2 and 3 of the

study. Furthermore, they are relevant for distinguishing nonbinding instruments from treaty law, but can also help to distinguish among nonbinding instruments. [Part 1, C.]

Seven parameters could be identified which appear to be particularly relevant: nonbinding status, norm characteristics, origin and norm development process, non-state actor involvement, addressees, adaptability and follow-up mechanisms. Nonbinding status provides the clearest distinction from treaty law, and plays a role for both utility, effectiveness and for legitimacy as shown throughout this study. The second parameter norm characteristics refers to the different types of provisions of a particular instrument may contain, i.e. for example whether it contains only general principles and objectives or prescribes concrete actions and measures, as for instance not to use certain nets for fishing or follow a specific procedure of prior informed consent. Third, origin of norms asks by whom and how a particular instrument was developed and how it was adopted and approved. One must in this respect distinguish instruments stemming from public international institutions and instruments developed and adopted by non-state actors. Fourth, non-state actor involvement in norm development and implementation matters is often particularly strong with non-binding instruments if compared to treaty law, but there also exist tremendous differences from one instrument to another. The parameter addressees captures that nonbinding instruments are often addressed to state actors just like treaty law, but often also or exclusively contain norms addressed directly at private non-state actors. Sixth, adaptability is often a particular advantage of nonbinding instruments over most treaty law instruments. Finally, it matters greatly for the effectiveness of a particular instrument whether there it is supported by institutionalised follow-up mechanisms or not.

5. The multi-level analysis in Part 2 illustrated the role and impact of nonbinding instruments on the international, regional, domestic and private level. To fully grasp their function, this analysis was combined with a closer look at the interplay of nonbinding instruments with binding law and the interplay between various institutions and instruments at each level and across various levels.

a) At the international level, nonbinding instruments gain impact through their interplay with international law and through their recognition by other international institutions and nonbinding instruments. [Part 2, A]

(1) Regarding the relationship with treaty law or customary law processes, one can distinguish precursory and supplementary roles of nonbinding instruments.

First, nonbinding instruments often play a precursory role. They shape future customary law or general principles, and provide a model for future treaty making. They are especially useful to pave the way for future treaty law where learning processes are needed due to prevailing uncertainty on the underlying problem and the adequate response. The success of the precursory function depends on the degree of institutionalisation and continuous discourse, established for example by means of reporting mechanisms. Customary law development is hampered in two instances: namely if the instrument contains an explicit disclaimer that it is only of voluntary nature, or when it prescribes norms primarily addressed to private actors. [Part 2, A.I.1.]

Second, nonbinding instruments often gain authority by supplementing treaty law. They do so for example as concretisations and interpretations of general provisions of treaty law. Some nonbinding instruments are also recognised through references in treaties, including such important ones as UNCLOS (e.g. fisheries and environmental provisions] and the WTO Agreement (e.g. Article XX GATT, TBT and SPS Agreement). The proposed new rules on subsidies for fisheries of the WTO Agreement on Subsidies and Countervailing Measures which directly link the acceptability of fisheries subsidies under WTO law to compliance with binding and nonbinding international fisheries instruments would not only mark a step forward for sustainable fisheries and the coordination of environmental and trade issues, but also represent a prime example for the hardening of nonbinding instruments through references in treaty law. [Part 2, A.I.2.]

(2) Nonbinding instruments contribute to the coordination and cooperation between international institutions at the international level. Nonbinding instruments were found to include a great number of references to other nonbinding or treaty instruments. In this manner, these instruments establish cross-cutting standards which integrate the products of various institutions or of other norm development processes, sometimes also from different sectors such as trade and human rights. By establishing such cross-cutting standards for a certain field which are recognised by other institutions, the nonbinding instruments contribute to the coordination of the work of the various institutions. Where the established norms cut across sectoral boundaries, they also achieve at least some integration of different issue areas – most notably social, economic and environmental issues – and in this way serve the

integrative goal of the principle of sustainable development. [Part 2, A.II.1.]

Inter-institutional cooperation is also achieved to the extent that nonbinding instruments influence the norm making and enforcement activities of other international and regional institutions. For example, the norms of the FAO Code of Conduct have been seen to influence norm making processes at the negotiations of the WTO on new rules on subsidies. And the recognition of the OECD Guidelines as a standard of assessment for listings of corporations by the UN Security Council or the recognition of FAO Guidelines in the operational procedures of the World Bank are but two examples that illustrate how these instruments provide standards on which other institutions rely in their own enforcement activities. Regional organisations such as fisheries management organisations or the European Union aim at enhancing compliance not only with treaty, but also with norms of nonbinding instruments such as the FAO Code of Conduct for Responsible Fisheries. Through these linkages, the work of the institution that has issued a particular instrument and that of the recipient institution are coordinated to the extent they both rely on the nonbinding instrument. Where the recipient institution has the competence to issue binding norms or has enforcement capabilities, the potential of the norms to induce behavioural change is greatly enhanced. [Part 2, A.II.2.]

b) In order to assess the impact of nonbinding instruments on the domestic state level, the study first looked at factors that enhance compliance by states, and then assessed the pathways of influence in national legal systems through which they are implemented by states. [Part 2, B.]

(1) International institutions increasingly undertake various efforts to enhance compliance with nonbinding instruments. One means that is often employed is reporting on implementation and progress. Even voluntary reporting mechanisms help keeping the respective issues on the agenda of national and international decision-making processes. This triggers interaction and broad discourse over the norms in question, exerting implementation pressure at least on those states concerned about their reputation. Capacity building programmes, if installed by organisations with significant financial and human resources such as the FAO or the World Bank, can over time help to implement norms at the national or local level and generally provide incentives for compliance. Institutionalised discourse over such norms may trigger persuasion and socialisation processes which ultimately shape interests and behaviour of states. NGOs use international nonbinding norms as pressure tools and a basis for independent monitoring. Due to these

mechanisms and depending on the specific characteristics of each instrument, compliance with nonbinding norms may thus be as good (or bad) than that induced by binding norms. [Part 2, B.I.]

(2) The analysis of this study shows that nonbinding instruments have a specific contribution to make in inducing behavioural changes of state actors, in particular through impacting domestic legislative and administrative processes, but also court decisions. The study finds some exemplary evidence which indicates that nonbinding instruments may have considerable impact on national policies, legislation as well as administrative decision-making. National policy makers can contribute in international processes and then better convince other constituencies nationally of the necessary legal or administrative changes. Absent specific legislative approval and incorporation, the instruments may be implemented *de facto* through other administrative or judicial means. Although not guaranteed, implementation without parliamentary approval is possible and takes place in practice. However, as neither legislative implementation nor enforcement in courts can be guaranteed in the case of nonbinding instruments, it can be said that they lack reliability. [Part 2, B.II.]

c) Regarding the private level, it is remarkable that international institutions develop a new regulatory approach by seeking directly to prescribe norms of behaviour for private actors such as corporations. These endeavours should not be derided as *per se* fruitless. Fear of reputational losses in a competitive global business environment and rising public criticism force companies at least to appear willing to comply. The work of industry associations and their efforts to force members to comply with international norms can be seen as evidence of some impact. Where international institutions establish international mediation and complaint mechanisms, these processes of compliance pressure are significantly strengthened. Mediation and complaint procedures enhance the possibilities for monitoring through NGOs. They may also have preventative effects. Apart from institutionalised procedures, international nonbinding instruments issued by international organisations provide tools on which NGOs can draw for their own campaigning, standard setting and labelling programmes. (e.g. Marine Stewardship Council). [Part 2, C.]

6. In a final assessment, nonbinding instruments have specific strengths, but also clear limitations. Although they have a specific and unique role to play, they therefore hardly ever appear as useful alternatives or substitutes for treaty law in the long run.

a) Nonbinding norms directed at state actors may under certain conditions constitute an important complementary strategy to regulation through treaty law in a regulatory division of labour.

They provide tools to mitigate some of the limitations of treaty law, by allowing states and international institutions to move forward even in the absence of political consensus or despite the lack of competence to issue binding rules. Their use therefore often allows integrating reluctant actors into common endeavours and continued norm-based discourse. The nonbinding form allows states to avoid lengthy and cumbersome treaty making processes. In particular in politically sensitive issue areas, this can reduce the risk of low ratification numbers or that an instrument does not enter into force at all. At the same time, one must not ignore that modern treaty regimes increasingly provide for flexibility such as escape clauses, sunset and exit provisions, framework-protocol approaches or – though still rare – majority decision making. In matured treaty regimes that provide for sufficient flexibility through these means, nonbinding instruments generally play a less prominent role.

Nonbinding instruments furthermore provide useful tools to establish norms at the international level and to guide – often through detailed legislative guidance or best practices – legislative development and administrative decision making at the national level. In particular in those developing countries where the legal system and administrative capabilities are less advanced, nonbinding instruments are often readily implemented and used as arguments against domestic constituencies or other parts of governments opposed to the changes.

Part of the particular strengths of nonbinding instruments is that their characteristics respond to some of the specific exigencies of environmental law. Environmental law is defined by a high level of uncertainty, the necessity to act in a preventive manner, rapid technological advancement and the related need for flexibility and learning. The evolutionary and process-based character of most nonbinding instruments delivers important flexibility able to accommodate the necessary learning processes. Thus, a nonbinding option is preferable to address issue areas that are not yet well understood and defined by high scientific uncertainty and scientific complexity. Binding norms could if necessary be adopted once the nonbinding ones are sufficiently tested and widely accepted by state and non-state actors. However, a prerequisite for a successful nonbinding learning strategy is the establishment of follow-up processes which secure continuous discourse and feed-back processes.

b) The ability of nonbinding instruments to induce compliance of state actors is however inherently limited. They therefore cannot serve as alternatives to treaty making. A number of factors related to the characteristics of nonbinding instruments prohibit relying on nonbinding instruments as long-term alternatives: [see also Part 2, B.III.]

(1) The first one is nonbinding status. Contrary to some suggestions in scholarship, the concept of bindingness remains an important, even if not the only factor for inducing compliance with norms. The binding status is a prerequisite for legal certainty, and therefore a nonbinding status seriously constrains the utility of nonbinding instruments in that respect. Building on a specific promise that a binding obligation will be obeyed and implemented, as well as broadly accepted law-making procedure, the authority of binding norms is distinct from that of nonbinding ones. This particular authority ultimately increases the credibility attached to binding promises. Given the particular credibility and authority of binding norms, states risk higher reputational costs in breaking legally binding rules. In consequence of the particular limitation of nonbinding instruments in providing for legal certainty, this study suggests that nonbinding instruments cannot replace international law, since one function of law is to provide legal certainty which is in turn the prerequisite for stable international relations. Nonbinding instruments are therefore not suited for long-term cooperation in areas that require a high degree of reassurance and legal certainty over time. Thus, it comes as no surprise that states that seriously wish to address environmental issues often propagate binding instruments for the allocation of common resources and goods such as fisheries or the allocation of emission rights. Legal certainty is also highly relevant for establishing market mechanisms such as emission trading, as demonstrated by the adoption of the first and second commitment periods under the Kyoto Protocol. It is also an essential element for long-term incentive mechanisms such as the access and benefit sharing system which build on the promise that environmental protection and allowing access to resources will pay off eventually in terms of benefits received from the use of genetic resources. In accordance with these considerations, states have eventually adopted legally binding instruments in these areas, as demonstrated by the adoption of the Nagoya Protocol on Access to Genetic Resources and the Fair and Equitable Sharing of Benefits Arising from their Utilization to the Convention on Biological Diversity in 2010.

(2) Another inherent limitation is connected to compliance mechanisms. If compared to compliance mechanisms which are increasingly employed for modern Multilateral Environmental Agreements, the

means usually employed to enhance compliance with nonbinding instruments only have limited potential. To be sure, this is not meant to imply that nonbinding instruments could not – theoretically – be equipped with strong compliance mechanisms. However, this study has revealed that most often, compliance mechanisms of nonbinding instruments lack some fundamental elements that make a compliance mechanism work.

First, reporting mechanisms based on voluntary norms often lack individualised compliance assessment and meaningful monitoring. If states do not need to justify their individual action in the follow-up to reporting, the effectiveness is constrained considerably.

Second, only binding treaty obligations can meet with the necessary legal certainty the need for financial support, capacity building and technological transfer that constitute the backbone of compliance management strategies. Often states only reliably provide resources for purposes of financial mechanisms and capacity building under binding treaty obligations. It is therefore not surprising that stronger compliance mechanisms and better financial mechanisms were two of the main reasons why states and NGOs favoured a shift from the voluntary to a binding PIC system. In practice, however, agreement on these additional elements is increasingly difficult to achieve, as demonstrated by the difficulties of establishing such mechanisms under the Stockholm and PIC Conventions.

(3) With respect to enforcement measures or sanctions, one might be tempted to neglect the difference between binding treaty law and nonbinding instruments for the simple reason that enforcement is in any case the exception in international environmental treaty law. However, one should not deduce from this that it is not desirable to develop international law further, with a view to strengthen enforcement possibilities through judicial and other means. Enforcement mechanisms adapted to the exigencies of environmental problems such as restrictions of market access can be highly effective, as demonstrated by the success of CITES or the Montreal Protocol and as envisioned in the Fish Stock Agreement. Such mechanisms are hardly ever based on nonbinding norms, except in rare cases. And it appears that states, when realising the necessity for enforcement as in the case of port state enforcement to control IUU fishing, resort to treaty making as a remedy. As suggested in particular by the case of the FAO Code of Conduct for Responsible Fisheries, voluntary standards that lack enforcement do not seem to be sufficient in areas faced with fierce economic competi-

tion, danger of free-riding and generally strong countervailing market conditions.

(4) One should also not forget that informality carries a potential equity problem, especially if one were to rely only on nonbinding norms for regulation. The increase in informality that can generally be associated with nonbinding instruments carries the danger of giving an advantage to powerful actors to the detriment of less powerful ones, who would lose the protection afforded to them by the principle of sovereign equality which is safeguarded by the requirement of consent.

(5) Accepting nonbinding norms as alternatives would remove pressure from negotiations for binding international law and for developing further international dispute settlement procedures and compliance mechanisms. Where nonbinding instruments can only be of limited effectiveness, as in areas of strong countervailing market forces and free-riding, relying solely on nonbinding solutions could thus actually inhibit effective action through legal development.

c) Nonbinding norms directed at private actors can represent a temporary solution in a field where international law is largely absent. And they can possibly prevent some extreme cases of corporate abuse of environmental and social rights through raising public pressure and by instigating naming and shaming by NGOs. They are useful in complementing international and national law in stimulating social and market-driven factors for achieving social and environmental objectives.

d) With respect to private actor regulation, nonbinding instruments are also of some utility, but cannot be considered as alternative to international and national law due to some inherent limitations.

(1) Nonbinding norms directed at private actors can represent a temporary solution in a field where international law is largely absent. And they can possibly prevent some extreme cases of corporate abuse of environmental and social rights through raising public pressure and by instigating naming and shaming by NGOs. They are useful in complementing international and national law in stimulating social and market-driven factors for achieving social and environmental objectives.

(2) But nonbinding instruments have clear limits in respect of private actors. First, one important limitation relates to the content of standards. Where substantive standards are promulgated, they often tend to demand little, since their success depends on their acceptability to business actors. Mere hortatory or aspirational statements of principles lack the precision and clarity which is a precondition for a standard to provide guidance to corporations and to be subjected to meaningful exter-

nal evaluation and monitoring. Further, as in the case of the OECD Guidelines, a large number of norms in these instruments only prescribe environmental management systems but not stringent substantive performance standards. Compliance therefore does not necessarily require a particularly environment-friendly business practice.

(3) Another limitation arises from the limited means of ensuring compliance. Sanctions or penalties for violation as well as external monitoring of nonbinding instruments are most often absent. Nonbinding instruments thus depend on public attention, continuing NGO pressure and consequently NGO resources, as well as the reputational sensitivity of business actors. Their effectiveness will therefore always remain limited to particular sectors, particular issues and particular time periods: they are not suited as a general regulatory tool.

(4) As a consequence, nonbinding instruments cannot absolve states from their duty to develop international treaty law that defines responsibilities of states in how to regulate multinational corporations. In particular in areas of corporate social responsibility and investment law, the move to nonbinding codes of conduct arguably deflected attention from developing the binding treaty law and develop national law necessary to control the social and environmental impact of activities of multinational corporations. The approach to regulate private actors through nonbinding international codes of conduct resembles in many ways self-regulatory approaches known from the domestic level. This approach is suited as a complement but not a substitute to traditional command-and-control or incentive based mechanisms as it better responds to the need for flexibility and innovation in complex and uncertain regulatory areas such as environmental law.

7. In Part 3, this study explored the question of the legitimacy of the use of nonbinding instruments. Legitimacy becomes an issue in light of the finding of this study that, at least in the environmental field, nonbinding activities of international institutions have a considerable effect on the international legal and institutional environment as well as direct or indirect effects on states and private actors. International institutions acting through nonbinding instruments not only provide fora that facilitate cooperation by states, they also increasingly act as influential autonomous actors. They also provide platforms for NGOs and industry representatives to influence and participate in international norm-setting and implementation processes. As much as this enhances the governance potential of international institutions, care must be taken to link these processes to sources of legitimacy.

The challenges outlined in the study cannot however be compared to those legitimacy problems triggered by the law-making activities of international institutions. Cases of international legislation or international administration raise more significant legitimacy problems and need for legal protection. A differentiated analysis is needed that reflects the specificities of the instruments under consideration. This study has therefore discussed some of the challenges that specifically and typically appear in the case of nonbinding instruments.

[Part 3, B.]

a) One challenge appears on the domestic level. The move to nonbinding instruments increases tendencies of deparliamentarisation and increase of executive power, in particular given the possibility to adopt nonbinding instruments on the international level and implement them at the national level without parliamentary approval. This becomes particularly problematic when it concerns issues formerly in the realm of domestic legislatures. The increase in executive power could be mitigated through national procedures that enhance the control of executives by the parliament in respect of the development, adoption and implementation of nonbinding instruments. [Part 3, B.III. and C.II.]

b) Cooperation through nonbinding instruments also affects traditional chains of legitimation from the national to the international level. First, the nonbinding form facilitates dynamic treaty development and institutional developments where the link to the original consent to the base treaty is attenuated. The nonbinding nature often allows for more flexible and autonomous decision making by international subsidiary bodies, international civil servants and expert bodies than would be the case if the norms concerned were binding. The mentioned actors often take decisions in the absence of specific institutional law that delineates and conditions their actions. These challenges can be addressed through strengthening the principles of legality and delegation. Specific institutional law that is – ideally – subject to judicial scrutiny must be developed for activities related to the adoption and implementation of nonbinding instruments. This should include binding procedural law enabling and controlling access to information and participation of civil society actors at the international level to improve the link between public discourse and norm-making processes which is otherwise increasingly strained. [Part 3, B.II. and C.III.]

c) One must not, however, ignore the need for administrative discretion, flexibility and informality. In fact, the need for more informal and nonbinding processes is why states and other actors resort to nonbinding instruments in the first place. Legal formalisation should not force these actors into even more informal and less transparent processes. As in national law, procedural law must and can address the balance between flexibility and effectiveness on the one hand and the long-term legitimacy of norms and institutions on the other hand, for instance by defining discretionary leeway. But only procedural law can turn arbitrariness into the legitimate exercise of authority by strengthening chains of delegation and securing equitable means of participation. However, the balance between flexibility and stability, between broad delegation to experts and political control, between confidential informal negotiations and public accessibility and accountability will be difficult to achieve and will require particular solutions for each regime and instrument in order to safeguard legitimacy of international institutions over time. [Part 3, D]

8. This study confirms that any dichotomous view of nonbinding versus binding, or soft law versus hard law, of soft enforcement versus hard enforcement is too simplistic. International relations today display an intricate and complex interdependency and interplay of nonbinding instruments with binding instruments. Nonbinding instruments have certain strengths which they develop through the interplay and linkage of various institutions and actors at private, national, regional and international levels, but they also have clear limitations where treaty law is needed. As can be observed in the case of EU and domestic law, mature legal systems therefore increasingly resort both to informal and formal, binding and nonbinding forms of ordering. In that sense, the use of nonbinding instruments may not necessarily be taken as a sign of weakness of the international system, but perhaps as a sign for a more complex and stronger international legal system that makes use of a variety of complementary regulatory tools. However, it remains essential not to take nonbinding instruments as an excuse to divert attention from the need to further develop international treaty law, its means of ensuring compliance and dispute settlement systems. In this sense, there remains a lot to be done in international environmental law. Both aspirational norms prescribing the need for change, often expressed in informal instruments, and clear obligations prescribed through formal binding law and supported by dispute settlement are part of a functioning legal system. Informality and nonbinding instruments provide additional flexibility, while formal binding law serves as a guarantor of legal security and stability. It is an ongoing task for legal scholarship and practitioners to find the right balance between both.

Bibliography

Abbott, Kenneth W./Snidal, Duncan, "Hard and Soft Law in International Governance", *International Organization* 54 (2000), 421-456.

Abi-Saab, Georges, "Fragmentation or unification: some concluding remarks", *New York University Journal of International Law & Politics* 31 (1999), 919-933.

Adede, Andrónico O., *The System for Settlement of Disputes under the United Nations Convention on the Law of the Sea*, 1987 (Dordrecht, Martinus Nijhoff Publishers).

Alam, S. M. Nazmul/C. Kwei Lin/Yakupitiyage, Amararatne/Demaine, Harvey/Phillips, Michael J., "Compliance of Bangladesh shrimp culture with FAO code of conduct for responsible fisheries a development challenge", *Ocean and Coastal Management* 48 (2005), 177-188.

Alvarez, José E., "The WTO as linkage machine", *The American Journal of International Law* 96 (2002), 146-158.

Alvarez, José E., *International Organizations as Law-makers*, 2005 (Oxford, Oxford University Press).

Alvarez, José E., "Do states socialize?", *Duke Law Journal* 54 (2005), 961-974.

Alvarez, José E., "International Organizations: Then and Now", *American Journal of International Law* 100 (2006), 324-347.

Assadourian, Erik, "The State of Corporate Responsibility and the Environment", *Georgetown International Environmental Law Review* 18 (2006), 571-594.

Aston, Jurij Daniel, *Sekundärgesetzgebung internationaler Organisationen zwischen mitgliedstaatlicher Souveränität und Gemeinschaftsdisziplin*, 2005 (Berlin, Duncker & Humblot).

Austin, John, *The province of jurisprudence determined: the uses of the study of jurisprudence*, 1954 (London, Weidenfeld & Nicolson).

Baade, Hans W., "The Legal Effects of Codes of Conduct for Multinational Enterprises", *German Yearbook of International Law* 22 (1979), 11-52.

Barnhizer, David, "Waking from Sustainability's "Impossible Dream": The Decisionmaking Realities of Business and Government", *Georgetown International Environmental Law Review* 18 (2006), 595-690.

Bassiouni, Cherif M., "A Functional Approach to 'General Principles of International Law'", *Michigan Journal of International Law* 11 (1990), 768-818.

Beermann, Jack M., *Administrative law*, 2006 (2nd ed., New York, Aspen Publ.).

Benvenisti, Eyal, "Public Choice and Global Administrative Law: Who's Afraid of Executive Discretion?", IILJ Working Paper 2004/3 (Global Administrative Law Series), available at www.iilj.org.

Benvenisti, Eyal/Downs, George W., "National Courts, Domestic Democracy, and the Evolution of International Law", *European Journal of International Law* 20 (2009), 59-72.

Benzing, Markus, "International Organizations or Institutions, Secondary Law", in: Wolfrum, Rüdiger (ed.), *Max Planck Encyclopedia of Public International Law* 2010, online at www.mpepil.com.

Bernstorff, Jochen von, "Zivilgesellschaftliche Partizipation in Internationalen Organisationen: Form globaler Demokratie oder Baustein westlicher Expertenherrschaft?", in: Brunkhorst, Hauke (ed.), *Demokratie in der Weltgesellschaft*, 2009 (Baden-Baden, Nomos), 277-302.

Bernstorff, Jochen von, "Procedures of Decision-Making and the Role of Law in International Organizations", in: Bogdandy, Armin von/Wolfrum, Rüdiger/Bernstorff, Jochen von/Dann, Philipp/Goldmann, Matthias (eds.), *The Exercise of Public Authority by International Institutions: Advancing International Institutional Law*, 2010 (Heidelberg, Springer-Verlag), 777-806.

Beyerlin, Ulrich, "State Community Interests and Institution-Building in International Environmental Law", *Zeitschrift für ausländisches öffentliches Recht und Völkerrecht* 56 (1996), 602-627.

Beyerlin, Ulrich/Marauhn, Thilo, *Rechtsetzung und Rechtsdurchsetzung im Umweltvölkerrecht nach der Rio-Konferenz 1992*, 1997 (Berlin, Erich Schmidt Verlag).

Beyerlin, Ulrich, *Umweltvölkerrecht*, 2000 (München, Beck Verlag).

Beyerlin, Ulrich/Marauhn, Thilo, *International Environmental Law*, 2011 (Oxford, Hart Publishing; München, Beck Verlag).

Beyerlin, Ulrich/Stoll, Peter-Tobias/Wolfrum, Rüdiger (eds.), *Ensuring compliance with multilateral environmental agreements: a dialogue between practitioners and academia*, 2006 (Leiden, Martinus Nijhoff Publishers).

Beyerlin, Ulrich, "Different Types of Norms in International Environmental Law: Policies, Principles and Rules", in: Brunnée, Jutta (ed.), *Oxford Handbook of International Environmental Law*, 2007 (Oxford, Oxford University Press), 425-448.

Birnie, Patricia/Boyle, Alan, *International law and the environment*, 2002 (2nd ed., Oxford, Oxford University Press).

Blanco-Bazán, Agustín, "IMO Interface with the Law of the Sea Convention", in: Nordquist, Myron H./Moore, John N. (eds.), *Current Maritime Issues and the International Maritime Organization*, 1999 (The Hague, Martinus Nijhoff Publishers), 268-287.

Blanpain, Roger, *The OECD Guidelines for Multinational Enterprises and Labour Relations 1976 – 1979*, 1979 (Deventer, Kluwer).

Blanpain, Roger, "Guidelines for Multinational Enterprises, for Ever?: The OECD Guidelines, 20 years later", *The International Journal of Comparative Labour Law and Industrial Relations* 14 (1998), 337-348.

Bleckmann, Albert, "Sittenwidrigkeit wegen Verstoßes gegen den ordre public international", *Zeitschrift für ausländisches öffentliches Recht und Völkerrecht* 34 (1974), 112-132.

Blom, Gaby G.R., "Institutional and legal aspects of the Food and Agriculture Organization of the United Nations (FAO)", *Revue de Droit International* 74 (1996), 227-332.

Blüthner, Andreas, "Ein Globalisierungspakt über Werte und Effizienz", in: Hobe, Stephan (ed.), *Kooperation oder Konkurrenz internationaler Organisationen*, 2001 (Baden-Baden, Nomos), 155.

Bodansky, Daniel, "Customary (and not so customary) international environmental law", *Indiana Journal of Global Legal Studies* 3 (1995), 105-131.

Bodansky, Daniel, "The Legitimacy of International Governance: A Coming Challenge for International Environmental Law?", *American Journal of International Law* 93 (1999), 596-624.

Bodansky, Daniel, "Legitimacy", in: Bodansky, Daniel/Brunnée, Jutta/Hey, Ellen (eds.), *Oxford Handbook of International Environmental Law*, 2007 (Oxford, Oxford University Press), 704-723.

Bodansky, Daniel, "The Concept of Legitimacy in International Law", in: Wolfrum, Rüdiger/Röben, Volker (eds.), *Legitimacy in International Law*, 2008 (Berlin/Heidelberg, Springer), 309-317.

Bodansky, Daniel, *The art and craft of international environmental law*, 2010 (Cambridge, Mass., Harvard University Press).

Böckenförde, Ernst-Wolfgang, "Die Zukunft politischer Autonomie: Demokratie und Staatlichkeit im Zeichen von Globalisierung, Europäisierung und Individualisierung", in: Böckenförde, Ernst-Wolfgang (ed.), *Staat, Nation, Europa: Studien zur Staatslehre, Verfassungstheorie und Rechtsphilosophie*, 1999 (Frankfurt a.M., Suhrkamp), 103-126.

Böckenförde, Markus, *Grüne Gentechnik und Welthandel: das Biosafety-Protokoll und seine Auswirkungen auf das Regime der WTO*, 2004 (Berlin/Heidelberg [u.a, Springer).

Böhmer, Alexander, "The Revised 2000 OECD Guidelines for Multinational Enterprises – Challenges and Prospects after 4 Years of Implementation", *Policy Papers on Transnational Economic Law* 3 (2004).

Bogdandy, Armin von, "Demokratie, Globalisierung, Zukunft des Völkerrechts – eine Bestandsaufnahme", *Zeitschrift für ausländisches öffentliches Recht und Völkerrecht* 63 (2003), 853-877.

Bogdandy, Armin von, "Lawmaking by International Organisations: Some Thoughts on Non-Binding Instruments and Democratic Legitimacy", in: Wolfrum, Rüdiger/Röben, Volker (eds.), *Developments of International Law in Treaty Making*, 2005 (Berlin/Heidelberg, Springer), 171-182.

Bogdandy, Armin von/Wolfrum, Rüdiger/ Bernstorff, Jochen von/ Dann, Philipp/Goldmann, Matthias (eds.), *The Exercise of Public Authority by International Institutions: Advancing International Institutional Law*, 2010 (Heidelberg, Springer).

Boisson de Chazournes, Laurence, "Policy Guidance and Compliance: The World Bank Operational Standards", in: Shelton, Dinah (ed.), *Commitment and Compliance: The Role of Non-Binding Norms in the International Legal System*, 2000 (Oxford, Oxford University Press), 281-302.

Boisson de Chazournes, Laurence, "Technical and Financial Assistance and Compliance: the Interplay", in: Beyerlin, Ulrich/Stoll, Peter-Tobias/Wolfrum, Rüdiger (eds.), *Ensuring Compliance with Multilateral Environmental Agreements*, 2006 (Leiden, Martinus Nijhoff Publishers), 273-300.

Boisson de Chazournes, Laurence "Technical and Financial Assistance", in: Bodansky, Daniel/Brunnée, Jutta/Hey, Ellen (eds.), *Oxford Handbook of International Environmental Law*, 2007 (Oxford, Oxford University Press), 947-973.

Bothe, Michael, "Legal and Non-Legal Norms – A Meaningful Distinction in International Relations?", *Netherlands Yearbook of International Law* 11 (1980), 65-95.

Bothe, Michael, "The Evaluation of Enforcement Mechanisms in International Environmental Law: An Overview", in: Wolfrum, Rüdiger (ed.), *Enforcing environmental standards: economic mechanisms as viable means?*, 1996 (Heidelberg, Springer), 13-38.

Boutros Ghali, Boutros, *An agenda for democratization*, 1996 (New York, United Nations).

Boyle, Alan, "Dispute Settlement and the Law of the Sea Convention: Problems of Fragmentation and Jurisdiction", *International and Comparative Law Quarterly* 46 (1997), 37-54.

Boyle, Alan E., "Marine Pollution under the Law of the Sea Convention", *American Journal of International Law* 79 (1985), 347-372.

Boyle, Alan E., "Some Reflections on the Relationship of Treaties and Soft Law", *International and Comparative Law Quarterly* 48 (1999), 901-913.

Boyle, Alan E., "Problems of Compulsory Jurisdiction and the Settlement of Disputes Relating to Straddling Fish Stocks", in: Stokke, Olaf S. (ed.), *Governing High Seas Fisheries: the interplay of global and regional regimes*, 2001 (Oxford, Oxford University Press), 91-120.

Bradley, Curtis A., "International delegations, the structural constitution, and non-self-execution", *Stanford Law Review* 55 (2003), 1557-1596.

Brakman Reiser, Dana/Kelly, Claire R., "Linking NGO accountability and the legitimacy of global governance", *Brooklyn Journal of International Law* 36 (2011), 1011-1073.

Breitmeier, Helmut/Young, Oran R./Zürn, Michael, "The International Regimes Database: Architecture, Key Findings, and Implications for the Study of Environmental Regimes", in: Jacob, Klaus/ Biermann, Frank/Busch, Per-Olof/Feindt, Peter H. (eds.), *Politik und Umwelt*, 2007 (Wiesbaden, Verlag für Sozialwissenschaften), 41-59.

Breitmeier, Helmut, *The legitimacy of international regimes*, 2008 (Aldershot, Ashgate).

Breuer, Rüdiger, "Zunehmende Vielgestaltigkeit der Instrumente im deutschen und europäischen Umweltrecht – Probleme der Stimmigkeit und des Zusammenwirkens", *Neue Zeitschrift für Verwaltungsrecht* 16 (1997), 833-845.

Brown, Edward D., *The International Law of the Sea*, 1994 (Aldershot, Dartmouth Publishing).

Brown Weiss, Edith (ed.), *International Compliance with Nonbinding Accords*, 1997 (Washington, D.C., American Society of International Law).

Brown Weiss, Edith, "Introduction", in: Brown Weiss, Edith (ed.), *International Compliance with Nonbinding Accords*, 1997 (Washington D.C., The American Society of International Law), 1-20.

Brown Weiss, Edith (ed.), *Engaging countries: strengthening compliance with international environmental accords*, 1998 (Cambridge, Mass., MIT Press).

Brown Weiss, Edith, "Conclusions: Understanding Compliance with Soft Law", in: Shelton, Dinah (ed.), *Commitment and Compliance: The Role of Non-binding Norms in the International Legal System*, 2000 (Oxford, Oxford University Press), 535-556.

Brown Weiss, Edith, "Understanding compliance with international environmental agreements: the Baker's dozen myths", *University of Richmond Law Review* 32 (1999), 1555-1589.

Brownlie, Ian, "Legal effects of codes of conduct for MNEs: Commentary", in: Horn, Norbert (ed.), *Legal Problems of Codes of Conduct for Multinational Enterprises*, 1980 (Deventer, Kluwer), 39-43.

Brownlie, Ian, *Principles of public international law*, 2008 (7th ed., Oxford, Oxford University Press).

Brunkhorst, Hauke, *Solidarität: von der Bürgerfreundschaft zur globalen Rechtsgenossenschaft*, 2002 (Frankfurt am Main, Suhrkamp Verlag).

Brunnée, Jutta/Toope, Stephen J., "International law and constructivism: elements of an interactional theory of international law", *Columbia Journal of Transnational Law* 39 (2000), 19-74.

Brunnée, Jutta/Toope, Stephen J., *Legitimacy and legality in international law: an interactional account*, 2010 (Cambridge, Cambridge University Press).

Brunnée, Jutta, "COPing with Consent: Law-Making under Multilateral Environmental Agreements", *Leiden Journal of International Law* 15 (2002), 1-52.

Brunnée, Jutta/Toope, Stephen J., "Persuasion and enforcement: explaining compliance with international law", *The Finnish Yearbook of International Law* 13 (2002), 273-295.

Brunnée, Jutta, "The Kyoto Protocol: Testing Ground for Compliance Theories?", *Zeitschrift für ausländisches öffentliches Recht und Völkerrecht* 63 (2003), 255-280.

Brunnée, Jutta, "The Stockholm Declaration and the Structure and Processes of International Environmental Law", in: Nordquist, Myron H./Moore, John N./Mahmoudi, Said (eds.), *The Stockholm Declaration and Law of the Marine Environment*, 2003 (The Hague, Kluwer Law International), 67-84.

Brunnée, Jutta, "Of Sense and Sensibility: Reflections on International Regimes as Tools for Environmental Protection", *International and Comparative Law Quarterly* 53 (2004), 351-367.

Brunnée, Jutta, "The Stockholm Declaration and the Structure and Processes of International Environmental Law", in: Chircop, Aldo (ed.), *The future of ocean regime building: essays in tribute to Douglas M. Johnston*, 2009 (Leiden, Martinus Nijhoff Publishers), 41-62.

Brunnée, Jutta, "International Legislation", in: Wolfrum, Rüdiger (ed.), *Max Planck Encyclopedia of Public International Law* (2010), online at: www.mpepil.com.

Bryde, Brun-Otto, *Internationale Verhaltensregeln für Private – Völkerrechtliche und verfassungsrechtliche Aspekte*, 1981 (Frankfurt am Main, Alfred Metzner Verlag).

Bryde, Brun-Otto, "Konstitutionalisierung des Völkerrechts und Internationalisierung des Verfassungsrechts", *Der Staat* 42 (2003), 61-75.

Buchanan, Allan/Keohane, Robert O., "The Legitimacy of Global Governance Institutions", in: Wolfrum, Rüdiger/Röben, Volker (eds.), *Legitimacy in International Law*, 2008 (Berlin/Heidelberg, Springer), 25-62.

Buchanan, Allen E., *Justice, legitimacy, and self-determination: moral foundations for international law*, 2003 (Oxford, Oxford University Press).

Burgstaller, Markus, "Amenities and Pitfalls of a Reputational Theory of Compliance with International Law", *Nordic Journal of International Law* 76 (2007), 39-71.

Butt, Mark, *Die Ausweitung des Rechts auf Umweltinformation durch die Aarhus-Konvention*, 2001 (Stuttgart, ibidem-Verlag).

Carr, Christopher J., "Recent developments in compliance and enforcement for international fisheries", *Ecology Law Quarterly* 24 (1997), 847-860.

Carstensen, Nils Christian, *Das Verhältnis des Seerechtsübereinkommens der Vereinten Nationen von 1982 zu fischereirechtlichen Übereinkommen und deren Streitbeilegungsvorschriften*, 2005 (Frankfurt am Main, Peter Lang Verlag).

Cassese, Sabino, "Il diritto amministrativo globale: una introduzione", *Rivista trimestrale di diritto pubblico* 55 (2005), 331-357.

Charlesworth, Hilary/Coicaud, Jean-Marc (eds.), *Fault Lines of International Legitimacy*, 2010 (Cambridge, Cambridge University Press).

Charney, Jonathan I., "Universal International Law", *American Journal of International Law* 87 (1993), 529-551.

Chayes, Abram/Handler Chayes, Antonia, *The New Sovereignty: Compliance with International Regulatory Agreements*, 1995 (Cambridge, Mass., Harvard University Press).

Chen, Chen-Ju, *Fisheries subsidies under international law*, 2010 (Heidelberg, Springer Verlag).

Cheng, Bin, "Custom: The Future of General State Practice in a Divided World", in: Macdonald, Ronald J. (ed.), *The Structure and Process of International Law: Essays in Legal Philosophy Doctrine and Theory*, 1983 (The Hague, Martinus Nijhoff Publishers), 513-554.

Chinkin, Christine, "The Challenge of Soft Law: Development and Change in International Law", *International and Comparative Law Quarterly* 38 (1989), 850-866.

Chinkin, Christine, "Normative Development in the International Legal System", in: Shelton, Dinah (ed.), *Commitment and Compliance: the Role of Non-binding Norms in the International Legal System*, 2000 (Oxford, Oxford University Press), 21-42.

Churchill, Robin R./Lowe, Vaughan, *The law of the sea*, 1999 (3rd ed., Manchester, Manchester University Press).

Churchill, Robin R., "International Tribunal for the Law of the Sea the Southern Bluefin Tuna cases (New Zealand v. Japan; Australia v. Japan): order for provisional measures of 27 August 1999", *The International and Comparative Law Quarterly* 49 (2000), 979-990.

Churchill, Robin R., "The Jurisprudence of the International Tribunal for the Law of the Sea Relating to Fisheries: Is There Much in the Net?", *The International Journal of Marine and Coastal Law* 22 (2007), 383-424.

Clark, Ian, *International Legitimacy and World Society*, 2007 (New York, Oxford University Press).

Coicaud, Jean-Marc/Heiskanen, Veijo (eds.), *The legitimacy of international organizations*, 2001 (Tokyo, The United Nations University Press).

Coicaud, Jean-Marc, "The Evolution of the International Order and Fault Lines of International Legitimacy", in: Charlesworth, Hilary/Coicaud, Jean-Marc (eds.), *Fault Lines of International Legitimacy*, 2010 (Cambridge, Cambridge University Press), 87-114.

Contini, Paolo/Sand, Peter H., "Methods to expedite environment protection: international ecostandards", *American Journal of International Law* 66 (1972), 37-59.

Cooper, Tracy, "Picture this: promoting sustainable fisheries through eco-labeling and product certification", *Ocean and Coastal Law Journal* 10 (2004), 1-49.

D'Amato, Anthony A., *The concept of custom in international law*, 1971 (Ithaca, Cornell University Press).

D'Amato, Anthony A., "Softness in International Law: A Self-Serving Quest for New Legal Materials: A Reply to Jean d'Aspremont", *European Journal of International Law* 20 (2009), 897-910.

D'Aspremont, Jean, "Softness in International Law: A Self-Serving Quest for New Legal Materials", *European Journal of International Law* 19 (2008), 1075-1093.

D'Aspremont, Jean, "Softness in International Law: A Self-Serving Quest for New Legal Materials; a rejoinder to Tony D'Amato", *European Journal of International Law* 20 (2009), 911-917.

Dahm, Georg/Delbrück, Jost/Wolfrum, Rüdiger, *Völkerrecht, Band I/1, Die Grundlagen. Die Völkerrechtssubjekte*, 1989 (2. völlig neu bearb. Auflage, Berlin, de Gruyter).

Dahm, Georg/Delbrück, Jost/Wolfrum, Rüdiger, *Völkerrecht, Band I/3, Die Formen völkerrechtlichen Handelns, die inhaltliche Ordnung der internationalen Gemeinschaft*, 2002 (2. völlig neu bearb. Auflage, Berlin, de Gruyter).

Dahm, Georg/Delbrück, Jost/Wolfrum, Rüdiger, *Völkerrecht, Band I/2, Der Staat und andere Völkerrechtssubjekte, Räume unter internationaler Verwaltung*, 2002 (2. völlig neu bearb. Auflage, Berlin, de Gruyter).

Dalton, Penelope D., "Implementing the International Code of Conduct for Responsible Fisheries", in: Nordquist, Myron H./ Moore, John N. (eds.), *Current Fisheries Issues and the Food and Agriculture Organization of the United Nations*, 2000 (The Hague, Kluwer Law International), 331-340.

Danilenko, Gennadij M., "Sources of International Law in a Changing International Community: Theory and Practice", in: Butler, William E. (ed.), *Perestroika and International Law*, 1990 (Dordrecht, Martinus Nijhoff Publishers), 61-80.

Dashwood, Hevina S., "Corporate Social Responsibility and the Evolution of International Norms", in: Kirton, John J./Trebilcock, Michael J. (eds.), *Hard Choices, Soft Law: Voluntary Standards in Global Trade, Environment and Social Governance*, 2004 (Aldershot, Ashgate), 189-201.

Delbrück, Jost, "Exercising Public Authority Beyond the State: Transnational Democracy and/or Alternative Legitimation Strategies?", *Indiana Journal of Global Legal Studies* 10 (2003), 29-43.

Dieckert, Ulrich, *Die Bedeutung unverbindlicher Entschließungen internationaler Organisationen für das innerstaatliche Recht der Bundesrepublik Deutschland*, 1993 (Dissertation Universität Hamburg).

Diller, Janelle M., "Social conduct in transnational enterprise operations: the role of the International Labour Organization", in: Blanpain, Roger (ed.), *Multinational enterprises and the social challenges of the XXIst Century*, 2000 (The Hague, Kluwer Law International), 17-28.

Dine, Janet, "Multinational enterprises: international codes and the challenge of sustainable development", *Non-State actors and international law* 1 (2001), 81-106.

Dinham, Barbara, "The Success of a Voluntary Code in Reducing Pesticide Hazards in Developing Countries", *Green Globe Yearbook of International Co-operation on Environment and Development* 3 (1996), 29-36.

Doehring, Karl, *Völkerrecht*, 2003 (Heidelberg, C.F. Müller).

Doulman, David J., "Code of Conduct for Responsible Fisheries: Development and Implementation Considerations", in: Nordquist, Myron H./Moore, John Norton (eds.), *Current Fisheries Issues and the Food and Agriculture Organization of the United Nations*, 2000 (The Hague, Martinus Nijhoff Publishers), 307-330.

Downs, George W./Rocke, David M./Barsoom, Peter N., "Is the good news about compliance good news about cooperation?", *International Organization* 50 (1996), 379-406.

Downs, George W., "Enforcement and the evolution of cooperation", *Michigan Journal of International Law* 19 (1998), 319-344.

Downs, George W./Danish, Kyle W./Barsoom, Peter N., "The transformational model of international regime design: triumph of hope or experience?", *Columbia Journal of Transnational Law* 38 (2000), 465-514.

Driesen, David M., "Is Emission Trading an Economic Incentive Program?", *Washington and Lee Law Review* 55 (1998), 289-350.

Dross, Miriam/Wolff, Franziska, "Do We Need a New Access and Benefit sharing Instrument?", *Yearbook of International Environmental Law* 15 (2004), 95-118.

Dupuy, Pierre-Marie, "Soft law and the international law of the environment", *Michigan Journal of International Law* 12 (1990-1991), 420-435.

Dupuy, Pierre-Marie, "Formation of customary international law and general principles", in: Brunnée, Jutta/Bodansky, Daniel/Hey, Ellen (eds.), *Oxford Handbook of International Environmental Law*, 2007 (Oxford, Oxford University Press), 449-466.

Dupuy, Pierre-Marie/Vierucci, Luisa (eds.), *NGOs in international law: efficiency in flexibility?*, 2008 (Cheltenham, Elgar).

Dupuy, René-Jean, "Declaratory Law and Programmatory Law: From Revolutionary Custom to 'Soft Law'", in: Akkerman, Robert J./Krieken, Peter J. van/Pannenborg, Charles O. (eds.), *Declarations on principles: a quest for universal peace. Liber Röling*, 1977 (Leyden, A. w. Sijthoff), 247-257.

Ebbesson, Jonas, "Public Participation", in: Bodansky, Daniel/Brunnée, Jutta/Hey, Ellen (eds.), *Oxford Handbook of International Environmental Law* 2007 (Oxford, Oxford University Press), 681-703.

Ebenroth, Carsten-Thomas, *Code of Conduct – Ansätze zur vertraglichen Gestaltung internationaler Investitionen*, 1987 (Konstanz, Universitätsverlag Konstanz).

Edeson, William, "Closing the Gap: The Role of 'Soft' International Instruments to Control Fishing", *Australian Yearbook of International Law* 20 (1999), 83-104.

Edeson, William, "The International Plan of Action on Illegal Unreported and Unregulated Fishing: The Legal Context of a Non-Legally Binding Instrument", *The International Journal of Marine and Coastal Law* 16 (2001), 603-623.

Edeson, William, "Soft and Hard Law Aspects of Fisheries Issues: Some Recent Global and Regional Approaches", in: Nordquist, Myron H./Moore, John N./Mahmoudi, Said (eds.), *The Stockholm Declaration and Law of the Marine Environment*, 2003 (The Hague, Kluwer Law International), 165-182.

Edeson, William, "The Role of Technical Bodies", in: Wolfrum, Rüdiger/Röben, Volker (eds.), *Developments of International Law in Treaty Making*, 2005 (Berlin/Heidelberg, Springer), 63-91.

Eifert, Martin, "Umweltinformation als Regelungsinstrument", *Die öffentliche Verwaltung* 47 (1994), 544-552.

Eifert, Martin, "Legitimationsstrukturen internationaler Verwaltung", in: Trute, Hans-Heinrich/Groß, Thomas/Röhl, Hans C./Möllers, Christoph (eds.), *Allgemeines Verwaltungsrecht – zur Tragfähigkeit eines Konzepts*, 2007 (Tübingen, Mohr Siebeck), 307-331.

Engel, Christoph, *Völkerrecht als Tatbestandsmerkmal deutscher Normen*, 1989 (Berlin, Duncker und Humblot).

Erbguth, Wilfried/Schink, Alexander, *Gesetz über die Umweltverträglichkeitsprüfung: Kommentar*, 1996 (2., vollständig überarbeitete Auflage, München, C.H. Beck).

Esty, Daniel C., "Next generation environmental law", *Capital University Law Review* 29 (2001), 183-204.

Esty, Daniel C., "Good Governance at the Supranational Scale: Globalizing Administrative Law", *Yale Law Journal* 115 (2006), 1490-1562.

Falk, Richard A./Strauss, Andrew, "On the creation of a global peoples assembly: legitimacy and the power of popular sovereignty", *Stanford Journal of International Law* 36 (2000), 191-220.

Fastenrath, Ulrich, "Relative Normativity in International Law", *European Journal of International Law* 4 (1993), 305-340.

Fatouros, Arghyrios A., "On the implementation of international codes of conduct: an analysis of future experience", *American University Law Review* 30 (1980-1981), 941-972.

Fatouros, Arghyrios A., "Les Principes directeurs de l'OECD à l'intention des entreprises multinationales: perspectives actuelles et possibilités futures", in: Reymond, Claude (ed.), *Études de droit international en l'Honneur de Pierre Lalive*, 1993 (Basel, Helbing & Lichtenhahn), 231-240.

Ferrell, Jessica K., "Controlling flags of convenience: one measure to stop overfishing of collapsing fish stocks", *Lewis & Clark Law School Environmental Law* 35 (2005), 323-390.

Fischer-Lescano, Andreas/Teubner, Günther, "Regime collisions: the vain search for legal unity in the fragmentation of global law", *Michigan Journal of International Law* 25 (2004), 999-1046.

Fischer-Lescano, Andreas/Liste, Philip, "Völkerrechtspolitik. Zu Trennung und Verknüpfung von Politik und Recht der Weltgesellschaft", *Zeitschrift für internationale Beziehungen* 12 (2005), 7-48.

Fitzmaurice, Malgosia, *Contemporary issues in international environmental law*, 2009 (Cheltenham, Elgar).

Franciose, Christopher N., "A critical assessment of the United States' implementation of the OECD guidelines for multinational enterprises", *Boston College International and Comparative Law Review* 30 (2007), 223-236.

Franck, Thomas M., "Legitimacy in the international system", *American Journal of International Law* 82 (1988), 705-759.

Franck, Thomas M., *The power of legitimacy among nations*, 1990 (New York, Oxford University Press).

Franck, Thomas M., *Fairness in international law and institutions*, 1995 (Oxford, Clarendon Press).

Franck, Thomas M., "The Power of Legitimacy and the Legitimacy of Power: International Law in an Age of Power Disequilibrium", *American Journal of International Law* 100 (2006), 88-106.

Freestone, David A./Makuch, Zen, "The new international environmental law of fisheries: the 1995 United Nations Straddling Stocks Agreement", *Yearbook of International Environmental Law* 7 (1996), 3-51.

Friedmann, Wolfgang, *The Changing Structure of International Law*, 1964 (London, Stevens & Sons).

Friedrich, Jürgen, "Carbon capture and storage: a new challenge for international environmental law", *Zeitschrift für ausländisches öffentliches Recht und Völkerrecht* 67 (2007), 211-227.

Friedrich, Jürgen, "Codes of Conduct", in: Wolfrum, Rüdiger (ed.), *Max Planck Encyclopedia of Public International Law* (2008), online at: www.mpepil.com.

Friedrich, Jürgen, "Environment, Private Standard Setting", in: Wolfrum, Rüdiger (ed.), *Max Planck Encyclopedia of Public International Law* (2010), online at: www.mpepil.com.

Friedrich, Jürgen, "Legal Challenges of Nonbinding Instruments: The Case of the FAO Code of Conduct for Responsible Fisheries", in: Bogdandy, Armin von/Wolfrum, Rüdiger/Bernstorff, Jochen von/Dann, Philipp/Goldmann, Matthias (eds.), *The exercise of public authority by international institutions: advancing international institutional law* (2010), 511-540.

Fritz, Simone, *Integrierter Umweltschutz im Völkerrecht*, 2009 (Berlin, Erich Schmidt Verlag).

Frowein, Jochen Abr., "Der Beitrag der internationalen Organisationen zur Entwicklung des Völkerrechts", *Zeitschrift für ausländisches öffentliches Recht und Völkerrecht* 36 (1976), 147-167.

Fuentes Véliz, Juan A., "L'évolution du rôle des organisations non gouvernementales dans le droit de l'environnement", *Revue Européenne de droit de l'Environnement* 4 (2007), 401-430.

Gaines, Sanford E., "Reflexive Law as a Legal Paradigm for Sustainable Development", *Buffalo Environmental Law Journal* 10 (2002-2003), 1-24.

Gehring, Markus W. (ed.), *Sustainable development in world trade law*, 2005 (The Hague, Kluwer Law International).

Goldmann, Matthias, "Inside Relative Normativity: From Sources to Standard Instruments for the Exercise of International Public Authority", in: Bogdandy, Armin von/Wolfrum, Rüdiger/Bernstorff, Jochen von/Dann, Philipp/Goldmann, Matthias (eds.), *The Exercise of Public Authority by International Institutions: Advancing International Institutional Law*, 2010 (Heidelberg, Springer-Verlag), 661-711.

Goldsmith, Jack L./Posner, Eric A., *The limits of international law*, 2005 (Oxford, Oxford University Press).

Goodman, Ryan/Jinks, Derek, "How to influence states: socialization and international human rights law", *Duke Law Journal* 54 (2004), 621-703.

Grosse, Robert, "Codes of Conduct for Multinational Enterprises", *Journal of World Trade Law* 16 (1982), 414-433.

Gruchalla-Wesierski, Tadeusz, "A Framework for Understanding "Soft Law"", *McGill Law Journal* 30 (1984-1985), 37-88.

Gündling, Lothar, "Compliance Assistance in International Environmental Law: Capacity-Building Through Financial and Technology Transfer", *Zeitschrift für ausländisches öffentliches Recht und Völkerrecht* 56 (1996), 796-809.

Gupta, Joyeeta, "The role of non-state actors in international environmental affairs", *Zeitschrift für ausländisches öffentliches Recht und Völkerrecht* 63 (2003), 459-486.

Guzmán, Andrew T., "A Compliance-Based Theory Of International Law", *California Law Review* 90 (2002), 1823-1887.

Guzmán, Andrew T., "Reputation and international law", *Georgia Journal of International and Comparative Law* 34 (2006), 379-391.

Guzmán, Andrew T., *How international law works: a rational choice theory*, 2008 (Oxford, Oxford University Press).

Haas, Peter, "Epistemic Communities", in: Bodansky, Daniel/Brunnée, Jutta/Hey, Ellen (eds.), *The Oxford Handbook of International Environmental Law*, 2007 (Oxford, Oxford University Press), 791-806.

Habermas, Jürgen, *Faktizität und Geltung: Beiträge zur Diskurstheorie des Rechts und des demokratischen Rechtsstaats*, 1998 (Frankfurt am Main, Suhrkamp Taschenbuch Wissenschaft).

Hahn, Anja von, "Implementation and Further Development of the Biodiversity Convention", *Zeitschrift für ausländisches öffentliches Recht und Völkerrecht* 63 (2003), 295-312.

Hailbronner, Kay, "Völkerrechtliche und staatsrechtliche Überlegungen zu Verhaltenskodizes für transnationale Unternehmen", in: Münch, Ingo von (ed.), *Staatsrecht – Völkerrecht – Europarecht: Festschrift für Hans-Jürgen Schlochauer*, 1981 (Berlin, Walter de Gruyter), 329-362.

Handl, Günther, *Multilateral Development Banking: Environmental Principles and Concepts Reflecting General International Law and Public Policy*, 2001 (The Hague, Kluwer Law International).

Handl, Günther, "A Hard Look at Soft Law", *American Society of International Law Proceedings* 82 (1988), 371-393.

Hart, Herbert L.A., *The concept of law*, 1961 (Oxford, Clarendon Press).

Held, David, *Democracy and the global order: from the modern state to cosmopolitan governance*, 1995 (Stanford, Stanford University Press).

Henkin, Louis, *How nations behave: law and foreign policy*, 1979 (2nd ed., New York, Columbia University Press).

Herberg, Martin, "Private Authority, global governance, and the law: The case of environmental self-regulation in multinational enterprises", in: Winter, Gerd (ed.), *Multilevel Governance of Global Environmental Change: Perspectives from Science, Sociology and the Law*, 2006 (Cambridge, Cambridge University Press), 149-178.

Herwig, Alexia, "The Contribution of Global Administrative Law to Enhancing the Legitimacy of the Codex Alimentarius Commission", in: O. Dilling/M. Herberg/G. Winter (eds.), *Transnational administrative rule making: performance, legal effects and legitimacy*, 2011 (Oxford, Hart Publishing), 171-212.

Herzog, Roman, "Art. 20 GG", in: Maunz, Theodor/Dürig, Günther (eds.), *Grundgesetz: Kommentar*, 2009 (München, Beck).

Hestermeyer, Holger P., *Human rights and the WTO: the case of patents and access to medicines,* 2007 (Oxford, Oxford University Press).

Heun, Werner, "Artikel 45a", in: Dreier, Horst (ed.), *Grundgesetz-Kommentar, Bd. 2*, 2006 (2. Auflage, Tübingen, Mohr Siebeck).

Heusel, Wolfgang, *"Weiches" Völkerrecht: Eine vergleichende Untersuchung typischer Erscheinungsformen*, 1991 (Baden-Baden, Nomos Verlagsgesellschaft).

Hewison, Grant J., "The Legally Binding Nature of the Moratorium on Large-Scale High Seas Driftnet Fishing", *Journal of Maritime Law and Commerce* 25 (1994), 557-579.

Hey, Ellen, "Sustainable development, normative development and the legitimacy of decision-making", *Netherlands Yearbook of International Law* 34 (2004), 3-53.

Hey, Ellen, "International Institutions", in: Bodansky, Daniel/Brunnée, Jutta/Hey, Ellen (eds.), *Oxford Handbook of International Environmental Law*, 2007 (Oxford, Oxford University Press), 749-769.

Higgins, Rosalyn, *The development of international law through the political organs of the United Nations*, 1963 (London, Oxford University Press).

Higgins, Rosalyn, "The Role of Resolutions of International Organizations in the Process of Creating Norms in the International System", in: Butler, William E. (ed.), *International Law and the International System*, 1987 (Dordrecht, Martinus Nijhoff Publishers), 21-30.

Hillgenberg, Hartmut, "A Fresh Look at Soft Law", *European Journal of International Law* 10 (1999), 499-515.

Hilpold, Peter, "EU Law and UN Law in Conflict: The Kadi Case", *Max Planck United Nations Yearbook* 13 (2009), 141-181.

Hinojosa Martínez, Luis M., "The legislative role of the Security Council in its fight against terrorism: legal, political and practical limits", *The International and Comparative Law Quarterly* (2008), 333-359.

Hobe, Stephan/Kimminich, Otto, *Einführung in das Völkerrecht*, 2004 (8. Auflage, Tübingen, Francke).

Hohmann, Harald, *Präventive Rechtspflichten und -prinzipien des modernen Umweltvölkerrechts*, 1992 (Berlin, Duncker und Humblot).

Holstein, Christoph, *Der Umweltschutz in der Tätigkeit der Weltbankgruppe: Instrumente, rechtliches Mandat und Bedeutung für das internationale Umweltrecht*, 2001 (Frankfurt am Main, Lang).

Horn, Norbert, "Codes of conduct for MNEs and Transnational Lex Mercatoria: An international Process of Learning and Law Making", in: Horn, Norbert (ed.), *Legal Problems of Codes of Conduct for Multinational Enterprises*, 1980 (Deventer, Kluwer), 45-81.

Horn, Norbert (ed.), *Legal Problems of Codes of Conduct for Multinational Enterprises*, 1980 (Deventer, Kluwer).

Horn, Norbert, "International Rules for Multinational Enterprises: the ICC, OECD, and ILO initiatives", *American University Law Review* 30 (1980-1981), 923-940.

Hosch, Gilles/Ferraro, Gianluca/Failler, Pierre "The 1995 FAO Code of Conduct for Responsible Fisheries: Adopting, implementing or scoring results?", *Marine Policy* 35 (2011), 189-200.

Hunter, David/Salzman, James/Zaelke, Durwood, *International Environmental Law and Policy*, 2002 (2nd ed., New York, Foundation Press).

Hurd, Ian, "Legitimacy and Authority in International Politics", *International Organization* 53 (1999), 379-408.

Jacobson, Harold K./Brown Weiss, Edith, "A Framework for Analysis", in: Jacobson, Harold K./Brown Weiss, Edith (eds.), *Engaging Countries: Strengthening Compliance with International Environmental Accords*, 1998 (Cambridge, Mass., MIT Press), 1-18.

Jarass, Hans D., *Bundes-Immissionsschutzgesetz: Kommentar*, 2005 (6. Auflage, München, C.H. Beck).

Johnston, Douglas M., "Book Review of Commitment and Compliance: The Role of Non-Binding Norms in the International Legal System", *American Journal of International Law* 95 (2001), 709-714.

Kadelbach, Stefan, "Die parlamentarische Kontrolle des Regierungshandelns bei der Beschlußfassung in internationalen Organisationen", in: Geiger, Rudolf (ed.), *Neuere Probleme der parlamentarischen Legitimation im Bereich der auswärtigen Gewalt*, 2003 (Baden-Baden, Nomos-Verlag), 41-57.

Kadelbach, Stefan/Kleinlein, Thomas, "Überstaatliches Verfassungsrecht. Zur Konstitutionalisierung im Völkerrecht", *Archiv des Völkerrechts* 44 (2006), 235-266.

Kahl, Wolfgang, "Parlamentarische Steuerung der internationalen Verwaltungsvorgänge", in: Trute, Hans-Heinrich/Groß, Thomas/Röhl, Hans C./Möllers, Christoph (eds.), *Allgemeines Verwaltungsrecht – Zur Tragfähigkeit eines Konzepts*, 2008 (Tübingen, Mohr Siebeck), 71-106.

Keller, Helen, "Codes of Conduct and their Implementation: the Question of Legitimacy", in: Wolfrum, Rüdiger/Röben, Volker (eds.), *Legitimacy in International Law*, 2008 (Heidelberg, Springer), 219-298.

Kellman, Barry, "Protection of Nuclear Materials", in: Shelton, Dinah (ed.), *Commitment and compliance: the role of non-binding norms in the international legal system*, 2000 (Oxford, Oxford University Press), 486-505.

Keohane, Robert O./Nye, Joseph S., "Transgovernmental Relations and International Organizations", *World Politics* 27 (1974), 39-62.

Keohane, Robert O./Haas, Peter M./Levy, Marc A., "The Effectiveness of International Environmental Institutions", in: Haas, Peter M./Keohane, Robert O./Levy, Marc A. (eds.), *Institutions for the earth: sources of effective international environmental protection*, 1993 (2nd print, Cambridge, Mass., M.I.T. Press), 3-24.

Khan, Rahmatullah, "The Anti-Globalization Protests: Side-show of Global Governance or Law-making on the Streets?", *Zeitschrift für ausländisches öffentliches Recht und Völkerrecht* 61 (2001), 323-355.

Kingsbury, Benedict, "Sovereignty and Inequality", *European Journal of International Law* 9 (1998), 599-625.

Kingsbury, Benedict, "The concept of compliance as a function of competing conceptions of international law", *Michigan Journal of International Law* 19 (1998), 345-372.

Kingsbury, Benedict/Krisch, Nico/Stewart, Richard, "The Emergence of Global Administrative Law", *Law & Contemporary Problems* (2004-2005), 15-62.

Kirchner, Christian, "Bedingungen interstaatlicher Institutionalisierung von wirtschaftlichen Prozessen", in: Hauff, Wilhelm (ed.), *Handbuch der Wirtschaftsethik, Band 2: Ethik wirtschaftlicher Ordnungen*, 1999 (Gütersloh, Gütersloher Verlagshaus), 376-389.

Kirgis, Frederic L., "Specialized law-making processes", in: Joyner, Christopher C. (ed.), *The United Nations and International Law*, 1997 (Cambridge, Cambridge University Press), 65-94.

Kirton, John J./Trebilcock, Michael J. (eds.), *Hard Choices, Soft Law: Voluntary Standards in Global Trade, Environment and Social Governance*, 2004 (Aldershot, Ashgate).

Kirton, John J./Trebilcock, Michael J., "Introduction: Hard Choices and Soft Law in Sustainable Global Governance", in: Kirton, John J./Trebilcock, Michael J. (eds.), *Hard Choices, Soft Law: Voluntary Standards in Global Trade, Environment and Social Governance*, 2004 (Aldershot, Ashgate), 3-29.

Kiss, Alexandre/Sicault, Jean-Didier, "La Conférence des Nations Unies sur l'Environnement", *Annuaire Français du Droit International* 18 (1972), 603-628.

Kiss, Alexandre/Doumbé-Billé, Stéphane, "La Conférence des Nations-Unies sur l'Environnement et le Développement", *Annuaire Français du Droit International* 38 (1992), 823-843.

Kiss, Alexandre C./Shelton, Dinah, *International environmental law*, 2004 (3rd ed., Ardsley, Transnational Publishers).

Klabbers, Jan, "The Redundancy of Soft Law", *Nordic Journal of International Law* 65 (1996), 167-182.

Klabbers, Jan, *The concept of treaty in international law*, 1996 (The Hague, Kluwer Law International).

Klabbers, Jan, "The Undesirability of Soft Law", *Nordic Journal of International Law* 67 (1998), 381-391.

Klabbers, Jan, "The changing image of international organizations", in: Coicaud, Jean-Marc/Heiskanen, Veigo (eds.), *The Legitimacy of International Organizations*, 2001 (Tokyo, United Nations University Press), 221-255.

Klabbers, Jan, *An Introduction to International Institutional Law*, 2002 (Cambridge, Cambridge University Press).

Klabbers, Jan, "Compliance Procedures", in: Bodansky, Daniel/Brunnée, Jutta/Hey, Ellen (eds.), *The Oxford Handbook of International Environmental Law*, 2007 (Oxford, Oxford University Press), 995-1009.

Klabbers, Jan, "Goldmann Variations", in: von Bogdandy, Armin/Wolfrum, Rüdiger/von Bernstorff, Jochen/Dann, Philipp/Goldmann, Matthias (eds.), *The Exercise of Public Authority by Inter-*

national Institutions, 2010 (Heidelberg, Springer-Verlag), 713-725.

Knauff, Matthias, *Der Regelungsverbund: Recht und Soft Law im Mehrebenensystem*, 2010 (Mohr Siebeck, Tübingen).

Kocher, Eva, "Selbstverpflichtungen von Unternehmen zur sozialen Verantwortung: Erfahrungen mit sozialen Verhaltenskodizes in der transnationalen Produktion", *Recht der Arbeit* 1 (2004), 27-31.

König, Doris, "The Protection of Marine Living Resources – The 1995 Agreement on Straddling Fish Stocks and Highly Migratory Fish Stocks ", in: Zhenghua, Tao/Wolfrum, Rüdiger (eds.), *Implementing International Environmental Law in Germany and China*, 2001 (The Hague, Kluwer Law International), 75-84.

Koh, Harold H., "Why do nations obey international law?", *The Yale Law Journal* 106 (1997), 2599-2659.

Koh, Harold H., "Internalization through socialization", *Duke Law Journal* (2005), 975-982.

Koskenniemi, Martti, "Breach of Treaty or Non-Compliance? Reflections on the Enforcement of the Montreal Protocol", *Yearbook of International Environmental Law* 3 (1992), 123-162.

Koskenniemi, Martti, *Fragmentation of international law: difficulties arising from the diversification and expansion of international law. Report of the Study Group of the International Law Commission*, 2007 (Helsinki, University of Helsinki).

Kratochwil, Friedrich, *Rules, norms and decisions: on the conditions of practical and legal reasoning in international relations and domestic affairs*, 1989 (Cambridge, Cambridge University Press).

Kummer, Katharina, "Prior Informed Consent for Chemicals in International Trade: The 1998 Rotterdam Convention", *Review of European Community and International Environmental Law* 8 (1999), 323-330.

Kunig, Philip, "The Relevance of Resolutions and Declarations of International Organizations for Municipal Law", in: Tunkin, Grigory I./Wolfrum, Rüdiger (eds.), *International Law and Municipal Law*, 1988 (Berlin, Duncker & Humblot), 59-78.

Kunig, Philip, "Deutsches Verwaltungshandeln und Empfehlungen internationaler Organisationen", in: Hailbronner, Kay/Ress, Georg/Stein, Torsten (eds.), *Staat und Völkerrechtsordnung – Fest-*

schrift für Karl Doehring, 1989 (Heidelberg, Springer-Verlag), 529-551.

Kunzmann, Katharina, "The Non-legally Binding Instrument on Sustainable Management of All Types of Forests – Towards a Legal Regime for Sustainable Forest Management?", *German Law Journal* 9 (2008), 981-1005.

Lang, Winfried, "Die Verrechtlichung des internationalen Umweltschutzes: Vom "soft law" zum "hard law"", *Archiv des Völkerrechts* 22 (1984), 283-305.

Leive, David M., *International regulatory regimes: case studies in health, meteorology, and food*, 1976 (Lexington, Mass., Lexington Books).

Lindblom, Anna-Karin, *Non-governmental organisations in international law*, 2005 (Cambridge, Cambridge University Press).

Lipson, Charles, "Why are some International Agreements Informal", *International Organization* 45 (1991), 495-538.

Louka, Elli, *International Environmental Law: Fairness, Effectiveness and World Order*, 2006 (Cambridge, Cambridge University Press).

Luhmann, Niklas, *Legitimation durch Verfahren*, 1989 (2. Auflage, Frankfurt am Main, Suhrkamp).

Lutz, Robert E./Aron, George D., "Codes of Conduct and Other International Instruments", in: Handl, Günther/Lutz, Robert E. (eds.), *Transferring Hazardous Technologies and Substances: The International Legal Challenge*, 1989 (London, Graham and Trotman), 129-151.

MacGibbon, Iain, "Means for the Identification of International Law: General Assembly Resolutions: Custom, Practice and Mistaken Identity", in: Cheng, Bin (ed.), *International law: teaching and practice*, 1982 (London, Stevens), 10-26.

Magraw, Daniel B./Hawke, Lisa D., "Sustainable Development", in: Bodansky, Daniel/Brunnée, Jutta/Hey, Ellen (eds.), *Oxford Handbook of International Environmental Law*, 2007 (Oxford, Oxford University Press), 614-638.

Malanczuk, Peter, *Akehurst's modern introduction to international law*, 1997 (7th rev., London, Routledge).

Mattli, Walter/Büthe, Tim, "Setting International Standards: Technological Rationality or Primacy of Power", *World Politics* 56 (2003), 1-42.

Matz-Lück, Nele/Wolfrum, Rüdiger, "Article XX lit. g GATT", in: Wolfrum, Rüdiger/Stoll, Peter-Tobias/Seibert-Fohr, Anja (eds.), *WTO – Technical Barriers and SPS Measures*, 2007 (Leiden, Martinus Nijhoff Publishers), 141-157.

Matz, Nele, "Environmental financing: function and coherence of financial mechanisms in international environmental agreements", *Max Planck Yearbook of United Nations Law* 6 (2002), 473-534.

Matz, Nele, *Wege zur Koordinierung völkerrechtlicher Verträge: völkervertragsrechtliche und institutionelle Ansätze*, 2005 (Berlin/Heidelberg, Springer).

Matz, Nele, "Financial and Other Incentives for Complying with MEA Obligations", in: Beyerlin, Ulrich/Stoll, Peter-Tobias/Wolfrum, Rüdiger (eds.), *Ensuring Compliance with Multilateral Environmental Agreements* 2006 (Leiden, Martinus Nijhoff Publishers), 301-318.

Maurer, Hartmut, *Allgemeines Verwaltungsrecht*, 2009 (17. Auflage, München, C.H. Beck).

Mechlem, Kerstin, "Food and Agriculture Organization of the United Nations (FAO)", in: Wolfrum, Rüdiger (ed.), *Max Planck Encyclopedia of Public International Law* (2008), online at: www.mpepil.com.

Meessen, Karl M., "Internationale Verhaltenskodizes und Sittenwidrigkeitsklauseln", *Neue Juristische Wochenschrift* 34 (1981), 1131-1132.

Mehri, Cyrus, "Prior Informed Consent: An Emerging Compromise for Hazardous Exports", *Cornell International Law Journal* 21 (1988), 365-389.

Mekouar, Mohamed A., "Pesticides and Chemicals: The Requirement of Prior Informed Consent", in: Shelton, Dinah (ed.), *Commitment and Compliance: The Role of Non-Binding Norms in the International Legal System*, 2000 (Oxford, Oxford University Press), 146-163.

Metaxas, Spyro A., Entreprises Transnationales et Codes de Conduite: Cadre Juridique et Questions d'Effectivité, 1988 (Zürich, Schulthess Polygraphischer Verlag).

Mitchell, Ronald B., "Compliance Theory: compliance, effectiveness, and behaviour change in international environmental law", in: Bodansky, Daniel/Brunnée, Jutta/Hey, Ellen (eds.), *Oxford Handbook of International Environmental Law*, 2007 (Oxford, Oxford University Press), 893-921.

Möllers, Christoph, "Transnationale Behördenkooperation: Verfassungs- und völkerrechtliche Probleme transnationaler administrativer Standardsetzung", *Zeitschrift für ausländisches öffentliches Recht und Völkerrecht* 65 (2005), 351-389.

Möllers, Christoph, *Gewaltengliederung: Legitimation und Dogmatik im nationalen und internationalen Rechtsvergleich*, 2005 (Tübingen, Mohr Siebeck).

Moore, Gerald, "The Code of Conduct for Responsible Fisheries", in: Hey, Ellen (ed.), *Developments in International Fisheries Law*, 1999 (The Hague, Kluwer Law International), 85-105.

Morgera, Elisa, "From Stockholm to Johannesburg: From Corporate Responsibility to Corporate Accountability for the Global Protection of the Environment", *Review of European Community and International Environmental Law* 13 (2004), 214-222.

Morgera, Elisa, "An Environmental Outlook on the OECD Guidelines for Multinational Enterprises: Comparative Advantages and Legitimacy in relation to Other International Initiatives", *Georgetown International Environmental Law Review* 18 (2006), 751-777.

Morgera, Elisa, *Corporate accountability in international environmental law*, 2009 (New York, Oxford University Press).

Morrison, Jason/Roht-Arriaza, Naomi, "Private and Quasi-Private Standard Setting", in: Bodansky, Daniel/Brunnée, Jutta/Hey, Ellen (eds.), *Oxford Handbook of International Environmental Law*, 2007 (Oxford, Oxford University Press), 498-527.

Muchlinski, Peter, *Multinational enterprises and the law*, 1995 (Oxford, Blackwell).

Muchlinski, Peter, "Human Rights, social responsibility and the regulation of international business: The development of international standards by intergovernmental organisations", *Non-State Actors and International Law* 3 (2003), 123-152.

Muchlinski, Peter T., "Global Bukowina' Examined: Viewing the Multinational Enterprise as a Transnational Law-making Community", in: Teubner, Gunther (ed.), *Global Law without a state*, 1996 (Aldershot, Dartmouth Publishing), 79-108.

Müller, Friedrich, *Demokratie zwischen Staatsrecht und Weltrecht: nationale, staatlose und globale Formen menschenrechtsgestützter Demokratisierung*, 2003 (Berlin, Duncker & Humblot).

Murphy, Sean D., "Taking multinational corporate codes of conduct to the next level", *Columbia Journal of Transnational Law* 43 (2005), 389-433.

Murray, Jill, "A new phase in the regulation of multinational enterprises: the role of the OECD", *The Industrial Law Journal* 30 (2001), 255-270.

Nanda, Ved P./Bailey, Bruce C., "Nature and scope of the problem", in: Handl, Günther/Lutz, Robert E. (eds.), *Transferring Hazardous Technologies and Substances*, 1989 (London, Graham and Trotman), 3-39.

Naudé Fourie, Andria, *The World Bank Inspection Panel and quasi-judicial oversight: in search of the 'judicial spirit' in public international law*, 2009 (Utrecht, Eleven International Publishing).

Neuhold, Hanspeter, "The Inadequacy of Law-Making by International Treaties: "Soft Law" as an Alternative", in: Röben, Volker (ed.), *Developments of International Law in Treaty Making*, 2005 (39-52).

Nordquist, Myron H. (ed.), *The Stockholm declaration and law of the marine environment*, 2003 (The Hague, Martinus Nijhoff Publishers).

Nowrot, Karsten, "Die UN-Norms on the Responsibility of Transnational Corporations and Other Business Enterprises with Regard to Human Rights – Gelungener Beitrag zur transnationalen Rechtsverwirklichung oder das Ende des Global Compact?", *Beiträge zum Transnationalen Wirtschaftsrecht* 21 (2003), 1-27.

Nowrot, Karsten, "The New Governance Structure of the Global Compact – Transforming a "Learning Network" into a Federalized and Parliamentarized Transnational Regulatory Regime", *Beiträge zum Transnationalen Wirtschaftsrecht* 47 (2005), 1-38.

Nowrot, Karsten, *Normative Ordnungsstruktur und private Wirkungsmacht: Konsequenzen der Beteiligung transnationaler Unternehmen an den Rechtssetzungsprozessen im internationalen Wirtschaftssystem*, 2006 (Berlin, BWV Berliner Wiss.-Verl.).

Oberthür, Sebastian/Buch, Matthias/Müller, Sebastian/Pfahl, Stephanie/Tarasofsky, Richard G./Werksman, Jacob/Palmer, Alice, *Participation of non-governmental organisations in international environmental co-operation: legal basis and practical experience*, 2002 (Berlin, Schmidt-Verlag).

Oberthür, Sebastian/Gehring, Thomas (eds.), *Institutional interaction in global environmental governance: synergy and conflict among international and EU policies*, 2006 (Cambridge, MA, MIT Press).

Oeter, Stefan, "Inspection in international law – Monitoring compliance and the problem of implementation in international law", *Netherlands Yearbook of International Law* 28 (1997), 101-169.

Oeter, Stefan, "Souveränität – ein überholtes Konzept?", *Tradition und Weltoffenheit des Rechts* (2002), 259-290.

Oeter, Stefan, "The openness of international organisations to transnational rule-making", in: O. Dilling/M. Herberg/G. Winter (eds), *Transnational administrative rule making: performance, legal effects and legitimacy*, 2011 (Oxford, Hart Publishing), 235-252.

Oeter, Stefan, "Towards a richer institutionalism for international law and policy", *University of Illinois Law Review* (2008), 61-70.

Orlitzky, Marc/Schmidt, Frank L./Rynes, Sara L., "Corporate Social and Financial Performance: A Meta-analysis", *Organization Studies* 24 (2003), 403-441.

Orrego Vicuña, Francisco, "In Memory of Triepel and Anzilotti: The Use and Abuse of Non-Conventional Lawmaking", in: Wolfrum, Rüdiger/Röben, Volker (eds.), *Developments of International Law in Treaty Making*, 2005 (Berlin/Heidelberg, Springer), 497-506.

Orts, Eric W., "Reflexive Environmental Law", *Northwestern University Law Review* 89 (1995), 1227.

Oster, Jan, "Das informell-kooperative Verwaltungshandeln im Umweltrecht: Begriffliche Abgrenzung, Erscheinungsformen und rechtliche Bewertung", *Natur und Recht* 30 (2008), 845-850.

Ott, Hermann E., *Umweltregime im Völkerrecht: eine Untersuchung zu neuen Formen internationaler institutionalisierter Kooperation am Beispiel der Verträge zum Schutz der Ozonschicht und zur Kontrolle grenzüberschreitender Abfallverbringungen*, 1998 (Baden-Baden, Nomos).

Oxman, Bernhard H., "The Duty to Respect Generally Accepted International Standards", *New York University Journal of International Law and Politics* 24 (1991-1992), 109-159.

Paarlberg, Robert L., "Managing Pesticide Use in Developing Countries", in: Haas, Peter M./Keohane, Robert O./Levy, Marc A. (eds.), *Institutions for the Earth: Sources of Effective International Environmental Protection*, 1993 (Cambridge, MIT Press), 309-350.

Pallemaerts, Marc, "Developments in International Pesticide Regulation", *Environmental Policy and Law* 18 (1988), 62-69.

Pallemaerts, Marc, "International Transfer of Restricted or Prohibited Substances, Regulation of Chemicals", *Yearbook of International Environmental Law* 2 (1991), 170-175.

Pallemaerts, Marc, "International Transfer of Restricted or Prohibited Substances, Regulation of Chemicals", *Yearbook of International Environmental Law* 3 (1992), 281-287.

Parker, Joan/Doulman, David/Collins, Jean, "Citation analysis for the 1995 FAO Code of Conduct for Responsible Fisheries", *Marine Policy* 34 (2010) 139–144.

Pauwelyn, Joost, "WTO compassion or superiority complex?: What to make of the WTO waiver on "conflict diamonds"", *Michigan Journal of International Law* 24 (2002-2003), 1177-1207.

Pauwelyn, Joost, *Conflict of norms in public international law: how WTO law relates to other rules of international law*, 2003 (Cambridge, Cambridge University Press).

Pearse, Andrew, *Seeds of Plenty, Seeds of Want: Social and Economic Implications of the Green Revolution*, 1980 (Oxford, Oxford University Press).

Perez, Oren, *Ecological sensitivity and global legal pluralism: rethinking the trade and environment conflict*, 2004 (Oxford, Hart Publishing).

Pernice, Ingolf, "Gesetzesvorbehalt für besondere Verträge (Art. 59 II 1 GG)", in: Dreier, Horst (ed.), *Grundgesetz-Kommentar Band II*, 2006 (2nd, Tübingen, Mohr Siebeck).

Peters, Anne/Pagotto, Isabella, "Soft Law as a New Mode of Governance: A Legal Perspective", *New Modes of Governance Project* 04/D11 (2006), 1 et seq.

Petersen, Niels, *Demokratie als teleologisches Prinzip: Zur Legitimität von Staatsgewalt im Völkerrecht*, 2009 (Berlin/Heidelberg, Springer).

Pitcher, Tony J./Kalikoski, Daniela/Pramod, Ganapathiraju (eds.), *Evaluations of Compliance with the FAO (UN) Code of Conduct for Responsible Fisheries*, 2006 (Vancouver University of British Columbia).

Pitcher, Tony J./Kalikoski, Daniela/Pramod, Ganapathiraju/Short, Katherine, "Not honouring the Code", *Nature* 457 (2009), 658-659.

Plaine, Daniel J., "The OECD Guidelines for Multinational Enterprises", *International Lawyer* 11 (1977), 339-346.

Protopsaltis, Panayotis M., "La mise en oeuvre des Principes directeurs de l'OCDE à l'intention des entreprises multinationales: Réflexions sur le nouveau mandat des Points de contact nationaux", *International Law FORUM du droit international* 7 (2005), 251-260.

Puhl, Thomas, "Entparlamentarisierung und Auslagerung staatlicher Entscheidungsverantwortung", in: Isensee, Josef/Kirchhof, Paul (eds.), *Handbuch des Staatsrechts der Bundesrepublik Deutschland, Band III*, 2005 (3. Auflage, Heidelberg, C.F. Müller), 639-682.

Rabkin, Jeremy A., *Law without nations?: Why constitutional government requires sovereign states*, 2005 (Princeton, N.J., Princeton University Press).

Ratner, Stephen R., "Business", in: Bodansky, Daniel/Brunnée, Jutta/Hey, Ellen (eds.), *The Oxford Handbook of International Environmental Law*, 2007 (Oxford, Oxford University Press), 807-828.

Raustiala, Kal, "The "participatory revolution" in international environmental law", *The Harvard Environmental Law Review* 21 (1997), 537-586.

Raustiala, Kal, "States, NGOs, and International Environmental Institutions", *International Studies Quarterly* 41 (1997), 719-740.

Raustiala, Kal/Victor, David G., "Conclusions", in: Victor, David G./Raustiala, Kal/Skolnikoff, Eugene B. (eds.), *The Implementation and Effectiveness of International Environmental Commitments*, 1998 (Cambridge, MIT Press), 659-708.

Raustiala, Kal, "Compliance & effectiveness in international regulatory cooperation", *Case Western Reserve Journal of International Law* 32 (2000), 387-440.

Raustiala, Kal, "The Architecture of International Cooperation: Transgovernmental Networks and the Future of International Law", *Virginia Journal of International Law* 43 (2002-2003), 1-92.

Raustiala, Kal, "Form and Substance in International Agreements", *American Journal of International Law* 99 (2005), 581-614.

Rayfuse, Rosemary, "The interrelationship between the global instruments of international fisheries law", in: Hey, Ellen (ed.), *Developments in international fisheries law*, 1999 (The Hague, Kluwer Law International), 107-158.

Redgwell, Catherine, "Regulating Trade in Dangerous Substances: Prior Informed Consent under the 1998 Rotterdam Convention", in: Kiss, Alexandre/Shelton, Dinah/Ishibashi, Kanami (eds.), *Economic Globalization and Compliance with International Environmental Agreements*, 2003 (The Hague, Kluwer Law International), 75-88.

Rehbinder, Eckard, "Das neue Pflanzenschutzgesetz", *Natur und Recht* 9 (1987), 68-71.

Reinisch, August, *International organizations before national courts*, 2000 (Cambridge, Cambridge University Press).

Riedel, Eibe, "Standards and Sources. Farewell to the Exclusivity of the Sources Triad in International Law?", *European Journal of International Law* 2 (1991), 58-84.

Riedel, Eibe, "The Development of International Law: Alternatives to Treaty-Making? International Organizations and Non-State Actors", in: Wolfrum, Rüdiger/Röben, Volker (eds.), *Developments of International Law in Treaty Making*, 2005 (Berlin/Heidelberg, Springer), 301-318.

Riedinger, Sonja, *Die Rolle nichtstaatlicher Organisationen bei der Entwicklung und Durchsetzung internationalen Umweltrechts*, 2001 (Berlin, Duncker & Humblot).

Röben, Volker, "Proliferation of Actors", in: Wolfrum, Rüdiger/Röben, Volker (eds.), *Developments of International Law in Treaty Making*, 2005 (Berlin/Heidelberg, Springer), 511-542.

Röben, Volker, *Außenverfassungsrecht: eine Untersuchung zur auswärtigen Gewalt des offenen Staates*, 2007 (Tübingen, Mohr Siebeck).

Rojahn, Ondolf, "Artikel 59 II", in: Münch, Ingo von/Kunig, Philip (eds.), *Grundgesetz-Kommentar (Art. 20 bis Art. 69)*, 2001 (5. Auflage, München, Beck), 1250 et seq.

Romano, Cesare P.R. Romano, "International dispute settlement", in: Bodansky, Daniel/Brunnée, Jutta/Hey, Ellen (eds.), *The Oxford Handbook of International Environmental Law*, 2007 (Oxford, Oxford University Press), 1036-1056.

Rosemann, Nils, "Business Human Rights Obligations – The "Norms of the Responsibility of Transnational Corporations and Other Business Enterprises with Regard to Human Rights", *Nordic Journal of Human Rights* 23 (2005), 47-62.

Ross, Jenifer, "Legally Binding Prior Informed Consent", *Colorado Journal of International Environmental Law and Policy* 10 (1999), 499-529.

Rothwell, Donald R., "The General Assembly Ban on Driftnet Fishing", in: Shelton, Dinah (ed.), *Commitment and Compliance: The Role of Non-Binding Norms in the International System*, 2000 (Oxford, Oxford University Press), 121-146.

Rothwell, Donald R., "The Contribution of ITLOS to Oceans Governance through Marine Environmental Dispute Resolution", in: Ndiaye, Tafsir M./Wolfrum, Rüdiger (eds.), *Law of the Sea, Environmental Law and Settlement of Disputes*, 2007 (Leiden/Boston, Martinus Nijhoff Publishers), 1007-1024.

Rubin, Seymour J., "Transnational Corporations and International Codes of Conduct: a study of the relationship between international legal cooperation and economic development", *American University Law Review* 30 (1980-1981), 907-921.

Ruggie, John G., "Current Development: Business and Human Rights: The Evolving International Agenda", *American Journal of International Law* 101 (2007), 819-839.

Saladin, Claudia/VanDyke, Brennan, "Implementing the principles of the Public Participation Convention in international organizations", *Yearbook of Human Rights & Environment* 1 (2001), 1-96.

Salzman, James, "Decentralized Administrative Law in the Organization for Economic Cooperation and Development", *Law and Contemporary Problems* 68 (2005), 189-224.

Sand, Peter H., "Institution-Building to Assist Compliance with International Environmental Law: Perspectives", *Zeitschrift für ausländisches öffentliches Recht und Völkerrecht* 56 (1996), 774-795.

Sand, Peter H., *Transnational Environmental Law: Lessons in Global Change*, 1999 (The Hague, Kluwer Law International).

Sand, Peter H., "Sticks, Carrots, and Games", in: Bothe, Michael/Sand, Peter H. (eds.), *La politique de l'environnement: De la réglementation aux instruments économique*, 2003 (The Hague, Martinus Nijhoff Publishers), 3-36.

Sand, Peter H., "Compensation for environmental damage from the 1991 Gulf War", *Environmental Policy and Law* 35 (2005), 244-249.

Sander, Gerald G, "International Organization for Standardization (ISO)", in: Wolfrum, Rüdiger (ed.), *Max Planck Encyclopedia of Public International Law* (2008), online at: www.mpepil.com.

Sanders, Pieter, "Implementing International Codes of Conduct for Multinational Enterprises", *American Journal of Comparative Law* 30 (1982), 241-254.

Sands, Philippe, "International courts and the application of the concept of "sustainable development"", *Max Planck Yearbook of United Nations Law* 3 (1999), 389-405.

Sands, Philippe, *Principles of International Environmental Law*, 2003 (2nd ed., Cambridge, Cambridge University Press).

Sands, Philippe, "Litigating Environmental Disputes: Courts, Tribunals and the Progressive Development of International Environmental Law", in: Ndiaye, Tafsir Malick/Wolfrum, Rüdiger (eds.), *Law of the Sea, Environmental Law and Settlement of Disputes*, 2007 (Leiden/Boston, Martinus Nijhoff Publishers), 313-325.

Schachter, Oscar, "The Twilight Existence of Nonbinding International Agreements", *American Journal of International Law* 71 (1977), 296-304.

Schachter, Oscar, "International Law in Theory and Practice: General Course in Public International Law", *Recueil des Cours de l'Académie de Droit International de La Haye* 178 (1982), 9-396.

Scharpf, Fritz W., *Regieren in Europa: effektiv und demokratisch?*, 1999 (Frankfurt/Main, Campus-Verlag).

Schermers, Henry G./Blokker, Niels M., *International Institutional Law: Unity within diversity*, 2003 (4th ed., Leiden, Martinus Nijhoff Publishers).

Schmidt-Aßmann, Eberhard, "Die Herausforderung der Verwaltungsrechtswissenschaft durch die Internationalisierung der Verwaltungsbeziehungen", *Der Staat* 3 (2006), 315-338.

Schroeder, Marcus, *Die Koordinierung der internationalen Bemühungen zum Schutz der Umwelt*, 2005 (Berlin, Duncker & Humblot).

Schwebel, Stephen M., *The legal effect of resolutions and codes of conduct of the United Nations*, 1986 (Deventer, Kluwer).

Scott, Joanne, "International Trade and Environmental Governance: Relating Rules (and Standards) in the EU and the WTO", *European Journal of International Law* 15 (2004), 307-354.

Seibert-Fohr, Anja/Wolfrum, Rüdiger, "Die einzelstaatliche Durchsetzung völkerrechtlicher Mindeststandards gegenüber transnationalen Unternehmen", *Archiv des Völkerrechts* 43 (2005), 153-186.

Seidl-Hohenveldern, Ignaz, "International Economic "Soft Law"", *Recueil des Cours de l'Académie de Droit International* 163 (1979), 169-246.

Senden, Linda, *Soft law in European Community law*, 2004 (Oxford, Hart Publishing).

Senti, Richard, *Die WTO im Spannungsfeld zwischen Handel, Gesundheit, Arbeit und Umwelt: geltende Ordnung und Reformvorschläge*, 2006 (Baden-Baden, Nomos).

Shelton, Dinah, "Compliance with international human rights soft law", *Studies of Transnational Legal Policy* 29 (1997), 119-143.

Shelton, Dinah (ed.), *Commitment and Compliance: The Role of Non-Binding Norms in the International Legal System*, 2000 (Oxford, Oxford University Press).

Shelton, Dinah, "Law, Non-Law and the Problem of 'Soft Law'", in: Shelton, Dinah (ed.), *Commitment and Compliance: The Role of Non-Binding Norms in the International Legal System*, 2000 (Oxford, Oxford University Press), 1-18.

Simma, Bruno/Alston, Philip, "The Sources of Human Rights Law: Custom, Jus Cogens, and General Principles", *Australian Yearbook of International Law* 12 (1992), 82-108.

Simma, Bruno/Heinemann, Andreas, "Codes of Conduct", in: Korff, Wilhelm et al. (eds.), *Handbuch der Wirtschaftsethik, Band 2: Ethik wirtschaftlicher Ordnungen*, 1999 (Gütersloh, Gütersloher Verlagshaus), 403-418.

Skala, Astrid K., *Internationale technische Regeln und Standards zum Umweltschutz – ihre Entstehungsarten und rechtlichen Wirkungen*, 1982 (Köln, Carl Heymanns Verlag).

Skjaerseth, Jon B./Stokke, Olav S./Wettestad, Jørgen, "Soft Law, Hard Law, and Effective Implementation of International Environmental Norms", *Global Environmental Politics* 6 (2006), 104-120.

Slaughter, Anne-Marie, *A New World Order*, 2004 (Princeton, Princeton University Press).

Sohn, Louis B., "'Generally accepted' International Rules", *Washington Law Review* 61 (1986), 1073-1080.

Spiro, Peter J., "Non-governmental organizations and civil society", in: Bodansky, Daniel/Brunnée, Jutta/Hey, Ellen (eds.), *Oxford Handbook of International Environmental Law*, 2007 (Oxford, Oxford University Press), 770-790.

Stein, Torsten, "Demokratische Legitimierung auf supranationaler und internationaler Ebene", *Zeitschrift für ausländisches öffentliches Recht und Völkerrecht* 64 (2004), 563-570.

Stephens, Tim, *International Courts and Environmental Protection*, 2009 (Cambridge, Cambridge University Press).

Stewart, Richard B., "A new generation of environmental regulation?", *Capital University Law Review* 29 (2001), 21-182.

Stoecker, Felix W., *NGOs und die UNO: die Einbindung von Nichtregierungsorganisationen (NGOs) in die Strukturen der Vereinten Nationen*, 2000 (Mannheim, Peter Lang Verlag).

Stokke, Olav S., "The Interplay of International Regimes: Putting Effectiveness Theory to Work", *FNI Report 14/2001* (2001), 1 et seq.

Stokke, Olav S./Coffey, Clare, "Institutional Interplay and Responsible Fisheries: Combating Subsidies, Developing Precaution", in: Oberthür, Sebastian/Gehring, Thomas (eds.), *Institutional interaction in global environmental governance: synergy and conflict among international and EU policies*, 2006 (Cambridge, Mass., MIT Press), 127-155.

Stoll, Peter-Tobias, "Hazardous Substances and Technologies", in: Morrison, Fred L./Wolfrum, Rüdiger (eds.), *International, Regional and National Environmental Law*, 2000 (The Hague, Kluwer Law International), 437-451.

Stoll, Peter-Tobias/Strack, Lutz, "Article XX lit. b GATT 1994", in: Wolfrum, Rüdiger/Stoll, Peter-Tobias/Seibert-Fohr, Anja (eds.), *WTO – Technical Barriers and SPS Measures*, 2007 (Leiden, Boston, Koninklijke Brill NV), 96-120.

Streinz, Rudolf, "Art. 59 [Völkerrechtliche Vertretungsmacht]", in: Sachs, Michael (ed.), *Grundgesetz: Kommentar*, 2009 (5. Auflage, München, C.H. Beck).

Techera, Erika J., *Marine environmental governance: from international law to local practice*, 2012 (London, Routledge).

Teubner, Gunther, "'Global Bukowina': Legal Pluralism in the World Society", in: Teubner, Gunther (ed.), *Global Law without a state*, 1996 (Aldershot, Dartmouth Publishing), 3-28.

Thürer, Daniel, "The emergence of non-governmental organizations and transnational enterprises in international law and the changing role of the state", *Non-state actors as new subjects of international law* (1999), 37-58.

Tietje, Christian, "The Changing Structure of International Treaties as an Aspect of an Emerging Global Governance Architecture", *German Yearbook of International Law* 42 (1999), 27-55.

Tietje, Christian, *Internationalisiertes Verwaltungshandeln*, 2001 (Berlin, Duncker und Humblot).

Tomuschat, Christian, "Die Charta der wirtschaftlichen Rechte und Pflichten der Staaten: Zur Gestaltungskraft von Deklarationen der UN-Generalversammlung", *Zeitschrift für ausländisches öffentliches Recht und Völkerrecht* 36 (1976), 444-491.

Tomuschat, Christian, "Der Verfassungsstaat im Geflecht der internationalen Beziehungen", *Veröffentlichungen der Vereinigung der Deutschen Staatsrechtslehrer* 36 (1978), 7-63.

Tomuschat, Christian, "Obligations Arising for States Without or Against Their Will", *Recueil des Cours de l'Académie de Droit International* 241 (1994), 195-374.

Tomuschat, Christian, "International law: ensuring the survival of mankind on the eve of a new century", *Recueil des cours de l'Académie de Droit International de La Haye* 281 (2001), 13-438.

Toope, Stephen J., "Formality and Informality", in: Bodansky, Daniel/ Brunnée, Jutta/Hey, Ellen (eds.), *The Oxford Handbook of International Environmental Law*, 2007 (Oxford, Oxford University Press), 107-124.

Trachtman, Joel P., "Institutional linkage: transcending "trade and ..."", *The American Journal of International Law* 96 (2002), 77-93.

Trebilcock, Michael J./Howse, Robert, *The regulation of international trade*, 2005 (3rd ed., London, Routledge).

Treves, Tullio, "A handbook on the new law of the sea", in: Dupuy, René-Jean/Vignes, Daniel (eds.), *A Handbook on the New Law of the Sea*, 1991 (Dordrecht, Martinus Nijhoff Publishers), 835-976.

Treves, Tullio, "The Settlement of Disputes According to the Straddling Stocks Agreement of 1995", in: Freestone, David (ed.), *International law and sustainable development: past achievements and future challenges*, 1999 (Oxford, Oxford University Press), 253-269.

Treves, Tullio, "Customary International Law", in: Wolfrum, Rüdiger (ed.), *Max Planck Encyclopedia of Public International Law* (2008), online at: www.mpepil.com.

Tsioumani, Elsa, "International Treaty on Plant Genetic Resources for Food and Agriculture: Legal and Policy Questions from Adoption to Implementation", *Yearbook of International Environmental Law* 15 (2004), 119-144.

Ulfstein, Geir, "Comment on J. Brunnée's "Reweaving the Fabric of International Law? Patterns of Consent in Environmental Framework Agreements"", in: Röben, Volker (ed.), *Developments of International Law in Treaty Making*, 2005 (Berlin/Heidelberg, Springer), 145-153.

van Hoof, Godefridus J.H., *Rethinking the sources of international law*, 1983 (Deventer, Kluwer).

van Reenen, Willem, "Rules of reference in the new Convention on the Law of the Sea, in particular in connection with the pollution of the sea by oil from tankers", *Netherlands Yearbook of International Law* 12 (1981), 3-44.

Venema, Siebren C., "The FISHCODE Project: Achievements", in: Nordquist, Myron H./Moore, John N. (eds.), *Current Fisheries Issues and the Food and Agriculture Organization of the United Nations*, 2000 (The Hague, Kluwer Law International), 341-359.

Vendzules, Sarah F., "The Struggle for Legitimacy in Environmental Standards Systems: The OECD Guidelines for Multinational Enterprises", 21 *Colorado Journal of International Environmental Law & Policy* 21 (2010), 451-489 (488).

Verdross, Alfred, "Entstehungsweisen und Geltungsgrund des universellen völkerrechtlichen Gewohnheitsrechts", *Zeitschrift für ausländisches öffentliches Recht und Völkerrecht* 29 (1969), 635-653.

Verdross, Alfred, *Die Quellen des universellen Völkerrechts*, 1973 (Freiburg, Rombach).

Verdross, Alfred/Simma, Bruno, *Universelles Völkerrecht: Theorie und Praxis*, 1984 (Berlin, Duncker & Humblot).

Victor, David G., "The Use and Effectiveness of Nonbinding Instruments in the Management of Complex International Environmental Problems", *American Society International Law Proceedings* 91 (1997), 241-250.

Victor, David G., ""Learning by Doing" in the Nonbinding International Regime to Manage Trade in Hazardous Chemicals and Pesticides", in: Victor, David G./Raustiala, Kal/Skolnikoff, Eugene B. (eds.), *The Implementation and Effectiveness of International Environmental Commitments: Theory and Practice*, 1998 (Laxenburg, MIT Press), 221-281.

Victor, David G./Raustiala, Kal/Skolnikoff, Eugene B., "Introduction and Overview", in: Victor, David G./Raustiala, Kal/Skolnikoff, Eugene B. (eds.), *The Implementation and Effectiveness of International Environmental Commitments: Theory and Practice*, 1998 (Laxenburg, MIT Press), 1-30.

Victor, David G./Raustiala, Kal/Skolnikoff, Eugene B. (eds.), *The Implementation and Effectiveness of International Environmental Commitments*, 1998 (Laxenburg, MIT Press).

Victor, David G., "The Sanitary and Phytosanitary Agreement of the World Trade Organization: an assessment after five years", *New York University Journal of International Law & Politics* 32 (2000), 865-937.

Vigneron, Giselle, "Compliance and International Environmental Agreements: A Case Study of the 1995 United Nations Straddling Fish Stocks Agreement", *Georgetown International Environmental Law Review* 10 (1998), 581-624.

Vignes, Daniel, "La valeur juridique de certaines règles, normes ou pratiques mentionnées au TNCO comme "généralement acceptées"", *Annuaire Français de Droit International* 25 (1979), 712-718.

Vitanyi, Béla, "Les positions doctrinales concernant le sens de la notion de 'principes généraux de droit reconnus par les nations civilisées'", *Revue générale de droit international public* 86 (1982), 48.

Vöneky, Silja, "The Liability Annex to the Protocol on Environmental Protection to the Antarctic Treaty", in: König, Doris/Stoll, Peter-Tobias/Röben, Volker/Matz-Lück, Nele (eds.), *International Law Today: New Challenges and the Need for Reform?*, 2008 (Berlin/Heidelberg, Springer), 165-197.

Vogelaar, Theo W., "The OECD Guidelines: Their Philosophy, History, Negotiation, Form, Legal Nature, Follow-Up Procedures and Review", in: Horn, Norbert (ed.), *Legal Problems of Codes of Conduct for Multinational Enterprises*, 1980 (Deventer, Kluwer), 127-140.

Vukas, Budislav, "Generally Accepted International Rules and Standards", in: Soons, Alfred H.A. (ed.), *Implementation of The Law of the Sea Convention Through International Institutions, 23 Law of the Sea Institute Proceedings*, 1990 (Honolulu, The Law of the Sea Institute), 405-421.

Weber, Max, *Wirtschaft und Gesellschaft*, 2002 (5., rev. Auflage, Nachdruck Tübingen, Mohr).

Weil, Prosper, "Towards Relative Normativity in International Law", *American Journal of International Law* 77 (1983), 413-442.

Weiler, Joseph H.H., "The geology of international law – governance, democracy and legitimacy", *Zeitschrift für ausländisches öffentliches Recht und Völkerrecht* 64 (2004), 547-562.

Weilert, Katharina, "Transnationale Unternehmen im rechtsfreien Raum? Geltung und Reichweite völkerrechtlicher Standards", *Zeitschrift für ausländisches öffentliches Recht und Völkerrecht* 69 (2009), 883-917.

Weiß, Wolfgang, "Allgemeine Rechtsgrundsätze des Völkerrechts", *Archiv des Völkerrechts* 39 (2001), 394-431.

Weissbrodt, David/Kruger, Maria, "Norms on the Responsibilities of Transnational Corporations and Other Business Enterprises with Regard to Human Rights", *American Journal of International Law* 97 (2003), 901-922.

Wellens, Karel, *Remedies against international organisations*, 2002 (Cambridge, Cambridge University Press).

Westfield, Elisa, "Globalization, governance, and multinational enterprise responsibility: corporate codes of conduct in the 21st century", *Virginia Journal of International Law* 42 (2002), 1075-1108.

Wettestad, Jørgen, "Monitoring and Verification", in: Bodansky, Daniel/Brunnée, Jutta/Hey, Ellen (eds.), *The Oxford Handbook of International Environmental Law*, 2007 (Oxford, Oxford University Press), 974-993.

Wilkie, Christopher, "Enhancing Global Governance: Corporate Social Responsibility and the International Trade and Investment Framework", in: Kirton, John J./Trebilcock, Michael J. (eds.), *Hard Choices, Soft Law: Voluntary Standards in Global Trade, Environment and Social Governance*, 2004 (Aldershot, Ashgate), 288-322.

Wirth, David A., "Reexamining Decision-Making Processes in International Environmental Law", *Iowa Law Review* 79 (1994), 769-802.

Wirth, David A., "Compliance with Non-Binding Norms of Trade and Finance", in: Shelton, Dinah (ed.), *Commitment and Compliance: The Role of Non-Binding Norms in the International Legal System*, 2000 (Washington, Oxford University Press), 330-344.

Wolfrum, Rüdiger, *Die Internationalisierung staatsfreier Räume: die Entwicklung einer internationalen Verwaltung für Antarktis, Weltraum, Hohe See und Meeresboden*, 1984 (Berlin, Springer).

Wolfrum, Rüdiger, "Konsens im Völkerrecht", in: Hattenhauer, Hans (ed.), *Mehrheitsprinzip, Konsens und Verfassung*, 1986 (Heidelberg, Müller Verlag), 79-91.

Wolfrum, Rüdiger, "The protection of the marine environment after the Rio Conference: progress or stalemate?", in: Beyerlin, Ulrich/Bothe, Michael/Hofmann, Rainer/Petersmann, Ernst-Ulrich (eds.), *Recht zwischen Umbruch und Bewahrung*, 1995 (Berlin/Heidelberg, Springer), 1003-1017.

Wolfrum, Rüdiger, "The Convention on Biological Diversity: Using State Jurisdiction as a Means of Ensuring Compliance", in: Wolfrum, Rüdiger (ed.), *Enforcing Environmental Standards: Economic Mechanisms as Viable Means?*, 1996 (Berlin/Heidelberg, Springer), 373-393.

Wolfrum, Rüdiger (ed.), *Enforcing environmental standards: economic mechanisms as viable means?*, 1996 (Berlin/Heidelberg, Springer).

Wolfrum, Rüdiger, "Kontrolle der auswärtigen Gewalt", *Veröffentlichungen der Vereinigung der Deutschen Staatsrechtslehrer* 56 (1997), 38-66.

Wolfrum, Rüdiger, "Means of Ensuring Compliance with and Enforcement of International Environmental Law", *Recueil des cours de l'Académie de Droit International de La Haye* 272 (1998), 13-154.

Wolfrum, Rüdiger, "IMO Interface with the Law of the Sea Convention", in: Nordquist, Myron H./Moore, John Norton (eds.), *Current Maritime Issues and the International Maritime Organization*, 1999 (The Hague, Kluwer Law International), 223-234.

Wolfrum, Rüdiger, "Die Entwicklung des Seerechts zum Recht der marinen Umwelt", in: Ehlers, Peter/Erbguth, Wilfried (eds.), *Aktuelle Entwicklungen im Seerecht*, 2000 (Baden-Baden, Nomos Verlagsgesellschaft), 69-81.

Wolfrum, Rüdiger, "The Role of the International Tribunal for the Law of the Sea", in: Nordquist, Myron H./Moore, John N. (eds.), *Current Fisheries Issues and the Food and Agriculture Organization of the United Nations*, 2000 (The Hague, Boston, London, Martinus Nijhoff Publishers), 369-385.

Wolfrum, Rüdiger/Matz, Nele, "The Interplay of the United Nations Convention on the Law of the Sea and the Convention on Biological Diversity", *Max Planck Yearbook of United Nations Law* 4 (2000), 445-480.

Wolfrum, Rüdiger/Röben, Volker/Morrison, Fred L., "Preservation of the Marine Environment", in: Morrison, Fred L./Wolfrum, Rüdiger (eds.), *International, Regional and National Environmental Law*, 2000 (The Hague, Kluwer Law International), 225-283.

Wolfrum, Rüdiger, "Vorbereitende Willensbildung und Entscheidungsprozeß beim Abschluß multilateraler völkerrechtlicher Verträge", in: Ipsen, Jörn/Schmidt-Jortzig, Edzard (eds.), *Recht-Staat-Gemeinwohl: Festschrift für Dietrich Rauschning*, 2001 (Köln, Heymanns), 407-418.

Wolfrum, Rüdiger, "American-European Dialogue: Different Perceptions of International Law – Introduction", *Zeitschrift für ausländisches öffentliches Recht und Völkerrecht* 64 (2001), 255-262.

Wolfrum, Rüdiger, "Article 55 (a) and (b)", in: Simma, Bruno (ed.), *The Charter of the United Nations: A Commentary*, 2002 (2nd ed., Oxford, Oxford University Press), 897-917.

Wolfrum, Rüdiger/Matz, Nele, *Conflicts in international environmental law*, 2003 (Berlin/Heidelberg, Springer).

Wolfrum, Rüdiger, "Die EG und das Meer: Versuch einer Neubewertung", *Archiv des Völkerrechts* 42 (2004), 67-79.

Wolfrum, Rüdiger, "Introduction", in: Wolfrum, Rüdiger/Röben, Volker (eds.), *Developments of International Law in Treaty Making*, 2005 (Berlin/Heidelberg, Springer), 1-13.

Wolfrum, Rüdiger/Langenfeld, Christine/Minnerop, Petra (eds.), *Environmental Liability in International Law: Towards a Coherent Conception*, 2005 (Berlin, Erich Schmidt Verlag).

Wolfrum, Rüdiger/Friedrich, Jürgen, "The Framework Convention on Climate Change and the Kyoto Protocol", in: Beyerlin, Ulrich/Stoll, Peter-Tobias/Wolfrum, Rüdiger (eds.), *Ensuring Compliance with Multilateral Environmental Agreements: A Dialogue between Practitioners and Academia*, 2006 (Leiden, Martinus Nijhoff Publishers), 53-68.

Wolfrum, Rüdiger, "Legitimacy in International Law from a Legal Perspective: Some Introductory Considerations", in: Wolfrum, Rüdiger/Röben, Volker (eds.), *Legitimacy in International Law*, 2008 (Berlin/Heidelberg, Springer), 1-24.

Wolfrum, Rüdiger, "The protection of the environment through international courts and tribunals", in: Fischer-Lescano, Andreas (ed.), *Frieden in Freiheit: Festschrift für Michael Bothe zum 70. Geburtstag*, 2008 (Baden-Baden, Nomos), 807-817.

Wolfrum, Rüdiger/Röben, Volker (eds.), *Legitimacy in International Law*, 2008 (Berlin/Heidelberg, Springer).

Wolfrum, Rüdiger, "Legitimacy of International Law and the Exercise of Administrative Functions: The Example of the International Seabed Authority, the International Maritime Organization (IMO) and International Fisheries Organizations", in: Bogdandy, Armin von/Wolfrum, Rüdiger/Bernstorff, Jochen von/Dann, Philipp/Goldmann, Matthias (eds.), *The Exercise of Public Au-*

thority by International Institutions: Advancing International Institutional Law, 2010 (Berlin/Heidelberg, Springer), 917-940.

Wolfrum, Rüdiger, "Obligation of Result Versus Obligation of Conduct: Some Thoughts About the Implementation of International Obligations", in: Arsanjani, Mahnoush H./Cogan, Jacob K./ Sloane, Robert D./Wiessner, Siegfried (eds.), *Looking to the Future: Essays on International Law in Honor of W. Michael Reisman*, 2010, Chapter 20.

Wouters, Patricia K., "Trade Measures in Multilateral Environmental Agreements", in: Bothe, Michael/Sand, Peter H. (eds.), *La politique de l'environnement: De la réglementation aux instruments économique*, 2003 (The Hague, Martinus Nijhoff Publishers), 159-189.

Young, Oran R., *Compliance and public authority*, 1979 (Baltimore, The Johns Hopkins University Press).

Zahedi, Nancy S., "Implementing the Rotterdam Convention: the challenges of transforming aspirational goals into effective controls on hazardous pesticide exports to developing countries", *Georgetown International Environmental Law Review* 11 (1999), 707-739.

Zerk, Jennifer A., *Multinationals and Corporate Social Responsibility: Limitations and Opportunities in International Law*, 2008 (Cambridge, Cambridge University Press).

Zumbansen, Peer, "Lex mercatoria: Zum Geltungsanspruch transnationalen Rechts", *Rabels Zeitschrift für ausländisches und internationales Privatrecht* 67 (2003), 637-682.

Zürn, Michael, "Global Governance and Legitimacy Problems", *Government and Opposition* 39 (2004), 260-287.

Index

Max-Planck-Institut für ausländisches öffentliches Recht und Völkerrecht

Beiträge zum ausländischen öffentlichen Recht und Völkerrecht

Hrsg.: A. von Bogdandy, A. Peters

Bde. 27–59 erschienen im Carl Heymanns Verlag KG Köln, Berlin (Bestellung an: Max-Planck-Institut für Völkerrecht, Im Neuenheimer Feld 535, 69120 Heidelberg); ab Band 60 im Springer Heidelberg New York Dordrecht London

---

247 Jürgen *Friedrich*: **International Environmental "soft law".** 2014. XXI, 503 Seiten. Geb. € 94,99

246 Anuscheh *Farahat*: **Progressive Inklusion.** 2014. XXIV, 429 Seiten. Geb. € 94,99

245 Christina *Binder*: **Die Grenzen der Vertragstreue im Völkerrecht.** 2013. XL, 770 Seiten. Geb. € 119,99

244 Cornelia *Hagedorn*: **Legitime Strategien der Dissensbewältigung in demokratischen Staaten.** 2013. XX, 551 Seiten. Geb. € 99,99

243 Marianne *Klumpp*: **Schiedsgerichtsbarkeit und Ständiges Revisionsgericht des Mercosur.** 2013. XX, 512 Seiten. Geb. € 94,99

242 Karen *Kaiser* (Hrsg.): **Der Vertrag von Lissabon vor dem Bundesverfassungsgericht.** 2013. XX, 1635 Seiten. Geb. € 199,99

241 Dominik *Steiger*: **Das völkerrechtliche Folterverbot und der „Krieg gegen den Terror".** 2013. XXX, 821 Seiten. Geb. € 139,99

240 Silja *Vöneky*, Britta *Beylage-Haarmann*, Anja *Höfelmeier*, Anna-Katharina *Hübler* (Hrsg.): **Ethik und Recht - Die Ethisierung des Rechts/Ethics and Law - The Ethicalization of Law.** 2013. XVIII, 456 Seiten. Geb. € 94,99

239 Rüdiger *Wolfrum*, Ina *Gätzschmann* (eds.): **International Dispute Settlement: Room for Innovations?** 2013. XIV, 445 Seiten. Geb. € 94,95 zzgl. landesüblicher MwSt.

238 Isabel *Röcker*: **Die Pflicht zur rahmenbeschlusskonformen Auslegung nationalen Rechts.** 2013. XXIII, 410 Seiten. Geb. € 89,95

237 Maike *Kuhn*: **Die Europäische Sicherheits- und Verteidigungspolitik im Mehrebenensystem.** 2012. XIII, 325 Seiten. Geb. € 79,95

236 Armin *von Bogdandy,* Ingo *Venzke* (eds.): **International Judicial Lawmaking.** 2012. XVII, 509 Seiten. Geb. € 94,95 zzgl. landesüblicher MwSt.

235 Susanne *Wasum-Rainer,* Ingo *Winkelmann,* Katrin *Tiroch* (eds.): **Arctic Science, International Law and Climate Change.** 2012. XIX, 374 Seiten. Geb. € 84,95 zzgl. landesüblicher MwSt.

234 Mirja A. *Trilsch*: **Die Justiziabilität wirtschaftlicher, sozialer und kultureller Rechte im innerstaatlichen Recht.** 2012. XIX, 559 Seiten. Geb. € 99,95

233 Anja *Seibert-Fohr* (ed.): **Judicial Independence in Transition.** 2012. XIII, 1378 Seiten. Geb. € 169,95 zzgl. landesüblicher MwSt.

232 Sandra *Stahl*: **Schutzpflichten im Völkerrecht - Ansatz einer Dogmatik.** 2012. XXX, 505 Seiten. Geb. € 94,95

231 Thomas *Kleinlein*: **Konstitutionalisierung im Völkerrecht.** 2012. XLII, 940 Seiten. Geb. € 149,95

230 Roland *Otto*: **Targeted Killings and International Law.** 2012. XVIII, 661 Seiten. Geb. € 109,95 zzgl. landesüblicher MwSt.

229 Nele *Matz-Lück*, Mathias *Hong* (Hrsg.): **Grundrechte und Grundfreiheiten im Mehrebenensystem - Konkurrenzen und Interferenzen.** 2012. VIII, 394 Seiten. Geb. € 89,95

228 Matthias *Ruffert,* Sebastian *Steinecke:* **The Global Administrative Law of Science,** 2011. IX, 140 Seiten. Geb. € 59,95 zzgl. landesüblicher MwSt.

227 Sebastian *Pritzkow*: **Das völkerrechtliche Verhältnis zwischen der EU und Russland im Energiesektor.** 2011. XXIV, 304 Seiten. Geb. € 79,95

226 Sarah *Wolf*: **Unterseeische Rohrleitungen und Meeresumweltschutz.** 2011. XXIII, 442 Seiten. Geb. € 94,95

225 Clemens *Feinäugle*: **Hoheitsgewalt im Völkerrecht.** 2011. XXVI, 418 Seiten. Geb. € 89,95

224 David *Barthel*: **Die neue Sicherheits- und Verteidigungsarchitektur der Afrikanischen Union.** 2011. XXV, 443 Seiten. Geb. € 94,95

223 Tilmann *Altwicker*: **Menschenrechtlicher Gleichheitsschutz.** 2011. XXX, 549 Seiten. Geb. € 99,95

222 Stephan *Bitter*: **Die Sanktion im Recht der Europäischen Union.** 2011. XV, 351 Seiten. Geb. € 84,95

221 Holger *Hestermeyer*, Nele *Matz-Lück*, Anja *Seibert-Fohr*, Silja *Vöneky* (eds.): **Law of the Sea in Dialogue.** 2011. XII, 189 Seiten. Geb. € 69,95 zzgl. landesüblicher MwSt.

220 Jan *Scheffler*: **Die Europäische Union als rechtlich-institutioneller Akteur im System der Vereinten Nationen.** 2011. XXXV, 918 Seiten. Geb. € 149,95

219 Mehrdad *Payandeh*: **Internationales Gemeinschaftsrecht.** 2010. XXXV, 629 Seiten. Geb. € 99,95

218 Jakob *Pichon*: **Internationaler Strafgerichtshof und Sicherheitsrat der Vereinten Nationen.** 2011. XXVI, 399 Seiten. Geb. € 89,95

217 Michael *Duchstein*: **Das internationale Benchmarkingverfahren und seine Bedeutung für den gewerblichen Rechtsschutz.** 2010. XXVI, 528 Seiten. Geb. € 99,95

216 Tobias *Darge*: **Kriegsverbrechen im nationalen und internationalen Recht.** 2010. XXXV, 499 Seiten. Geb. € 94,95

215 Markus *Benzing*: **Das Beweisrecht vor internationalen Gerichten und Schiedsgerichten in zwischenstaatlichen Streitigkeiten.** 2010. L, 846 Seiten. Geb. € 139,95

214 Urs *Saxer*: **Die internationale Steuerung der Selbstbestimmung und der Staatsentstehung.** 2010. XLII, 1140 Seiten. Geb. € 169,95

213 Rüdiger *Wolfrum*, Chie *Kojima* (eds.): **Solidarity: A Structural Principle of International Law.** 2010. XIII, 238 Seiten. Geb. € 69,95 zzgl. landesüblicher MwSt.

212 Ramin S. *Moschtaghi*: **Die menschenrechtliche Situation sunnitischer Kurden in der Islamischen Republik Iran.** 2010. XXIII, 451 Seiten. Geb. € 94,95

211 Georg *Nolte* (ed.): **Peace through International Law. The Role of the International Law Commission.** 2009. IX, 195 Seiten. Geb. € 64,95 zzgl. landesüblicher MwSt.

210 Armin *von Bogdandy*, Rüdiger *Wolfrum*, Jochen *von Bernstorff*, Philipp *Dann*, Matthias *Goldmann* (eds.): **The Exercise of Public Authority by International Institutions.** 2010. XIII, 1005 Seiten. Geb. € 149,95 zzgl. landesüblicher MwSt.

209 Norman *Weiß*: **Kompetenzlehre internationaler Organisationen.** 2009. XVIII, 540 Seiten. Geb. € 99,95

208 Michael *Rötting*: **Das verfassungsrechtliche Beitrittsverfahren zur Europäischen Union.** 2009. XIV, 317 Seiten. Geb. € 79,95

207 Björn *Ahl*: **Die Anwendung völkerrechtlicher Verträge in China.** 2009. XIX, 419 Seiten. Geb. € 89,95

206 Mahulena *Hofmann*: **Von der Transformation zur Kooperationsoffenheit?** 2009. XIX, 585 Seiten. Geb. € 99,95

205 Rüdiger *Wolfrum*, Ulrike *Deutsch* (eds.): **The European Court of Human Rights Overwhelmed by Applications: Problems and Possible Solutions.** 2009. VIII, 128 Seiten. Geb. € 59,95 zzgl. landesüblicher MwSt.

204 Niels *Petersen*: **Demokratie als teleologisches Prinzip.** 2009. XXVII, 280 Seiten. Geb. € 79,95

203 Christiane *Kamardi*: **Die Ausformung einer Prozessordnung sui generis durch das ICTY unter Berücksichtigung des Fair-Trial-Prinzips.** 2009. XVI, 424 Seiten. Geb. € 89,95

202 Leonie F. *Guder*: **The Administration of Debt Relief by the International Financial Institutions.** 2009. XVIII, 355 Seiten. Geb. € 84,95 zzgl. landesüblicher MwSt.

201 Silja *Vöneky*, Cornelia *Hagedorn*, Miriam *Clados*, Jelena *von Achenbach*: **Legitimation ethischer Entscheidungen im Recht.** 2009. VIII, 351 Seiten. Geb. € 84,95

200 Anja Katarina *Weilert*: **Grundlagen und Grenzen des Folterverbotes in verschiedenen Rechtskreisen.** 2009. XXX, 474 Seiten. Geb. € 94,95

199 Suzette V. *Suarez*: **The Outer Limits of the Continental Shelf.** 2008. XVIII, 276 Seiten. Geb. € 79,95 zzgl. landesüblicher MwSt.

198 Felix *Hanschmann*: **Der Begriff der Homogenität in der Verfassungslehre und Europarechtswissenschaft.** 2008. XIII, 370 Seiten. Geb. € 84,95

197 Angela *Paul*: **Kritische Analyse und Reformvorschlag zu Art. II Genozidkonvention.** 2008. XVI, 379 Seiten. Geb. € 84,95

FSC
www.fsc.org
MIX
Papier | Fördert gute Waldnutzung
FSC® C083411

Kolibri 360 GmbH
Ferdinand-Jühlke-Straße 7
99095 Erfurt, Deutschland
produktsicherheit@kolibri360.de